Integrated Circuits and Systems

For further volumes, go to
http://www.springer.com/series/7236

Hoi-Jun Yoo · Chris van Hoof
Editors

Bio-Medical CMOS ICs

Springer

Editors
Hoi-Jun Yoo
Department of Electrical Engineering
Korea Advanced Institute of Science and Technology
Guseong-dong, Daejeon 305-701
Republic of Korea
hjyoo@ee.kaist.ac.kr

Chris van Hoof
Katholieke Universiteit Leuven
Interuniversity Microelectronics Center (IMEC)
Kapeldreef 75, Leuven B-3001
Belgium
chris.vanhoof@imec.be

ISSN 1558-9412
ISBN 978-1-4419-6596-7 e-ISBN 978-1-4419-6597-4
DOI 10.1007/978-1-4419-6597-4
Springer New York Dordrecht Heidelberg London

Library of Congress Control Number: 2010936189

Printed on acid-free paper

Springer is part of Springer Science+Business Media (www.springer.com)

Preface

A major societal challenge for the decades to come will be the delivery of effective medical services while at the same time curbing the growing cost of healthcare. It is expected that new concepts-particularly electronically assisted healthcare will provide an answer. This will include new devices, new medical services as well as networking. On the device side, impressive innovation has been made possible by micro- and nanoelectronics or CMOS Integrated Circuits. Even higher accuracy and smaller form factor combined with reduced cost and increased convenience of use are enabled by incorporation of CMOS IC design in the realization of bio-medical systems. The compact hearing aid devices and current pacemakers are good examples of how CMOS ICs bring about these new functionalities and services in the medical field. Apart from these existing applications, many researchers are trying to develop new bio-medical solutions such as Artificial Retina, Deep Brain Stimulation, and Wearable Healthcare Systems. These are possible by combining the recent advances of bio-medical technology with low power CMOS IC technology.

CMOS IC design alone is a challenging discipline and needs a long-term education to master it. Bio-medical services also require long experience and deep knowledge about the physiology, pathology, as well as the psychology of the patients. Bio-medical CMOS IC is different from other CMOS ICs in that it will be used in very intimate contact with the body with the bio-medical services in mind. Biological effects and interactions related to the specific application should therefore be considered before the chip design and after the chip fabrication. Recently, electronic detection has contributed even to psychological monitoring by sensing stress and emotions through vital-sign monitoring, and this has been enabled by bio-medical IC design. The obvious interdisciplinary nature of bio-medical IC design is very challenging. One of the editors has taught "Bio-Medical IC Design" course to graduate students in KAIST, and feels the necessity of a text book or a good reference book for students explaining the basic principles and current trends of such a wide range of topics in relatively simple and clear terms.

The purpose of this book is therefore to provide the readers with a complete overview of how to design and apply CMOS ICs for bio-medical applications. Even though the importance of CMOS ICs is well-known, there is hardly any literature devoted to this subject. Many journal papers and lectures at conferences have dealt with the bio-medical ICs, but they were not comprehensive and generally only one circuit or a specific application was covered. This bio-medical IC design book in contrast covers the basic circuits and systems of the current leading-edge researches and products. The editors and the authors have been involved in bio-medical IC design for many years, and have tried to write a book showing how to design bio-medical ICs specifically targeting those readers that have limited experience in CMOS IC design. The book is also suitable for experienced engineers who would like to be introduced to recent trends in bio-medical ICs. For this purpose, we start every chapter with a brief introduction of a basic principle behind the circuits and systems. Also, we tried to write the contents with simple and plain expressions. In addition, where appropriate, most chapters provide a summary of current trends and an overview of the future directions.

This book would not have been possible without intensive help from many people. First of all, we would like thank all authors for their invaluable contributions. We would like to also thank Prof. Anantha Chandrakasan, the series editor, for his encouragement to write this book, and all members of the Springer team, especially Alex Greene and Ciara Vincent, for their support. Last but not least, we give our sincere thanks to graduate students of Semiconductor Systems Lab of KAIST, IMEC, and the Holst Centre for their help and support during the preparation of the book.

Daejeon, Korea Hoi-Jun Yoo
Leuven, Belgium Chris van Hoof

Contents

1 **Introduction to Bio-Medical CMOS IC** 1
Hoi-Jun Yoo and Chris van Hoof

Part I Vital Signal Sensing and Processing

2 **Introduction to Bioelectricity** 13
Yong Jeong

3 **Biomedical Electrodes For Biopotential Monitoring and Electrostimulation** 31
Eric McAdams

4 **Readout Circuits** 125
R. Firat Yazicioglu

5 **Low-Power ADCs for Bio-Medical Applications** 157
J. Craninckx and G. Van der Plas

6 **Low Power Bio-Medical DSP** 191
Hyejung Kim and Hoi-Jun Yoo

Part II Bio-Medical Wireless Communication

7 **Short Distance Wireless Communications** 219
Robert Puers and Jef Thoné

8 **Bio-Medical Application of WBAN: Trends and Examples** 279
Julien Penders, Chris van Hoof, and Bert Gyselinckx

9 **Body Channel Communication for Energy-Efficient BAN** 303
Seong-Jun Song and Hoi-Jun Yoo

Part III Examples of Bio-Medical ICs

10 **Wearable Healthcare System** 339
Jerald Yoo and Hoi-Jun Yoo

11 Digital Hearing Aid and Cochlear Implant 371
Sunyoung Kim and Hoi-Jun Yoo

12 Cardiac Rhythm Management IC's 421
Erno Klaassen

13 Neurostimulation Design from an Energy and Information Transfer Perspective 453
David A. Dinsmoor, Robert W. Hocken, Jr., Wesley A. Santa, Jalpa S. Shah, Larry Tyler, and Timothy J. Denison

14 Artificial Retina IC 481
Jun Ohta

Index 515

Contributors

J. Craninckx Interuniversity Microelectronics Center (IMEC), Leuven, Belgium, jan.craninckx@imec.be

Timothy J. Denison Medtronic Neuromodulation Integrated Circuit Technology, Columbia Heights, MN 55421, USA, timothy.denison@medtronic.com

David A. Dinsmoor Medtronic Neuromodulation Integrated Circuit Technology, Minneapolis, MN 55432, USA, david.a.dinsmoor@medtronic.com

B. Gyselinckx Interuniversity Microelectronics Center (IMEC), Eindhoven, The Netherlands

Robert W. Hocken, Jr. Medtronic Neuromodulation Integrated Circuit Technology, Minneapolis, MN 55432, USA

Yong Jeong Department of Bio and Brain Engineering, Korea Advanced Institute of Science and Technology, Gwahangno, Yuseong-gu, Daejeon 305-701, Republic of Korea, yong@kaist.ac.kr

Hyejung Kim Ultra Low Power and Extreme Electronics Group IMEC, Kapeldreef 75, B-3001 Leuven, Belgium, hyejung@imec.be

Sunyoung Kim Interuniversity Microelectronics Center (IMEC), Leuven, Belgium, sunyoung@imec.be

Erno Klaassen St. Jude Medical, Cardiac Rhythm Management Division, Sunnyvale, CA, USA, eklaassen@sjm.com

Eric McAdams Biomedical Sensors Group, INSA (National Institute of Applied Science) of Lyon, Léonard de Vinci Building, 20 Avenue Albert Einstein, 69621 Villeurbanne Cedex, France, et.mcadams@ulster.ac.uk

Jun Ohta Graduate School of Materials Science, Nara Institute of Science and Technology, Nara 630-0192, Japan, ohta@ms.naist.jp

Julien Penders Holst Centre, IMEC, Eindhoven, The Netherlands, julien.penders@imec-nl.nl

Robert Puers Katholieke Universiteit Leuven, Kasteelpark Arenberg 10, 3001 Leuven, Belgium, robert.puers@esat.kuleuven.be

Wesley A. Santa Medtronic Neuromodulation Integrated Circuit Technology, Minneapolis, MN 55432, USA

Jalpa S. Shah Medtronic Neuromodulation Integrated Circuit Technology, Minneapolis, MN 55432, USA

Seong-Jun Song Samsung Electronics Co., LTD, Suwon 443-742, Republic of Korea, sj33.song@samsung.com

Jef Thoné Katholieke Universiteit Leuven, Kasteelpark Arenberg 10, 3001 Leuven, Belgium, jef.thone@esat.kuleuven.be

Larry Tyler Medtronic Neuromodulation Integrated Circuit Technology, Minneapolis, MN 55432, USA

G. Van der Plas Interuniversity Microelectronics Center (IMEC), Leuven, Belgium, geert.vanderplas@imec.be

Chris van Hoof Katholieke Universiteit Leuven, Interuniversity Microelectronics Center (IMEC), Kapeldreef 75, B-3001 Leuven, Belgium, chris.vanhoof@imec.be

R. Firat Yazicioglu Interuniversity Microelectronics Center (IMEC), Leuven, Belgium, firat@imec.be

Hoi-Jun Yoo Department of Electrical Engineering, Korea Advanced Institute of Science and Technology, Guseong-dong, Daejeon 305-701, Republic of Korea, hjyoo@ee.kaist.ac.kr

Jerald Yoo Microsystems Engineering, Masdar Institute of Science and Technology (MIST), Abu Dhabi, United Arab Emirates, jyoo@masdar.ac.ae

Chapter 1
Introduction to Bio-Medical CMOS IC

Hoi-Jun Yoo and Chris van Hoof

1.1 Introduction to Bio-Medical CMOSIC

Living a healthy life without threatening or debilitating illness is our primary hope and the medical systems and services continue to strive to guarantee it. Nevertheless, in developed countries, the per capita cost of care keeps rising and this is a threat to sustained healthcare. Recently, many promising technological advances in IT are about to change our concept about healthcare, as well as the provision of medical cares. For example, the telemedicine, e-hospital, and ubiquitous healthcare are enabled by emerging wireless broadband communication technology. While initially becoming main-stream for portable devices such as notebook computers and smart phones, wireless communication is evolving towards wearable solutions and even implantable solutions are being introduced. Such healthcare devices rely on WSN (Wireless Sensor Network) and BSN (Body Sensor Network). Before the advent of these technologies, healthcare provided centralized care (at the doctor's office or at the hospital) which came with a penalty in cost and time. With the help of IT technology and the deployment of wearable healthcare, people are largely breaking free from these limitations and will be able to monitor their health condition anywhere and at any time while being able to get the expert's help whenever needed. This will enable a more individualized and more proactive healthcare, and these new concepts and systems are expected to change our daily lives as well as medical profession and its industry. "E-Healthcare" or "U-Healthcare" is a recent term for healthcare supported by information and communication technology in general, and by small personal wireless monitoring devices in particular. An important emerging example is remote and continuous wireless vital signs monitoring. The combination of two technologies, ultra-low power sensor technology and ultra-low power wireless communication technology, enables long-term continuous monitoring and feedback to medical professionals wherever needed.

H.-J. Yoo (✉)
Department of Electrical Engineering, Korea Advanced Institute of Science and Technology, Guseong-dong, Daejeon 305-701, Republic of Korea
e-mail: hjyoo@ee.kaist.ac.kr

H.-J. Yoo, C. van Hoof (eds.), *Bio-Medical CMOS ICs*,
Integrated Circuits and Systems, DOI 10.1007/978-1-4419-6597-4_1,

With the improvement of the standard of living, the life expectancy is increasing and the resulting aging population leads to an exploding cost of healthcare. This cost explosion should be curbed by new technologies apart from lowering the cost of healthcare, E-Healthcare is also expected to change our daily lives. The focus of the healthcare is shifting from illness management to health management, from healing to prevention and from centralized care to delocalized individual care as in Fig. 1.1. Seeing how Nintendo Wii Fit and Nike+ are having a large impact on lifestyle already, further and even more significant lifestyle changes are possible thanks to this added digitization and wireless communication of vital signal. This information can even be integrated with other types of data such as food and calorie intake, physical exercise information to overall improve wellness.

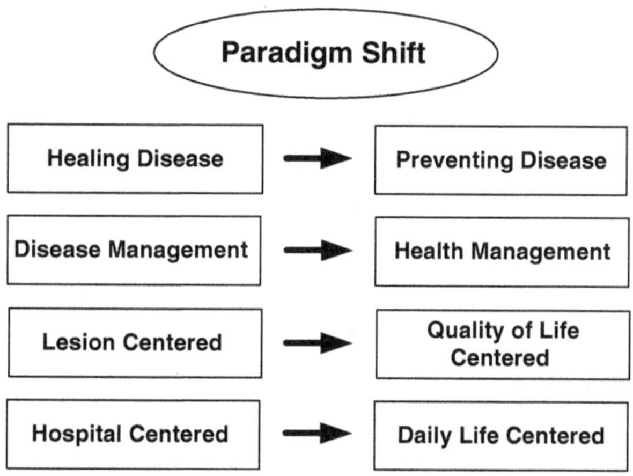

Fig. 1.1 Recent paradigm shift in bio-medicine

Historically, most of the medical devices have been very bulky and were usually connected to the body by many electrical leads wires. Their cost significantly exceeded that of consumer electronics partly due to the conservative medical industry which requires long pre-testing and many regulations. However, by using advanced semiconductor technology, it is possible to implement the same functionality in a much smaller form factor with even higher performance and much lower cost. The recent "Continua Healthcare Consortium" is a good example for this trend [7]. Many IT companies such as Intel, Samsung and Nokia formed a consortium to make industrial standards for the next generation healthcare systems. It is expected that the medical landscape will radically change: Where previously only a few bio-medical system companies occupied the strongly regulated (and therefore conservative) bio-medical market, the increased government stimulation will enable the implementation of many new technologies in the market of consumer and personal health.

Bio-medical applications cover medical device applications, healthcare/lifestyle applications and life sciences (see Fig. 1.2). While the third category specifically needs advances in biology and biochemistry (which could be termed "bioware"),

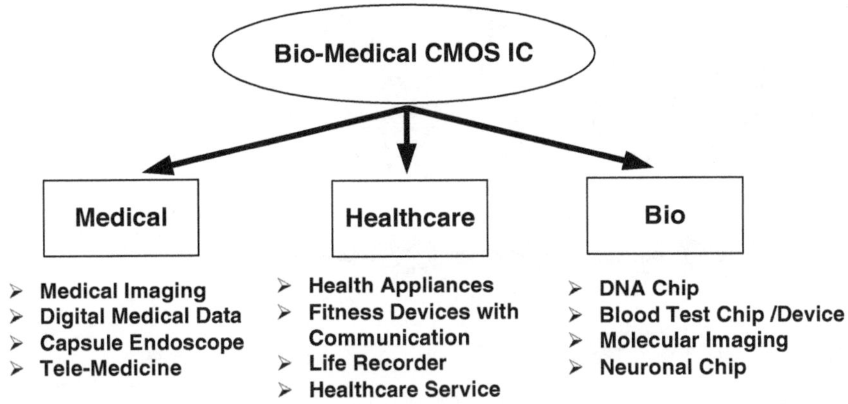

Fig. 1.2 3 major categories of bio-medical IC

all areas strongly benefit from semiconductor and innovations (both hardware and software).

The silicon microelectronics technology represented by "Moore's Law" has enabled the integration of more than a billion transistors on a single chip, and has enabled a digital age. These advances have not been limited to digital technology. Also ultra-low power CMOS IC technology is becoming available enabling implantable applications with a battery life time exceeding more than 10 years. Furthermore, CMOS circuit solutions for of PAN (Personal Area Network) and BAN (Body Area Network) technologies such as Bluetooth, UWB, Zigbee, and their emerging Ultra-Low-Power counterparts, as well as Body Channel Communication, enable low power versatile short distance communication. Concurrently, IC-centric (i.e., CMOS or MEMS technology based) sensor technologies have been established. They can now be integrated monolithically or heterogeneously with the CMOS circuits to reduce the form factor as well as the power consumption. In addition to the low power communication circuits for PAN/BAN, and the sensor technologies mentioned above, other specific considerations are necessary for achieving bio-medical CMOS IC design. For example, design of high sensitivity mixed-signal biopotential front-ends that are impedance matched to on-the-body signal electrodes, and compensation circuits to mitigate motion artifacts and other noises sources due to human motion.

These are some examples that will be elaborated in the following chapters in more detail.

1.2 Architecture of Sensor Systems with Bio-medical CMOS IC

A representation of the generic architecture of a bio-medical CMOS IC based sensor system is shown in Fig. 1.3. Irrespective of their form factor, integration and packaging method, and, deployment around or in the body, most bio-medical systems contain electrodes, ADC, Processor, and Communication blocks. For specific

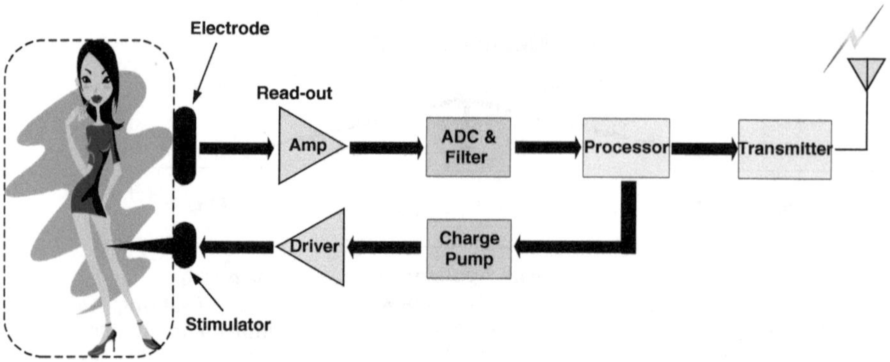

Fig. 1.3 Typical architecture of bio-medical IC subsystem

applications where electrical stimulation is needed, also DAC and driver blocks are needed. Compared with other CMOS ICs, the ultra-low energy consumption is the stringent requirement in the design of the bio-medical CMOS ICs. For example, if a battery is used with an energy capacity of 625 mAh at 1.5 V, a CMOS IC with 100 mW power consumption can only last for nine and a half hours. Most of implantable bio-medical devices in contrast should last more than 10 years and this limits the average system power consumption (when using the same battery) to10 μW. With this extremely low energy budget, bio-medical devices should still be able to detect complex vital signals, with wide dynamic range (μV–mV) covering low frequency band (<10 kHz). These signals concern EEG, ECG, and EMG measurements but also electrical conversion signals of Heart Rates and respiration rates, as illustrated in Fig. 1.4. Recently, other applications are emerging such as the capsule endoscopy and the neural recording that require more than >2 Mbps data transmission, taxing the power budget even more. The overall functionality of those systems should be carefully selected to ensure a match between the system requirements and the ultra-low energy consumption. In addition, its building blocks shown in Fig. 1.3 should be optimized for the proper performance under the given energy budget.

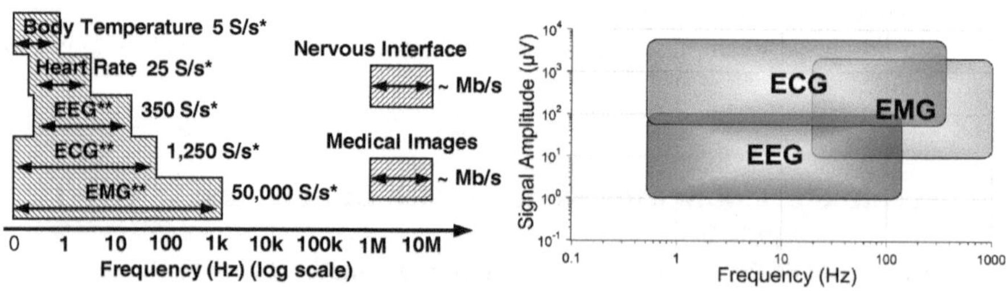

Fig. 1.4 Frequency and voltage ranges of on-the-body biopotential signals

In this application, the first interface to the human body consists of electrodes to sense the electrical signals from the human body. While other sensors may be used to sense such biopotential signals, metal-based electrodes most often in conjunction with a wet gel are the most common interfaces. The next circuit block is the Read-out Amplifier or Analog Front End. Traditionally, differential instrumentation amplifiers have been used as a Front End. For robust sensing of these weak signals, the instrumentation amplifier should have high CMRR.

In the Analog-to-Digital Converter (ADC) the amplified analog signal is converted in a bits-tream allowing digital signal processing. The type, speed, resolution, and dynamic range of the ADC block should be selected carefully based on the overall system requirements because these parameters significantly impact on the power consumption. Among the Nyquist-rate ADCs, where the sampling frequency is inversely proportional to the resolution, SAR ADCs of 6–8 bit resolutions are usually chosen for the consideration of the low power dissipation [1]. In the next signal processing stage, the digital bio signals are processed to remove for example noise, detect any anomalies, and compress the data to reduce either the required memory or the amount of data communication. Low power digital signal processing technology has received much attention in the last two decades and this has low power high performance CPUs [2]. Their on-chip integration allows fine-tuning of the system by modification of the embedded software, by personalization of the device (e.g., setting thresholds for the alarm), and by encryption of the vital data to ensure data security and privacy.

The next block in the system is the transmitter which can communicate the processed digital signals by wireline or wirelessly. Most devices are using wireless communication schemes such as NFC, PAN, BAN, or new technology like Body Channel Communication(BCC) [3]. They can be grouped into two, Near Field Inductive Coupling and Far Field Electromagnetic Communication. The inductive coupling is effective for the short distance "through body" communication connecting for example implanted devices with on-body monitoring devices. Applications of this kind are Implanted Cardiac Defibrillators (ICDs), cochlear implants, and artificial retinas. However, far-field communication based on PAN and BAN is most widely used in connecting the on-the-body sensors for sensor networking. In this case, the power consumption of the wireless transmitter typically dominates the system power. It is therefore necessary to use a very low power PAN such as Zigbee or to tightly control the duty cycle to confine its turn-on time or its energy consumption. Lower-Power BAN radio architectures are being developed and the decision for an IEEE standard (802.15.6) is nearly taken. Body Channel Communication is a promising alternative because it can transmit $\sim$10 Mbps signal with very low energy consumption, but its long term effect on the human body is not verified so far [4]. It requires electrical coupling to the body, but most electrode-based vital sign measurements are doing this anyway for signal extraction (Fig. 1.5).

In therapeutic applications, electrical stimulation is required to achieve "closed loop" healthcare. From the nervous stimulation in artificial cochlea applications to the high voltage stimulation of ICD, stimulation occurs when the on-board process decides this action is required: the digital processor checks the input data and

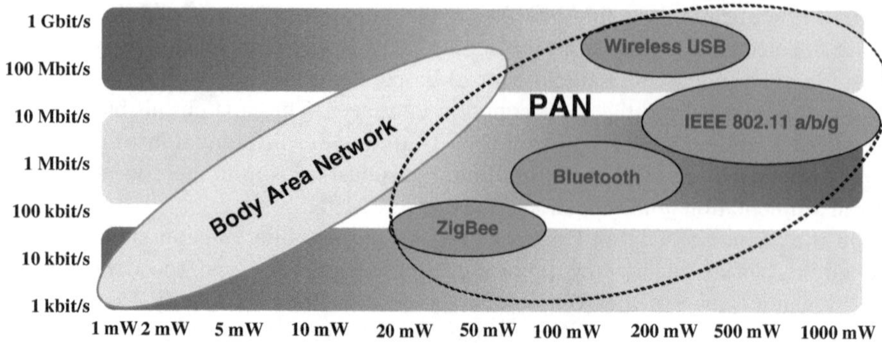

Fig. 1.5 Bandwidth and communication distance of BAN and PAN

triggers the charge pump to build up the energy required for the stimulation. The stimulation level is controlled by the charge quantity injected as a pulsed current, rather than a pulsed voltage. This is because it is difficult to derive an exact current level from a voltage pulse due to the unpredictably varying impedance of the body. Usually, a series of biphasic current pulses are used for the stimulation so as to achieve overall zero net charge and avoiding charge accumulation. Table 1.1 shows the most relevant electrical parameters, such as stimulating voltage and current for stimulators. Note that most of stimulators require more than 10 V of operation. Special circuitry is therefore necessary in order to boost such high voltage from a normal battery voltage, and this also requires a high voltage CMOS technology in these circuits.

Table 1.1 Relevant electrical parameters for different stimulators

		Voltage	Current	Impedance	Duration	Frequency
Retinal Stimulator		±10 V	100 μA-1 mA	~10 KΩ	100 μs–1 ms	10–120 Hz
Deep Brain Stimulator		1–10.5 V	~100 μA	~KΩ	60–450 μs	2–185 Hz
Neural Stimulator (FES)		5–10 V	~mA	5 KΩ	~100 μs	10–100 Hz
Defibrillator	Implantable	10–70 V	–	10–25 Ω	100 ms	1-pulse
	External	10–750 V	–	200–300Ω	5–50 ms	1-pulse

1.3 Applications and Future Trends

The healthcare or health-assisting devices, as well as medical care enabled by these device will enable a level of unprecedented care during each person's life from cradle to grave. For achieving practical and convenient, personalized healthcare, these devices should be ultra-portable. With the help of the bio-medical IC developments, their sizes can be drastically reduced from the current notebook size systems used today in the hospital or ambulatory care to smart phone type embodiments. Such smart phones (Fig. 1.6a) are used to check personal health data and act as health

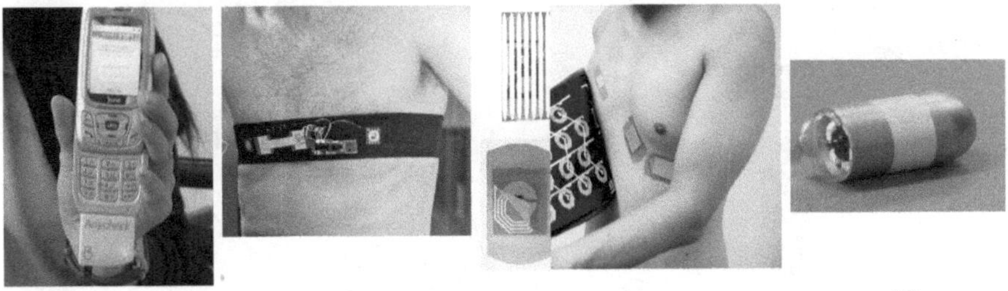

(a) (b) (c) (d)

Fig. 1.6 (**a**) Hand-held (cell phone). (**b**) Chest band. (**c**) Bandage. (**d**) Capsule

coaches or early warning devices. Moreover, they can connect to the healthcare professional in real time when necessary. Furthermore, portability is further improved by turning them into wearable form-factors like Fig. 1.6b, c. In that case, they are embedded in shirts, vest, belt, straps, wristbands, and head bands enabling unobtrusive monitoring during daily physical activity. Further miniaturization makes the capsule-like bio-medical devices possible as shown in Fig. 1.6d. These are hermetically sealed and stay inside the human body either temporarily (for gastrointestinal transit after swallowing) or for an extended period of time after surgical implantation. The implantable types may require surgery for replacing the battery unless wireless recharging or powering is possible.

Continuous monitoring of physiological parameters is beneficial for chronic diseases, and it is being enabled by bio-medical IC and wearable electronics technology. The collected data are intended to detect the onset of a medical condition and take the preventive or therapeutic measures. Long-term data collection also assists a more exact diagnosis. Fall detection is another bio-medical IC application enabling or assisting longer independent living of elderly with minimal invasion of privacy. Implantable diabetic care is still one of the holy grails where a tiny capsule can monitor blood glucose and dispense insulin when necessary. For non-chronic illnesses, bio-medical ICs can assist rehabilitation of patients. For patients recovering from stroke from impediments in limb motion, wrist-band type sensors can quantitatively monitor limb motion and inform the medical expert as well as the patient about their recovery. Graphics displays on the sensor can even help the patient to correct his/her right posture. The closed cycle involving monitoring-preventive-diagnosis-therapeutic-rehabilitation is illustrated in Fig. 1.7.

On top of these healthcare and therapeutic applications, these new devices can be used for emerging applications such as wellbeing, sports, lifestyle and, entertainment as depicted in Fig. 1.7. The significant amount of data gathered by the so-called personal health-coach devices can be utilized to analyze the owner's daily nutrition, personal habits and life styles, and assist him/her to live "a healthier life". This technology can evidently be exploited for entertainment as well. For example, the wristband sensor developed for the rehabilitation can be used for the input device to enable new mode of interaction between gamer and game machine. Another

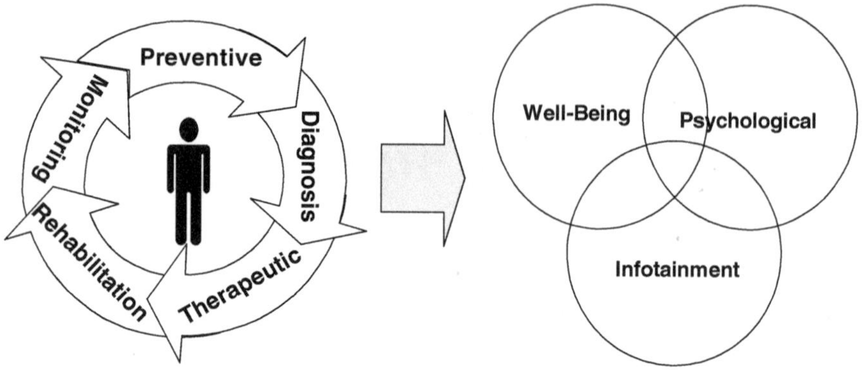

Fig. 1.7 Healthcare and emerging applications of the new bio-medical devices

interesting application area is the monitoring of stress and emotion (e.g., personal psychological analysis) [5] as well as social computing (Group psychology analysis) [6]. If a correlation between the measured data and psychological variables can be made, it can assist reduction of stress level improve the productivity of a team.

It is expected that these new bio-medical devices will be able to enhance our sensing ability beyond the conventional 5 senses, and can also provide prosthetic functions if required. Cochlear implants are widely used to restore hearing to hearing impaired. Artificial retina and artificial noses are under intensive study. These devices concern the ear, the eye, and the nose. For the heart, pacemakers and ICD exist and for the brain, Deep Brain Stimulation (DBS) is used and brain-computer interfaces (BCI) are under development. Beyond these known applications sensing of EM waves is a possibility as well, which you may call the transcendence of the 5 senses. These will extend our limits and as a result, our thought enabling us to live a new life.

1.4 Organization of the Book

This book consists of 14 chapters which are grouped into three parts. The first part is composed of 5 chapters for sensing and processing of the bio signals. It covers the bio-electricity, the bio-electrodes, Analog read-out circuits, Low power ADC, and Bio Signal Processors, following the signal flow from the human skin to sensor's signal processor. These chapters will introduce the basic principles and some practical bio-medical application examples. The second part of this book has three chapters covering the bio-medical wireless communication. Different types of wireless PAN/BAN will be explained in detail with emphasis on the bio-medical applications. Next, the new wireless communication scheme, body channel communication, will be introduced for the possible application to bio-medical area. Three commercial examples of WPAN and WBAN will be summarized and their

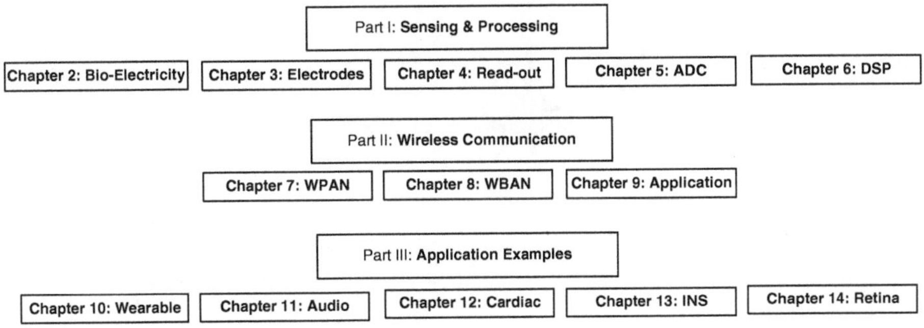

Fig. 1.8 Organization of the chapters

use in bio-medical applications will be illustrated. The third part covers interesting, cutting-edge bio-medical device examples. It consists of five chapters, the Wearable Healthcare Systems, Digital Hearing Aids and Cochlear Implants, Cardiac Rhythm Management ICs, Neuro-stimulation Designs, and Artificial Retina ICs. These chapters contain the most recent research results as well as the basic operation principles and architectures, and update the readers with detail explanation of advanced research (Fig. 1.8).

References

1. Długosz R, Iniewski K (April 2007) Flexible architecture of ultra-low-power current-mode interleaved successive approximation analog-to-digital converter for wireless sensor networks. VLSI Design. 2007(2), Article 2
2. Chandrkasan A, Verma N, Dally D (2008) Ultralow-power electronics for biomedical applications. Annu Rev Biomed Eng. 10:247–274
3. Ashok R, Agrawal A (2003) Next-generation wearable networks. IEEE Computer. Nov 31–39
4. Yoo H-J et al (2007) Low energy on-body communication for BSN. IFMBE Proceedings (BSN 2007), March 2007, Aachen Germany, pp 15–20
5. Picard R, Vyzas E, Healey J (Oct. 2001) Toward machine emotional intelligence: analysis of affective physiological state. IEEE T Pattern Anal 23(10):1175–1191
6. Yano K et al (Feb. 2008) Life thermoscope: integrated microelectronics for visualizing hidden life rhythm. Proceeding of ISSCC 2008, pp 136–137
7. Marc Holland, Lynne A. Dunbrack (2006) The Continua Health Alliance: Will Remote Health Monitoring Create a Paradigm Shift in Healthcare? Health Industry Insights Newsflash. IDC. Jun. 2006

Part I
Vital Signal Sensing and Processing

Chapter 2
Introduction to Bioelectricity

Yong Jeong

2.1 Introduction

It can be said that the use of electricity by biological systems as a signal between the nerves and muscles was first discovered in 1789 in a frog leg when the Italian physicist Luigi Galvani touched an exposed sciatic nerve with a charged metal scalpel and observed the dead frog's leg flex as if it were alive. This finding provided the basis for the current understanding that electrical energy is the impetus behind muscle movement and also the driving force in other systems. This work was reported in the *Proceedings of the Bologna Academy* in 1791. At that time, Galvani believed that the muscular contractions were due to electrical energy emanating from the animal. However, Allesandro Volta was convinced that the electricity in Galvani's experiments originated from the presence of the dissimilar metals. Both of these interpretations represent the two different aspects of electrical potential in biological system, the action potential and the steady source of electrical potential [1, 2].

Bioelectricity is the electrical phenomenon of life processes. The basic unit of this phenomenon is a cell which is polarized by certain processes using energy. Specialized classes of cells that have electrically excitable membranes such as neurons or muscle cells have additional capabilities of developing action potentials. Many biomedical instruments such as electroencephalography, electrocardiography or electromyography measure the compounds of these action potentials from the brain, heart, and muscle, respectively.

In this chapter, the basic biological mechanisms behind bioelectricity and their applications will be introduced.

Y. Jeong (✉)
Department of Bio and Brain Engineering, Korea Advanced Institute of Science and Technology, Gwahangno, Yuseong-gu, Daejeon 305-701, Republic of Korea
e-mail: yong@kaist.ac.kr

H.-J. Yoo, C. van Hoof (eds.), *Bio-Medical CMOS ICs*,
Integrated Circuits and Systems, DOI 10.1007/978-1-4419-6597-4_2,

2.2 Electrical Properties of the Human body

The electrical properties of the human body follow the laws of nature. The main chemical components of our body is water, salts (or their ions) and some organic chemicals such as proteins, lipids and so on. These components show different electrical characteristics, however, we can build a circuit model using a combination of resistors, capacitors and/or generators. The main difference between this model from the actual human body is that current flow is caused not by the movement of electrons but by that of ions dissolved in water.

The human body exhibits many levels of structural complexity. It is composed of various organ systems which can be grouped into functional units. These are the integumentary system, skeletal system, muscular system, nervous system, endocrine system, cardiovascular system, respiratory system, digestive system, urinary system, and reproductive system. An organ is a structure composed of several tissue types, which performs specific functions. Again, tissues consist of a group of similar cells that have a common function. Cells are the smallest units of all living things. Thus, even if different systems or organs perform different function, they consist of the same building blocks, the cell and cement (the extracellular matrix).

This means that each organ shares common features including electrical properties. Some variations exist in terms of the chemical composition in the cytoplasm (intracellular compartment), the cell membrane and extracellular components depending on cell or tissue type. These variations, however, has little effect on electrical properties. The main factor which determines the electrical properties of the tissue is the distribution pattern of ion channels on the cell membrane. The amount and composition of lipids can also influence the electrical properties of human body by acting as a capacitor but the capacitance value is almost the same across the cell.

2.2.1 Cell Membrane

Cells are basic building blocks of living organisms. The boundary of animal cells is a plasma membrane composed of thin lipid bilayer and proteins embedded in it (Fig. 2.1). The main role of the cell membrane is to regulate the exchange of chemical substances. Both the lipid bi-layer and some of the embedded proteins play critical roles in the electrical properties of the cell using this exchange. The plasma (fluid in inner space of cells) and interstitial fluids (fluid in outer space) are composed of ions or electrolytes of different species which are unequally distributed across the membrane. This membrane prevents water molecules and ions from diffusing across it. The most common electrolytes are Na^+ (sodium), K^+ (potassium), Cl^- (chloride) and Ca^{++} (calcium). Other components such as H^+ (hydrogen), HCO_3^- (bicarbonate), NH_4^+ (ammonium) or phosphate ions contribute minimally to membrane potential. Protein components endow the membrane with a selective permeability to some ions. The driving force of this exchange is initially a difference in concentration between the inside and outside of the cell. This difference or concentration gradient is maintained mainly by Na^+/K^+ pumps that move Na^+ out of the cell and K^+ into the cell using energy. Selective permeability is also determined

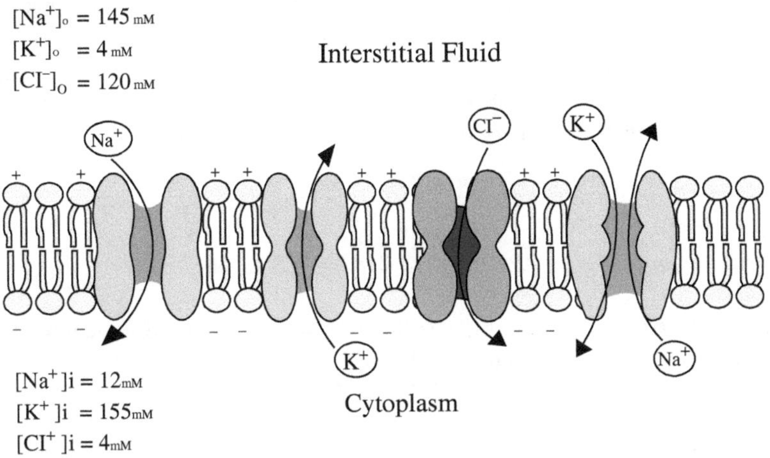

Fig. 2.1 Structure of the plasma membrane. Membrane is composed of a lipid bilayer with embedded proteins. The proteins shown in the figure are ion channels and Na^+/K^+ pumps as illustrated, however there are other proteins embedded in the membrane that act as receptors, transporters and serve other functions

by ion channels. Ions pass through channels that are selective for that specific ion. There are four major ion channels, Na^+ channel, K^+channel, Cl^- channel, and Ca^{++} channel. Figure 2.1 represents a plasma membrane with ion channels and a Na^+/K^+ pump with typical ion concentrations of each side of the membrane. The flow of ions in response to concentration gradients is limited by the selectively permeable cell membrane and the resultant electrical field [3, 4, 5].

2.2.2 Membrane Potential

To understand membrane potential better consider an environment with only one ion species, for example, K^+ is present in solution. If the solution is divided into two compartments by a membrane with K^+ channels, and the concentration of K^+ is higher in one compartment, then there will be a net flux from the compartment with higher concentration to the lower one (extracellular interstitial fluid). For this situation, the flux due to concentration gradient (diffusion) tends to push K^+ to compartment with low concentration (outside of the cell) and is given by

$$J_K\,(\text{diffusion}) = -Dd[\mathrm{K}^+]/dx$$

from Fick's law where D is the diffusivity constant.
The flux of K^+ will lead to a positive charge accumulation outside of the cell. The flux due to electrical field (drift) tends to push K^+ inside the cell and is given by

$$J_K\,(\text{drift}) = -\mu Z[\mathrm{K}^+]dv/dt$$

from Ohm's law where μ is mobility in m^2/sV, Z is ionic valence and v is voltage across the membrane.

Thus, the net flow is

$$J_K = J_K\,(\text{diffusion}) + J_K\,(\text{drift}) = -Dd[\text{K}^+]/dxJ_K - \mu\text{Z}[\text{K}^+]dv/dt$$

Using the Einstein relation, $D = \text{K}T\mu/q$, the total flow is given by

$$J_K = -RT/q\mu d[\text{K}^+]/dx - \mu Z[\text{K}^+]dv/dt$$

where R is Boltzmann's constant, T is the absolute temperature in degrees Kelvin, and q is the magnitude of the electric charge.

At equilibrium when the flow of K^+ into the cell is balanced by the flow out of the cell thus

$$RT/q\mu d[\text{K}^+]/dx - \mu\text{Z}[\text{K}^+]dv/dt = 0$$

Integrating this equation from outside the cell to inside

$$\int_{v_o}^{v_i} dv = -\frac{KT}{q}\int_{[\text{K}^+]_o}^{[\text{K}^+]_i} \frac{d[\text{K}^+]}{[\text{K}^+]}$$

where v_o and v_i are the voltages outside and inside the membrane and $[\text{K}^+]_o$ and $[\text{K}^+]_i$ are the concentrations of K^+ outside and inside the membrane. Thus,

$$vi - vo = -\frac{KT}{q}\text{In}\frac{[\text{K}^+]_i}{[\text{K}^+]_o} = \frac{KT}{q}\text{In}\frac{[\text{K}^+]_o}{[\text{K}^+]_i}$$

This equation is known as the Nernst equation, and we can obtain K^+ equilibrium potential,

$$E_K = v_i - v_o = \text{RT}\,\text{In}[\text{K}^+]_o/[\text{K}^+]_i\,\text{mV}$$

In the same way Na^+ and Cl^- equilibrium potentials are

$$E_{\text{Na}} = \text{RT}\,\text{In}[\text{Na}^+]_o/[\text{Na}^+]_i\,\text{mV}$$

$$E_{\text{Cl}} = -\text{RT}\,\text{In}[\text{C1}^-]_o/[\text{Cl}^+]_i\text{mV}, \text{respectively.}$$

In a real system, however, all ions coexist together and the membrane potential (V_m) is given by the Goldmann equation for K^+, Na^+, and Cl^- can be derived as

$$\text{V}m = \frac{KT}{q}\text{In}\left(\frac{P_\text{K}[\text{K}^+]_o + P_\text{Na}[\text{Na}^+]_\text{o} + P_\text{Cl}[\text{Cl}^-]_i}{P_\text{K}[\text{K}^+]_i + P_\text{Na}[\text{Na}^+]_i + P_\text{Cl}[\text{Cl}^-]_o}\right)$$

where P is the relative membrane permeability of each ion.

2.2.3 *Equivalent Circuit Model for the Plasma Membrane*

Developing of an equivalent circuit model of the cell membrane is helpful for understanding membrane potential. The membrane is a lipid bilayer that is embedded with different types of ion channels. Ion channels act as a resister however, since they are characterized as being open or closed, thus they are variable resisters. Equilibrium potential for each ion is the electrical potential difference across the channel and is modeled as batteries.

There is a steady outflow of K^+ ions and an inflow of Na^+ ions, thus when left alone, this would drive the membrane potential toward 0. To prevent this, Na^+/K^+ pumps are used in equal and opposite directions to these passive currents and can be modeled as generators.

The cytoplasm and interstitial fluid are the electrical conductors and they are separated by the lipid bilayer of the membrane which has an insulating property. This feature can be modeled as capacitor. Capacitance for a cell membrane is approximately 1 $\mu F/cm^2$.

By combining the above ion channels as resistors, Na^+/K^+ pumps as generator, and lipid bilayer as capacitors, one can develop an integrated model of the cell membrane as shown in Fig. 2.2.

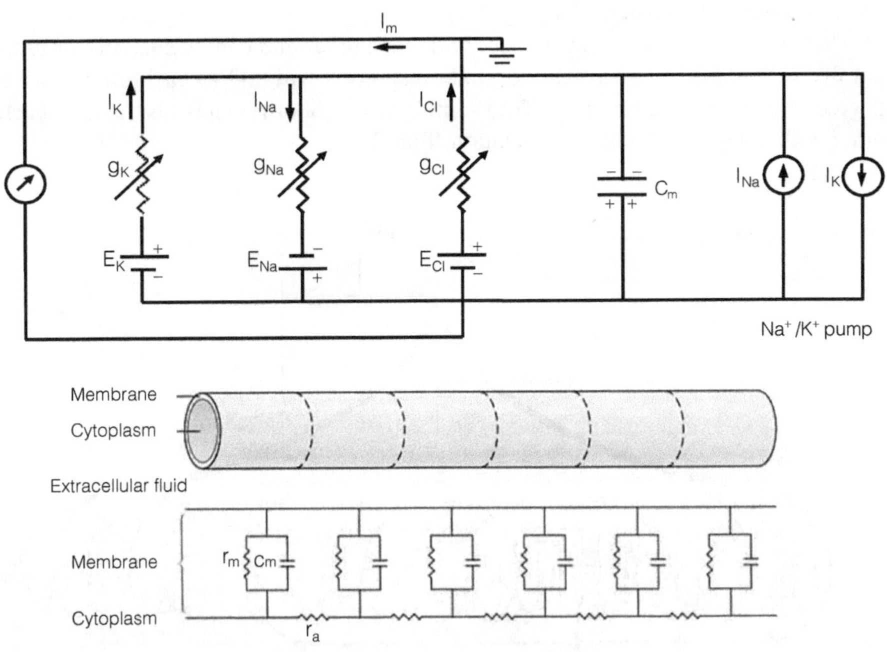

Fig. 2.2 Equivalent circuit model of the plasma membrane. Lipid bilayer acts as a capacitor and each ion channel acts as a variable resister. Na^+/K^+ pump acting as a generator is also illustrated

2.2.4 Graded Response of Membrane Potential

Changes in the open probability of sets of ion channels cause membrane potential change in limited areas. The change can depolarize or hyperpolarize the membrane potential. For example, either a decrease of K^+ permeability or an increase of Na^+ permeability will cause the membrane potential to move closer to the sodium equilibrium potential (depolarization). Opening either sodium channels or nonselective cation channels will lead inward movement of Na^+ thus causes depolarization by moving the membrane potential toward some value roughly midway between the sodium equilibrium potential and the potassium equilibrium potential. This graded potential spreads along the membrane by changing the charge on the membrane capacitance and by flowing through opened channels which are equivalent to a resistor. This structure is equivalent to a RC circuit and in this case a step change in current flow causes an exponential change in membrane potential. The time it takes for the membrane potential to reach about 63% ($1 - 1/e$) of its final value is the time constant (τ) of the membrane. The time constant can be calculated by multiplying the resistance and the capacitance of the membrane (Fig. 2.3).

As the current flows along the membrane, some of the current leaks through open channels in the neighboring areas. As a result the membrane potential progressively decreases with increasing distance from a current source. This spatial pattern is exponential and the distance where the voltage change to 37% ($1/e$) of its original value is the length constant (λ) (Fig. 2.4).

These grade responses can interact with each other and can be spatially or temporally summated. Two successive grade responses will add to each other with a degree of temporal summation. Two stimuli at neighboring sites also add to each other with a degree of spatial summation (Fig. 2.5).

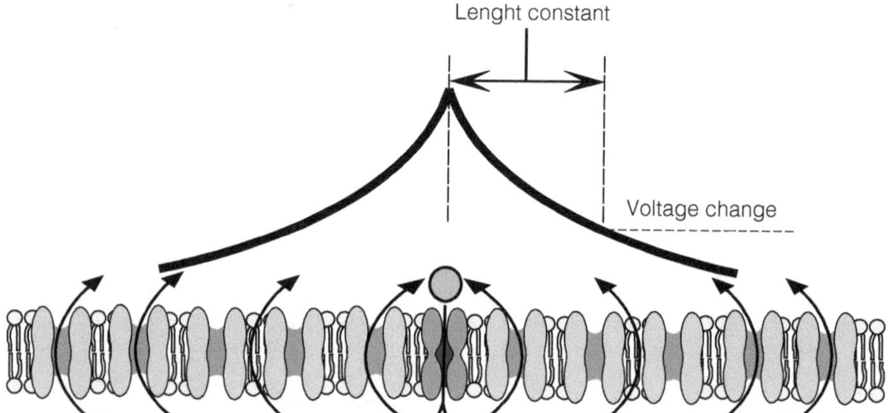

Fig. 2.3 An injection of current at one point causes a voltage decline exponentially along with distance. The distance where the voltage reaches about 37% of its original value is the length constant

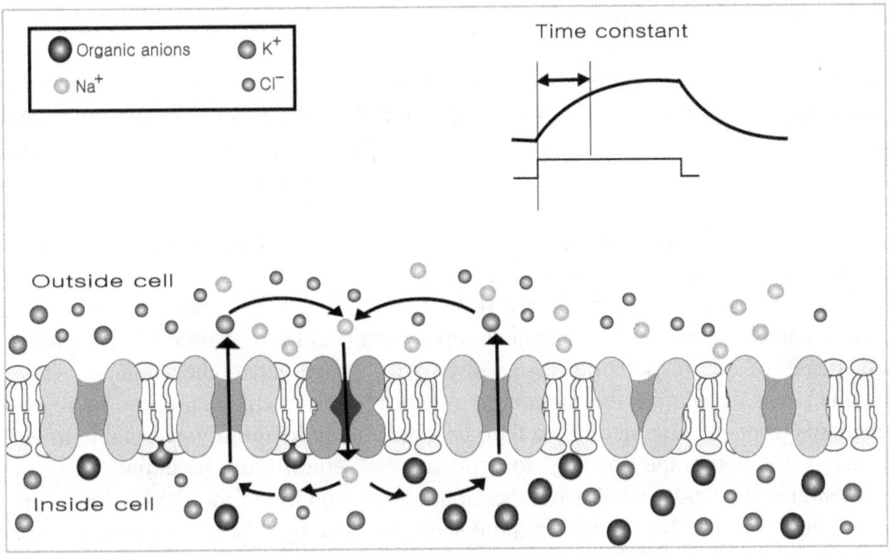

Fig. 2.4 An injection of square wave causes an exponential voltage change due to the parallel resistance and capacitance of the membrane. The time when the voltage reaches about 63% of its final value is the time constant

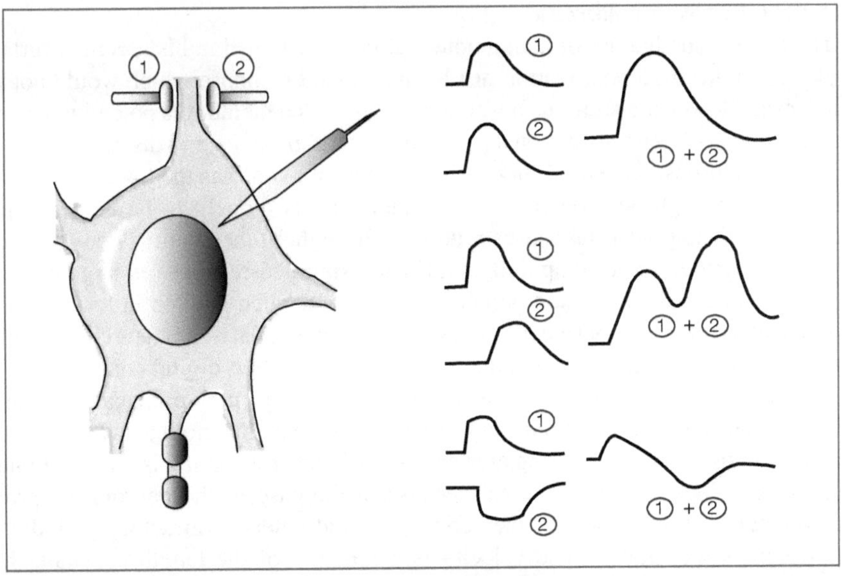

Fig. 2.5 Temporal and spatial summation. *Upper*: Simultaneous activation of two synapses results in spatial summation. *Middle*: Sequential activation of two synapses results in both spatial and temporal summation. *Low*: Spatial summation can occur with excitatory and inhibitory synaptic input

2.2.5 Action Potential

As most neurons or muscle cells are much longer than their length constants, the grade responses disappear when flowing along the cell, thus the responses cannot deliver signal from one end to the other in the cell. Excitable cells are distinguished by their ability to generate action potentials that can propagate without losing their amplitude.

The core structures for generating action potential are voltage-gated ion channels, mainly voltage-gated Na^+ and K^+ channels.

Depolarization by synaptic input or by receptor potential causes the opening of both channels; however Na^+ channels open faster and are responsible for the rising phase of action potential. Since they are voltage-gated, the opening of Na^+ channel will depolarize the membrane potential which will then cause more Na^+ channels to open. The membrane then becomes overwhelmingly permeable to Na^+ causing the membrane potential to approach Na^+ equilibrium potential. This phenomenon can be calculated using the Goldmann's equation described before. Once open, however, the Na^+ channels spontaneously close by inactivation gate and they cannot open again until the membrane potential returns to resting membrane potential. Closing of Na^+ channels causes the membrane potential to return to its resting level. In addition, K^+ channel start to open slowly and this facilitates the falling phase. The permeability of K^+ in this stage dominates and the membrane potential first approaches resting membrane potential then approaches K+ equilibrium potential (after hyperpolarization) (Fig. 2.6).

The all-or-none feature of action potential implies that stimulus less than certain level (threshold) of depolarization results in a graded response which would not be transferred. However a stimulus big enough to move the membrane potential beyond the threshold will generate action a potential that can propagate to distant regions of the cell. In neurons, the axon hillock (initial point of axon) has the lowest threshold with relatively high densities of Na^+ channels and is thought to be the principal trigger zone. The graded responses produced throughout the dendrites or cell body is summed spatially and temporally, and if the summed response is large enough to pass the threshold, an action potential will be generated at axon hillock. At this point the amount of response determines the frequency of action potential. Thus the function of the axon hillock is similar to that of an analogue-digital converter.

After generating an axon potential at trigger zone, it begin to propagate to neighboring segments of the membrane and depolarize them to threshold triggering action potentials in the next neighboring area and so on. This propagation is unidirectional, from axon hillock to axon terminal because in the case of the neuron, the proximal segment just traversed by the action potential enters a refractory period and thus becomes inexcitable. The velocity is a function of the length constant, that is, the longer the length constant the further an action potential can travel down axon segment before it decreases to subthreshold levels. To deliver the action potential faster, invertebrates have thick axon fibers, up to hundreds of micrometers in width, to increase the length constant. Vertebrates on the other hand have myelinated axons which allow rapid conduction with thin axons. Myelin acts as an insulating

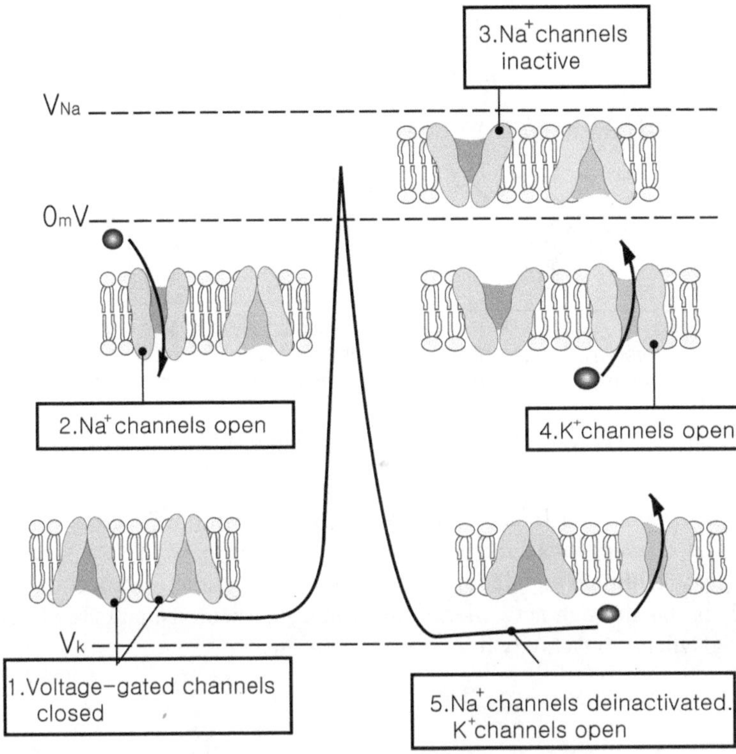

Fig. 2.6 Na^+ and K^+ channel opening and closing underlie action potential development. E_{Na}: sodium equilibrium potential, E_K: potassium equilibrium potential, V_m: membrane potential

sheath, allowing an action potential to spread along the axon until it gets to a node of Ranvier, which is a bare portion of axon without myelin. As a result, action potentials jump from one node to the next and so on. This conduction is called saltatory conduction. Saltatory conduction not only functions as a method of fast conduction of action potentials, but it also has the function of verifying information by ensuring that the frequency of action potential is correct (Fig. 2.7).

2.2.6 Synaptic Transmission

One of the main functions of neurons is transferring information form one neuron to another. The synapse is the place where this process occurs. There are two types of synaptic transmissions; electrical and chemical synapses transmissions. Electrical synapses refer gap junctions between two cells. These synapses separate two adjoining neurons by a few nanometers. The cytoplasm of one cell is connected to the next cell through channels named connexons. Current can flow through these channels either way, thus depolarization and hyperpolarization can spread from one cell

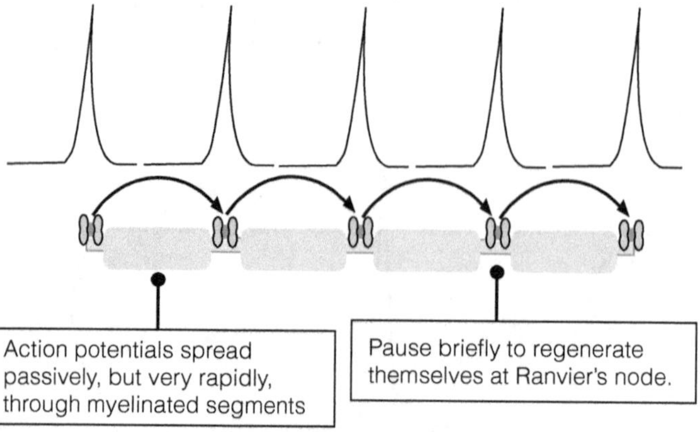

Fig. 2.7 Saltatory conduction of action potential along a myelinated axon

to next cell instantaneously and form an electrical syncitium. The electrical properties of this channel thus follow that of a graded response. These synapses are present only in a limited population of neurons. Gap junctions are good at spreading electrical signals through networks of interconnected neurons, and are effective in developing synchronic activity in clusters of neurons.

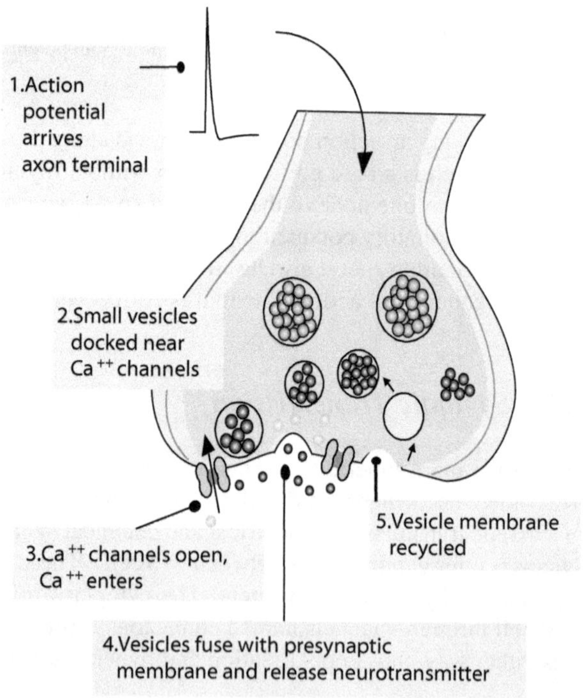

Fig. 2.8 Release of neurotransmitter

Chemical synapses are the main measure of inter-cellular communication and are different from electrical synapses. There is no continuity of cytoplasm at all and the direction of signal transmission is unidirectional (There are some exceptions. However, main frame is unidirectional). The gap between two cells is extracellular space and named the synaptic cleft. When the action potential reaches the axon terminal, intracellular Ca^{++} increases by opening of voltage dependent Ca^{++} channels. The increased Ca^{++} levels triggers fusion of synaptic vesicles wherein neurotransmitters are stored within neuronal membrane (Fig. 2.8). Glutamate, GABA, acetylcholine, dopamine, and several other chemicals are used as neurotransmitters. The released neurotransmitters bind to specific receptors of the post-synaptic cell membrane and change the membrane potential. Chemical synapses are similar to digital to analogue converter. They have much more signaling flexibility than electrical synapses thus allows for more room for plastic changes of synaptic transmission which is thought be the basis of learning and memory.

2.3 Equivalent Circuit Model of Tissues and Organs

The electrical properties of tissues and organs are basically the combination of the electrical properties of cells that make up the tissue or the organ. As shown in Fig. 2.9, one can make a model of a body segment, which contains skin, fat, muscle, bone and extracellular fluid (ECF) components. This model represents typical limbs [6]. Nerves and vessels components can be added, however their proportion is small in terms of the magnitude of resistance and capacitance. Modeling of the trunk portion of the human body is also possible by adding other components, for example an air component in case of a lung or gut. The values of resistance and capacitance of each component are sum of the composing cells.

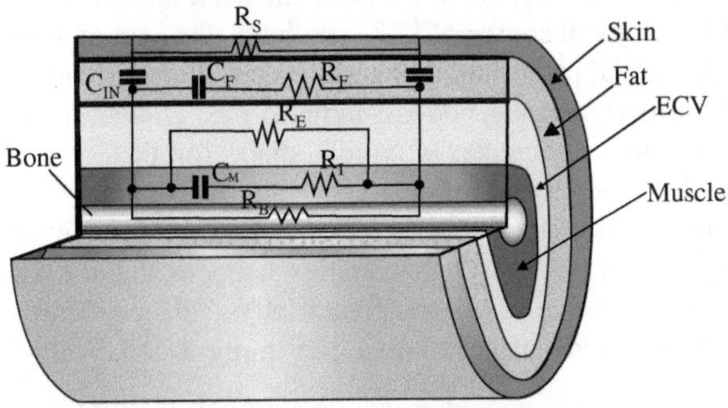

Fig. 2.9 The equivalent model of segmental body composition with the components of skin, fat, muscle, extracellular and intracellular fluid volume and bone. R_S: skin resistance, C_{In}: connect capacitance between skin and fat, C_F: capacitance of the fat, R_F: resistance of the fat, C_M: capacitance of the muscle cell membrane, R_I: resistance of the sarcoplasmic fluid, R_E: resistance of the extracellular fluid, R_M: resistance of muscle, R_B: resistance of bone (modified from Ref. [3])

When measuring the electrical properties of the body, the main problem is the impedance developed between electrodes and skin. This issue is beyond the scope of this chapter. Detailed information on this subject can be found elsewhere [7].

2.4 Biomedical Devices

The main sources of bioelectrical signals are generated by muscles, the heart or the brain. Chemical or mechanical signals also can be measured after being converted to electrical signals. The initial point of measuring is called the sensor. The outputs of the sensor are analog signals, which are amplified, filtered, conditioned, and converted to digital signals through a processor and digital converter. Once the analog signals are converted to digital signals, they can be stored and processed by many different methods [5].

Although the electrical signals that different instruments measure do not look the same, they share the same underlying mechanisms. All instrument systems also have an output display device, a storage device, a calibrating system and feedback elements. The display device enables users to view signals numerically or graphically in a discrete or continuous way. The storage device allows operators to carry out further analysis after obtaining the data.

This chapter will introduce basic mechanisms and applications of major biomedical instruments.

2.4.1 Electrocardiography

Electrocardiography (ECG or EKG) is routinely performed in clinical practices to measure the heart function. With ECG doctors can detect many clinical situations such as arrhythmia, cardiac hypertrophy, ischemic heart disease and even electrolyte imbalances. The biological source of ECG rhythm is the sum of action potentials in heart muscle cells. Serial propagation of action potential from sino-atrial node (pace maker) through atrio-ventricular node, Bundle of His, bundle branches and finally to ventricular muscle cells creates a typical shape for ECG rhythm (Fig. 2.10). The first wave, P is developed when the atrium is excited. The PR interval is when the action potentials stay for 100–200 ms to ensure time for blood flow from atrium to ventricle. The second QRS complex happens mainly when ventricle is excited and the third T wave is when the ventricle is relaxed. Thus, changes in the shape, timing or amplitude of each wave can imply a certain dysfunction in the heart.

Conventional ECG examination use 12 channels with 9 skin surface electrodes and a reference electrode. One electrode is attached to each limb and among those electrodes three are for picking up ECG signal and the remaining one (right leg) is a reference electrode. Six chest electrodes are connected on the thoracic surface at defined positions near the heart (Fig. 2.11).

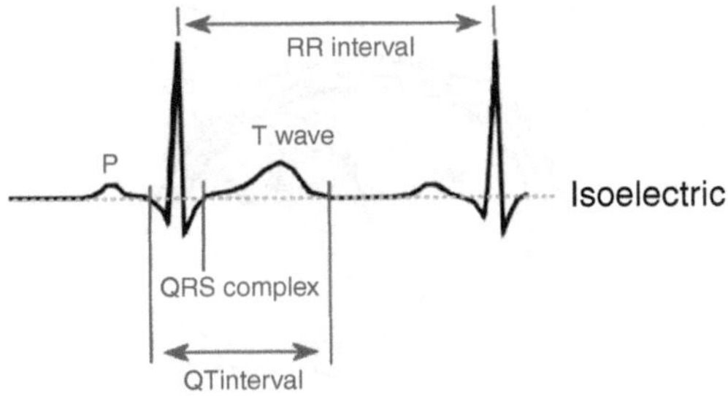

Fig. 2.10 Typical ECG rhythm

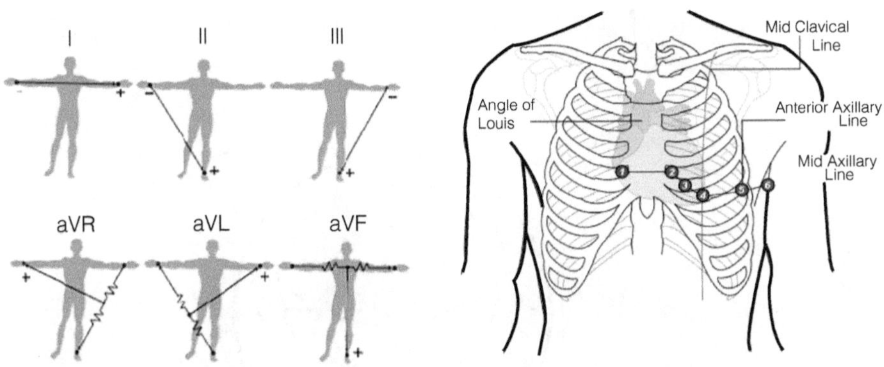

Fig. 2.11 Leads in electrocardiography recording. Six limb leads (*left*) and six chest leads are illustrated

Four electrodes are attached to each limb; right arm (RA), left arm (LA), and left leg (LL) for measuring the signal and right leg (RL) for reference. Three bipolar electrodes (I, II, III) measure the voltage difference between RA-LA in channel I, RA-LL in channel II, LA-LL in channel III. Einthoven, the inventor of the first practical ECG found the following difference among the electrodes: $u_{II} = u_I + u_{III}$. The three unipolar electrodes (aVR, aVL, aVF) are voltage differences between one limb and sum of the other two limbs. Signals from these electrodes have a higher voltage, and they are called augmented electrodes. Interestingly the vector leads of the augmented leads interlace the vector angle of lead I, II and III (0°, 60°, 120°) so that each 30° is covered. Six chest electrodes are for measuring the cardiac electrical activities in transverse plane. The electrodes are unipolar with reference of RA or the sum of three limb leads. With these 12 electrodes one can detect the cardiac vectors both in coronal and transverse plane.

2.4.2 Electroencephalography

Electroencephalography (EEG) measures the electrical activity of 10^{11} neurons underneath the scalp. In some situations, electrodes are placed on the surface of

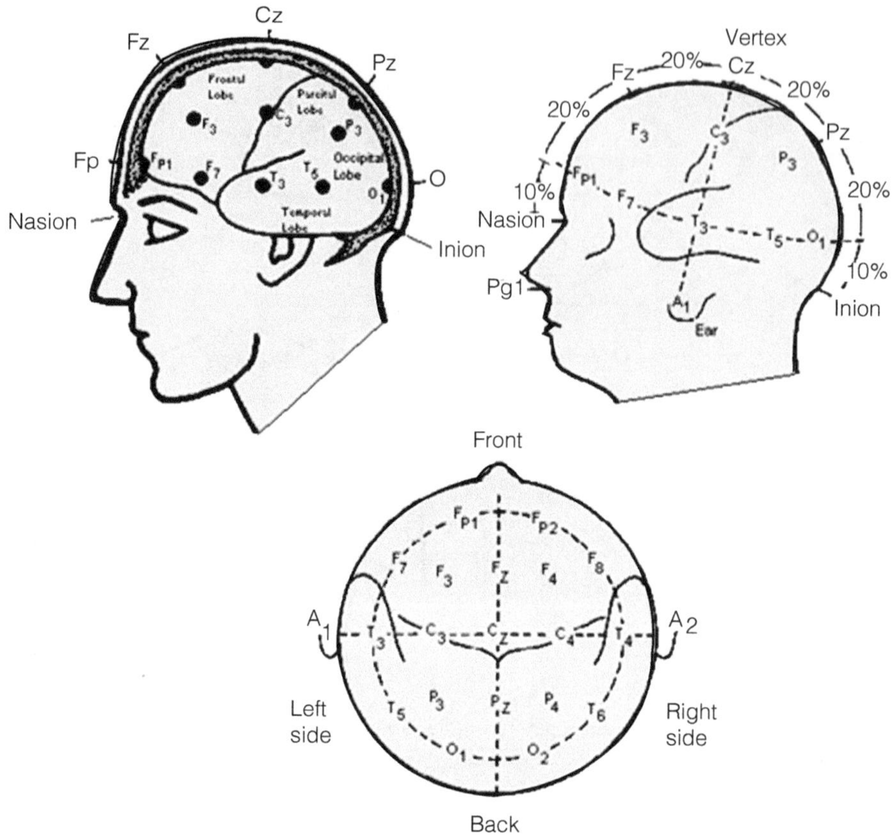

Fig. 2.12 Positioning of international 10–20 EEG electrode system. F: frontal lobe, P: parietal lobe, T: temporal lobe, O: occipital lobe

brain itself after or during surgery. In this case it is called Electrocorticogram (ECoG). Standardized EEG recording uses 21 electrodes (Fig. 2.12). The leads can be bipolar or unipolar according to the montage the operator selects. When recording, one can easily record oscillating electrical activity. The amplitude is on the order of 50 μV and the frequency between 1 and 50 Hz. The waves change markedly according to the status of brain, for example depending on the conscious level of the subject, different brain waves will be recorded by the EEG machine. Some of the changes in EEG waves are characteristic of specific abnormalities of the brain, such as epilepsy or brain death. Others are found in normal people and classified according to frequency level. Alpha waves are rhythmic waves occurring at a frequency between 8 and 13 Hz. They are recorded in almost all people when they are awake in a quiet and resting state. Beta waves in the frequency range from 13 to 30 Hz, and are recorded during mental activities and tension. Theta waves occur at frequencies between 4 and 7 Hz. It is prominent during drowsiness. Delta waves include all the

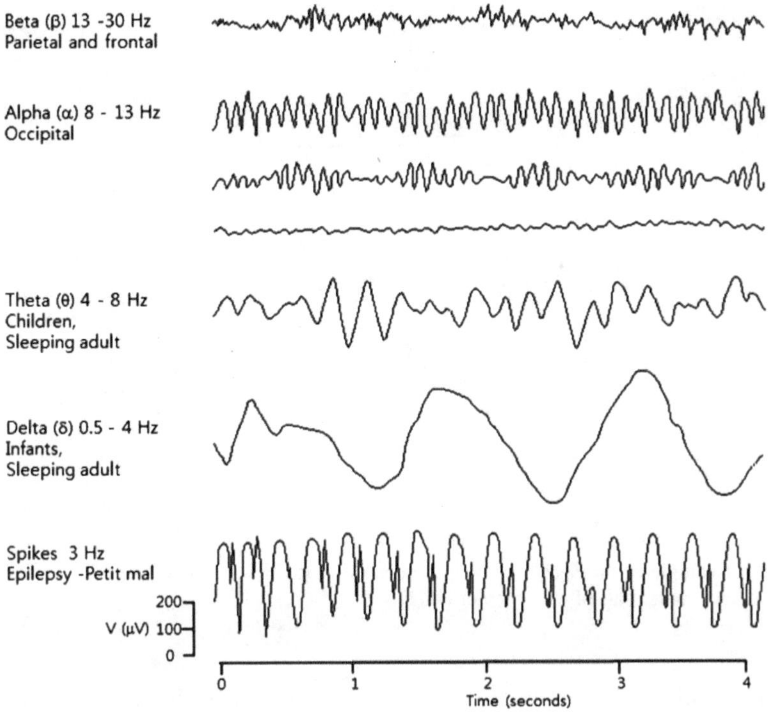

Fig. 2.13 Examples of EEG waves

waves below 3 Hz. It occurs in deep sleep, in babies and in serious organic brain disease (Fig. 2.13).

One application of EEG is sleep studies. When an individual becomes inattentive, drowsy and falls asleep, the alpha rhythm is replaced by slower, larger waves. In deep sleep, large, irregular delta wave appears with bursts of alpha-like activity called sleep spindles. The high-amplitude, slow waves are sometimes replaced by rapid low voltage irregular activity resembling that those obtained in alert subjects. This stage is called paradoxical sleep or rapid-eye-movement (REM) sleep. Other stages are nonrapid-eye-movement, NREM, or slow-wave sleep.

2.4.3 Electromyography

The functional unit of skeletal muscle is the motor unit which includes motor neurons, its axon and muscle fibers innervated by the motor neuron. Cross sectionally, however, the muscle fibers of a motor unit are interspersed with fibers of other motor units. Thus the muscle fiber component of a given motor unit constitutes a distributed electric source located in a volume conductor consisting of all other muscle fibers, both active and inactive. The potential from active fibers has a triphasic form

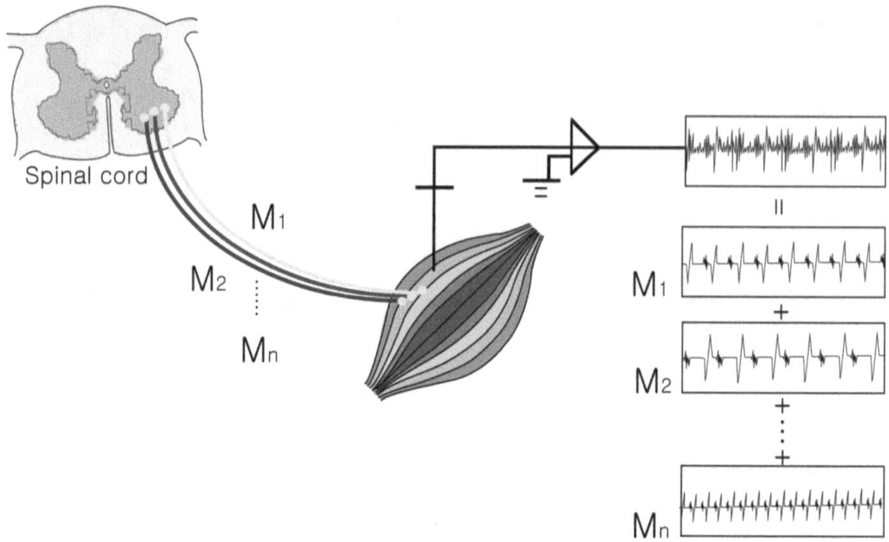

Fig. 2.14 Example of an electromyography recording. Recorded EMG waves are compounded from the action potentials of many motor units (M)

of short duration with an amplitude of 20–2000 μV, depending on the size of the motor unit (Fig. 2.14).

Surface electrodes are convenient but only can detect the activities of surface muscles and have less spatial resolution. Various types of monopolar, bipolar, and multipolar insertion-type electrodes are commonly used for recording the activity of deep muscles and from single motor units.

In clinical studies, the main purpose of EMG is to differentiate whether symptoms such as weakness are from dysfunction of neurons or from the muscles. With neuronal diseases, spontaneous activities and giant muscle unit potentials are observed. In contrast with muscle diseases the amplitude of each motor unit potential is decreased.

2.5 Current Research Trends in Biomedical Electrical Instruments

From the engineering perspective, the technology for biomedical instruments does not need to be either advanced or sophisticated. In terms of technology, the level of currently developed bioelectrical devices seems to have reached their plateau stage. Thus further advances or innovations should be found through different approaches.

One of the directions biomedical instrument device development is headed in is to provide users the interpretation of measured data. By using a certain algorithm, the ECG system provides not only simple data such as the heart rate or irregularity, it provides many possible clinical situations. Even though physician should verify the interpretation, the accuracy rate is impressive. The EEG system also provides brain mapping solutions thus allowing the physician to detect the source region of abnormal brain waves. This kind of approach is promising since it reduces the clinical burden to doctors, which may result in patients getting better medical service at less financial cost.

Medical instruments are also becoming miniaturized, portable, and designed to use less electrical energy. All these are becoming possible due to the development of microchips and better performing batteries. Also recently developed instruments can be connected to a network. This means patient data can be stored and delivered to a database using a personal computer or even cellular phones. One example of this technology is a hybrid of cell phone with a blood glucose tester. When a diabetic patient checks his or her blood glucose level with tester attached to the cell phone, the results are transmitted to the physician and family to be stored on a PC or database. By using this device the patient can get more dedicated attention and care, and therefore might have better prognosis. This idea can be applied to other devices that can monitor EKG or EEG. By creating hand-held devices that can monitor these things we can predict possible heart attacks or seizures.

Nowadays, all scientific and engineering fields emphasize multi-disciplinary approaches and fusion or converging technology. Electrical engineering for biomedical instruments is no exception. Above all, engineers should have a better understand of anatomy and the physiologic nature of biological systems and clinical needs. With this knowledge, engineer can identify the real problems and can provide novel solutions for unsolved problems. By converging information technology and biological technology we can expect the beginning of a new era in the field of biomedical devices.

References

1. Enderle J (2004) Bioelectric phenomenon. In: Introduction to biomedical engineering, 2nd edn. Elsevier, Oxford, UK, pp 627–692
2. Grimnes S, Martinsen OG (2008) Bioimpedance and bioelectricity basics, 2nd edn. Academic Press, Oxford, UK
3. Hille B (2001) Ion channels of excitable membranes, 3rd edn. Sinauer Associates, Sunderland, MA
4. Aidley DJ (1998) The physiology of excitable cells. Cambridge University Press, Cambridge, UK
5. Delmar M (2006) Bioelectricity. Heart Rhythm 3:114–119
6. Zhu F, Leonard EF, Levin NW (2005) Body composition modeling in the calf using an equivalent circuit model of multi-frequency bioimpedance analysis. Physiol Meas 26:S133–43
7. Geddes LA (1972) Electrodes and the measurement of bioelectric events. Wiley-Interscience, New York, NY
8. Reilly JP (1998) Applied bioelectricity. From electrical stimulation to electropathology. Springer, New York, NY

Chapter 3
Biomedical Electrodes For Biopotential Monitoring and Electrostimulation

Eric McAdams

3.1 Introduction

Biomedical electrodes are used in various forms in a wide range of biomedical applications including

(i) the detection of bio-electric events such as the electrocardiogram (E.C.G.)
(ii) the application of therapeutic impulses to the body e.g. cardiac pacing and defibrillation and transcutaneous electrical nerve stimulation (T.E.N.S.)
(iii) the application of electrical potentials in order to facilitate the transdermal delivery of ionized molecules for local and systemic therapeutic effect (iontophoresis) and
(iv) the a.c. impedance characterization of body tissues.

Good electrode design is not as simple and straight forward a matter as is often assumed and all electrode designs are not equal in performance [41]. One must therefore not simply chose an electrode with as conductive a metal plate as possible. Unfortunately this was and appears to still be the case in many designs. Probably due to this mistaken view, it would appear that the associated electronic systems are often first developed and the electrode design is left to the end, almost as an after thought. If the clinician is to properly diagnose the patient's cardiac problem (for example), it is imperative that the measured biosignal is clear, undistorted and artefact free. Unfortunately, monitoring bioelectrodes, if they are not chosen correctly, give rise to significant problems that make biosignal analysis difficult, if not impossible. Similarly, stimulation electrodes must be well chosen if they are to optimally supply the therapeutic waveforms without causing trauma to the patient.

E. McAdams (✉)
Biomedical Sensors Group, INSA (National Institute of Applied Science) of Lyon, Léonard de Vinci Building, 20 Avenue Albert Einstein, 69621 Villeurbanne Cedex, France
e-mail: et.mcadams@ulster.ac.uk

H.-J. Yoo, C. van Hoof (eds.), *Bio-Medical CMOS ICs*,
Integrated Circuits and Systems, DOI 10.1007/978-1-4419-6597-4_3,

Current or charge is carried by ions inside the patient's body and by electrons in the electronic device itself and in its leads. The "charge transfer" mechanism between current/charge carriers takes place at the electrode-patient interface and is of major importance in the design of an optimal electrode. Both the electrode-electrolyte interface and the skin under the electrode (collectively known as the "contact") give rise to potentials and impedances that can distort the measured biosignal or adversely affect the electrotherapeutic procedure.

Implanted electrodes are generally made from "inert" or "noble" materials which do not react with surrounding tissues. Unfortunately, as a consequence, they tend to give rise to large interface impedances and unstable potentials. Implanted biosignal monitoring electrodes, in particular, require stable potentials and low interface impedances to minimize biosignal distortion and artifact problems. External biosignal-monitoring electrodes can generally use high electrical performance "non-noble" materials such as silver-silver chloride with out fear of biocompatibility problems [53]. They do however have to address the additional and very significant problem of the skin with its sizeable impedance and unstable potential. Along with the desired biosignal, one amplifies the difference between the two contact potentials. If the contact potentials were identical (highly improbable), they would cancel each other out due to the use of a differential amplifier. If the potential mismatch were very large (several hundred mV) the amplifier would not be able to cope and would saturate. If the mismatch in contact potentials is small and stable, this mismatch will be amplified along with the biosignal and the biosignal will appear shifted up or down on the oscilloscope screen or printout paper. This would generally not a major problem as the additional voltage offset can be easily removed. What is a significant problem, however, is when the contact potentials fluctuate with time. Their mismatch therefore varies and the baseline of the biosignal is no longer constant. This leads to the problem termed "baseline wonder" or "base line drift" which makes analysis of some of the key features of the biosignal difficult. Filtering out of the drift is often not an option as the filtering often also removes key components of the desired biosignal.

Large mismatched contact impedances can cause signal attenuation, filtering, distortion and interference in biosignal-monitoring. If contact impedances are significant compared to the input impedance of the amplifier, they can give rise to signal attenuation as a result of the "voltage divider effect". Attenuation of the signal is not a major problem, after all we are going to use the amplifier to amplify the signal by a factor of around 1000 (in the case of an ECG). A significant problem arises, however, due to the fact that the contact impedance varies with frequency. The frequency-dependence of the contact impedance is a consequence of the presence of parallel capacitances at the electrode-electrolyte interface and/or at the skin under the electrode. At very high frequencies the contact impedances are very small and there is therefore no attenuation of the high frequency parts of the biosignal. At low frequencies the contact impedances *can* be very large and hence there can be significant attenuation of low frequency components of the biosignal. The overall signal is not only attenuated, it is also distorted with its low frequency components selectively reduced. The measurement system in effect acts as a high pass filter and the signal is differentiated. In the case of the ECG, the P, S and T waves are deformed, leading in particular to a modification of the S-T segment. The S-T segment is of vital importance to the electrocardiologist, hence the importance of avoiding such biosignal distortions.

50/60 Hz interference can be amplified along with any monitored biosignal due to the mismatch of the contact impedances. Displacement currents flow from power lines through the air to the monitor cables and then through the electrodes and the patient to ground. If the contact impedances are not identical, the displacement currents flowing through the two contact impedances connected to a differential amplifier will give rise to different voltages at the amplifier's inputs. This 50 Hz "offset" voltage will be amplified along with the desired biosignal and its amplitude is proportional to an electrode-skin impedance mismatch [Webster 1998].

Other applications such as electrical impedance plethysmography and electrical impedance tomography [23], do not monitor intrinsic biosignals emanating from the body, but inject small currents or voltages into the body and record the resultant voltages or currents. The electrical properties of the body or a body segment can then be calculated. In many of these applications, the magnitude and mismatch of contact impedances can give rise to significant errors or artifacts [94]. As relatively high frequencies are often involved in these techniques, even the series resistance of the gel pad (which is generally ignored) may become significant.

Although interface impedance and potential are generally less critical for implanted stimulation electrodes, many such electrodes (e.g. implanted pacing electrodes) are used to monitor biosignals as well as to deliver the required stimulation impulses. Even in the case of a purely stimulating electrode, a low interface impedance is required to minimize energy wastage and to prolong the life of the power source. Various techniques are therefore used to effectively decrease the otherwise large interface impedances of the "noble" or "inert" materials used for their "biocompatible" properties. Electrode material and high electrical performance is generally less critical for external stimulation electrodes such as TENS and external defibrillation electrodes. Current density distribution is of major importance in these applications in order to avoid electrical "hotspots" and resultant burns to the skin. In some applications such as TENS and external pacing, it is even sometimes advantageous to use a relatively resistive electrode material or gel as this has been found to optimize current density distribution under the stimulation electrode.

As in the above applications, the avoidance of current density hot spots is one of several key factors in iontophoretic, transdermal delivery [129]. An additional important constraint which is generally not relevant in other electrotherapies is the maintenance of the delicate electrochemical balance at the electrode/reservoir/skin interface. The electrode potential and impedance as well as the composition of the drug reservoir must generally remain within certain narrow ranges in order to avoid the deterioration of the electrode, the contamination of the drug reservoir and the irritation of the patient's skin.

The electrical properties of the electrode contacts are therefore of great importance in most applications. Ideally the contact with the patient should give rise to

Zero potential. Unfortunately this is not possible and a more realistic goal is to achieve a low, stable potential at each of the contacts

Zero Impedance. Unfortunately this too is not possible and a more realistic goal is to achieve impedances at the two contacts which are low and as similar as possible

The potentials and impedances of the electrode-electrolyte interface and the skin will therefore be studied in more depth in the following sections.

3.2 Electrical Properties of Electrode-Skin Interface

As briefly outlined above, the electrode-electrolyte interface and the skin under the electrode both give rise to potentials and impedances that can either distort any measured biosignal or give rise to problems during electrical stimulation.

3.2.1 The Electrode-Electrolyte Interface

3.2.1.1 The Electrode-Electrolyte Potential

When a metallic electrode comes in contact with an electrolyte (in body tissues or in an electrode gel), an ion-electron exchange occurs as a result of an electro-chemical reaction. There is a tendency for metal atoms M to loose n electrons and pass into the electrolyte as metal ions, M^{+n}, causing the electrode to become negatively charged with respect to the electrolyte. (Fig. 3.1). Reaction (1) is termed oxidation.

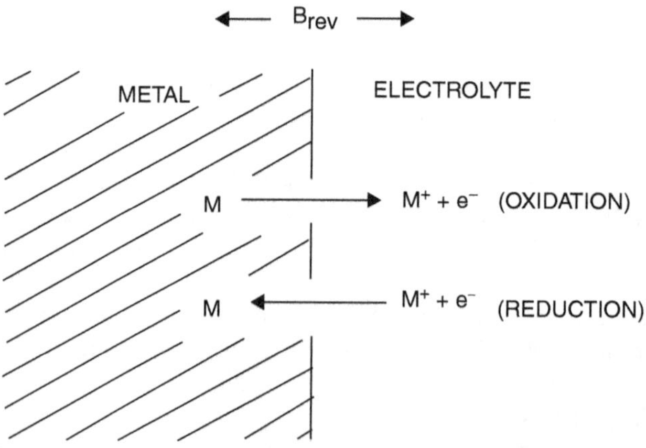

Fig. 3.1 The Electrode-electrolyte interface and reactions involved in generating its reversible or equilibrium potential

$$M \rightarrow M^{+n} + ne^{-} \tag{3.1}$$

Similarly, under equilibrium conditions, some of the ions in solution, M^{+n}, take n electrons from the metal and deposit onto the electrode as metal atoms, M. Electrode becomes positively charged with respect to the electrolyte. Reaction (2) is termed reduction

$$M^{+n} + ne^{-} \rightarrow M \tag{3.2}$$

The overall chemical reaction taking place at the interface is therefore

$$M \rightleftharpoons M^{n+} + ne \tag{3.3}$$

Under equilibrium conditions, the rate at which metal atoms lose electrons and pass into solution is exactly balanced by the rate at which metal ions in solution deposit onto the electrode as metal atoms. The current flowing in one direction, i_O, is equal to and cancels out the current flowing in the opposite direction. The electrode is said to be behaving reversibly and the common value of currents, i_O, is termed the "exchange current (density)". Although the net current flowing through the electrode interface is zero, a potential difference is found to exist between the electrode and the electrolyte and depends on the position of the equilibrium between the two processes (1) and (2). Generally the metal is negative relative to the electrolyte. The potential difference depends on the relative activities (or concentrations) of the ions present and on the electrode metal [9]. This potential has been termed the "equilibrium", "reversible" or "half-cell" (i.e. one interface only) potential in the literature.

When trying to measure the potential of a "half-cell" (i.e. one interface only), one is immediately faced with a problem as one requires two electrodes to make a potential measurement, thus effectively creating an electrochemical cell with two electrode-electrolyte interfaces. One therefore measures not only the potential of the electrode-electrolyte interface under study, but also that of the second electrode used to complete the circuit. If one uses the same metal for the second electrode to that used in the first, the potentials will be identical (in theory at least) and will cancel each other out. The measured potential will be (theoretically) equal to zero. [In practice however, slight differences in the composition of the metal used, the electrode surfaces and in the gel will result in differences in the two "half-cell" potentials]. If on the other hand one uses a different metal for the second electrode, the measured potential of the cell will be due to the combination of the potentials of the two "half-cells". It will be impossible to separate the potential of the "half-cell" under investigation.

In order to resolve this problem, early electrochemists decided to measure all electrode interface potentials with respect to a standardized electrode or reference electrode. The standard hydrogen electrode (SHE) was chosen to be the universal reference electrode and its "half-cell" potential was specified as zero. Other metal-to-ion interface potentials were then measured with reference to SHE and the entire measured offset voltage was attributed to electrode system being tested.

Hydrogen electrode consists of platinized plate submerged to one half its height HCl over which hydrogen gas at one atmosphere is bubbled. The "half-cell" potential of SHE depends on concentration of hydrogen ions in the solution and hence it is quite stable and reproducible. At the time that this decision was agreed, the necessary glass blowing and silver soldering were common skills and the SHE was thus easy and inexpensive to make. Although it is however no longer convenient for modern routine measurements as a reference electrode (the flowing hydrogen gas is potentially explosive), electrode potentials are standardized with respect to the SHE [9].

The "reversible", "equilibrium", or "half-cell" potential of a given electrode-electrolyte interface depends on the activity (almost synonymous with concentration) of the ions taking part in the reactions (see Table 3.1). This potential, E_{rev}, is given by the Nernst equation,

$$E_{rev} = E_O + [RT/nF]\ln[\text{activity of oxidized form/activity of reduced form}] \quad (3.4)$$

E_{rev} is the "reversible", "equilibrium", or "half-cell" potential
E_O is the standard half-cell potential (measured relative to the standard hydrogen electrode)
R the universal Gas constant,
n the number of electrons involved in reaction,
T the absolute temperature (°K),
F the Faraday constant.

Activity, $a = \gamma C$

Where C is concentration and γ, the activity coefficient, is a measure of the interaction between ions. When solution is infinitely dilute, $\gamma = 1$ and activity is equal to concentration.

Note the two components of E_{rev}. One is constant, E_O, while the other will vary due to slight variations in concentration etc., from one electrode to another. If two chemically identical electrodes make contact with the same electrolyte/body, the two interfaces should, in theory, develop identical half-cell potentials. When connected to a differential amplifier, the half-cell potentials of such electrodes would cancel each other out and the "offset voltage" would be zero. The electrode potentials would therefore make zero contribution to a biosignal they were being used to detect. Unfortunately, slight differences in electrode metal or gel result in the creation of offset voltages which can greatly exceed the physiological variable to be measured. Generally, a more significant problem is that the electrode offset voltage can fluctuate with time, thus distorting the monitored biosignal [6].

Table 3.1 Reversible potentials for common electrode materials at 25°C. The metal undergoing the reaction shown has the magnitude and polarity of standard half-cell potential, E_O, listed when the metal is referenced to the Standard Hydrogen Electrode. [Webster, 1995]

Metal and reaction	Potential E°, V
$Al \rightarrow Al^{3+} + 3e^-$	−1.706
$Zn \rightarrow Zn^{3+} + 2e^-$	−0.763
$Cr \rightarrow Cr^{3+} + 3e^-$	−0.744
$Fr \rightarrow Fe^{2+} + 2e^-$	−0.409
$Cd \rightarrow Cd^{2+} + 2e^-$	−0.401
$Ni \rightarrow Ni^{2+} + 2e^-$	−0.230
$Pb \rightarrow Pb^{2+} + 2e^-$	−0.126
$H_3 \rightarrow 2H^+ + 2e^-$	0.000 by definition
$Ag + Cl^- \rightarrow AgCl + e^-$	+0.223
$2Hg + 2Cl^- \rightarrow Hg_3Cl_2 + 2e^-$	+0.268
$Cu \rightarrow Cu^{2+} + 2e^-$	+0.340
$Cu \rightarrow Cu^+ + e^-$	+0.522
$Ag \rightarrow Ag^+ + e^-$	+0.799
$Au \rightarrow Au^{3+} + 3e^-$	+1.420
$Au \rightarrow Au^+ + e^-$	+1.680

Half-cell potentials for common electrode materials at 25°C.
The metal undergoing the reaction shown has the sign and potential E° when referenced to the hydrogen electrode.

3.2.1.2 The Electrode-Electrolyte Impedance

It has already been stated that in the electrode and the connecting lead, electrical charge is carried by electrons whereas in the gel and in the human body charge is carried by ions. There is a transition at the interface between the electrode and the electrolyte where charge is transferred from one kind of carrier to the other. In order for some of the ions in the electrode gel or in the body fluids to transfer their charge across the interface, many must first diffuse to the electrode-electrolyte interface under the influence of electrostatic attraction. Here they "stick" (or "adsorb" as it is termed in electrochemistry) to the electrode surface and form the outer Helmholtz plane (OHP) [9]. If the electrode has a negative charge relative to the electrolyte, positive ions will be attracted to the interface region and adsorb onto the electrode surface. As a consequence, there is a layer of negative charge on the metal surface and a layer of equal but opposite charge on the electrolyte side of the interface, both separated by a small distance across the OHP (see Fig. 3.2). A "double layer" of charge therefore exists at the interface and such a system behaves like a parallel plate capacitor. Not altogether surprisingly, the interface's capacitance is often termed the double layer capacitance, C_{dl}, and is connected in parallel to the charge transfer resistance in our simple equivalent circuit model.

Just in case one believed that the electrode interface was that simple, one must point out, for example, that as well as the cations electrostatically attracted to the negatively-charged electrode surface (coulombic adsorption) there may be anions that are adsorbed on the electrode surface and form the inner Helmholtz layer or

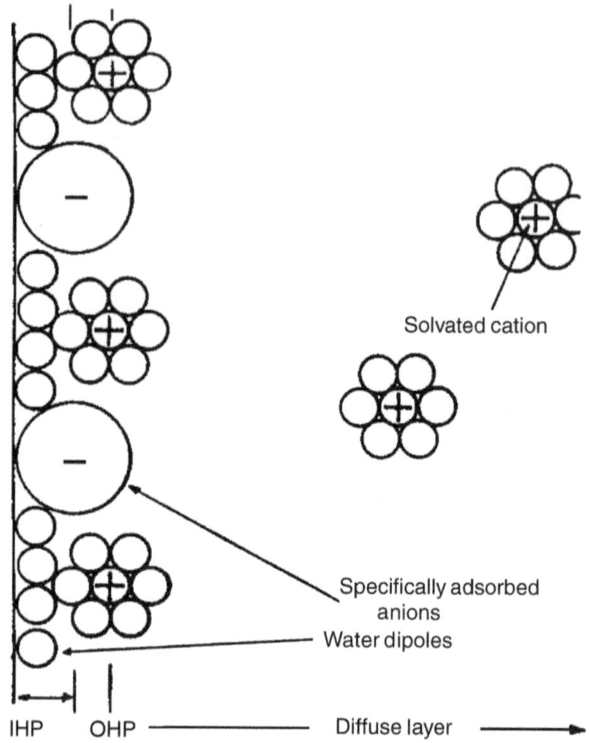

Fig. 3.2 The Electrode-electrolyte interface. A metal in an electrolyte forms a double layer of charge. Redrawn from Bard et al. [9]

plane (IHP). These anions have tended to loose their hydration sphere and consequently are in close contact with the electrode. As they are negative ions adsorbed onto a negative electrode surface, electrostatic forces can not be responsible. Some force *specific* to the ion (rather than its electrical charge) must be responsible, hence the use of the term "Specific Adsorption" to describe this phenomenon. Van der Waals or chemical forces are thought to be responsible [9].

In order to understand some aspects of the double layer capacitance, it is good to consider the basic equation for a parallel plate capacitor. If two identical conductive plates, each of area A cm^2 are separated by a distance d cm which is filled with a material of dielectric constant ε_O, then the capacitance of this parallel plate capacitor, C_{pp}, is given by:

$$C_{pp} = \varepsilon_O A/d \tag{3.5}$$

and the magnitude of the capacitive impedance, Z_{pp}, is given by

$$Z_{pp} = 1/2\pi f C_{pp} \tag{3.6}$$

where f is the frequency of the applied ac signal and π is a constant.

Some dc (or faradaic) current does however manage to leak across the double layer due to electrochemical reactions (1) and (2) taking place at the interface. These reactions experience a "charge transfer" resistance, R_{CT}, which can be thought of as shunting the non-faradaic, double layer capacitance and whose expression can be derived from the Butler-Volmer equation.

For small applied signal amplitudes, [9].

$$R_{CT} = \frac{RT}{nF}\frac{1}{i_o} \tag{3.7}$$

A good electrode, from an electrical point of view, will have a very low value of R_{CT}. Charge will be transferred across the interface almost unimpeded and little voltage will be dropped across the interface. One should note from (7) that R_{CT} is inversely proportional to i_O. i_O is the "exchange current", i.e. the current flowing across the interface (in both directions) under equilibrium conditions (no net current flow). Simplistically, if an interface can cope with large currents under equilibrium conditions, it will be able to cope well with currents under non-equilibrium conditions. A good electrode system will therefore be characterized by a large value of exchange current or a low value of R_{CT}.

The interface impedance should theoretically be well represented by an equivalent circuit model comprising the double layer capacitance in parallel with the charge transfer resistance, R_{CT}. Both are in series with R_{TOTAL}, the relatively small resistance due to the sum of the lead and electrolyte resistances.

3.2.1.3 Complex Impedance Plot

If, for each frequency of AC signal used to measure the impedance, the real part of the measured impedance (Z' or R_S) is plotted on the X axis and the imaginary part (Z'' or X_S) on the Y axis of a graph, one obtains a "Nyquist" or complex impedance plot. The impedance locus for the above simple equivalent circuit-model of the interface impedance is plotted on a complex impedance plot in Fig. 3.3. *Note:* Electrochemists plot -X_S vs R_S and not X_S vs R_S as electrode (and tissue) impedances tend to be capacitive and thus negative. It is generally found easier to look at the plots with the Z'' axis inverted. Low frequency data are on the right side of the plot and higher frequencies are on the left. This is generally the case for electrode interface data.

The impedance locus has the form of a semi-circle with high and low frequency intercepts with the real axis at 90° (due to the presence of C_{dl} in parallel with R_{CT}). At very low frequencies the impedance is equal to $R_{TOTAL}+R_{CT}$, the diameter of the semi-circle being equal to R_{CT}. At higher frequencies, the impedance is influenced by the value of the parallel capacitance C_{dl}. As the capacitive impedance decreases with increasing frequency, current therefore flows through it and the total impedance of the parallel combination decreases. The reactive component and the phase angle increases from zero, reaches a maximum value (which depends on the relative sizes of R_{TOTAL} and R_{CT}) and the decreases again towards zero (see Fig. 3.3 and 3.4).

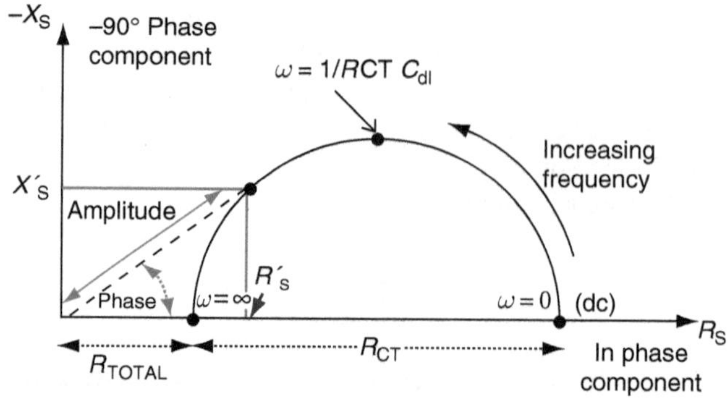

Fig. 3.3 Impedance plot for simple equivalent circuit model of the electrode-electrolyte interface. The impedance locus is semi-circular as a result of the parallel combination of C_{dl} (the double layer capacitance) and R_{CT} (the charge transfer resistance). Both of these are in series with R_{TOTAL}, the sum of the lead and electrolyte resistances

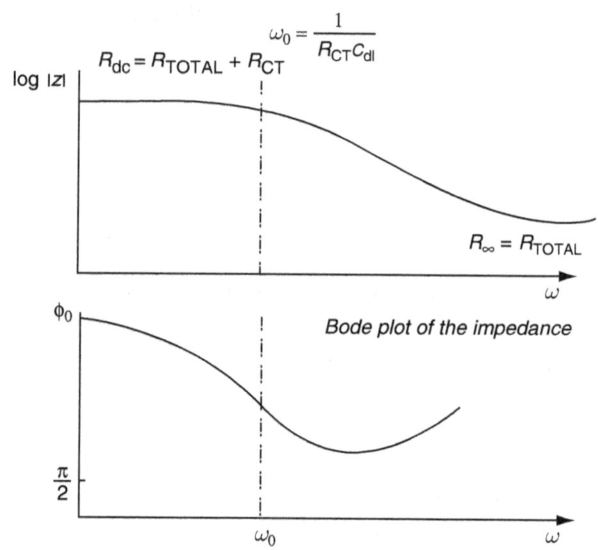

Fig. 3.4 *Bode plot for the simple "3-component" equivalent circuit model. The magnitude of the impedance decreases from its low frequency value* ($R_{TOTAL} + R_{CT}$) *to* R_{TOTAL} *at very high frequencies. The phase angle of the overall interface impedance increases from 0° at low frequencies, reaches a maximum which depends on the relative sizes of* R_{TOTAL} *and* R_{CT} *and then decreases again.*

The frequency at which the reactive component reaches its maximum value (ω_O) is given by $\omega_O = 1/R_{CT}C_{dl}$ (Fig. 3.3).

At high frequencies the impedance is determined by the series resistance R_{TOTAL}.

3.2.1.4 Bode Plot

Another popular method of presenting impedance data is the "Bode plot". The impedance is plotted with log frequency on the x-axis and both the absolute value of the impedance ($|Z| = [Z'^2 + Z''^2]^{1/2}$) and the phase-shift ($\phi = \tan^{-1}[Z''/Z']$)

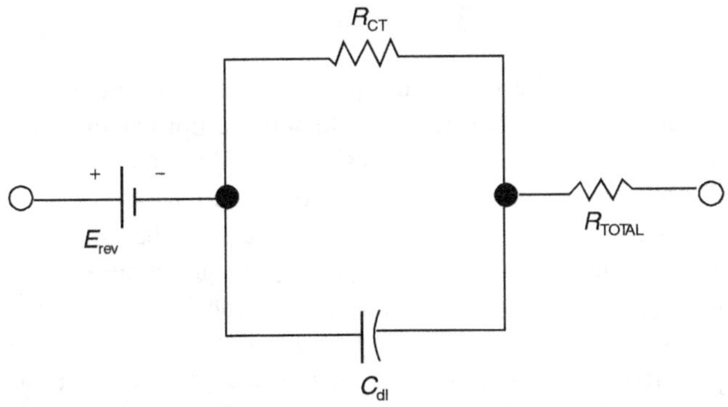

Fig. 3.5 Simple equivalent circuit model of the electrode-electrolyte interface. C_{dl} represents the double layer capacitance, R_{CT} the charge transfer resistance, R_{TOTAL} the sum of the lead and electrolyte resistances and E_{rev} represents the reversible or equilibrium potential

on the y-axis. Unlike the Complex Impedance plot, the Bode plot explicitly shows frequency information.

The Bode plot for the electric circuit of Fig. 3.5 is shown in Fig. 3.4.

As in the complex impedance plot, the magnitude of the impedance is equal to $R_{TOTAL} + R_{CT}$ at very low frequencies (R_{dc}). The phase angle is zero at this point as the impedance is purely resistive. At very high frequencies the magnitude of the impedance (R_{∞}) is equal to that of the series resistance R_{TOTAL}. At frequencies in between these two limits, the interface impedance is influenced by the value of the parallel capacitance C_{DL}. As the capacitive impedance decreases with increasing frequency, current therefore flows through it and the total impedance of the parallel combination decreases. The phase angle increases from zero, reaches a maximum value (generally less than 90° or $\pi/2$ radians) and then decreases again towards zero (see Fig. 3.4).

The above model can be used to explain most key aspects of the electrode-electrolyte interface. It must be pointed out, however, that the equivalent circuit is a gross approximation.

For example, diffusion of ions to the interface from the bulk of the electrolyte (gel or patient) takes place at a finite rate and thus gives rise to impedance to current flow, especially at low frequencies. The diffusion (often termed "Warburg") impedance is generally located in series with the charge transfer resistance, both of these being in parallel with the double layer capacitance. The diffusion impedance has been ignored in the above model as it tends not to be observed for many biomedical electrode systems over the range of frequencies typically used.

A further simplification is the use of a simple capacitance in the above model. Such ideal capacitive behavior is rarely observed with solid metal electrodes. Instead, an empirical "pseudo capacitance" or Constant Phase Angle impedance, Z_{CPA}, is often used which has a constant phase angle, much like a capacitor.

$$Z_{CAP} = K(j\omega)^{-\beta} \tag{3.8}$$

where K is a measure of the magnitude of Z_{CPA} and has units of $\Omega s^{-\beta}$, and β is constant such that $0 < \beta < 1$. The phase angle of this empirical circuit element ($\varphi = \beta\pi/2$ radians or 90β degrees) generally lies between 45° and 90° [80]. Typically, β has a value of 0.8 for many biomedical electrode systems

Fricke [40] used the term "polarization" to describe the constant phase angle impedance and postulated that it was due to "spontaneous depolarization" of the electrode. Although he did not enlarge on the hypothesis, many authors have used Fricke's terminology over the intervening years. The present author must concur with Cole and Curtis' observation that "the use of the term polarization for describing the unexplained effects occurring at the metal-electrolyte interface is only an admission of our ignorance" [26].

The two most likely causes of the observed constant phase angle impedance are specific adsorption and surface roughness effects [130]. With solid biomedical electrodes, the non-ideal behavior is probably due to the surface roughness of the electrodes [28]. This is supported by reports that roughing an electrode surface decreases the measured value of phase angle.

It is also naive to think that surface effects will only distort the non-faradaic impedance and will have no effect on R_{CT} as assumed in the above model. It is more realistic that surface effects will affect the parallel combination of C_{dl} and R_{CT} giving rise to skewed [29, 77] or distorted [81] arcs. The simple equivalent circuit used in this presentation is however a useful approximation which enables qualitative interpretation of much of the published data.

3.2.1.5 Polarization

Since the work of Fricke [40], the term "polarization" has been used to describe just about anything associated with the electrode-electolyte interface – frequency-dependence, non-linearity, noise, etc. Polarization has been defined as "the departure of the electrode potential from the reversible value upon the passage of faradaic current" [9].

Under equilibrium conditions the electrode potential, E, is equal to its reversible potential, E_{rev}. When a dc or faradaic current, i_{dc}, is applied to the electrode interface, it must flow through the resistance R_{CT} which is in parallel with C_{dl}. From Ohm's law, the voltage dropped across this charge transfer resistance will be equal to i_{dc} multiplied by R_{CT}. The electrode potential, E, is now given by: (see Fig. 3.6)

$$E = E_{rev} + i_{dc}R_{CT} \tag{3.9}$$

The electrode therefore is no longer operating at its equilibrium or reversible value, E_{rev}. This change in the electrode interface's potential from its equilibrium value is termed "polarization". The degree of polarization is measured by the "additional voltage" dropped across R_{CT}, (or "over-potential", η, as it is termed in electrochemistry) where: (see Fig. 3.7)

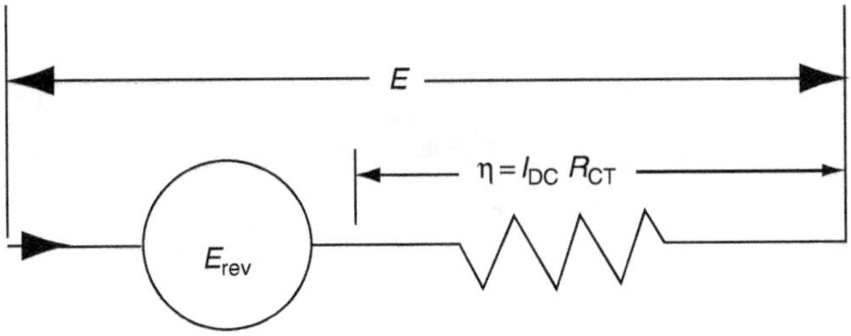

Fig. 3.6 Polarisation, the departure of the electrode potential from its reversible value upon the passage of faradaic current

$$\eta = /E - E_{rev}/ \tag{3.10}$$

An ideal "non-polarizable" electrode would have a value of R_{CT} equal to zero and hence there would be no resistance to faradaic current. The nonfaradaic impedance would effectively be shorted out and the total interface impedance would be zero. In this case, current would pass freely across the interface unimpeded. Measured biosignals, for example, would be unattenuated and undistorted. A perfect electrode system! The electrode potential would always remain constant at its reversible value.

A perfectly "polarizable" electrode would not permit the flow of any dc or faradaic current as the charge transfer resistance in this case is infinite. Such an electrode is sometimes termed a "blocking" electrode. No faradaic charge would cross the interface, even for large overpotentials, and the electrode couples capacitatively with the tissues/electrolyte in this extreme case.

Real electrodes are, however, neither perfectly polarizable nor perfectly non-polarizable. Any net current flow across an electrode-electrolyte interface will experience a finite faradaic impedance across which an overpotential will develop.

An electrode system which has a very low value of R_{CT} lets current traverse the interface almost unimpeded, there is little energy wasted at the interface, the overpotential is relatively small and the electrode system is relatively non-polarisable. Such electrode systems are highly sought after, especially when recording small biosignals from the body surface.

Electrodes made of noble metal come closest to behaving as perfectly polarizable electrodes. As these metals are "inert", they tend not to react chemically with the surrounding electrolyte or tissue. Noble metals are therefore generally used in the construction of implant electrodes where chemical reaction with surrounding tissues must be avoided in order to minimise tissue toxicity problems. Little steady current can pass in such cases as the charge transfer resistance for these electrodes is therefore very large. The small current that does pass represents the charging and discharging of the double layer capacitance. A problem therefore exists when designing implant electrodes. For a biocompatibility point of view one requires a

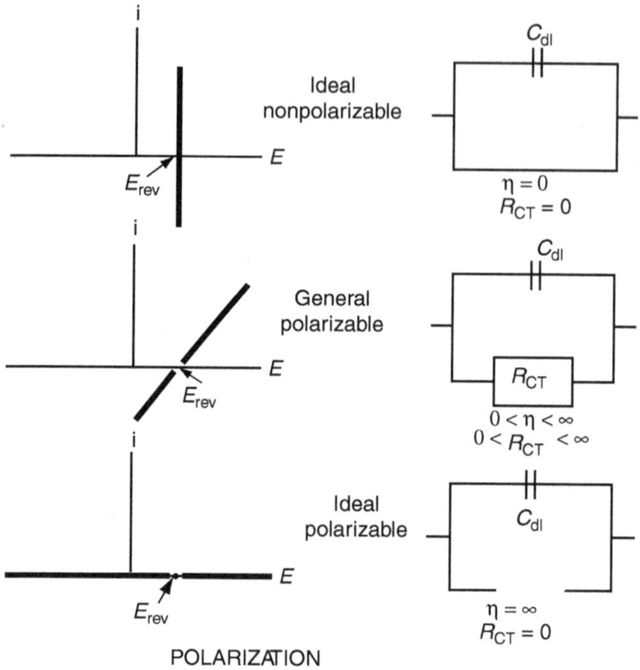

Fig. 3.7 DC Current-Voltage plots for "Ideal Non-polarizable", "General Polarizable" and "Ideal Polarizable" electrode interfaces

noble hence polarizable electrode system, whereas from an electrical performance point of view one requires a non-polarizable system. A compromise is achieved by using a polarisable electrode and roughening the surface of the electrode, thus decreasing the large interface impedance.

3.2.1.6 Transient Response and Tissue Damage

The response of the electrode system to sine waves of varying frequencies has been considered above (Complex Impedance and Bode plots) as this is a very useful tool in analysing circuits or, in our case, electrode systems. Equally relevant is the response of an electrode system to voltage and current steps or pulses as these will approximate therapeutic stimulation applications.

It must be borne in mind that the conversion from electrical to ionic current takes place at the electrode-tissue interface. Based on our simple equivalent circuit model, current can flow either through the parallel resistance or flow through the double layer capacitance

Current flowing through the parallel resistance involves faradaic charge transfer reactions. At the anode, the electrolysis of water and the oxidation of organic compounds can occur. The oxidation of the electrode itself can also occur, which results in the dissolution of metal. At the cathode, hydrogen ions are reduced to form hydrogen gas. This results in a change in pH near the electrode. The new chemical by-products in all of these reactions may lead to tissue damage and hence faradaic charge transfer reactions must be avoided [14, 17]. Current must not therefore be allowed to flow through R_{CT}.

For current "flowing through" the double layer capacitance, no charge actually crosses the electrode-tissue interface. Instead ions in the tissue are attracted or repelled by charges on the electrode, resulting in transient pulses of ionic current. As no net current flows through the interface and electrochemical reactions are not involved, capacitive current is relatively safe. One must therefore seek as far as possible to couple capacitively with tissue when seeking to stimulate tissue without causing trauma.

Voltage Response to a Step in Current

If one applies a pulse of current of amplitude I_{dc} at time $t = 0$, the voltage response of the electrode-interface equivalent circuit model and, it is believed, the electrode-patient system is as shown on Fig. 3.8.

$$V(t) = L_{dc}R_{TOTAL} + I_{dc}R_{CT}(1 - \exp[-t/R_{CT}C_{dl}]) \quad (3.11)$$

At $t = 0$ the applied current flows unopposed through the capacitor and hence only "sees" the series resistance R_{TOTAL}. The initial voltage response is therefore

$$V_o = I_{dc}R_{TOTAL} \quad (3.12)$$

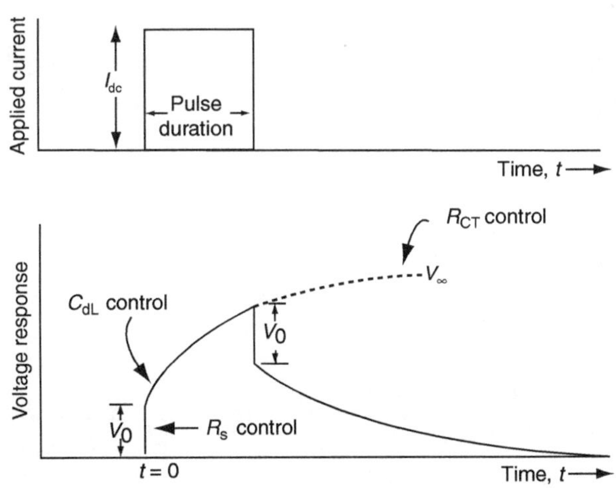

Fig. 3.8 Redo Voltage response to a pulse or step in current

The voltage response is then observed to gradually increase from V_o. The initial increase in voltage with time is inversely proportional to the magnitude of the capacitance.

For long pulse durations all of the current will flow through the resistances R_{CT} and R_{TOTAL}. The total resistance seen by the current is therefore

$$Z_{(t=\infty)} = R_{TOTAL} + R_{CT} \tag{3. 13}$$

and the limit voltage, V_∞ is given by

$$V_\infty = I_{dc}(R_{TOTAL} + R_{CT}) \tag{3. 14}$$

The voltage response will reach this limit value V_∞ in a time period of approximately five time constants, T, where $T = C_{dl}R_{CT}$.

Current Response to a Step in Voltage

If a perfect step in voltage, V_{dc}, is applied to the electrode system or the three-component model, the current response is as shown on Fig. 3.9 and given by the equation

$$I(t) = V_{dc}\left\{\left(\frac{1}{R_{TOTAL} + R_{CT}}\right) + \left(\frac{R_{CT}/R_{TOTAL}}{R_{CT} + R_{TOTAL}}\right)\exp\left[-\frac{R_{CT} + R_{TOTAL}}{R_{CT}R_{TOTAL}C_{dl}}t\right]\right\} \tag{3. 15}$$

At the beginning of the voltage step the resultant current jumps to a relatively large value I_o, where

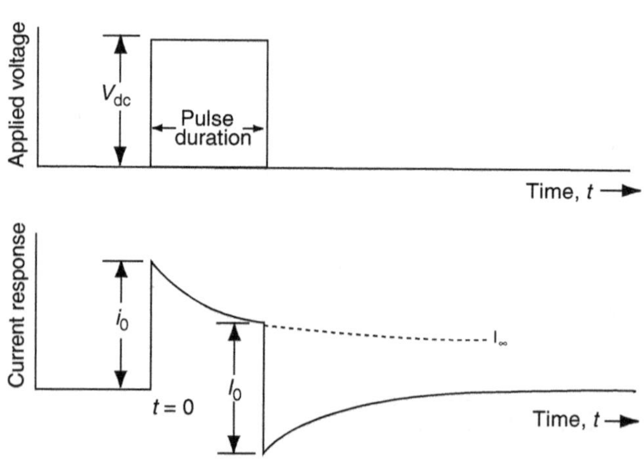

Fig. 3.9 Redo Current response to a pulse or step in voltage

$$I_o = V_{dc}/R_{TOTAL} \tag{3. 16}$$

As time passes, the resultant current decreases exponentially with an initial slope inversely proportional to the capacitance C_{dl}. Eventually the current will reach a limiting value of I_∞, where

$$I_\infty = \frac{V_{dc}}{R_{TOTAL} + R_{CT}} \tag{3. 17}$$

Tissue Damage

The rise in voltage response or the decrease in current response is often attributed in the literature to a mysterious phenomena called "polarization". It is in fact nothing more than the transient response of a simple "3-component" circuit model.

When one applies pulses to an electrode-tissue interface (either for implanted or surface stimulation), the applied or resultant current initially flows into the patient via the double layer capacitance. There are no reactions associated with the capacitive flow of current and hence there are few undesirable effects when using short duration pulses (i.e. with a duration less than one time constant). As the pulse duration is increased, however, progressively more current flows through the parallel charge transfer resistance and the patient is more at risk due to the reaction by-products. This is especially true if the pulse is applied for a length of time longer than five time constants giving the voltage or current response time to "level off" and reach its "steady state" value of V_∞ (or I_∞). At this point, most of the current is flowing through the charge transfer resistance and charge is therefore "injected" into the tissues via a faradaic process. The by-products of the electrochemical reactions involved in the charge transfer process will diffuse into the skin and/or tissue, causing chemical injury [16]. As the levelling off of the voltage response is indicative of pure charge transfer control, this feature must be avoided at all costs by either avoiding long duration pulses or by using electrode systems with very long time constants, $T = C_{dl}R_{CT}$. In both these cases, charge will be largely applied to the body via relatively safe capacitive processes.

It may not always be possible to use sufficiently short pulse durations to avoid tissue damage and still achieve therapeutic effect. It must also be noted that even when using "short" pulses where the current or voltage response has not "levelled off", some current still flows through R_{CT} and a faradaic charge transfer reaction takes place – with the associated, albeit reduced, problems [17]. Over extended periods of stimulation, the by-products will accumulate in the tissues. In the past, however, the applied waveforms found experimentally to minimize electrode and tissue damage are consistent with the basic goal of minimizing the flow of faradaic current across the electrode interface [35] and thus ensuring little, if any, net transfer of charge. Decreasing the duration, amplitude and rate of current pulses have all been suggested as each of these ensures C_{dl} control of the transients and thus avoids, to some extent at least, the undesirable faradaic processes.

An alternative to using short pulse durations is to use electrode systems with long time constants, i.e. with a large value of R_{CT} (faradaic charge transfer resistance) and/or C_{dl} (double layer capacitance)]. As we have seen, noble metals react with difficulty with surrounding tissues and thus have large values of R_{CT} [16]. For a given pulse duration, they tend to couple capacitively with tissue. The reactions on both anode and cathode are reversible, involving surface oxygen, and this reversibility explains the observed identical nature of anodic and cathodic waveforms [16]. As a result, noble electrodes are widely used for physiological stimulation. Unfortunately, noble metal give rise to large interface impedances. From a biocompatibility point of view one requires a noble (hence large impedance) electrode system, whereas from an electrical performance point of view one requires a low impedance system. A compromise is achieved by using such polarizable electrode materials and roughening the surface of the electrode, thus decreasing the large interface impedance. Roughening the electrode surface gives rise to a significant increase in the interface capacitance, thus increasing further the time constant ($T\uparrow = C_{dl}\uparrow R_{CT}$) and ensuring that the voltage or current response is even more dominated by the highly desirable capacitive processes while decreasing the interface impedance.

It would be a gross oversimplification to attempt to demystify biomedical electrode design by stating that all high performance biomedical electrodes simply have rough surfaces. There would however be a significant element of truth in the statement. For example, terms like "activated", "sintered", "porous" etc have been used to describe implant electrodes for cardiac pacing and indicate that the electrode fabrication process results, deliberately or otherwise, in a rough surfaced electrode. It could be argued that it is often the surface finish rather than the electrode metal that gives rise to the favorable electrical properties reported, especially the low interface impedances.

Most electrical stimulation electrodes rely to some extent on faradaic mechanisms at the interface between the metal and the tissue. Even in the case of noble metal electrodes and with short pulse durations, by-products of the electrochemical reactions involved will accumulate over time in the tissues when a signal is applied in one direction (monophasic) and will eventually give rise to irritation. With surface stimulation, for example, early designs of electrodes incorporated thick pads of electrolyte-impregnated lint in order to distance the patient's skin from the electrode-electrolyte interface and the undesirable by-products. Obviously, with implanted electrodes, that short term solution is not possible.

In particular, monophasic anodic pulses must be avoided as they will cause corrosion problems. Additionally, for most applications, cathodic stimulation has a lower threshold than anodic stimulation. Even in the case of monophasic cathodic pulses, however, current flows in only one direction and the chemical reactions at the interface are not reversed.

Biphasic waveforms are preferred in most electrotherapies as the by-products of the "forward" reaction are thought to be "recaptured" by the "reverse" reaction [71]. In using "charge balanced" waveforms, it is often believed that because there is no net charge transfer across the electrode-skin interface, there is also no net flow of potentially harmful by products into the skin. Unfortunately, the electrochemical reaction which occurs to enable the flow of current during the first phase is not necessarily that involved in the second. By-products of the first reaction are therefore not always "recaptured" and may "escape" from the interface into the patient [144]. However, it must be pointed out that the use of charge balanced biphasic waveforms does indeed greatly minimise the problem and hence its widespread use in a range of surface and implant applications. Additionally, surface biphasic stimulation is found to be more comfortable than monophasic.

3.2.1.7 Limit Voltages and Currents of Linearity

Although the non-linearity of the skin's electrical properties has been investigated under a range of conditions, the phenomenon is still far from well understood. "There appears to be, so far, no model available which accounts for both the linear and non-linear behavior of the electrodes in the frequency and time domains" [39]. It is an important feature of an electrode as "it appears...that electrodes often introduce non-linear characteristics which are erroneously ascribed to the biological system under study" [126].

Schwan proposed empirical relationships for the "Limit Current of Linearity" and the "Limit Current of Linearity"

Limit Current of Linearity

If has been observed that electrode-tissue interface impedance non-linear behavior is first evidenced at low frequencies. As the applied current amplitude is increased, progressively higher frequency points are affected. Schwan proposed a "Limit Current of Linearity", i_L. He observed that the relationship between the angular velocity of a given impedance point and the current amplitude required to drive it into non-linearity (deviate by more than 10% from its linear, small-signal value) was well expressed by the empirical relationship

$$i_L = B\omega^{\beta} \tag{3.18}$$

where B is a constant particular to the electrode system and β is the fractional power which appears in equation (3.8).

Schwan and others [54, 105, 125, 128] have observed that this empirical relationship is valid for many electrode systems over wide frequency ranges

The presence of β (a parameter describing the frequency dependence of the linear interface impedance) in a relationship describing the non-linearity of the system was found most intriguing.

The solution to this mystery is quite simple when it is approached from the right direction. Generally, researchers have assumed that the observed non-linear behaviour is attributable to the high-frequency Z_{CPA} impedance (Equation 3.8) which they observe under linear, small-signal conditions and over the limited frequency ranges they use. However, in parallel with Z_{CPA} is the "Charge Transfer" resistance R_{CT} which, in the linear range, has a very large value $R^{O}{}_{CT}$ where

$$R^{O}_{CR(o)} = \frac{RT}{nF}\frac{1}{i_o} \tag{3.19}$$

As a result, its contribution is either not observed or ignored.

The value of the "Charge Transfer" resistance can be derived from the Butler-Volmer equation and is very non-linear, decreasing rapidly with applied signal (ac or dc) amplitude. Compared to R_{CT}, Z_{CPA} is relatively linear. R_{CT} is therefore the source of the observed non-linear behaviour.

As the applied current amplitude is increased, the charge transfer resistance decreases rapidly causing the diameter of the impedance locus to decrease. As the low frequency end of the arc is dominated by the charge transfer resistance, the effects of such nonlinearity will be first evidenced at these frequencies. Low frequency points are therefore the first to deviate significantly (by more than 10%) from their small signal, linear values, as observed by Schwan and colleagues. As the applied signal amplitude is further increased, the diameter of the impedance locus decreases further and progressively higher frequencies are affected [81, 86].

Simplistically, it can be shown that the following approximations can be made over limited ranges of frequency and/or applied signal amplitude

- Approximate relationship between applied current and R_{CT}

$$i \propto 1/R_{CT} \tag{3.19a}$$

- Approximate relationship between R_{CT} and the frequency at which nonlinearity occurs

$$R_{CT} \propto \omega^{-\beta} \tag{3.19b}$$

Then, by cancelling $R_{CT,}$in the above two equations

- Approximate relationship between applied current and the frequency at which nonlinearity occurs

$$i_L \propto \omega^{\beta} \tag{3.19c}$$

as found by Schwan. The presence of β in the expression of an electrode system's non-linear behavior is therefore simply due to presence of a very nonlinear resistance in parallel with a relatively linear, frequency-dependent Z_{CPA}. A more accurate calculation based on the equivalent circuit model outlined above and the Butler-Volmer equation, was published [86].

Limit Voltage of Linearity

Schwan and his colleagues [105, 125] also postulated that the electrode-electrolyte interface impedance becomes nonlinear at a certain limit voltage, V_L, which they found to be independent of the frequency of the applied signal.

Using the Butler-Volmer equation and the equation for the impedance of the equivalent circuit model, the voltage limit of linearity can be calculated for a range of frequencies. [89]. It can be shown that the charge transfer resistance decreases pseudo exponentially with applied voltage amplitude, initially causing low frequency impedance points on the locus to deviate from their small signal values [84, 86].

At very low frequencies, such that $\omega \rightarrow 0$, the voltage limit of linearity, V_L, approximates to the voltage at which the charge transfer resistance decreases by 10% from its small signal value. This occurs at $V_L = 40/n$ mV, where n is the number of electrons per molecule oxidized or reduced [89].

As the applied voltage is increased above this low frequency limiting value, the charge transfer resistance further decreases and affects progressively higher frequency points (i.e. "become nonlinear"). The derived log (f) versus V_L plot is found to be a straight line over a wide range of frequencies. V_L is observed to increase only very slightly with frequency. This would agree qualitatively with Onaral and Schwan's results [1982], where V_L increased from 106 to only 129 mV over the frequency range of 10 mHz–100 Hz for platinum electrodes in saline. This would also explain why, in the past, V_L has been assumed constant and independent of the applied frequency.

3.2.1.8 Electrode Metals

As biocompatibility is of great importance in implants, implant electrode materials are generally confined to those that are essentially inert and do not react with the surrounding tissues. As cardiac pacing electrodes were among the first implanted and have had a long, generally successfully and well researched history, most conclusions drawn on the suitability of materials for implant electrodes are based on pacing electrodes.

Implant electrodes are and have been generally made from noble metals such as gold, platinum, iridium, rhodium and palladium. Platinum has been the most widely used as it has excellent corrosion resistance and produces relatively low polarization [133]. Platinum, however, is mechanically relatively soft and for many applications is alloyed with much harder iridium, producing platinum-iridium. Other noble metal alloys which have been used include gold-platinum-rhodium, platinum-rhodium and gold-palladium-rhodium.

Passive metals, such as titanium, tantalum, zirconium, tungsten and chromium have been successfully implanted. Titanium has been widely used because it forms a non-conducting oxide layer at the surface. This coating prevents charge transfer at the electrode interface. Titanium therefore exhibits a high resistance to corrosion. Stainless steel is similar in that it acquires a protective oxide layer which renders it inert. Although stainless steels were used in early pacing electrodes, they do not appear to have the required corrosion resistance for long-term use. Stainless steel pacing electrodes were discontinued after the 1960s because of unreliable corrosion resistance [132, 133].

Some early pacing electrodes were made of Elgiloy (an alloy of Fe, Ni, CO, Cr, and MO from Elgin Watch Co.) However, Elgiloy has marginal corrosion resistance and produces a relatively high polarization overvoltage. It was discontinued in the 1980s. Carbon is an inert, non-metallic element which has similar electrochemical characteristics to noble metals and continues to be used successfully as an implant electrode. Materials such as zinc, copper, mercury, nickel, lead, copper, silver, silver chloride, iron and mild steel have been found toxic to body tissues and are normally not used

Biocompatibility has been defined as the ability of a material to perform with an appropriate host response in a specific application [147]. Strictly speaking there is no such thing as a biocompatible material as an implant's "biocompatibility" will also depend on a range of variables including its shape and surface finish.

Stimulation threshold is a key parameter in implant stimulation electrode design. When activitated vitreous carbon electrodes were first introduced pacing electrodes they were found to have relatively low chronic thresholds. These thresholds were thought to be the result of the superior "biocompatibility" of the carbon electrode. Other researchers similarly interpreted the low thresholds observed for their new "exotic" materials such as indium oxide, titanium nitride, semi-metal ceramics. Stokes (1996) however concluded that "material selection appears to have little or nothing to do with threshold evolution – as long as the material is biocompatible and reasonably corrosion resistant. Thus our experiments with biocompatible materials such as carbon, titanium, platinum, iridium oxide, and many more have all produced about the same results when tested as polished electrodes, all other factors held equal". Stokes went on to point out "while the bulk properties of an electrode material are important, it is the electrode-tissue interface that determines the electrode's performance. In fact, the surface microstructure of the electrode is critical" [133]. It would appear that the microstructure of an electrode surface may affect cellular adhesion and activation, thus reducing the foreign body response. It is therefore the surface structure of many of the new materials (resulting from their fabrication process) that gives rise to the observed positive effect on threshold evolution over time, rather than the "biocompatibility" of the bulk material.

Another advantage of porous and microporous implant surfaces is their reduced interface impedance. Although interface impedance is generally less critical for implanted stimulation electrodes, many such electrodes (e.g. implanted pacing electrodes) are used to monitor biosignals as well as to deliver the required stimulation impulses. Decreased interface impedance helps in this regard.

Implanted biosignal monitoring electrodes require stable potentials as well as low interface impedances to minimize biosignal recording problems. These metals have high positive standard electrode potentials (E^o in equation 3.4) and are the lowest ones on the electromotive series. As "noble" metal electrodes do not tend to react chemically with the electrolyte, the Nernst equation is not defined and the measured potential is often influenced more by any traces of impurities on the surface than by the intrinsic properties of the metal itself. The electrode potential can drift randomly, especially immediately following implantation. It may fluctuate widely under apparently identical circumstances. This is an inherent disadvantage of "noble": materials

External biosignal-monitoring electrodes can generally use high electrical performance "non-noble" materials such as silver-silver chloride with out fear of biocompatibility problems [53]. Silver-silver chloride has been found to be an excellent electrode sensor material as, when it is in contact with a chloride gel it has

1. A low, stable electrode potential
2. Low level of intrinsic noise
3. A small value of charge transfer resistance, ie it is relatively nonpolarizable.
4. A small interface impedance

A silver – silver chloride electrode is generally made by the deposition of a layer of silver chloride onto a silver electrode. Silver chloride is a sparingly soluble salt and thus effectively provides the silver electrode with a saturated silver-chloride buffer which facilitates exchanges of charge between the silver electrode and the sodium chloride environment of the gel and human body. The system behaves as a reversible chloride-ion electrode and the Nernst potential in this case depends on the activity (which is closely related to concentration) of the environment chloride ions and not on that of the silver ions. The potential of this electrode is therefore quite stable (as well as small) when the electrode is placed in an electrolyte containing Cl^- as the principal anion – as is the case in the human body and electrode gels [53].

Electrical "noise" (potential fluctuations) can occur spontaneously at the electrode interface without any physiological input. Ag/AgCl electrodes have been shown to be particularly stable and resistant to noise [43].

A silver-silver chloride electrode has a relatively large value of exchange current density [53] (Equation 3.7) and hence a very low value of charge transfer resistance, R_{CT}. Charge is transferred across the interface with relative ease and little voltage is dropped across the interface. The electrode therefore operates close to its equilibrium or "reversible" potential. Ag-AgCl electrodes are therefore relatively non-polarisable.

When a smooth surfaced electrode is chlorided, the AgCl deposit can give rise to a very rough surface and thus to relatively very low interface impedances [27, 43]. K, the magnitude of the interface pseudo capacitance (Equation 3.8), is observed to decrease following the deposition of an AgCl layer [60]. However, although AgCl facilitates the interfacial electrochemistry, it is very resistive having a resistivity of around $10^5 - 10^6 \Omega$cm [53]. As the layer thickness increases, the series resistance, R_{TOTAL} will therefore increase. This series resistance dominates the very high frequency interface impedance and the latter will also increase with chloride deposit.

There is therefore an optimal layer thickness, for a given frequency, which decreases the interface impedance and yet does not significantly increase the series resistance, R_{TOTAL} [87]. The optimal silver chloride layer thickness consequently depends on the frequency range of interest [42].

Tin-stannous chloride, a material somewhat similar to silver – silver chloride, was used in some biosignal electrodes [48].

Electrode material and high electrical performance is generally less critical for external stimulation electrodes such as TENS and external pacing electrodes (current density distribution is the key concern). The majority of commercially available TENS electrodes are molded from an elastomer such as silicone rubber or a plastic such as ethylene vinyl acetate and loaded with electrically conductive carbon black. Mannheimer and Lampe (1987) pointed out that the only tangible disadvantage with having a large electrode interface impedance is that more power will be required from the stimulator to drive the stimulating current through the electrodes into the patient.

Graphite-loaded polyesters and similar materials are used in external pacing electrodes, for example. Some are constructed using tin as the metal layer. In early electrodes the combination of tin and the chloride based-gel gave rise to pitting of the metal. Improvements made to the gels and the use of high purity tin have effectively removed this problem.

Although silver-silver chloride has been and still is used in some external electrostimulation electrodes, these should be used with care. Silver chloride is deposited electrolytically and can therefore be either removed by the passage of current or a thicker, high resistance layer deposited, depending on the polarity of the electrode. This can be a significant problem in iontophoretic transdermal drug delivery and may cause problems in multi-function pads which include a silver – silver chloride layer to enable distortion free-monitoring of the ECG through electrodes designed to deliver the pacing or defibrillation impulses.

3.2.2 The Skin

3.2.2.1 Structure of the Skin

The skin is a multi-layered organ which covers and protects the body. It is made up of three principal layers – the epidermis, the dermis and the subcutaneous layer [Note: In the literature, there exist variations in the terminology used to denote these layers].

The epidermis, the outermost layer, is around 100 μm thick, depending on body site. It is the strongest layer, providing a protective barrier against the outside hostile environment. Unlike any other organ of the body, the epidermis renews itself continually. It can be subdivided into several layers with the "basal layer" forming the innermost layer and the "stratum corneum" the outermost layer. Cells in the basal layer constantly multiply and, as they are pushed up towards the skin's external surface, the cells undergo changes. Eventually layers of compacted, flattened, non-nucleated, dehydrated cells (called corneocytes) form the stratum corneum. These dead cells are continuously being shed and are replaced from the underlying epidermal layers. The intercellular spaces between corneocytes are occupied by arrays of bilaminar membranes with the morphological features of polar lipids [109]. This matrix appears to serve to bind the cells and the stratum corneum has been described in terms of corneocyte "bricks" surrounded by lipid "mortar" [15]. On average, the stratum corneum comprises around 20 cell layers thick and has a thickness of around 10–15 μm. Thickness will however vary with the number of cell layers making up the stratum corneum and the state of hydration. On some body areas it can be several hundred microns thick. The epidermal layer is traversed by numerous skin appendages such as hair follicles, sebaceous glands and sweat glands.

The underlying layers of the epidermis are, in contrast, a relatively aqueous environment. The transition from an essentially non-conductive, lipophilic membrane (the stratum corneum) to an aqueous tissue (viable epidermis and dermis) gives rise to the skin's barrier properties.

The dermis is the second layer of the skin and, with an approximate thickness of 2 mm, is considerably thicker than the epidermis. It is formed from a dense network of connective tissue made of collagen fibres, giving the skin much of its elasticity and strength. Embedded in the dermis are blood vessels, hair follicles, sebaceous and sweat glands and several types of sensory nerve endings.

The final layer of the skin (the Subcutaneus layer) is found beneath the dermis layer. It contains structures of connective tissues and enables the skin on most parts of the body to move freely across the underlying bone structures. It is one of the body's areas for fat storage and acts as a cushion to protect delicate organs lying beneath the skin.

3.2.2.2 Electrical Properties of the Skin

Skin Impedance

As the stratum corneum is relatively non-conductive, it presents a high impedance to the transmission of electrical currents. As a result, the impedance of the skin is the largest component of the overall inter-electrode impedance (Fig. 3.10). Nonetheless, due to the stratum corneum's dielectric properties and its thinness, it permits capacitive coupling between a conductive metal electrode placed on the skin surface and the underlying conductive tissues. One can imagine the relatively non-conductive stratum corneum sandwiched between the conductive electrode and the conductive tissues underlying the stratum corneum forming a parallel plate capacitor. The stratum corneum's electrical impedance is therefore often represented by a simple capacitor, C_{SP} [The subscript "SP" refers to "Skin" and "Parallel"].

Some ions do, however, manage to cross the stratum corneum via paracellular pathways and through the skin's appendages (hair follicles, sweat ducts, sebaceous glands, imperfections in the integrity of the skin). As skin appendages extend through the stratum corneum barrier they can act as shunts to the interior. The flow

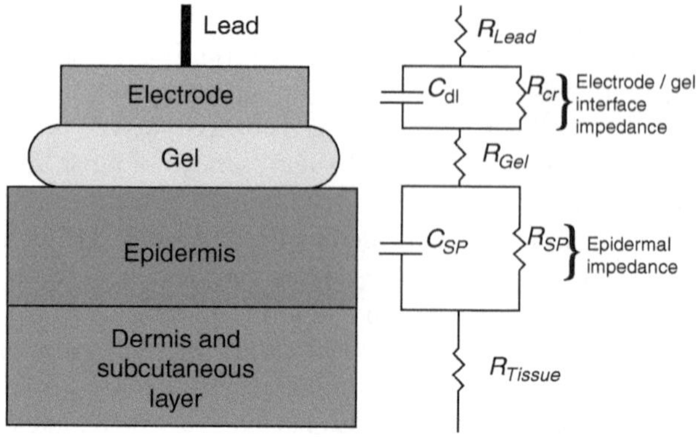

Fig. 3.10 Schematic representation of the skin and its equivalent circuit model

of ionic current can be represented electrically by a large resistance, R_{SP}, in parallel with C_{SP}.

The underlying layers of the epidermis are, in contrast, relatively conductive and can be collectively represented by a tissue resistance, R_{Tissue}.

A simple equivalent circuit model of the overall electrode-gel-skin system is therefore shown on Fig. 3.10 and includes the electrode lead resistance, R_{Lead}; the electrode-gel interface impedance (the double layer capacitance, C_{dl}, in parallel with the charge transfer resistance, R_{CT}); the gel resistance, R_{Gel}; the skin impedance (the parallel combination of a capacitance, C_{SP} and a resistance, R_{SP}) and the underlying tissue resistance, R_{Tissue}.

It must be borne in mind that this equivalent circuit model is a simplification of the rather complex electrical properties of the skin. For example, it has been found experimentally [114, 115] that the capacitance of the skin is better described by an empirical, constant phase angle impedance, Z_{CPA} where

$$Z_{CPA(S)} = K_S(j\omega)^{-\alpha} \tag{3.20}$$

[This is similar to the empirical expression for the "pseudo capacitance" often used to represent the non-ideal capacitive properties of the electrode interface's double layer capacitance (3.8)].

K has units of Ohm.sec$^{-\alpha}$. α is constant such that $0 \leq \alpha \leq 1$. The fractional power, α, of the capacitive impedance has been found to be related to the degree of hydration of the stratum corneum [Salter, 1980]. If the epidermal layer behaved as a simple capacitance, α would equal unity. The actual value of α, normally around 0.8–0.9, is a measure of the deviation from this ideal behaviour.

The use of C_{SP} will, however, be sufficient for this present review. The lead resistance, the gel resistance, the tissue resistance, and the electrode-gel interface impedance are all relatively small in comparison to the large skin impedance. Skin impedance therefore generally dominates and will therefore be studied in more depth.

3.2.2.3 The Skin's Parallel Capacitance, C_{SP}

It was suggested above that the electrode-skin interface can be approximated by a capacitor with the stratum corneum forming the dielectric layer sandwiched between the electrode and the underlying tissues which form the conductive plates. It can be seen from (3.5) that the skin's capacitance will increase as the thickness of the stratum corneum decreases, its dielectric constant increases or the area of the electrode increases.

The number of cell layers in the stratum corneum can range from 12 to 30 [61]. Epidermal thickness can therefore vary greatly for different body sites within the range of about 10 μm to well over 100 μm [111]. The stratum corneum can be, for example, as thick as 400–600 μm in the palm and plantar areas and as little as 10–20 μm on the back, legs and abdomen [24]. The value of the capacitance of the skin is related to the thickness and composition of the stratum corneum and has a typical value in the range 0.02–0.06 $\mu F/cm^2$ when measured using electrodes with "wet" electrolyte gels several minutes following electrode application [36, 151]. As the stratum corneum is typically at least 10 times as thick on the palms of the hands and soles of the feet as compared with other body areas, the skin capacitance at these points is considerably smaller than at other sites on the body. The stratum corneum on the face and scalp is not as thick as on other body parts and is characterised by large capacitance values.

Dark-skinned subjects have stratum corneum layers which are more dense and contain more layers of cells than fair-skinned subjects [61]. Not surprisingly, they are characterised by skin capacitances which are much lower (skin impedances (3.6) much higher) than those for fair-skinned subjects. One should therefore take care when assessing a new electrode system or associated device that they are tested on a range of subjects and skin sites. What may work well on a subject with low skin impedance in a warm and humid environment may be found later to fail on a high impedance subject, especially in a cold or dry environment.

3.2.2.4 The Skin's Parallel Resistance, R_{SP}

Although the stratum corneum does not easily allow foreign substances to traverse it, some current, carried by ions, manages to flow through it. The difficulty or resistance, this current experiences in passing through the skin is represented in the equivalent circuit (Fig. 3.10) by the parallel resistance, R_{SP}.

The skin's resistance is highly dependant on the presence and activity of sweat glands and on the presence of other appendageal pathways. An average human skin surface is believed to contain between 200 and 250 sweat ducts on every square centimetre [25]. The density of sweat glands varies greatly over the body surface with a value of approximately 370 per cm^2 on the palms of the hands and the soles of the feet and a value of approximately 160 per cm^2 on the forearm [111]. The diameter of the ducts can range from 5 to 20 μm. It is therefore not surprising that R_{SP} is reported to vary greatly from patient to patient, from body site to body site and with time. The measured values of R_{SP} are much smaller on areas with high densities of sweat glands, such as the palms of the hands (in spite of the thicker stratum corneum layer), especially when the glands are active in response to thermal or psychophysiological stimuli.

An average human skin surface is reported to contain between 40 and 70 hair follicles per square centimetre [24]. The presence of a high density of hair follicles (which act as low resistance shunts) gives rise to a very low value of skin parallel resistance, R_{SP}. However, this observation is counterbalanced by the difficulty in making firm mechanical and electrical contact to hirsute body sites or patients. In such cases, the skin impedance is very large at best. Generally the electrodes fall off and hence require the shaving of the skin site prior to electrode application.

Observed inter-site and inter-patient variations in skin impedance tend to be due to large variations in R_{SP}. In the low frequency range, dominated by R_{SP}, regional differences in skin impedance were observed by Rothman (1956), Lawler et al. [66] and Rosell et al. [116]. Low frequency skin impedance was observed to decrease in the following order – thumb, forearm, abdomen and, smallest of all, forehead. Similarily, Almasi and Schmitt [5] observed the low frequency skin (10 Hz) impedance to decrease in the order- outer forearm, leg, inner forearm, back, chest, earlobes and forehead. The forehead appears to have a very low skin impedance value [47], presumably as a result of the stratum corneum on the face and scalp being thinner than that on other body parts [61] and the presence of a high density of sweat glands. Almasi and Schmitt [5] plotted their average impedance values for the body sites on a complex impedance plot and found that most of the points lay along a "smooth common locus of monotonically increasing phase angle and impedance magnitude". This behaviour was successfully interpreted by McAdams and Jossinet [84] who showed that such frequency loci were formed when the skin's parallel resistance varied greatly from site to site while the skin's capacitance remained relatively constant. Two body sites did not fit the locus and hence the physical explanation; these were the palm and fingertips. These body sites have much larger epidermal thicknesses and hence have skin capacitance values much smaller than other body sites.

One must be very careful when assessing different electrode designs or gels. Testing different electrodes on different patients is certain to give misleading results due to the inter-subject variations, unless of course large numbers of subjects are used and statistically significant differences are observed.

R_{SP} varies greatly over time due to a number of parameters including room temperature and psychophysiological stimuli. The latter effect is exploited in so called "lie detectors". Schmitt and Almasi [123] reported that there is a considerable daily variation in a given subject, and seasonal changes have also been reported [150]. Testing a range of electrodes on the same subject but on different days is therefore not optimal either as day-to-day variations in skin impedance, especially fluctuations in R_{SP}, will nullify the validity of this approach. For example, Searle and Kirkup [127] found that the diurnal variations on a given subject for a given electrode was much larger than any difference between the range of electrode designs they tested in any one recording session. It should be further noted that the electrode test sites should be allowed to recover for several days between experiments to enable the skin to recover. For example, pealing off an adhesive electrode will remove some of the underlying stratum corneum. Any electrode subsequently tested on the site will benefit from this prior "skin stripping" (see below).

Electrode designs must therefore be compared in vivo by testing them at the same time on the same subject. One must still bear in mind the significant differences in skin impedance that exist over the subject's body, as outlined above. Even testing the electrodes at the same time on a limb of a given subject remains problematic. The different skin sites involved, even if located close together, will give rise to significant differences in the measured electrode-skin impedances which may be wrongly attributed to the electrode designs or gels under test. Searle and Kirkup [127], for example, showed that testing a range of dry electrode metals on the inner forearm gave rise to potentially very misleading results. Electrodes placed closer to the wrist gave rise to lower impedances due to the presence of a higher concentration of sweat glands.

Electrodes must therefore be repeatedly tested at the same time, under the same conditions, varying their relative positions, in order to clearly establish their relative performances. McAdams and colleagues developed a four channel impedance monitoring system to enable the "simultaneous" comparison of electrode designs/gels [93].

Skin Potential and Motion Artifact

A potential difference E_S, given by the Nernst equation, exists across the epidermis as a result of ionic concentration differences. This potential varies from patient to patient, site to site, and depends on gel composition (if used) and skin condition.

The skin surface is normally negative with respect to the inside of the body. Skin potential becomes more negative when sweat glands are active and palmer and plantar surfaces, with their higher sweat gland concentrations, are the most negative. Increasing gel concentrations of NaCl or KCl also render the site more negative. E_S has a typical value of 15–30 mV [31].

The dependence of the skin potential on the thickness of the epidermal layer is important to many ECG recording applications. If the thickness of the layer is changed by stretching or pressing down on the skin, the skin potential can vary by 5–10 mV compared to, for example, the 1 or 2 mV ECG signal. As these fluctuations generally result from patient movement they are termed "motion artefact". Motion or skin-deformation artefact is a serious problem during exercise cardiac stress testing of patients on treadmills or exercise bicycles, during ambulatory monitoring and while monitoring patients lying in bed [136] (Fig. 3.11).

Abrading or puncturing the skin is often used in "stress testing" to remove or bypass the problem source. Although skin potential increases with gel concentration, artefact gradually decreases with time as the conductive electrode gel soaks into the skin and renders the stratum corneum more conductive. High concentration gels are often used for short-term diagnostic applications where the risk of skin irritation is outweighed by the need for clear traces.

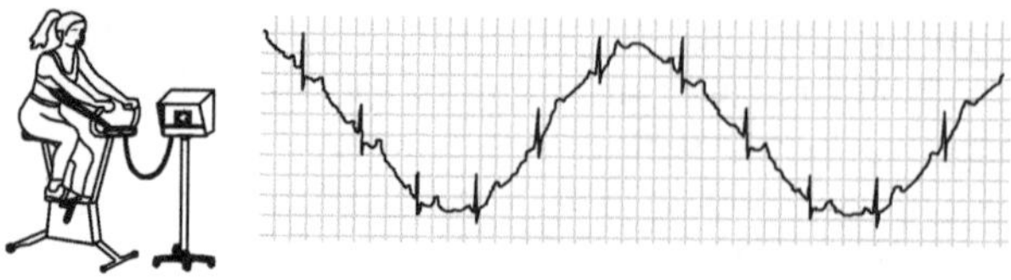

Fig. 3.11 Disturbance of Biosignal due to patient movement. Redrawn from Zinc [152]

In general, hydrogel-based electrodes (see below) should not be used for stress testing or in other monitoring applications which are likely to suffer motion artefact problems. Hydrogels tend not to hydrate the skin and hence do not actively "attack" the source of the problem. The same comment applies for dry (gel-less) electrodes.

In stress testing and ambulatory event monitoring, modified electrode locations are used to avoid muscular or flabby areas of the body and thus minimise skin-deformation (and EMG) artefact. "Stress loops" are formed in the connecting leads which are taped to the patient and used to avoid direct pull on electrodes and the underlying skin. The use of foam-backed electrodes tends to absorb any pull on the electrode and minimises artefact.

Electrode Gels and Their Effects

Dry electrodes are successfully used in some monitoring applications. Suitably designed gel-less electrodes have advantages when used in the home environment where the patient may not remember or have the time to apply gel electrodes prior to use [95].

For many home-based monitoring applications, electrodes are manufactured from non-corroding materials such as stainless steel which can be repeatedly washed and reused. Unfortunately, such polarisable materials give rise to poor electrical performances. In order to ensure good, stable electrode potentials, silver-silver chloride electrodes should be used (see below).

Jossinet and McAdams [56] demonstrated that the impedance of a dry electrode decreases pseudo exponentially due to the gradual build up of sweat under an occlusive, gel-less electrode and the resultant progressive hydration of the underlying skin. Searle and Kirkup [127] reported that the decrease in skin impedance of dry electrodes is polyexponential and requires two time constants, one very short (approximately 45 s) and the other almost 10 times longer (approximately 450 s), possibly indicating two different processes at work.

Given that the surface of the skin is irregular, a flat dry electrode will initially only make contact with a few "peaks" on the skin surface. There is therefore a smaller effective contact area than one would otherwise expect. However, as sweat builds up under the occlusive, "dry" electrode, a better contact with the skin will result in a relatively rapid increase in the measured value of C_{SP}. Human sweat contains a small amount of sodium chloride (around 0.1–0.4 % NaCl [111]) and hence serves as a weak electrolyte. It is suggested by the author that this accounts for the shorter time constant. [The longer time constant is probably indicative of the progressive hydration on the underlying skin resulting in a gradual decrease in R_{SP}]. As will be outlined below, R_{SP} is observed to decrease with a time constant of around 10 min in the presence of an electrolyte gel. This agrees quite well with the 7.5 min observed by Searle and Kirkup [127] .

Before leaving "gel-less" electrodes it should be pointed out that in certain applications which employ very high frequency signals such as Electrical Impedance Tomography (EIT) the use of a gel pad may not be needed as it will contribute a small but significant contact resistance to the desired measurement [94]. In such instances the use of a very thin "spray" of moisture onto the electrode surface prior to its firm application to the patient's skin may be all that is required. Profiled "dry" electrodes firmly pressed onto the skin may also be adequate for certain home-based biosignal applications. If skin impedance is a problem with standard "button" electrode designs, this can be addressed by increasing the electrode area in the non-critical axis. For example, long narrow dry electrodes are used for precordial ECG recording which enable a large contact area while ensuring sufficient inter-electrode distances on the chest [95].

Electrode gels serve (i) to ensure a good electrical contact between the electrode and the patient's skin, (ii) to facilitate the transfer of charge at the electrode-electrolyte interface between the two kinds of charge carrier (electrons in the electrode and ions in the gel) and (iii) to decrease the large impedance of the stratum corneum.

There exist two main types of electrode gel, viz. "wet" gels (often described as pastes, creams or jellies) and hydrogels.

"Wet" gels are generally composed of water, a thickening agent, a bactericide/fungicide, an ionic salt and a surfactant [19]. The ionic salt is present to achieve the appropriate electrical conductivity of the gel and this will depend on the specific application. As the major portion of ions present in tissue fluids and sweat are sodium, potassium and chloride (Cl^-), in order to insure bio-compatibility the ionic salts most commonly used in electrode gels are NaCl (sodium chloride) and KCl (potassium chloride). High concentrations of these salts tend to be better tolerated by the body than other salts. The ions in the gel serve not only to ensure electrical conductivity of the gel but to decrease the skin impedance by diffusing into the skin due to the existing concentration gradient. A relatively high concentration of electrolyte will also decrease the value of the charge transfer resistance (thus rendering the electrode more non-polarisable).

When a standard pre-gelled "wet" electrode is applied to the skin, the gel rapidly fills up the `troughs' on the electrode and skin surfaces, thus ensuring maximum effective contact area. The skin capacitance, C_{SP}, is therefore observed to initially increase rapidly in value following electrode application and then to remain relatively constant [84]. [A similar effect was probably noticed by Searle and Kirkup [127] as a result of sweat accumulation under a "dry" occlusive electrode]. Although C_{SP} does not exhibit a strong time dependence, it does vary with the electrolyte composition and concentration [41, 111], increasing with increasing concentration [103].

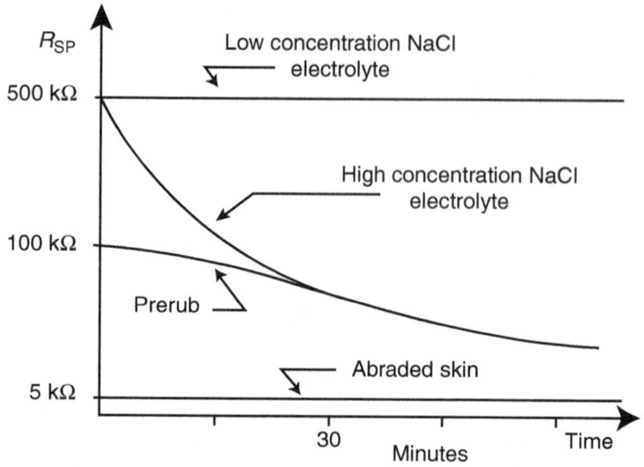

Fig. 3.12 Variation of skin resistance with time for a range of gel concentrations and skin preparation techniques [41]

Following electrode application, the skin's parallel resistance, R_P, generally decreases with time in a pseudo exponential manner as the ions in the gel diffuse through the skin rendering it more conductive [84, 104] see Fig. 3.12. It has been observed, however, that when a cold gel is applied to a skin site the measured value of R_{SP} is observed to initially increase [84, 116]. This is attributed to the cold gel causing the sweat pores to contract. Once the gel and skin has warmed up, the value of R_{SP} is observed to decrease as the electrolyte ions diffuse through the epidermal layer.

The skin temperature effect should be borne in mind when assessing a range of electrode designs. If, for example, the patient/subject removes his/her shirt just before the tests, the electrodes tested at the start of the experiment will have an advantage (i.e. smaller skin impedance) over those tested later as the uncovered skin sites will gradually cool down with time following removal of the shirt. Meaningful in vivo assessment of electrodes is not straightforward and wrong conclusions can very easily be made by the unaware or the unscrupulous.

The time constant for the skin's parallel resistance, R_{SP}, appears to be inversely proportional to the concentration of the gel (see Fig. 3.13). The decay has a time constant of around 10 min [41], thus indicating that it takes almost 1 h for the electrode-skin impedance to decrease to its lowest value. For example, 50/60 Hz interference, linked to mismatch of electrode-skin impedances, is often observed experimentally to decrease with time. One should therefore, where possible, apply the electrodes to the patient first, for example, before setting up the rest of the measurement system, to enable the skin impedance to decrease as much as possible.

High salt concentrations give rise to a more rapid diffusion of ions into the skin and a more rapid decrease in the skin's parallel resistance, R_{SP} [41, 84] (Fig. 3.12). Such aggressive gels tend to be used in short term biosignal monitoring applications such as "stress testing" where "instant", high quality traces are required

[82]. Biological tissues can not tolerate long term exposure to salt concentrations which depart significantly from physiological levels (approximately 0.9% NaCl for body fluids and around 0.1–0.4 % NaCl for human sweat [111]). Aggressive gels (>> 5 % NaCl) should not therefore be used, for example, for the long term monitoring of bed-ridden patients or for the monitoring of neonates. In the latter case, the incompletely formed skin is very susceptible to skin irritation problems. In any monitoring application, aggressive gels should not be used in combination with skin abrasion (see below). It is especially to be avoided in longer term monitoring applications where the removal of the body's defensive barrier coupled with the long term exposure to an aggressive gel will lead to sever discomfort to the patient.

The second kind of electrode gel commonly used in electrode systems are hydrogels. Hydrogel-based electrodes have recently become popular for numerous biomedical applications including resting ECG. Hydrogels are "solid" gels which originally incorporated natural hydrocolloids (e.g. Karaya gum a polysaccharide obtained from a tree found in India) [19]. The use of natural hydrocolloids give rise to variable performances and, in some cases, an unattractive colour. Synthetic (e.g. Polyvinyl pyrrolidone) hydrocolloids are now widely used.

The use of such "solid" gels entails numerous advantages when they are used in conjunction with screen printing or similar technologies. The use of an adhesive hydrogel pad dispenses with the need of the standard gel-impregnated sponge, gel-retaining ring and surrounding disc of adhesive foam that are used in wet-gel electrode designs. It is possible to construct thin, light-weight, highly-flexible electrode arrays with accurately defined electrode/gel areas, shapes and inter-electrode distances [90, 91].

Hydrogels also tend to cause less skin irritation compared to "wet" gels. A simplistic explanation of the advantageous/disadvantageous features of hydrogels is that hydrogel serves principally to ensure a good electrical contact between the skin and the electrode and that they do not significantly affect (compared to "wet" gels) the properties of the stratum corneum.

The impedance of the gel layer can be represented by a simple resistance in series with the impedances of the skin and the electrode plate – electrolyte interface. The magnitude of the gel resistance will depend on the composition and concentration of the gel and on the dimensions of the gel layer. Hydrogels are generally more resistive than "wet" gels. Typical resistivities for "wet" gels are of the order of 5–500 Ohm.cm (the higher the salt concentration, the lower the resistivity) compared to 800–8,000 Ohm.cm for hydrogels (the higher resistivity hydrogels tend to be used in cardiac pacing electrodes). "Wet" ECG electrodes, for example, have a gel layer thickness of around 0.3 cm and typical areas of 3 cm^2. The resistance of a "wet" gel layer is therefore generally in the range 0.5–50 Ohms. Although hydrogels have higher resistivities, this disadvantage is generally compensated for by the use of larger gel areas. This need not necessarily entail the use of a larger overall electrode area as the adhesive hydrogel may not require the use of a large surrounding disc of adhesive foam. Another way to compensate for hydrogel's inherent disadvantage is to decrease the gel layer thickness, a variable generally ignored in electrode design even though it can have a significant effect on electrical performances. Many commercial hydrogels used in biosignal monitoring electrodes have layer thicknesses of around 1 mm (compared to around 3 mm for pre-gelled "wet" electrodes) and that coupled with larger areas of around 7 cm^2 can lead to hydrogel pad resistances in the range 10–100 Ohms [94].

It is suggested that further improvements can be made to the performances of hydrogel electrodes (and "wet" electrodes) by the use of even thinner gel layers. It must be borne in mind that gel layer resistance is not solely determined by the dimensions and properties of the gel pad. When a large area gel pad is used in conjunction with a small area sensor, the dimensions of the smaller sensor will largely determine the magnitude of the gel layer resistance, the overlapping section of gel pad carrying relatively little current. This is important in both biosignal monitoring and electrostimulation applications.

Hydrogels, being hydrophilic, are used for wound dressings in order to absorb exudate. They are therefore poor at hydrating the skin and will even absorb surface moisture. With hydrogel electrodes, R_{SP} is observed to fluctuate with sweat gland activity and the subject's state of mental arousal, decreasing during increased activity and gradually increasing again as the hydrogel absorbs the excess surface moisture [83, 92]. In contrast, C_{SP} remains relatively constant after a slight initial increase [84].

Hydrogels are therefore not only more resistive than "wet" gels, but they hydrate the skin less effectively and give rise to higher skin impedances (i.e. higher values of R_{SP} and lower values of C_{SP}). Typical values of R_{SP} for hydrogels can be as high as 15 MΩ cm^2 compared to a high of 5 MΩ cm^2 for "wet" gels [92]. Once again, this disadvantage can be overcome, at least partially, by the use of larger hydrogel pad areas. An additional way of increasing the value of C_{SP} is the use of thinner hydrogel pads [84].

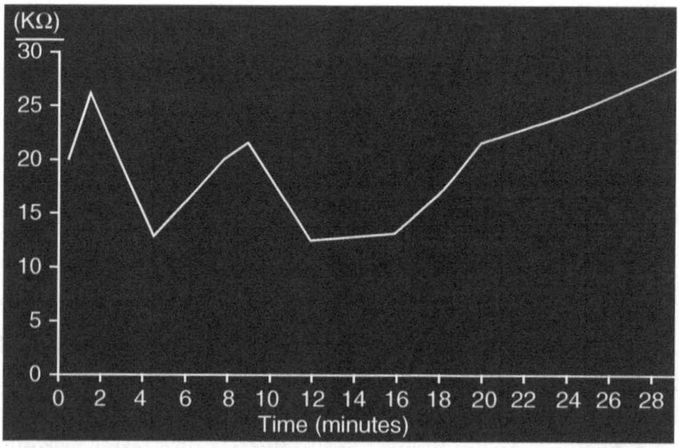

Fig. 3.13 Variation of skin resistance with time for a hydrogel based electrode. Fluctuations due to variations in subject's state of relaxation/arousal [84]

Skin Preparation Techniques

In the clinical environment the skin site is often degreased using an alcohol wipe prior to electrode application. This probably removes some of the loose, outermost cells of the stratum corneum and the poorly conducting lipid substances from the surface of the skin [66]. However, the use of alcohol wipes may initially increase the impedance of the skin by dehydrating the outer layers of the skin [20]. Motion artefact also may increase initially following application of alcohol to the skin [136]. When "wet" gel electrodes are applied to the skin to alcohol-wiped skin, the gel will eventually penetrate the degreased skin more readily once the electrode has been on the skin for several minutes, leading to a more rapid decrease in skin impedance and possibly to a decrease in motion artefact. This may not be the case, however, in the case of hydrogel electrodes which do not actively hydrate the skin. The use of an alcohol wipe accompanied by vigorous rubbing should result in low initial impedances due to the additional mild abrasion.

A related method of rapidly decreasing skin impedance is to prerub the skin site with a high-concentration electrolyte, thus forcing the gel into the outer layers of the skin. This can result in a significant decrease in R_{SP} (Fig. 3.12) and an increase in C_{SP}, especially when accompanied by vigorous rubbing. Arbo-prep® cream is supplied for this purpose and it is claimed to reduce skin resistance by up to 90% (from 40 or 50 kΩ to 4 or 5 kΩ according to an advertisement). Some commercial gels such as Hewlett-Packard's Redox® paste contain abrasives such as crushed quartz, which, when rubbed into the skin prior to electrode application, greatly reduce skin impedance. Such aggressive gels should only be used in short term biosignal monitoring applications such as "stress testing" where high quality traces are required.

The outer layers of the stratum corneum can also be removed by rubbing the skin with abrasive pads especially designed for this purpose. This can give rise to a major decrease in R_{SP} (Fig. 3.12) and an increase in C_{SP}.

Unomedical, for example, market a small disposable skin preparation abrasive pad which, when adhered to the finger tip, can be used to dramatically reduce skin impedance. A Skin Rasp® which resembles a strip of Velcro® is marketed by Medicotest for this purpose. The Quinton Quick-Prep Applicator® rotates the abrasive centre of the Quick-Prep® electrodes, causing a marked decrease in skin impedance. ECG electrodes are often supplied with abrasive pads built into the electrode release backing.

In skin "stripping", the stratum corneum is progressively removed by repeatedly applying and removing adhesive tape to and from the skin [66, 149]. Skin stripping can greatly decrease skin impedance as a consequence of a dramatic decrease in the value of R_{SP} and an increase in C_{SP}. As the outermost layers of the stratum corneum are the most resistive, the most significant decrease in skin impedance is achieved with the first few strippings [149]. There is therefore no need for the complete removal of the stratum corneum which would obviously be clinically unacceptable due to the discomfort (pain, bleeding and/or irritation) caused to the patient during and following the recording. The more the skin is abraded for a given gel composition, the sooner discomfort develops and the more severe the irritation. The level of irritation also varies with the salt concentration and the additives present in the gel.

As pointed out above, abrading or stripping the skin is often used in "Stress testing" to decrease motion artifact [31] as well as the 50/60 Hz noise induced by any mismatch of the contact impedances. High concentration gels are also often used for such demanding applications, rapidly soaking the skin and thus effectively removing the source of the problem.

The use of both skin abrasion/stripping and an aggressive gel will however maximise the potential for severe skin irritation problems and these approaches should not be used together. Even with the use of mild gels, it is probably unwise to abrade the skin for long term monitoring applications. The increased length of exposure of the abraded skin to the gel will be conducive to skin irritation. Somewhat surprisingly, long term monitoring electrodes are sometimes commercially supplied with integral abrasive pads. This is not only risky but probably unnecessary as the use of a suitable mild gel would eventually decrease the skin impedance without the need for skin abrasion.

3.3 Electrode Design

3.3.1 External Biosignal Monitoring Electrodes

3.3.1.1 Historical Background

In 1887 Augustus Waller, using Etienne Jules Marey's modification of the capillary electrometer, obtained *surface* ECGs (as opposed to recording directly from the exposed heart of an animal) of one of his patients, "Jimmy". The patient turned out to be his pet dog. Waller used two buckets of saline to measure the canine ECG, one for the front paws and one for the hind paws [42, 139, 140].

Waller eventually succeeded in recording the first human ECG in 1887 using the capillary electrometer (Fig. 3.14). However, he initially concluded "*I do not imagine that electrocardiography is likely to find any very extensive use in the hospital.* It can at most be of rare and occasional use to afford a record of some rare anomaly of cardiac action." [10]

It was therefore left to a more visionary and tenacious Dutchman, Willem Einthoven, to establish the clinical relevance of this strange new trace and to develop and commercialise a clinically acceptable system based on the string galvanometer. Einthoven's achievement was truly awesome. However, it must be pointed out that he did build (very significantly, it is conceded) on the work of earlier pioneers. The electrode system used, for example, was Waller's bucket electrode while the moving photographic plate recording technique was originated by Marey [118].

The input impedance of Einthoven's galvanometer was such that very low contact impedances were necessary, hence the very large bucket electrodes (Fig. 3.15). Obviously, the range of applications was somewhat limited.

Fig. 3.14 Human subject connected to capillary electrometer via large area "bucket" electrodes [141]

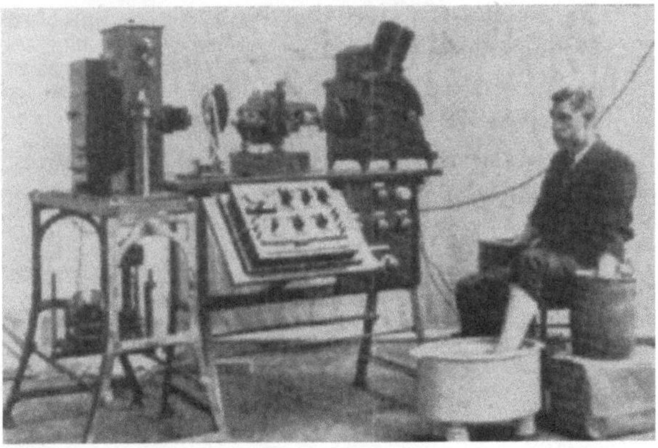

Fig. 3.15 Early commercial ECG machine and electrodes [12]

Realistically, only one's limbs could be conveniently placed in to the buckets. Hence the use of limb "leads" in electrocardiograpghy, even to the present day. It is therefore important to note several points. Present "state of the art" is often based on historical quirks rather than on a profound scientific basis. The monitoring device and amplifier determined the electrode size, design and location of the electrodes which in turn determined the clinical application and the presentation of the physiological data.

Einthoven's device in its early form could not be used for the monitoring of bedridden patients or for ambulatory monitoring. These applications had to wait for improvements to be made to the amplifiers which then enabled the use of smaller electrodes which could be more conveniently attached in other anatomical locations. However, the early monitoring locations and the form of the signals observed became accepted as "standard" and there is often considerable resistance to novel monitoring scenarios (e.g. "smart" clothing) which require or are based on different "lead" systems and which present physiological data in a different format to that familiar to the clinician.

In the 1920s vacuum-tubes were used to amplify the electrocardiogram instead of the mechanical amplification of the string galvanometer. This lead to smaller, more rugged systems which were transportable (Fig. 3.16). The input impedances of the new ECG monitors were larger, and the large metal buckets could be replaced by smaller metal plate electrodes (still large compare to present day electrodes) [118]. These advances enabled bedside monitoring and by the 1930s some ECG devices could be carried to the patient's home. Not unsurprisingly, the new plate electrodes were attached to the limbs, both for historical and practical reasons. The metals used were chosen for their availability and ease of machining (Fig. 3.17). They included German silver, nickel-silver and nickel plated steel. The foil plates were used in conjunction with moistened pads of paper toweling, lint, cotton gauze or sponge and were generally held in place with rubber straps. Around 1935, conductive gels were developed to replace the soaked pads. A wide range of gel ingredients were assessed and it was noticed that the presence of an abrasive in the gel greatly reduced the skin impedance [13, 44].

They also noted that slightly abrading the skin before applying the electrolyte helped achieve very low skin impedances.

A dry version of the plate electrode was reported by Lewes [68]. The "multipoint" stainless plate electrode resembled a large nutmeg grater which penetrated

Fig. 3.16 Mobile ECG used to monitor bed-ridden patient in hospital ward circ. 1920. [12]

Fig. 3.17 Metal plate limb electrode

the skin when firmly strapped on to the limb or applied to the skin with a slight rotary movement, thus resulting in a very significant reduction in skin impedance.

Modern versions of the limb plate electrode still exist. Some have a convenient spring clip mechanism which dispenses with the need for the rubber strap.

In the 1930s clinicians, some using electrodes held on the chest by the patient himself or by another member of clinical staff, experimented with precordial leads and established their clinical value [148]. In 1938 the American Heart Association and the Cardiac Society of Great Britain defined the standard positions and wiring of chest leads V1 – V6 [11] .

Research then focused on the development of electrodes that could be conveniently attached to the chest to enable convenient routine clinical measurements. Several designs involved a rubber "bulb" which was used to create suction sufficient to hold the metal electrode on the chest. One of the first suction electrodes was developed by Rudolph Burger in 1932 for the precordial leads [18]. The suction electrode shown in Fig. 3.18a is one developed by Ungerleider [138]. Another more recent system incorporated the "multi-point" electrode of Lewes [65] into the suction head (Fig. 3.18b). The most popular suction electrode design, widely used around the world and still in use today, was developed by Welch [145] and often called the "Welch" or "Suction" cup/bulb electrode (Fig. 3.19). It consists of a hollow, metallic, cylindrical electrode that makes contact with the skin at its base. A rubber suction bulb fits over the other end of the cylinder. The suction bulb was squeezed while the electrode was held against the skin. Upon releasing the bulb, the electrode is held in place. The suction electrode can be used any where on the chest and can even be used on hairy subjects. A single electrode can, if necessary, be used to take a measurement at a given location and then moved to another site.

Although the "Welch cup" electrode became widely used as a precordial electrode, it could only be realistically used for resting (supine) diagnostic ECG recording. The weight and bulk of the electrode generally rules out its use on upright, ambulatory and/or clothed subjects. [Since then, more suitable, lightweight, low-profile suction electrodes have been developed which are pneumatically connected to remote vacuum pumps (Geddes, 1972). Some arrays of suction electrodes are commercially available, for example, for use with exercise bicycles for cardiac "stress testing" [90].

A method had to be invented to attach small disks of suitable metal and their conductive gel coating to a patients chest (in the case of ECG) or to other body parts

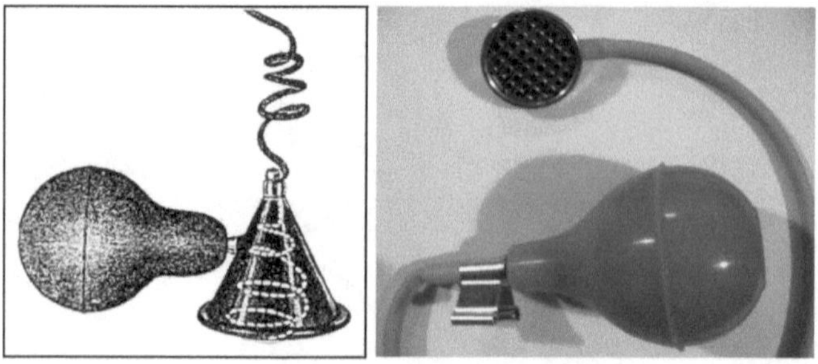

Fig. 3.18 Early designs of suction precordial electrode (**a**) Ungerleider [138]. b Lewes [65]

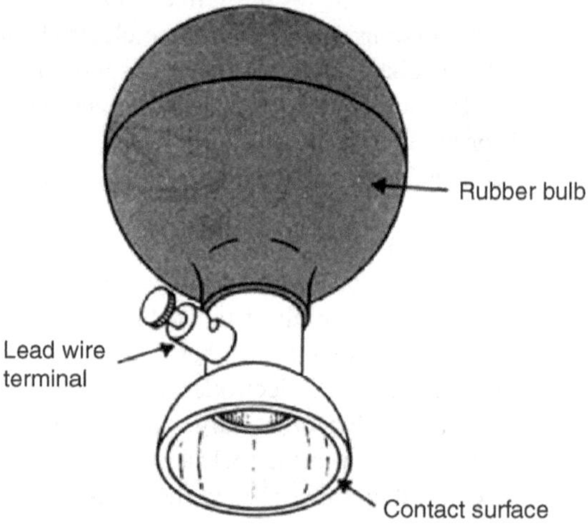

Fig. 3.19 A metallic suction electrode is often used as a precordial electrode on clinical electrocardiographs [142]

in the case of other biosignal applications such as EEG and EMG. Simply taping a metal disc to the skin site with a sandwiched gel layer was a method often used [55].

Conically-formed metal disc electrodes were and often still are used for EEG recordings (see Fig. 3.20). The base of the metal cone is attached to the patient's scalp using elastic bandages or wire mesh or, more recently, using a strong adhesive such as colloidon. There is aperture in the apex of the cone to enable the introduction into the recessed electrode of electrolyte gel or to enable the abrasion of the underlying skin by means of a blunt hypodermic needle. The cone electrodes were often made of gold as it has high conductivity and inertness, desirable in reusable electrodes. More recently, Ag/AgCl is used.

Early plate electrode designs were presumably very messy and gave rise to considerable artefact problems. The observed artefacts were attributed to disturbance of the double layer region at the electrode/skin (or more precisely electrode/electrolyte) interface (termed the electrokinetic effect by Khan and Greatbatch [57]). When the electrode moves with respect to the electrolyte, the distribution of the double layer of charge on electrode interface was thought to change and cause transient fluctuations in the half cell potential or give rise to a "streaming potential".

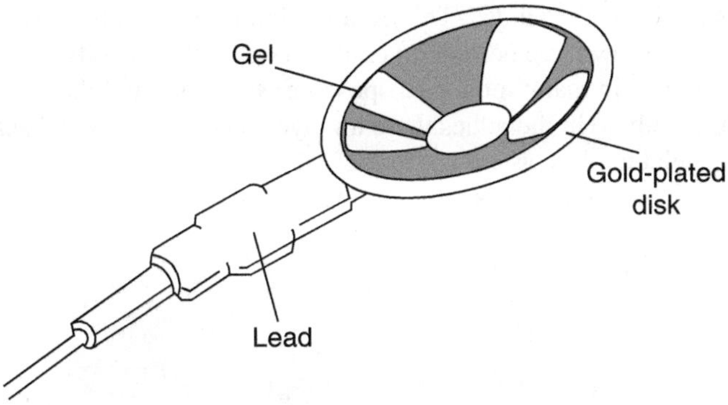

Fig. 3.20 Conically-formed metal disc EEG electrodes

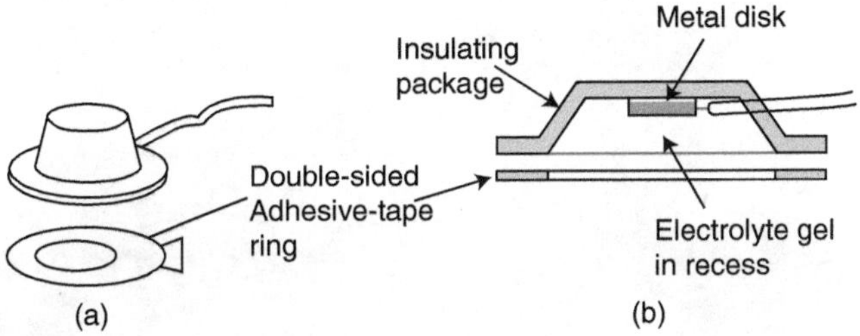

Fig. 3.21 Examples of a "floating" or "recessed" biosignal electrode. (**a**) Recessed electrode with "top-hat" structure. (**b**) Cross-sectional view of the electrode in (**a**). Webster [142]

Recessed or "floating" electrodes were introduced in an effort to protect the electrode-gel interface from such mechanical disturbance and resultant movement artefact. A metal disk was recessed in a plastic housing which was filled with electrolyte gel prior to application to the patient. The "top hat" shaped container was adhered to the skin by means of an annulus of double sided adhesive tape. (Fig. 3.21). Later a gel-impregnated sponge was used to ensure good electrical contact between the electrode disk and the skin surface. The electrode disk was therefore not in direct contact with the skin and this was found to reduce "motion artifact". At first various metal plates were used as the electrode conductor, then a sintered Ag/AgCl disc with preattached wire ensured better performances for more demanding applications.

3.4 Modern Disposable Electrodes

The "top hat" housing was eventually replaced with a smaller "retaining ring" or plastic cup and the electrode was held in place by means of a surrounding disc of adhesive foam. The plastic cup holds the gel-impregnated sponge in place and stops the gel from spreading beyond the set boundary, either during storage or use on the patient. Low-cost Ag/AgCl-plated plastic "eyelets" (part of a snap fastener) are used in these disposable electrodes and the leads are connected to the electrodes via the electrodes' snap fastener "studs". The rigid retaining ring was however uncomfortable as it did not allow the electrode to conform optimally to body contours. It was eventually removed in many modern disposable electrodes and the "recess" is now often formed by a hole in the adhesive foam layer. The backing label serves to hold the snap and eyelet in place as well as to present the company's logo (Fig. 3.22). The resultant electrode structure is much more flexible and more comfortable to wear.

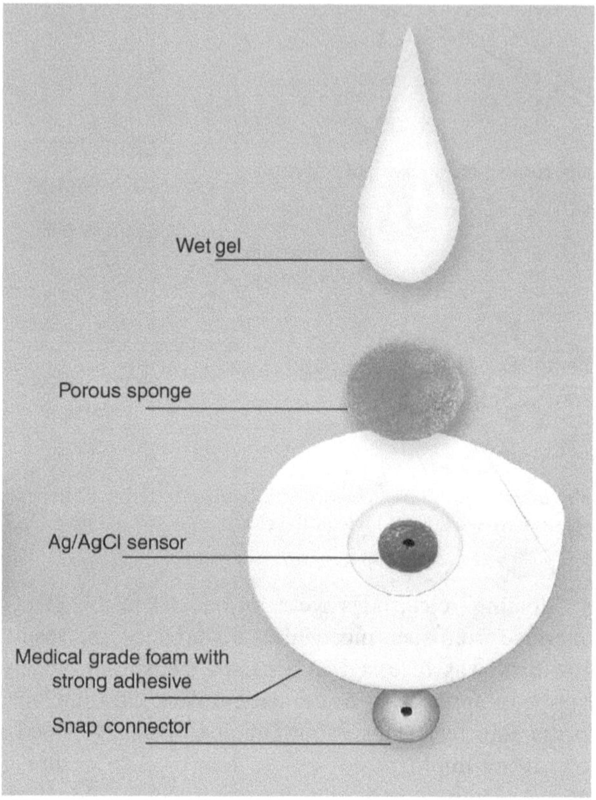

Fig. 3.22 Modern wet gel disposable Electrode (Courtesy of Unomedical A/S)

The use of a snap fastener style connection in disposable electrodes has one significant drawback for certain applications. The male stud protruding from the back of the electrode and the female connector required on the connecting lead results in a relatively heavy, large-profile electrode/connector interface which is less than optimal for applications such as neonatal and pediatric monitoring. The use of such electrodes in long term monitoring of bedridden patents could lead to considerable discomfort and the heavy connection could also give rise to significant motion artifact problems. The "integrated lead" design seeks to over come these disadvantages. A thin, highly flexible lead wire is bonded directly to the back of a specially designed Ag-AgCl coated eyelet. This results in a very low-profile, light-weight

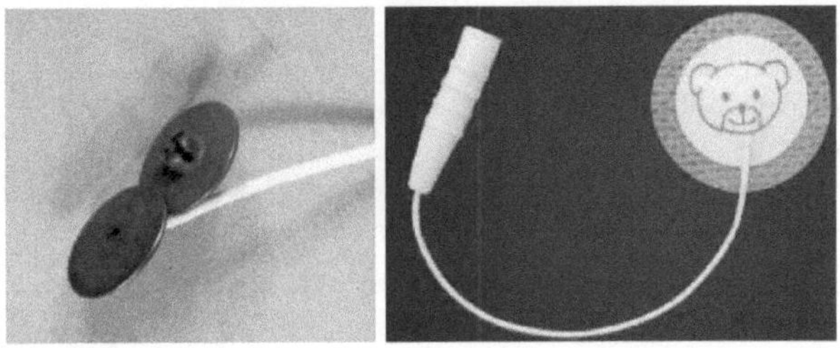

Fig. 3.23 (**a**) Specially designed Ag-AgCl coated eyelet with bonded lead. (**b**) Low profile, light weight paediatric electrode with bonded lead (Courtesy of Unomedical A/S)

electrode-connector system much used in neonatal and pediatric monitoring and attractive for long term monitoring applications (Fig. 3.23).

In an effort to decrease motion artifact, many electrode designs feature an "offset centre". The connector, often in the form of a snap fastener, is separated from the gelled sensor by a strip of metal or similar conductive layer. The connector is thus one or two centimeters away from the metal-gel-skin interface and it is possible to connect the lead to the electrode or to pull on the connector without pulling directly on the gelled skin site and thus causing artifact problems. This design appears well suited for stress testing applications though arguably less so for long term monitoring of bedridden patients due to the bulky connector. The invention was patented by Manley [75] and the concept has been commercially exploited very successfully by Ambu A/S.

More recently, many other manufactures supply electrodes with offset connectors (Fig. 3.24). Leadlock have developed an electrode which incorporates a slit in the foam backing. Once the electrode has been applied to the patient and the lead connected to it, part of the foam backing is used tape down the lead, locking in place and minimizing direct pull on the connector and underlying electrode/skin interface.

A wide range of backing materials now exists and some are better suited for specific monitoring applications, types of patients, skin types, etc. Given the great variety of materials, adhesives and designs used, the following comments are generalizations.

Open cell foam layers, made from a plastic such as polyethylene, are much used and have thicknesses typically in the rage 1 to 2 mm. They have a 10 to 100 μm coating of a "pressure sensitive adhesive", generally a polymeric hydrophobic substance [19, 63]. Adhesive foams generally give rise to firm adhesion, are resistant to liquids and tend to cushion lead pull thus giving rise to less artifact. They are therefore generally well suited to cardiac stress testing and similar applications. However, as they are occlusive and generally have a relatively aggressive adhesive, they can give rise to more skin irritation and they must be used with caution for neonatal and long term monitoring.

Porous, breathable layers, such as nonwoven clothes or tapes, have the advantages of being soft, stretchable and conformable with the skin. [Note: The term "micropore", although sometimes used loosely to describe any breathable backing

Fig. 3.24 An example of an "off-set connector" electrode (Courtesy of Unomedical A/S)

material, is strictly a 3M product]. Porous layers tend to cause less mechanically-induced trauma to the skin which can occur with more rigid materials and is due to shearing of underlying skin layers. As they are highly air permeable and use milder adhesives, porous "tapes" cause less skin irritation and are well suited for Long Term Monitoring. Larger backing areas tend to be required. The gelled centre can however pull away from the skin as a result of the stretchable backing. Ambu A/S use a central ring of adhesive around the gelled eyelet to minimize this problem.

As we have already seen, "wet" (as opposed to "solid" gels) vary in composition and concentration depending on the application. Aggressive gels with higher concentrations of electrolyte and/or including abrasive particles are used for short term, demanding monitoring applications such as cardiac stress testing. Mild gels are used in pediatric and neonatal applications due to the increased vulnerability of the patient's skin. It such be noted that no matter how hypo-allergenic a gel or an adhesive is claimed to be, there will always be some patients who will experience some form of skin reaction to one of the components.

3.5 Solid Conductive Adhesive Electrodes

The growing monitoring market has led to the development of even lower cost disposable electrodes. "Solid conductive adhesive" or "hydrogel" electrodes were first introduced by LecTec Corp around 1980 [52]. Hydrogels are composed of a hydrocolloid, alcohol, a conductive salt, water and a preservative. The hydrocolloid use can be either natural (e.g. Karaya gum) or synthetic (e.g. Polyvinyl pyrrolidone) [19]. Early hydrogel electrodes were based on the natural hydrocolloid, Karaya, which comes from the bark of a tree. The rather unaesthetic appearance of these early gels and the variations in their electrical and mechanical properties limited their widespread acceptance. The use of synthetic hydocolloids, with their more attractive appearance and performances, has led to the recent revolution in electrode design.

Solid adhesive gels reduce the number of electrode parts required, dispensing with the need of a gel-impregnated sponge or a surrounding disc of adhesive tape. This gives rise to small area, low profile electrodes suitable for neonatal monitoring, especially when coupled with integrated leads as discussed above (Fig. 3.23b).

"Tab" solid adhesive electrodes are now widely used for many biosignal monitoring (and stimulation) applications. Thin, highly-flexible metallic/conductive foils or printed conductive ink layers are laminated with solid, adhesive hydrogels. A section of the foil or printed layer is left uncovered. Once the electrodes are cut out, the exposed conductive "tab" acts as a means of connection, the leads being connected via alligator clips (Fig. 3.25). Electrode design is therefore very simple and manufacturing costs low. These flexible, low profile electrodes are best used for short-term, resting diagnostic monitoring. Tab electrodes are not suitable for ambulatory or longterm monitoring as the tab connection will cause the electrode to peal off quite easily when pulled from any angle other than directly "downwards". Also, hydrogels are hydrophilic and tend to absorb moisture and loose their adhesive properties over time and will falling off the patient if an additional adhesive backing is not used. Hydrogels, being "solid", do not leave a messy residue on the skin requiring cleaning. Tab electrodes are also repositionable and are reuseable (on the same patient!) in certain home monitoring applications.

When used with an adhesive backing layer, the hydrophilic hydrogels tend to be relatively non-drying (a significant problem with pre-gelled "wet" electrodes) and their electrical properties may even improve as they absorb moisture. As they don't actively hydrate or otherwise affect the skin they tend to be relatively non-irritating compared to "wet" gels.

There are some disadvantages however associated with hydogels. Hydrogels are more resistive than "wet" gels and hence the gel pad resistance will be higher. This can be compensated for by using larger hydrogel pad areas and thinner layer thicknesses as compared to those used with "wet" gels. Although the area of the solid adhesive gel in a "tab" electrode, for example, is considerably larger for this reason than that in a standard disposable "wet" gelled electrode, the absence of a surrounding adhesive layer results in the "tab" electrode having a smaller over all area.

Hydrogels, being hydrophilic, are poor at hydrating the skin and may even absorb surface moisture. They therefore give rise to larger skin impedances. This disadvantage can also be overcome, at least partially, by the use of larger hydrogel pad areas. Hydrogels are also more expensive than "wet" gels but generally lead to less expensive electrodes due to the simpler designs involved.

Hydrogels are more sensitive to motion artifact as they do not actively hydrate the skin. They are therefore not well suited for stress testing.

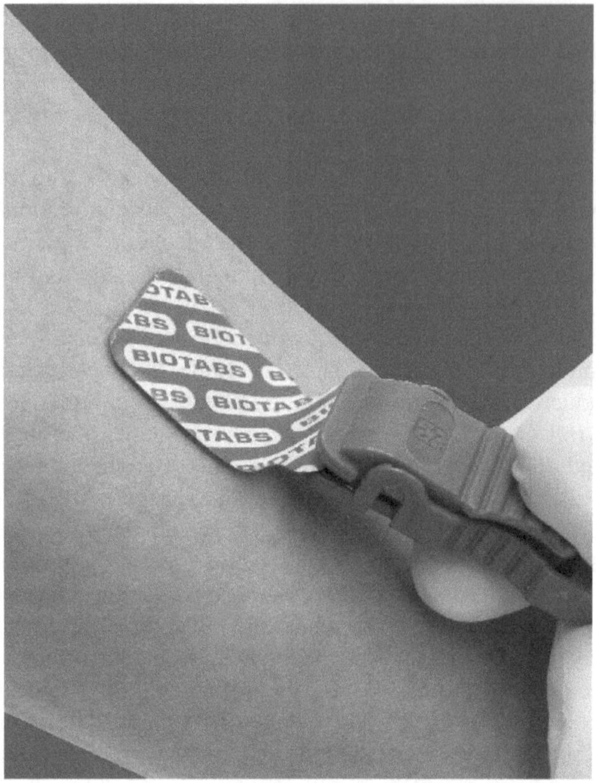

Fig. 3.25 Hydrogel-based tab electrode with connector (Courtesy of Unomedical A/S)

The use of such "solid" gels entails numerous advantages when they are used in conjunction with screen printing technology [30], especially for body surface mapping and similar applications. It is possible to construct thin, light-weight, highly-flexible electrode arrays with accurately defined electrode/gel areas, shapes and inter-electrode distances for a wide range of novel stimulation and biosignal recording applications. As the solid gel will not spread between electrodes, it is possible to position electrodes very close together without electrical shorting (Fig. 3.26).

3.6 "Wearable" Electrodes for "Personalized Health"

The recent and continuous trend towards home-based and ambulatory monitoring for "Personalized Health Care", although exciting and potentially leading to a revolution in health care provision, necessitates even more demanding performance criteria for the monitoring sensors (Andreas and Dittmar). Many groups around the world are seeking to incorporate electrodes into clothing in order to monitor

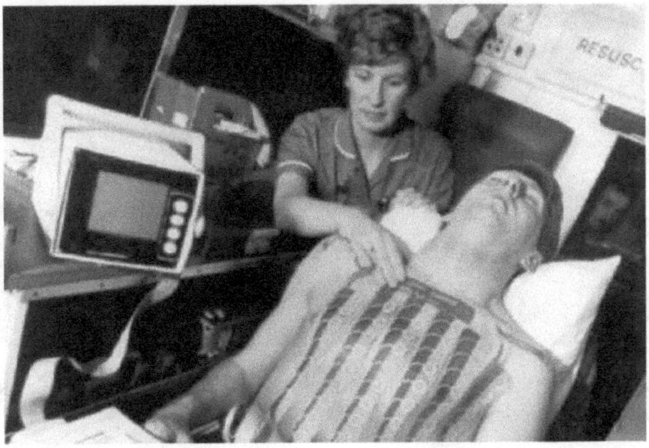

Fig. 3.26 Cardiac mapping electrode harness

military personnel, firefighters and eventually the average citizen who wishes to monitor his or her health. Systems already exist on the market, e.g. Life Shirt, which resemble waistcoats into which one plugs-in standard ECG electrodes and other sensors. These sensors are removed and replaced periodically by the subject and hence require the knowledgeable involvement of the motivated wearer, presently military personnel, athletes, rescue workers, etc.

For the more widespread use of "wearable" monitoring systems, especially by the average citizen, the system must be very easy and comfortable to use and require no preparation – literally as simple as putting on their shirt. Electrodes must therefore (i) require no "prepping" (ii) be located in the correct location once the "smart" garment is put on, (iii) make good electrical contact with the skin (iv) not give rise to motion artifact problems (v) not cause discomfort or skin irritation problems (vi) be reusable and machine-washable. Although much work has been carried out in this novel area, it is not surprising given the above list of required performance criteria that the electrodes/sensors tend to form the "bottle-neck" in the success of the overall monitoring systems. One must therefore not simply choose an electrode with as conductive a metal element as possible. Unfortunately, it would often appear that the associated electronic systems are first developed and the electrode design is left to the end, almost as an after thought. The author would therefore suggest that researchers start with the desired biosignal and establish the optimal body site(s) and electrode design for the given application before developing the rest of the monitoring system. This may involve the use of novel "lead" or "montage" electrode positions in order to conveniently pick up artifact-free signals. Although this will necessitate clinicians interpreting non-traditional waveforms, it will at least enable feasible monitoring and, as it involves novel body sites and electrode designs, it may well be patentable. After all, if it is not patented and commercialized, it will not benefit the patient.

One of the most promising "smart" garments is that developed under a European Fifth Framework programme called WEALTHY (Wearable Health Care System)

Fig. 3.27 **a**: The WEALTHY physiological monitoring vest with integrated sensors **b**: An early version of the "fabric" electrodes

(Fig. 3.27). WEALTHY is a wearable, fully integrated system, able to monitor a range of physiological parameters including electrocardiogram, respiration, posture, temperature, and a movement index. "Fabric" electrodes are made using conductive fibres woven into the stretchable yarn of the body-contour hugging garment and connections are integrated into the fabric structure (Fig. 3.27b). Various membranes are being assessed to ensure optimal electrode-skin contact and minimize skin irritation. The garment is comfortable and can be worn during everyday activities. It is washable and easy to put on.

3.6.1 External Electrostimulation Electrodes

3.6.1.1 Historical Background

The evolution of external stimulation electrode design shares some of the key land marks as the development of biosignal monitoring electrode and hence this section will be somewhat shorter.

From the mid 1700s when electrostatic generators were used to deliver arguably therapeutic impulses to various parts of the body, hand held (by the practitioner) electrodes had to be designed capable of delivering the impulses to the patient without shocking the practitioner who was holding them against the body part in question. The electrodes used tended therefore to be simply long metal rods insulated with wooden handles (Fig. 3.28). Although the electrodes were initially terminated in a simple metallic sphere, more exotic terminations were soon invented as these were observed to lead to different therapeutic effects on the body by means of the variations in the "streams of the electric fluid". A modern parallel would be the use of different shaped probes in Electrosurgery for different effect.

Around 1800 saw the discovery of Galvanism (DC current) and Voltaic Piles (early batteries). There are numerous examples of practitioners using their hand held probe for localized effect and the second contact to the patient was made by means of a container of water into which the patient put a hand or foot (Fig. 3.29).

Fig. 3.28 Early electrostatic generator and hand-held electrodes [1]

Fig. 3.29 "Container" electrode. Used on the dead and the living [4]

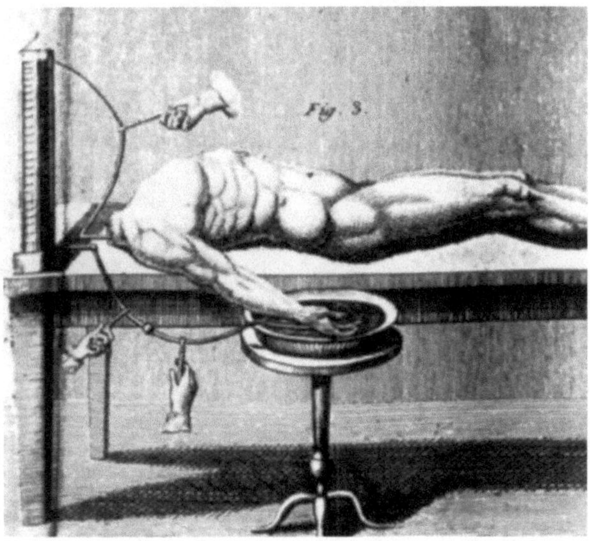

Following the discoveries of self and mutual induction (circ.1830), Guillaume Duchenne made great contributions to the clinical application of the new "Faradic" current. At that time there was much interest in the localization of what became known as motor points. It was common to combine the prevailing interest in acupuncture and use needles to stimulate muscles and nerves under the skin, termed "electropuncture" [118]. Duchenne was not happy with this approach and developed his own electrodes for "localized electrization". His electrodes were in various shapes (discs, spheroids and cones) covered with leather moistened with salt water prior to application [33] (Fig. 3.30).

Fig. 3.30 Duchenne's moistened conical electrodes for "localized electrization" [34]

During the 1900s, as with biosignal monitoring electrodes, electrostimulation electrodes involved the use of simple metal buckets or receptacles, filled with water or another electrolyte, into which the subject introduced their foot or hand, especially in early iontophoretic applications. Obviously, the range of applications was somewhat limited. Metal probes were still manually pressed against skin for short term applications. The electrodes were either gelled before application or had moistened chamois coverings similar to those used by Duchenne. In the 1950s, early external pacing and defibrillator electrodes, termed "paddles" because of their shape, consisted of bare metal disks made of non-corrosive material and were simply pressed against the patient's chest [122] see Fig. 3.31.

Rigid metal plate electrodes which were eventually held in place with rubber straps on the limbs and even the thorax (Fig. 3.32). Some of these electrodes were and still are made with rigid stainless steel plates [114]. The foil plates were generally used in conjunction with moistened pads of paper toweling, lint, cotton gauze or

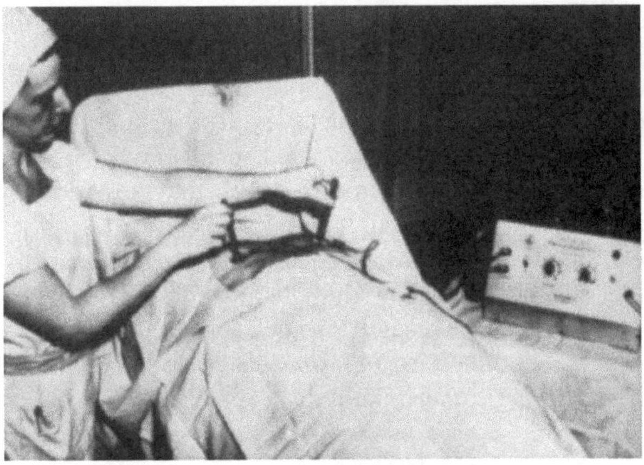

Fig. 3.31 Early Pacing equipment and hand-held electrodes [122]

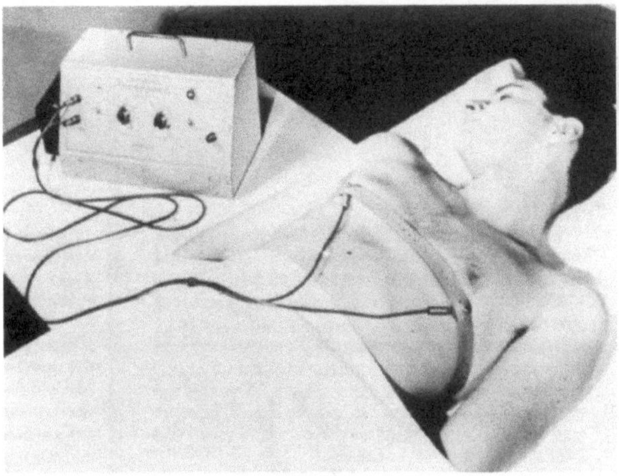

Fig. 3.32 Early Pacing equipment and metal plate electrodes [122]

sponge. The pads were moistened by the therapist prior to electrode application with water or electrolyte. Such electrodes could be easily reused by simply washing and regelling the electrode. Being rigid, however, these plate electrodes did not always make optimal contact with the body surface and gave rise to current density "hot spots". External cardiac pacing at this time, for example, was very painful [118].

Malleable metal foil electrodes were the next evolutionary step in electrode design. Malleable electrodes have been made using a range of metals including tin-plate lead and aluminium foils [73]. Such electrodes had the advantage of being able to conform, to some extent, with body contours, thus ensuring a better, more comfortable contact between the electrode and the patient than was the case with rigid plates. Wrinkles in malleable metal foil could however encourage preferential current flow through small areas of the gel and into the patient.

More convenient, disposable pregelled foil electrodes were then developed for a range of external electrostimulation applications. The metal foil was laminated onto an adhesive foam backing. A gel-impregnated sponge layer was located on top of the metal layer and the complete electrode is attached to the patient by means of the surrounding layer of adhesive backing foam.

Unfortunately, the "wet" gel in these disposable pregelled electrodes tended to pool to one side, depending on how they were stored, giving rise once again to current density hot spots. More recently, the gel-impregnated sponge layer has been replaced by a conductive adhesive gel layer as this does not have the potential for pooling to one area during storage and it does not squeeze out under pressure [19] (Fig. 3.33).

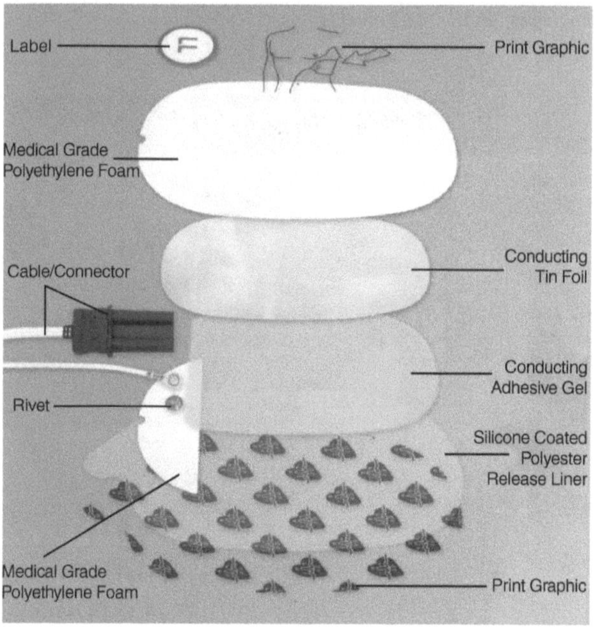

Fig. 3.33 Construction of a modern External Pacing or Defibrillation Electrode (Courtesy Unomedical A/S)

3.6.1.2 Current Density Considerations

The distribution of current density under an electrode is an important parameter when designing and using electrostimulation electrodes. In the simplest case, current density (the amount of current per unit of conduction area) is inversely proportional to the electrode/skin contact area. For a given current, the current density under a small area electrode will be higher and more localized than that under a large area electrode. There is generally an optimal electrode area for a given therapeutic application, based on a range of criteria including the anatomical position and size of the nerve/muscle/organ and the relative positions and sizes of the electrodes. Small electrodes are therefore well to target precisely known points such as motor points. If a small electrode is used in conjunction with a large area electrode, the effect is more pronounced under the smaller of the two. In such "monopolar" stimulation, the small electrode is often used as the "active" electrode to target the therapeutic effect. The larger electrode is simply used to complete the electrical circuit and is termed the "indifferent" or "dispersive" electrode. The use of two equally sized electrodes is termed "biopolar" stimulation. In TENS, for example, biopolar stimulation is often used to stimulate large muscle groups "sandwiched" between the (large) electrodes. Too large an area of electrode, however, may cause the current to spread to neighboring tissues.

High current densities can cause tissue injury due to, among other things, heating effects. Assuming uniform current density distribution under an electrode, it is possible to calculate the minimum area of electrode necessary to achieve therapeutic effect and avoid tissue trauma [76]. If, for whatever reason, a large portion of the applied current flows into the patient through a small portion of the electrode-skin interface, a high current density "hot spot" can occur which may cause considerable pain and trauma to the patient when applying apparently "safe" therapeutic impulses. At best in such cases, the applied current may have to be limited to less than therapeutic values due to the patient's discomfort [19]. There are many potential sources of "accidentally" high current densities. Wrinkles or breaks in the metal electrodes; gel squeezing out from under the electrode or drying out; electrodes partially peeling off the skin, poor electrode application, etc. can encourage preferential current flow through small areas of the gel and into the patient. However, current density "hot spots" can also occur due to poor electrode design and a considerable amount of research has been and is being spent investigating this important problem.

In this presentation, stimulation electrodes have been divided into "conductive" electrodes and "resistive" electrodes in order to facilitate the review of the various design features.

With highly conductive metal electrodes, such as those used for external cardiac pacing, defibrillation or electrosurgery, current density hot spots are observed to occur under the perimeter of the electrode, often evidenced in the past by annular shaped burns to the patient [111, 131].

Current density hot spot problems are now often studied using thermal imaging cameras. Thermograms of the patient's skin (or a substitute such as pig skin) are taken immediately following the application of a given series of pulses and the removal of the electrode under test (Fig. 3.33). Increases in skin temperature reflect the magnitude of the current density at a particular point [107].

Wiley and Webster [142] showed that current flow through a circular electrode placed on a semi infinite medium could be solved analytically. They found that for an electrode of radius, a, and total current, Io, into the electrode, the current density into the body as a function of radial distance from the centre, r, was given by:

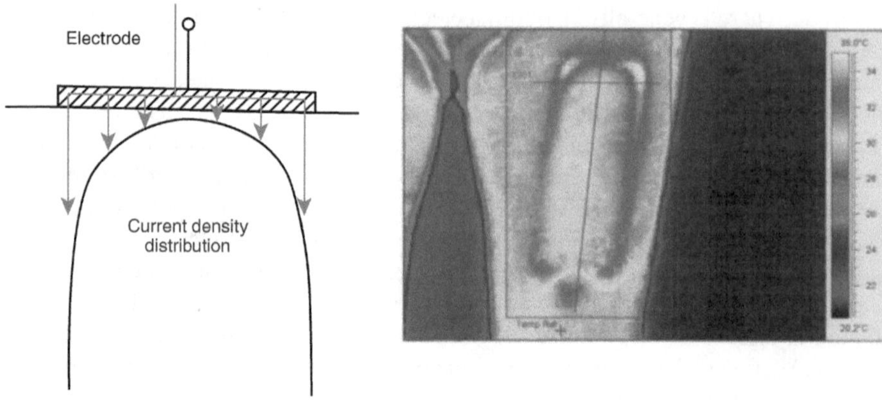

Fig. 3.34 (**a**) Schematic representation of the current density distribution under a "conductive" electrode plate. (**b**) Thermal image of the skin under an electrosurgical electrode following testing

$$J(r, o) = \frac{Jo}{2[1 - (r/a^2)]^{1/2}}, (A/cm^2) \tag{3.21}$$

where $Jo = I_0/\pi a^2$, i.e. a hypothetical uniform current density.

As a result of the approximations made in deriving this simple equation, the value of current density at the edge (when $r = a$) would theoretically approach infinity. More realistically, the current density at the perimeter can be around three times higher than that at the centre of the electrode [29] (Fig. 3.34). The above equation shows that the middle portion of the electrode is relatively ineffective in carrying the current as half the total current flows through an outermost annulus 0.14a wide or 1/7th of the radius.

Efforts in this area concentrate on "encouraging" more of the current to flow through the central portion of the electrode.

A main concern in the design of the "conductive" stimulation electrodes used for external cardiac pacing, defibrillation or electrosurgery is the decrease in the high current densities observed at the edges

In TENS, relatively resistive conductive rubber is often used and the opposite problem arises. When current is introduced into the conductive rubber (via a small metallic connector), it tends to flow into the skin immediately under the connector rather than laterally through the resistive electrode. Efforts in this area concentrate on "encouraging" the current to flow laterally through more of the electrode surface.

3.6.1.3 Modern Electrode Designs

"Conductive" Electrodes

Electrosurgery, external cardiac pacing and defibrillation share a common problem: electrodes tend to deliver or sink a substantial portion of the outgoing or incoming current through their peripheral area as opposed to providing a uniform current density along their surface. This is referred to in the literature as the fringe, edge or perimeter effect.

Many suggestions have been made to reduce this edge effect observed with metal electrodes. These include

1. Increase overall area of the electrode. Obviously an increase in electrode area will lead to a decrease in current density [59, 62]. However it is generally not practical to use very large electrodes as the applied electrical field must be sufficiently focused to stimulate the targeted tissues and them alone. Also, there is a strong commercial interest in decreasing the size of the electrodes – to save money and to facilitate packaging and storage of the electrodes.
2. Avoid sharp edges in the metal plate [59, 62]. It has long been observed that square or rectangular electrodes with angular edges concentrate the electrical field at their "corners" giving rise to current density hot spots in these locations. Using round electrodes or rectangular ones with rounded edges have been found advantageous in this regard.
3. Make the gel pad slightly larger than the electrode to enable the electric field lines to spread out before reaching the skin [59] (Fig. 3.35). Using a gel pad much larger than the size of the metal plate has less effect than would be expected as the perimeter of such a large gel pad will carry little current and the additional gel is electrically redundant. This arguably applies to some of the snap connection electrodes used in TENS which do not have an additional current dispersing element.

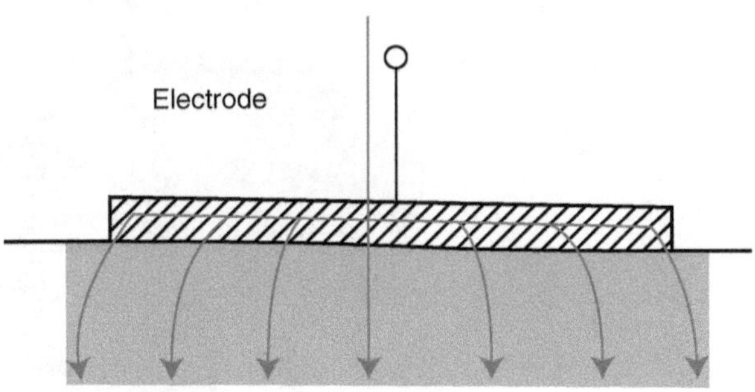

Fig. 3.35 Schematic representation of the current density distribution under an electrode plate coupled with a larger gel pad. Current is allowed to spread out beyond the boundaries of the metal plate, thus minimizing the edge effect

4. Increasing the overall resistance or thickness of the gel layer in order to give the current more "time" to spread out evenly through the gel [59, 62]. It is well known that in applications such as external cardiac pacing, the use of relatively resistive gels decreases the pain and burning to the patient's chest. [62] suggest that the use of a layer of intermediate resistivity, comparable to that of the

underlying tissues, optimally improves the distribution. However, in other applications such as external defibrillation, a high resistance gel pad would lead to energy wastage and a decrease in the desired therapeutic effect.

Taken to its logical conclusion, this approach results in the coating of the electrode metal plate surface with a dielectric film. Such "capacitive" electrodes have been shown to give rise to nearly uniform current densities [107].

5. Increase the resistance or thickness of the gel at the edges. Kim et al. [58] proposed covering the electrode metal with resistive gel of increasing resistivity as one moved out from the centre towards periphery, according to a specific relation with respect to the electrode radius. Although an intriguing concept, the commercial manufacture of such an electrode system is not as yet feasible.
6. Make electrode conductive plate progressively more resistive toward the peripheral edge of the electrode. Wiley and Webster [146] suggested subdividing the electrode plate into concentric segments and connecting external resistors to the individual segments. The connected resistors had progressively higher resistances towards the periphery in order to equalize the currents in the separate segments. A simpler system that has been successfully commercialized was patented by Netherly and Carim [102]. A resistive layer is deposited on the outer edge of the electrode conductive plate, thus forcing more current to flow through the central portion of the electrode (Fig. 3.36). Krasteva and Papazov [62] demonstrated theoretically that a high-resistivity perimeter ring decreased the maximum periphery current by 12% without increasing the total interface resistance, hence the resistance to the defibrillation current.

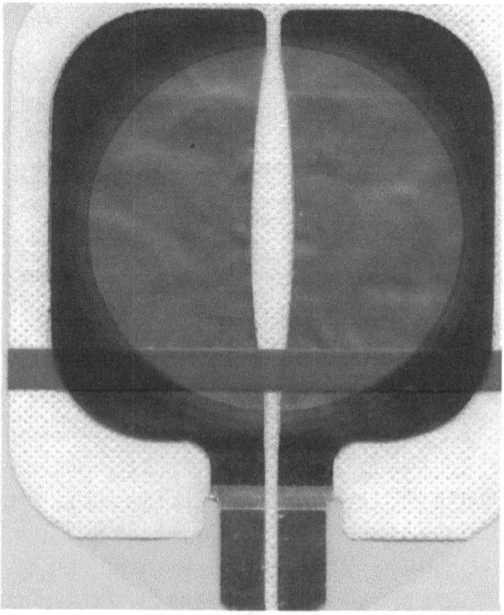

Fig. 3.36 3M's Electrosurgical dispersive Electrode. Note the green "lossy dielectric" material deposited around the peripheral edge of the electrode

Another related approach to this problem starts with a resistive graphite-based conductive layer and progressively builds up multilayered (at least two) coatings of more conductive silver/silver chloride towards the center of the electrode [38]. The perimeter resistance is approximately 200 times that of the centre. This is the technique exploited in Medtronic EDGE ^{TM} system electrodes for defibrillation, noninvasive pacing and ECG monitoring (Fig. 3.37). It is claimed that the design distributes the current density evenly over the entire surface area of the electrode, rather than concentrating it at the edges.

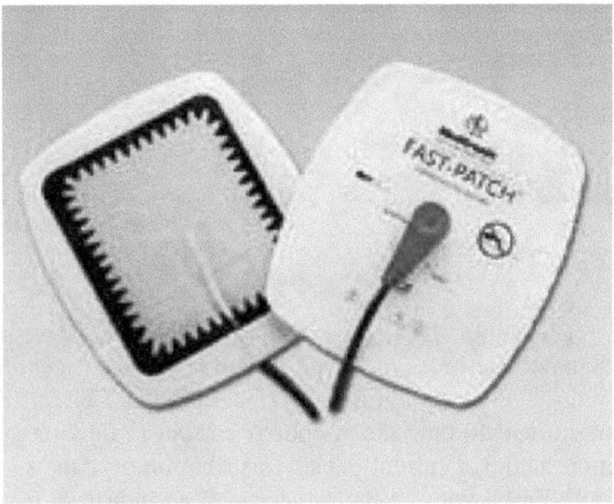

Fig. 3.37 Medtronic's EDGE ^{TM} system electrode Better Photo needed

7. Scallop or otherwise shape the edges of the metallic plate so that the length of the perimeter is increased and hence the peripheral current density is decreased. Over the years various designs have incorporated this concept. For example, it has been shown that using a figure-of-eight design rather than a rectangular metal plate reduced the maximum temperature (reflective of current density) by 30 to 50% [180]. An alternative design is shown in Fig. 3.38. Caution is advised with this approach as the formation of "fingers" in the metal layer may serve only to concentrate the current at the tips of the fingers and one could be effectively left with a reduced peripheral area.
8. Make holes in the central portion of the metal plate in order to provide "internal peripheral edges" to block the lateral flow of current. Some early claims were made that holes in the metal layer improved current density under the electrode. Presumably it was believed that the holes "blocked" the current from flowing from the connector to the edge of the metal plate, "forcing" it to flow into the patient at the "edges". It is the author's belief that such holes in the metal plate achieve little apart from further decreasing the area of the electrode and, if anything, increasing the current density at the edges. This impression appears to be confirmed by the work of Krasteva and Papazov [62] who investigated electrode structures with openings in the metal plate for skin "breathing".

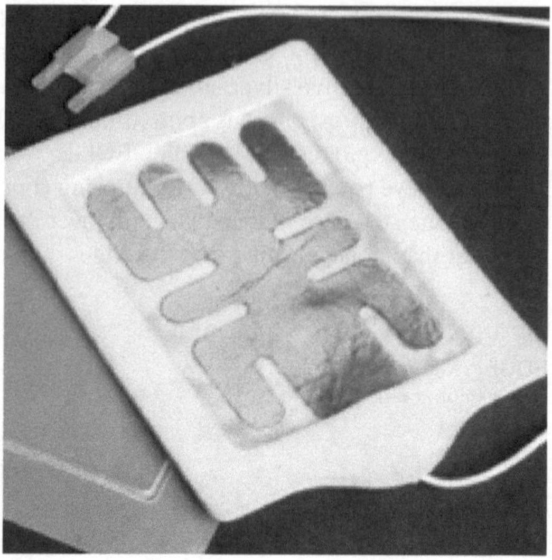

Fig. 3.38 An electrostimulation electrode with cut out metal plate in an effort to increase peripheral edge. The green sponge impregnated with gel has largely been removed to facilitate inspection of the underlying plate

The author has suggested that the use of concave slits in the metal layer rather than circular holes may well have a favorable effect on current density distribution with the concave "internal peripheral edges" effectively blocking the lateral flow of current, forcing the "trapped" current to flow into the gel and thus achieving a more uniform current density distribution over the surface of the conductive layer [88]. Early work on the project with an industrial partner appeared promising but the work was never completed.

"Resistive" Electrodes

A TENS electrode system appears relatively simple and generally comprises a conductive plate, an ion-containing gel, a means of attachment to the skin and a means of connection to the stimulator's lead. [76] pointed out, however, that of allof the component parts of the overall TENS system, the electrode-skin interface has probably been the least understood and the most problematic. In addition to influencing the effectiveness of the treatment, poor electrode design can give rise to electrically-, chemically- and mechanically-induced skin irritation and trauma to the patient.

Initially, electrodes originally designed for ECG and other biosignal monitoring applications were used with TENS units. Some still are. Larger, more suitable electrode designs were eventually developed in order to reduce the current densities under the electrodes, to reduce skin irritation problems and to increase stimulation comfort [135].

A large percentage of commercially available TENS electrodes are now molded from an elastomer (e.g. silicone rubber) or a plastic (e.g. ethylene vinyl acetate) and loaded with electrically conductive carbon black (Fig. 3.39). Very few irritation or allergic reactions have been reported for conductive rubber electrodes as they do not

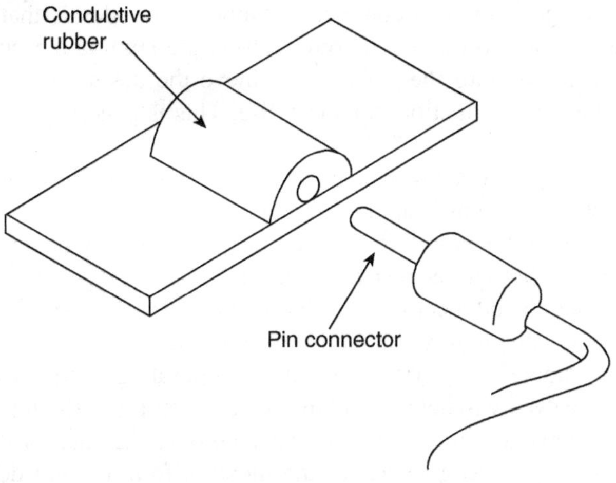

Fig. 3.39 Carbon-filled silicone rubber electrode [142]

generate the corrosion products often observed with metal electrodes [76]. The great advantage of such electrodes is that they can be molded into almost any size or shape and a wide range of choice exists in the market. They can be made sufficiently thin to have high flexibility and thus are able to conform with body contours, making them suitable for a wide range of TENS applications.

Conductive rubber electrodes are often used in conjunction with an electrolyte gel and attached to the patient using elastic straps or custom-cut discs or patches of adhesive "tape". Expanded polyester foam tends to give the most secure adhesion. However, as this backing is occlusive, the use of foam can give rise to skin irritation problems. Breathable, cloth-like fabrics allow the transmission of air and moisture and generally cause less skin irritation problems. Cloth-like materials tend to stretch, however, (an advantage when it accommodates skin stretching due to movement) and this can lead to the electrode working loose and making poor contact with the skin, possibly resulting in current density "hot spots".

"Wet" gels can squeeze out from under parts of the electrode and give rise to increased current densities in other areas. The use of hydrogel minimizes this problem source [19]. Conductive, adhesive pads of "solid" hydrogel help ensure firm electrical contact between the electrode and the skin, reduce the incidence of current density hot spots and often simplify the design of the electrode. As a large surrounding disc of adhesive tape is not required, the electrode size can be reduced to the "active" electrode area. These solid gel pads can be, depending on the application, replaced, refreshed or simply reused in various "semi"- or "totally"-reusable electrode systems. In some applications, the gel pads can be removed and the conductive rubber electrode cleaned and regelled with a fresh gel pad for further use. In other cases, the electrode can be intermittently reused, on the same patient, by re-hydrating the gel pad. Such reusable electrodes are ideal for home-based patient use.

One disadvantage with such conductive rubber electrodes is that they are relatively resistive. More power is required to drive the stimulating current through the resistive electrodes into the body and achieve the desired stimulation. There may therefore be some reduction in battery life. This is generally not a significant problem, however.

A more serious problem involves current density distribution under the resistive electrode. When current is applied through the conductive rubber (via a small metallic connector), it tends to flow into the skin immediately under the connector rather than laterally through the resistive electrode, thus giving rise to a current density hot spot under the connector. Effectively, this is the opposite problem to that encountered when using highly conductive electrodes.

Efforts to overcome this problem include incorporating conductive elements in the rubber to more evenly to help spread the current over the entire interface surface. Some electrodes have a thin metallic layer coated onto the back of the conductive rubber and these appear to give rise to the most uniform current density profiles [19].

The growing home-based market has led to the great variety of "low cost" disposable and reusable electrodes which are generally based on solid adhesive gels. Some electrodes are made using conductive cloth-like materials; thin metallic foils; aluminized carbon-filled mylar; or wire strands. Electrical connection is generally made to these electrodes via alligator clips, snap fasteners or pin connectors. Many of these hydrogel-based electrodes can be trimmed to the desired size or shape by simply cutting with a pair of scissors. The current density profiles under these electrodes will very much depend on the relative resistivities of the "metal" and gel layers as well as on the actual design.

Snap fastener designs resembling standard ECG electrodes, are available with hydrodel pads or sponge discs containing low chloride "wet" gels (to minimize skin irritation). These may require an additional "current dispersing" element to ensure that the current spreads out beyond the immediate confines of the eyelet electrode [19]. Axelgaard Ltd's UltraStim® Snap Electrodes feature a highly conductive grid pattern printed on to a conductive flexible layer and coated with a moderately conductive adhesive gel layer. The conductivity of the conductive pattern is controlled through the use of various grid designs with preselected line widths and spacing as well as thickness and ink compositions [7]. The pattern is thus used to control and optimize the spread of electrical current over the surface of the electrode with an intentional current "drop off" towards the edge of the electrode (Fig. 3.40).

It is interesting to note that while considering current density under "conductive" and "resistive" electrodes leads to a similar optimal design. To improve conductive electrodes one places a resistive layer between it and the patient. To improve a resistive electrode one puts a conducting layer either behind or in front of the resistive layer. Such "sandwich" electrode designs appear promising for a range of electrostimulation applications.

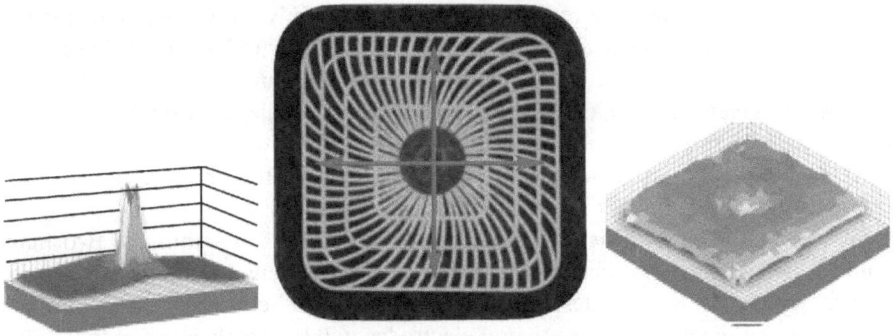

Fig. 3.40 (**a**) Current density distribution under a conventional snap electrode. (**b**) Axelguaard's UltraStim snap electrode with current controlling grid. (**c**) Current density distribution under Axelguaard's UltraStim snap electrode. (Courtesy of Axelgaard Ltd)

"Garment" Electrodes

A range of researchers in the TENS, FES and body toning areas of electrostimulation are endeavoring to incorporate electrodes in to "body hugging" garments to enable the convenient and accurate application of a (large) number of electrodes to the body part to be stimulated. The use of a large number of electrodes can enable, for example, several muscle groups to be stimulated together or sequentially in a coordinated manner to achieve a more natural movement of a limb. Garments are already on the market which resemble tight fitting cycling shorts/pants (what is the US term??) and have integrated wires and connectors for the attachment of standard TENS (or similar) snap electrodes prior to application. Other, more challenging designs include the integration of reusable electrodes into a stretchable garment.

3.7 Implant Electrodes

Implantable monitors/stimulators and their electrodes are used, or are being developed, for a wide range of applications including cardiac pacing and defibrillation; cochlear implants; urinary control; phrenic nerve stimulation for respiration control; functional electrical stimulation of limbs; vagal stimulation for control of epilepsy; spinal stimulation for chronic pain relief; deep brain stimulation for Parkinson's disease or depression; bonehealing and several visual neuroprostheses.

Implanted monitoring electrodes are used to more accurately pick up the desired signal while minimizing the contributions of extraneous signals. Implanted stimulation electrodes deliver the applied waveform more selectively to the targeted tissue, making the therapy more effective and, as the stimulation electrode is generally implanted away from cutaneous pain receptors and afferent nerve fibres, more comfortable for the patient. One significant draw back, however, is the greater potential for damage from improper electrode design, installation and use.

The design of an implant electrode will depend greatly on the anatomical structure it is to be implanted against, into or around. Electrodes can be/have been implanted in, on or near a given muscle; in, on or around a given nerve; in, on or around a given bone; in, on or around the spinal cord; and in or on the surface of the cerebellar cortex.

A review of all of these designs is beyond the scope of this chapter. The reader is referred to the appropriate chapters in this encyclopedia.

To facilitate this overview of some of the key design possibilities, two main application areas will be concentrated on; muscular and neural electrodes. Muscular (especially cardiac) electrodes, using more "traditional" electrode fabrication, are presented in a separate section. Neural electrodes will be largely covered in the section on "newer" microelectrodes constructed using thin-film and similar techniques. These categories are very loose and there is obviously a considerable degree of overlap between applications and the various electrode fabrication techniques. Once again, the reader is referred to the appropriate chapters in this encyclopedia for more detailed descriptions of electrodes and their fabrication for specific applications.

Cardiac electrodes are the most important example of muscular electrodes. As cardiac pacemakers and defibrillators have the longest and most successful track records as implantable devices, much of the science underpinning the newer (and future) implantable devices (muscular, neural and other) has been developed by the cardiac implant pioneers. Key contributions were made in the areas of implant electrodes, biomaterials and powersources, to name but a few. The present review of electrodes is therefore largely based on cardiac electrodes. However, ideas can be gleaned from this area and, once suitably customized, applied to others with success. The advantageous aspects of biphasic impulses (discovered in the 1950s, arguably earlier) is a good example of something being rediscovered in a range of stimulation applications over the past few decades.

3.7.1 Historical Background

Implant stimulation electrodes, or at least percutaneous stimulation electrodes, predate the earliest biosignal monitoring electrodes. In the early 1800s, there appeared a renewal of interest in acupuncture in Europe which had been introduced into Europe in the second half of the eighteenth century by Jesuit missionaries. In 1825 Sarlandière was the first to apply an electric (galvanic) current to thin metal needle electrodes (derived from acupuncture needles) thus creating "electropuncture for the application of current to specific points on or in the body [70, 121] ".

Electropuncture soon became the accepted method of stimulating muscles, nerves or organs beneath the skin [118]. Electropuncture of the heart was first attempted by Krimer in 1828 without recorded success (Fig. 3.41)[122]. This technique was then abandoned for several decades. Meanwhile, W. Morton successfully introduced the use of ether as an anesthetic in 1846. Eventually chloroform was found to be more suitable though cardiac arrest was a frequent complication of chloroform anesthesia in those early days. In 1871 Steiner over-anaesthetized horses,

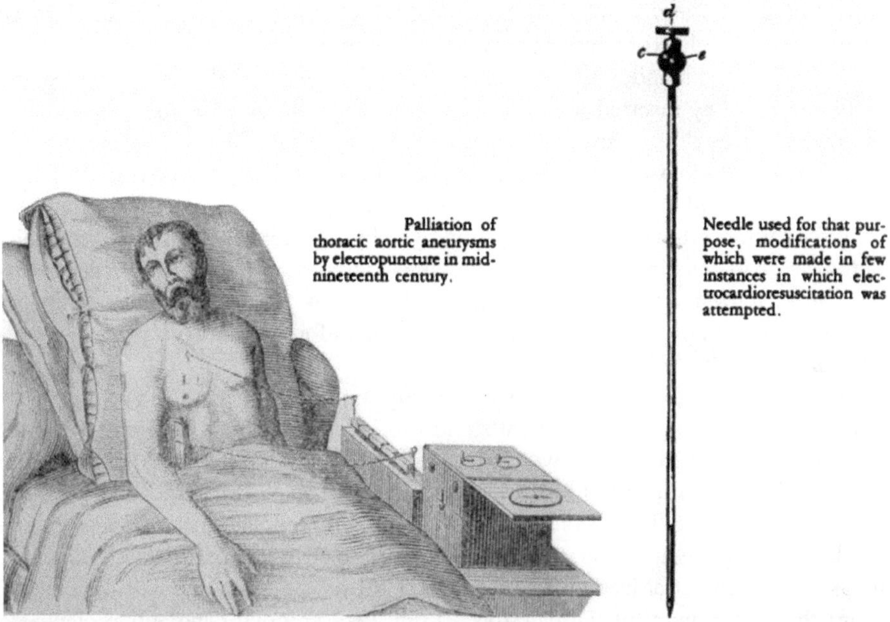

Fig. 3.41 Kimer's electropuncture of the heart [122]

dogs, cats and rabbits to produce cardiac arrest. He reported successfully applying an intermittent galvanic current to a percutaneous needle in the heart to evoke rhythmic contractions. Terms such as galvano-puncture and farado-puncture soon started to appear in the literature [122].

In the early 1900s, cardio-stimulating drugs such as epinephrine were injected directly into the heart of sudden death victims by means of a large needle inserted through the chest wall to restore automatic activity. It was eventually established that one of the key factor in the occasional success of these intracardiac injection procedures was the actual puncture of the heart wall rather than the medication administered. Based on this observation, Hymen went on to build the first hand-cranked, spring-driven "artificial pacemaker" [51]. He used transthoracic needle electrodes plunged into the atrium and even introduced the concept of using a biopolar needle arrangement as "in having the two electrodes so close together that only a small pathway is concerned in the electric arc established by the heart muscle, an irritable point is produced" [122].

In the 1950s, Lillehei, Weirich and others pioneered the use of cardiac pacing for the management of heart block accidentally resulting from cardiac surgery and for other emergency cardiac treatment. Slender wire electrodes were implanted into the myocardium before closing the chest with the connecting leads thus exiting through the chest wall. Pacing impulses could then be delivered through these wires for a week or so until the heart healed. Once the heart had recovered, the electrodes were pulled out. Early versions of these electrodes consisted of silver-plated braided copper wires insulated with polyethylene or Teflon [96, 122].

In 1958 Furman and Schwedel reported the first instance of transvenous pacing of the heart. They inserted a unipolar catheter electrode into the right ventricle of the patient through a superficial vein and paced the heart via the endocardial surface. The electrode used was a solid copper wire with a bare terminal tip [96]. The electrode was withdrawn once the patient's heart resumed its own idioventrcular rhythm. Although the cardiac pacing employed by Lillehei et al. and by Furman and Schwedel was much better tolerated than the external stimulation of Zoll, perielectrode infection was a major drawback as was the transport of the external pacemaker. The development of an implantable pacemaker therefore became the goal of several groups around the world.

In 1958, Senning and Elmqvist successfully implanted the first Pacemaker without leads emerging from the patient's chest to invite infection. The implanted unit was powered by cells which were recharged from outside the body using a line connected, vacuum tube radiofrequency generator. The electrodes/leads used were stainless steel wires. The second version of the unit failed due to a lead fracture one week following implantation. It was then decided to abandon pacemaker therapy for this patient until better leads were developed [96].

At that time, many of the electrodes used were unipolar. The active electrode tended to consist of the bared tip of an insulated wire implanted in the myocardium while the indifferent electrode was a similar wire implanted subcutaneously in the chest wall. Unfortunately, the stimulation threshold was observed to rise following implantation during longer term pacing. This increase in threshold was thought to be due to the development of scar tissue around the active electrode. Hunter and Roth developed a bipolar electrode system in 1959. This electrode consisted of two rigid, 0.5 cm long, stainless steel pins attached to a silicone rubber patch. The cathode-anode pins were positioned in to myocardial "stab wounds" surgically created for the purpose and the pad was then sutured to the epicardial surface [132]. The lead wire was a Teflon-coated, multistrand stainless steel wire with an outer sleeve made from silicone rubber tubing [100].

In 1960, Chardack, Gage and Greatbatch successfully produced a wholly implantable, battery powered pacemaker (Fig. 3.42a). Initially, they used a pair of multistrand stainless steel wires in a Teflon sleeve with the bare ends sutured to the myocardium [132]. Other metallic formulations were tried, such as solid wire, silver wire, stainless steel, orthodontic gold, and platinum and its alloys [46].

They eventually adopted the "Hunter-Roth" intramyocardial electrode (Fig. 3.42a, b). Considerable surgery was required as the pacemaker had to be implanted into the abdomen and the electrodes were sutured to the heart wall. The bipolar electrode did however dispense with the need of a dispersive "chest" electrode and the associated pain it caused [122]. Stimulation thresholds tended to stabilize at much lower levels with this electrode [96] which enabled successful pacing for many months.

Breakage of lead wires, due to metal fatigue, was a major concern. One of the main problem areas occurred at the point were the two metal components were

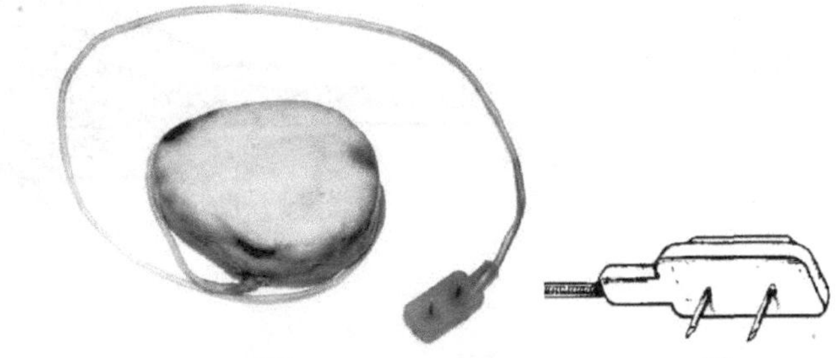

Fig. 3.42 (**a**) Chardack, Gage and Greatbatch's wholly implantable pacemaker and the "Hunter-Roth" bipolar intramyocardial electrode [118] (**b**) Diagram of an early Hunter-Roth lead with two bipolar myocardial pin electrodes [132]

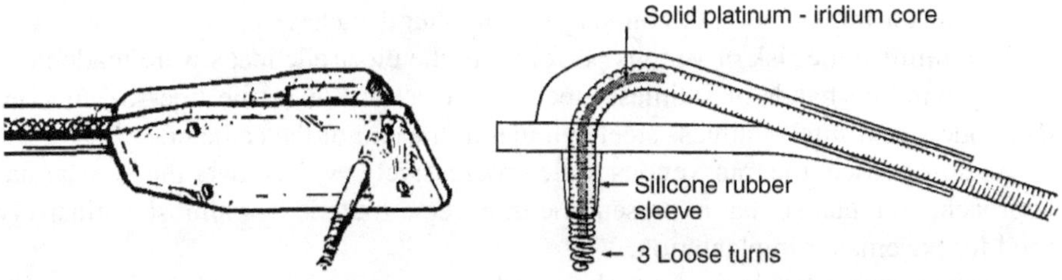

Fig. 3.43 The "Chardack" electrode (**a**) [132] (**b**) Chardack et al [21]

welded together [100]. Corrosion also occurred at the small area stainless steel anode, causing cessation of pacing within a few months.

Chardack and colleagues [22] devised a replacement for the Hunter-Roth electrode based on a continuous helical coil of platinum-iridium (Fig. 3.43). The electrode was simply a few turns of the coiled lead wire, exposed and extended to enable fibrous tissue to grow between the spirals and firmly anchor the electrode in place. The use of a helical coil greatly increased flexibility and decreased the number of fatigue failures, as did the use of one continuous wire (without a join) for both lead and electrode. The use of the same metal for lead and electrode also had the advantage of preventing corrosion from galvanic action. Additionally, platinum-iridium is more corrosion resistant than the metals used in many electrodes prior to Chardack's electrode.

The sutureless "screw-in" lead was later introduced by Hunter in 1973 [132]. The screw-in electrode was simply rotated into the myocardium and did not require a stab wound or sutures for insertion. The electrode was effectively the means of attachment. As this "corkscrew" electrode tended not to dislodge, it dominated pacing for a long time and is still used today for many epimyocardial implants (Fig. 3.44).

A thoracotomy was required to attach many of the above electrodes to the heart and this complicated surgical procedure resulted in a 10% early mortality [46]. The first so-called "modern" pacemaker, which combined an implanted generator

Fig. 3.44 The corkscrew myocardial electrode [132]

and a transvenous lead, was developed simultaneously in 1962 by Parsonnet and Lagergren [64, 106]. The endocardial catheter electrodes could be installed under local anesthesia and this approach virtually eliminated early mortality. As they did not require the opening of the chest cavity, the use of catheter leads opened the field of pacemaker implantation to non-surgeons in later decades.

To minimize the risk of venous perforation, the electrode leads were made flexible by winding bands of stainless steel around a core of textile fibers. [96]. The electrode was a small stainless steel cylinder at the end of the catheter.

As time passed, the transvenous route progressively evolved over the myocardial approach, so much so that at present the transverse route is that almost exclusively used for pacemaker implantation.

Cardiac pacing has been the earliest and most successful example of implanted electrodes and associated hardware. Many present and future developments in other implanted electrostimulation (and biosignal recording) areas are/will be based to a large extent on the pioneering work carried out in the pacing area.

3.7.2 Some Modern Electrode Designs

With the early transvenous leads, the stimulation threshold was observed to greatly increase if the electrode pulled away even slightly from the myocardium. A wide variety of active fixation devices was therefore invented. These included springs, deployable radiating needles, barbs, hooks, claws and screws designed to anchor the electrodes by actively penetrating the myocardium [132, 137]. The "Bisping" transvenous screw-in electrode is the most popular as it allows the screw helix to be extended from the tip once the lead has been successfully threaded through the vein and located against the desired part of the heart (Fig. 3.45). It can be used as a combined anchor and electrode. The screw can be retracted allowing for an easier extraction of the lead, when necessary. [Note: A similar design of electrode is used for detecting the fetal electrocardiogram during labor. The intracutaneous needles are screwed in to the fetus' presenting scalp. Similar designs are also used in EEG monitoring].

A wide variety of passive fixation devices were also invented. Various tines, flanges and other soft, pliant projections were formed at the distal end of the lead, generally as an extension of the silicone or polyurethane lead insulation, and

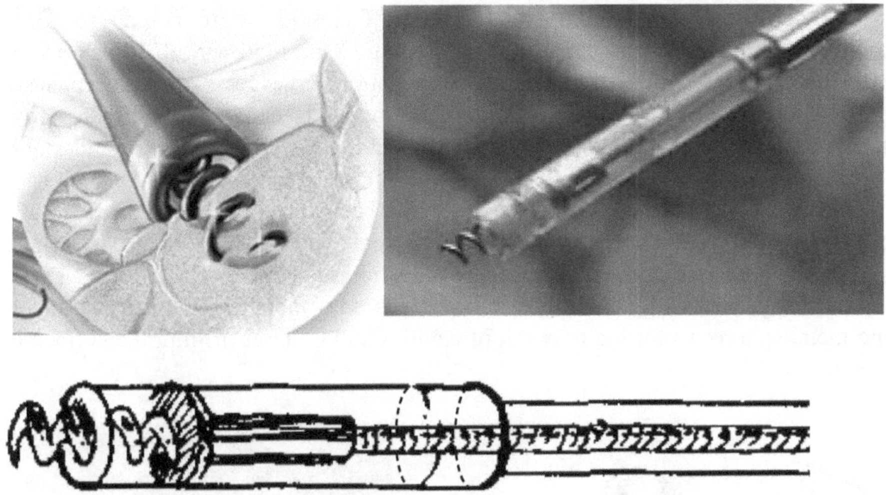

Fig. 3.45 The "Bisping" transvenous screw-in lead with the helical screw electrode extended. From S.S. Barold's the third decade of cardiac pacing [132]

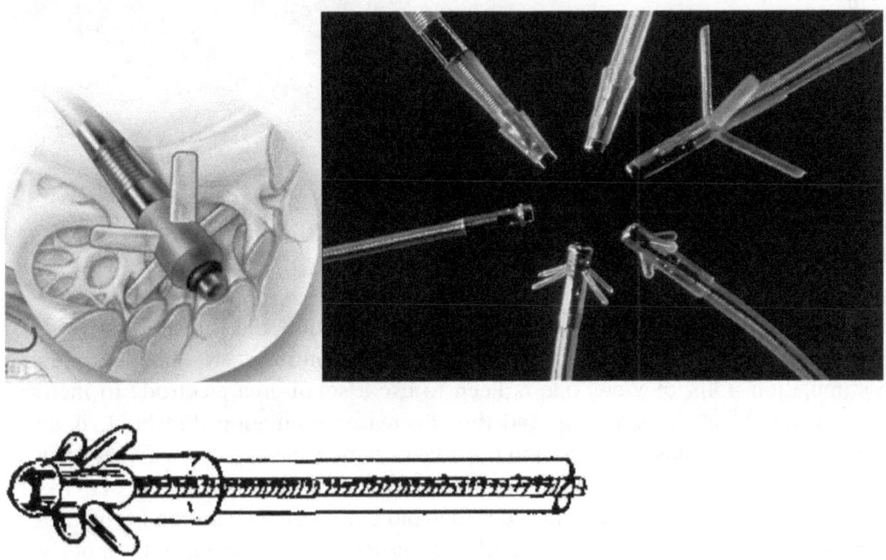

Fig. 3.46 Transvenous lead with silicone rubber tines [132]

designed to passively and atraumatically wedge the electrode between endocardial structures such as trabeculae (Fig. 3.46). In some designs, the electrode has the form of a closed loop helical coil which when twisted clockwise becomes lodged in the trabeculae [137].

Early electrodes had smooth metal surfaces. Techniques were then developed to roughen the surface in the hope of encouraging tissue ingrowth, thus locking the electrode in place, minimizing mechanical irritation and excessive fibrous encapsulation and ensuring low chronic stimulation thresholds. Studies found that porous electrodes did indeed achieve better fixation, thinner fibrous capsules and stable thresholds.

A variety of "porous" electrode tips have been developed including "totally porous" structures such as CPI's "meshed screen" electrode and electrodes whose surfaces had been textured using a range of techniques (Fig. 3.47). Porous surfaces have been generated by coating metal surfaces with metallic granules, by sintering metal spheres to form a network of cavities and by laser drilling the surface of electrodes [137].

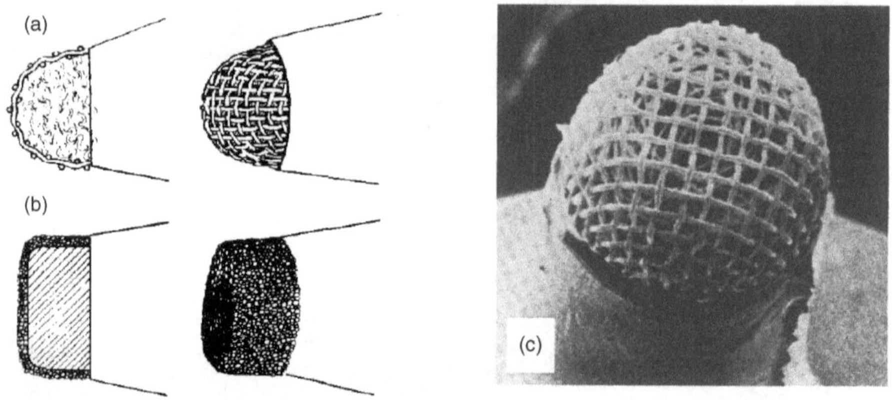

Fig. 3.47 (**a**) Cross-section of a totally porous electrode [132] (**b**) Cross-section of a porous surface electrode [132] (**c**) Photo of "totally porous" electrode [137]

Not only does roughening improve electrode fixation and threshold stability, it has been found to have a very advantageous effect on interface impedance. From a stimulation point of view, one is keen to use a small area electrode to increase current density at the small tip and thus decrease stimulation threshold. A small area electrode also has a high pacing impedance which can decrease current drain on the generator and thus prolong implant life [132]. However, one would also like to have a low sensing impedance in order to avoid excessive attenuation of the cardiac signal. There are two ways these conflicting criteria can be optimised – modifying the electrode surface or modifying its design.

Roughening the surface of a small area electrode increases its effective area without changing its "geometric" or "outer envelope" surface area [132]. Many electrode systems have been developed which incorporate this concept – electrodes with terms such as "activated", "porous", "sintered" in their company's description. These electrodes have been found to be effective in lowering the electrode interface impedance under small-signal sensing conditions. [Note: the electrode-electrolyte interface is very nonlinear and hence smaller under stimulation]. Unfortunately, the reduction in interface impedance has been erroneously interpreted as rendering the electrode non-polarizable. As stated previously, the word polarization appears to be used in a rather vague manner and has been used as the explanation of, among other things, the non-linearity of the interface impedance as well as its frequency and time dependence. The fact that the current or voltage response to a step in voltage or current is not a simple step has been attributed to "polarization". The observed transient responses are merely due to the presence of the double layer capacitance (see Figs. 3.8 and 3.9). Roughening the surface of an electrode effectively increases the area of the interface and the value of C_{dl}. This in turn results in an increase in the response's time constant ($T = R_{CT}C_{dl}$). The observed response thus looks "stretched out" along the time axis. This "flattened" response has been mistaken for that of a purely resistive, "non-polarizable" electrode. At any rate, roughening the surface of the electrode almost gives us the "best of both worlds" – a noble or inert electrode with a low interface impedance.

Another way of achieving a small simulation surface area (high current density) while ensuring a large sensing surface area (low interface impedance) is to modify the design of the electrode.

The porous electrode of Amundson involved a hemispherical platinum screen which enclosed a ball of compacted 20 μm diameter platinum-iridium fibers [2]. As electrolyte could penetrate this "3 dimensional" or "multi-layered" electrode, the design resulted in a major increase in effective surface area as well as promoting tissue ingrowth and long-term stability of thresholds. Lagergren et al. [65] introduced the "birdcage" design which also exploited some of these features [137].

One interesting example of a modified design by Parsonnet and colleagues involved the use of an electrolyte-filled hollow electrode, called a differential current density (DCD) electrode [69]. The actual stimulating electrode is the mouth of the electrolyte filled pore, which can be small to provide high current density at the point of contact with tissue (Fig. 3.48). The inside of the hollow electrode chamber has a large metallic surface (a helical coil forming a cylinder) and thus gives rise to a low electrode-electrolyte interface impedance.

This appears to be an electrode design that could readily be customised and used in a wide range of monitoring or stimulation applications. The electrode-electrolyte interface is effectively recessed and protected from any disturbance, a further advantage to those already listed above.

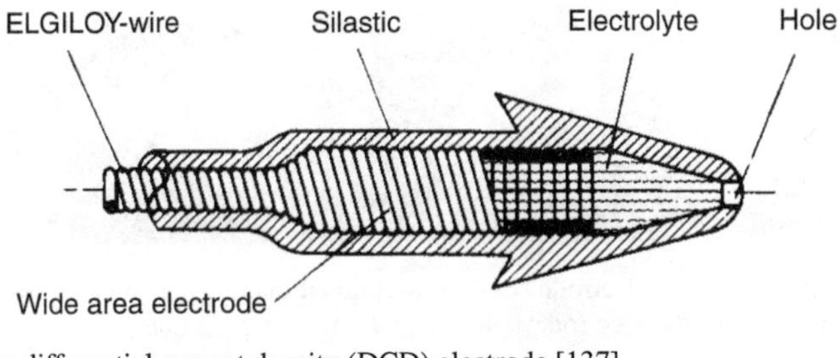

Fig. 3.48 The differential current density (DCD) electrode [137]

Several other designs exist which aim to achieve a similar effect by manipulating the current density distribution around an electrode tip. Electrodes with complex shapes have irregular patterns of current density with localized "hot spots" at points of greatest curvature [137]. It is possible to exploit these areas of high current density for stimulation purposes while the larger overall surface area gives rise to a low interface impedance [37]. A hollow, ring-tipped electrode (effectively similar to the DCD electrode) has a large current density at its annular mouth while having a large electrode-electrolyte interface area. Such electrodes are reported to have better stimulation thresholds and sensing characteristics than hemispherical designs and have proved popular. Several manufactures have combined this ring-tip design with increased surface porosity [137]. Other related designs include a dish-shaped electrode for edge-focusing of current (with laser-drilled pores for interface impedance reduction) and a grooved "hemispherical" platinum electrode coated with platinum black particles (Target-Tip electrode, Fig. 3.49) [37, 137].

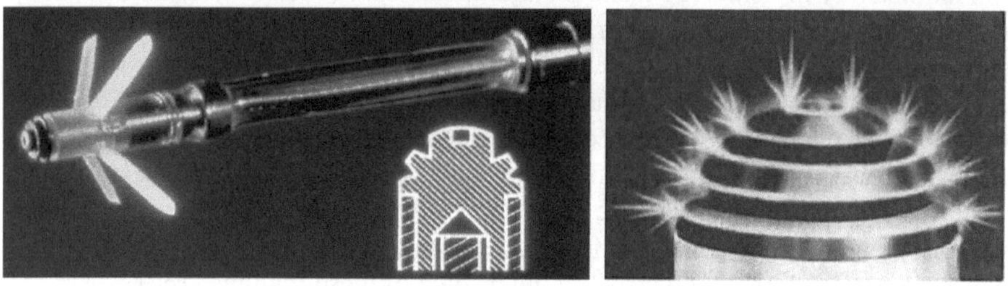

Fig. 3.49 The target tip electrode. Microporous, plantinized plantinum electrode. The target appearance is due to shallow grooves separated by peaks. (Courtesy Medtronic, Inc.)

Steroid eluting electrodes were introduced in 1983 in an effort to minimise the growth of connective tissue. The first generation electrode was made of titanium, with a platinum-coated porous titanium surface (Fig. 3.50). The electrodes incorporated a silicone core that was impregnated with a small quantity an anti-inflammatory corticosteroid [133]. Upon implant, the steroid is gradually eluted

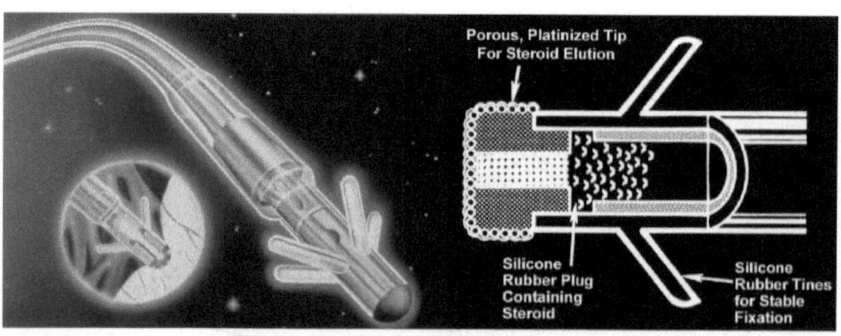

Fig. 3.50 (**a**) Steroid-eluting electrode (**b**) Cross-sectional diagram of an early design of Steroid eluting electrode. Behind the electrode is the silicone rubber plug compounded with (Courtesy Medtronic, Inc.)

into the interface between the lead electrode and the endocardium, reducing the inflammation and fibrosis that would normally occur. Steroid-eluting leads are characterised by a lower long-term capture threshold. Similar improvements in capture thresholds have been achieved [97, 98].

Most cardiac electrodes now involve the combination of drugs and complex surface structures at the "macro" and "micro" scales.

For newer applications, such as cochlear implant electrodes, an array of electrodes is involved. In some such multi-electrode applications, one may be interested not only in the current density profiles under the surfaces of the individual electrodes, but in the interplay between the electrical fields produced by the electrodes in the array in the hope of achieving more effective stimulation and/or imitate more effectively physiological stimulation. For example, the Clarion Hi-Focus electrode system of Advanced Bionics Corp incorporates 16 electrodes in a flexible array that are designed to deliver improved focused stimulation to the auditory nerve [67].

3.7.3 Microelectrodes

Over the past few years, exciting developments have taken place in areas of biomedical engineering which involve implantable devices for the recording and/or stimulation of the nervous system.

In the previous section, we saw the success in commercializing pacemakers. Other implant devices which have also reached the patient in clinical routine practice or research settings include - cochlear implants to restore hearing; deep brain stimulators to alleviate symptoms of Parkinson's Disease and depression; vagal nerve stimulators to minimize the effects of epilepsy as well as FES systems to restore or improve function in the upper extremity, lower extremity, bladder and bowel, and respiratory system [79, 108]. Other areas of research which are likely to come to fruition within the next few years include various visual prostheses to restore functional vision in the profoundly blind and the exploration of the brain-computer interface [79, 119, 143].

Much of the early research in these areas started around the 1960s [108]. Where possible and appropriate, surface and percutaneous electrodes were first used to establish the feasibility of the given recording/therapy. Early implant electrodes involved fine metallic wires or small discs placed near, in, on or around the targeted muscle or nerve. The fabrication of these electrodes was time consuming and the electrode properties were not very reproducible given the variations in areas, surfaces, inter-electrode distances etc. This was particularly a problem when several electrodes were to be used in an array. As the demands on human implantable diagnostic/stimulation devices increases, there is generally an increased need for a larger number of smaller area electrodes with well defined and reproducible surfaces and dimensions. Although, due to their high level of specificity, muscle-based electrodes will continue to be used, new electrode designs tend to concentrate more on direct nerve stimulation as this may provide more complete muscle recruitment and the same electrode may successfully recruit several muscles, thus reducing the number of electrode leads required [108]. Electrodes are therefore needed which can interface electrically with the neural system at the micrometer scale [119].

For example, the goal for a high-resolution retinal prosthesis is a 1000-electrode stimulating array in a 5 ×5 mm package [143]. If this area of research is to be clinically successful and if the other areas are to continue to improve, microelectrodes must be (and are being) manufactured using the thin-film technologies associated with the IC circuit industry. Microfabrication involves either material deposition or removal. Either rigid silicon wafers or flexible polyimide substrates act as platforms for the microelectrodes and associated circuitry. The deposited films (for connectors, leads, electrodes and/or insulation) are produced by Electroplating, Evaporation and Sputtering. The layers can be photo-patterned and etched to sub-micron resolutions and finally encapsulated in biomaterials such as diamond-like-carbon; bio-ceramic or a biocompatible polymer. Processes such as photo-lithography; Reactive Ion Etching (RIE); CMOS processing; MEMS processing; Focused Ion Beam patterning and AFM lithography can be used to achieve the desired micro-electrode design.

The benefits of a micro-fabrication approach include a high degree of reproducibility in physical, chemical and electrical characteristics. This is a high yield, low cost process once the design and processing sequence have been developed. Additionally, there is precise control of the spatial distribution of electrode sites and this may be of interest when seeking to optimally stimulate or record from a target site. A high packing density of electrode sites for a given implant volume is also readily achievable using photo-lithographic techniques. There is the possibility of incorporating the interface circuitry directly on the microsensor platform thus reducing the need for complex interconnections.

The widespread availability of silicon micro-manufacturing techniques has enabled the fabrication of a range of silicon-based "wedge-" or "needle-" shaped electrodes to allow penetration of the nervous tissue. 3-dimensional arrays of such structures have been developed for insertion into, for example, the cortex to detect local potentials [79].

One-dimensional arrays of electrodes are fabricated using lithographic patterning and deposition of thin-film metal leads and electrodes onto not only silicon, but also glass and even flexible polyimide substrates [119]. Much of the work on silicon-based microprobe fabrication has been pioneered at the Center for Integrated Sensors and Circuits at the University of Michigan.

A three-dimensional electrode array can be fabricated by assembling a range of one dimensional probes (such as those shown on Fig. 3.51.). As each probe has multiple recording sites along its length, the complete volume of the tissue under study can be assessed, giving rise to very dense sampling. The Michigan Probe has evolved a large number of single-shaft, multi-shaft, and 3-D stacked microelectrode arrays [119].

Recent improvements in silicon microtechnology have made it possible to create not only planar microelectrodes but also penetrating "brush" electrode structures for in vivo measurements. In contrast to the University of Michigan's planar devices,

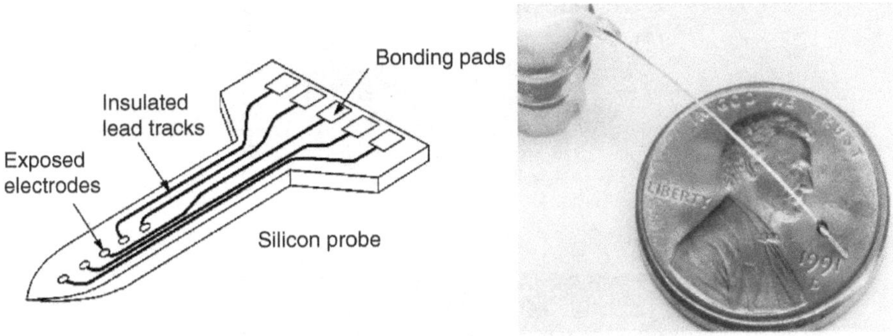

(a) (b)

Fig. 3.51 (**a**) Multielectrode silicon probe after Drake et al. [32] (**b**) Michigan micromachined multielectrode probe for recording and/stimulation of central nervous system

the two-, and three-dimensional cortical multi-microelectrode arrays developed at the University of Utah are fabricated out of a single solid block of silicon. Etching of the block results in a 10 × 10 array of needles, each 1.0–1.5 mm long, arranged on a 4.2 mm ×4.2 mm base. The metal and insulation layers are then applied, creating 35- to 75-micron-long platinum recording tips [79, 119, 134] (Fig. 3.52a).

This design has the advantage of placing a relatively large number of recording sites in a compact volume of the cortex. However, with a single recording site on the end of each "needle" set at a fixed depth into the cortex, this version of the Utah array is classified as a "2-D" array as all the electrode are in the same plane [119]. The Utah probe can achieve high-density sampling by spacing many needles close together but does not have multiple sites along each shaft. When the length of the needles in such an array is graded (the array is said to be "slanted", Fig. 3.52b) or the needles have some other distribution of lengths, these arrays are termed 3-D as the electrode tips are no longer in the same plane. These designs are thought to give the better spatial selectivity [79, 119].

Implanting such "needle" or "brush" electrode systems is obviously associated with damage of the tissue. Moreover, the stiffness of many systems may lead to damage of nervous tissue, especially if there is relative movement between the sharp needles and the delicate tissues. Breakage of the brittle needle is also a concern. Considerable efforts are therefore being directed at miniaturizing the width of the needles or at introducing more flexible materials.

For example, some versions of the Michigan Probe consist of four parallel, dagger-like probes connected to a micro-silicon ribbon cable. The ribbon cable is semiflexible and allows the probes to move up and down with the cortex as it pulses [134].

In the development of subretinal stimulating arrays using current silicon micro-manufacturing techniques, it has been pointed out that a planar, rigid implant is likely to mechanically damage the compliant, spherical retina [143]. Concerns have also been expressed regarding the use of penetrating microelectrodes, the relative

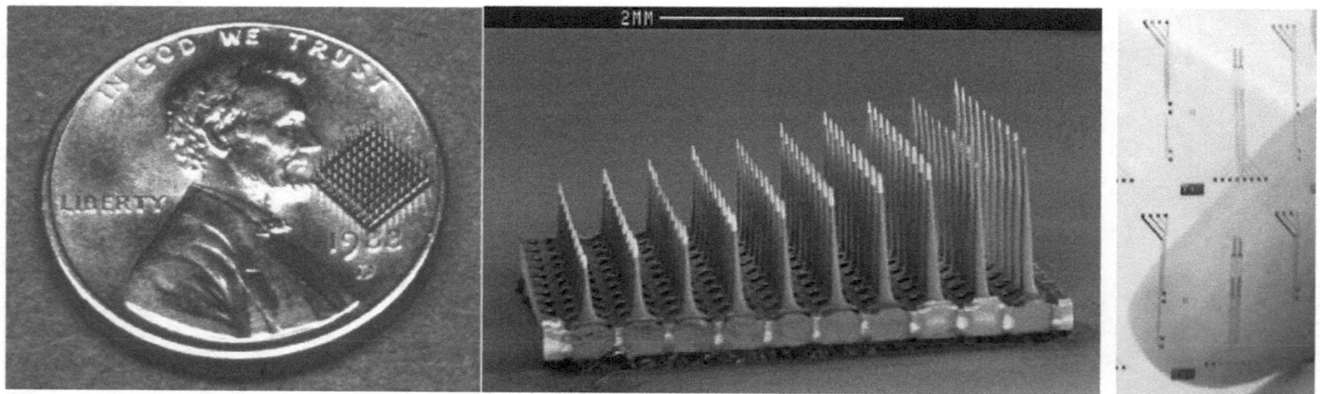

(a) (b)

Fig. 3.52 (**a**) Utah Electrode Array shown on a US penny to convey size. (**b**) Modified Utah Electrode Array in which the length of the needles is uniformly graded. [79]

micro-motion between the array and the retina potentially provoking mechanical damage and a significant encapsulation response [79]. The ideal retinal stimulating electrode would therefore have the flexibility to match the curvature of the retina and the next generation of electrode arrays are likely to be constructed on flexible substrates.

Microelectrode arrays on flexible substrate have been demonstrated in a range of applications including the European project "Microcard", Si-Based Multifunctional *Micro*system needle for Myo*card*ial Ischaemia Monitoring. Initially, work centred on silicon-based micro-probes to monitor the electrical impedance of tissue, tissue temperature, pH and local ionic concentrations of potassium, sodium, and calcium. These parameters were found to vary considerably when, for example, a heart undergoes an ischaemic phase, thus establishing the clinical value of the technique and device [3].

In the course of the silicon probe development it was foreseen that the brittle nature of silicon could make intact probe removal difficult. Additionally, the rigid "needle" could cause damage to the delicate tissues. The thrust of the project thus changed to the development of flexible, polymer-based probes.

Thin film devices for the measurement of tissue impedance and ion concentrations were manufactured on flexible polyimide substrates (Fig. 3.53a). Gold thin film electrodes were deposited using an improved photolithography process for 1micron resolution. Polyimide insulation layers were spin-coated onto the PTFE surface after suitable conditioning, and proved to be insulating and continuous.

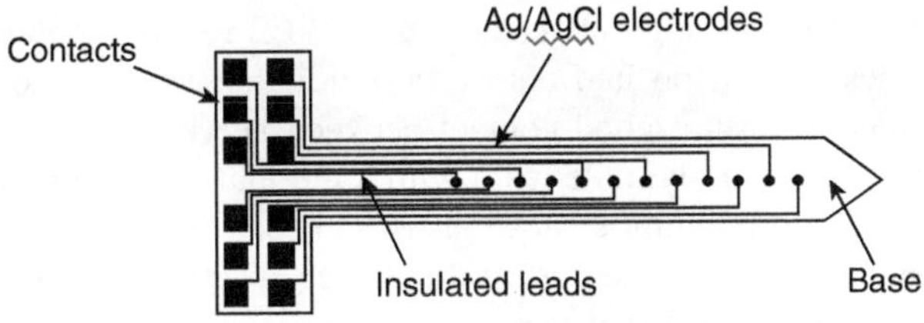

Fig. 3.53 (**a**) Micro-fabrication of sensors onto flexible substrates. Early prototype (**b**) Similar one-dimensional probe electrode array (after [78])

Electrochemical characterisation of the gold thin film impedance electrodes showed them to possess high interface impedance. Pt and IrO oxide coatings were electrochemically applied to the gold thin film surface and resulted in a drastic reduction in interface impedance for monitoring or stimulation applications [72].

Encircling neural electrodes may be of a cuff or spiral design. The term "`cuff electrodes" applies to those devices which engulf the entire circumference of a nerve First models which rather stiff, carried only one or two electrodes and they were made using a platinum foil electrodes which were located on the inside of a cylinder of silicone rubber which was wrapped around a nerve (Fig. 3.54b) [119]. It is generally recommended that the diameter of the cuff be 50% larger than the nerve diameter to avoid nerve compression and necrosis due to swelling and fibrous tissue in-growth. Cuff electrodes do however have a long and successful track record in a range of FES applications [99].

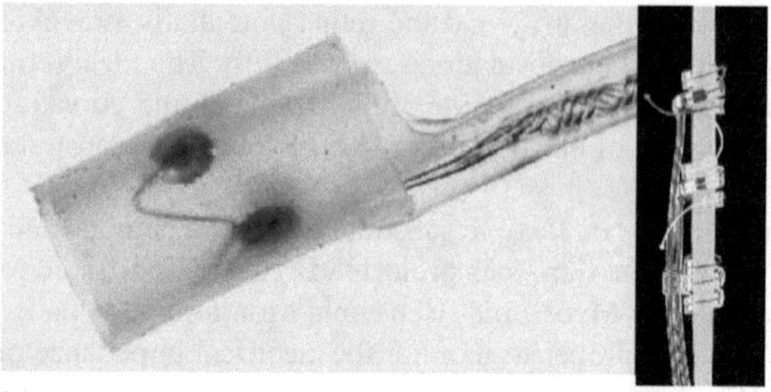
(a) (b)

Fig. 3.54 (**a**) Neural Cuff electrode Aalburg University (**b**) Spiral electrode (*Cyberonics*.)

The spiral electrode is a loose, open helix which is wound around the nerve [99]. The open design can accommodate swelling and is very flexible. A version of this electrode is marketed by Cyberonics for use with their vagus nerve stimulator (Fig. 3.54b). New designs of nerve cuff electrodes seek to reshape the geometry of the nerve to more selectively stimulate or record from particular nerve fascicles. Efforts are also directed at controlling the electrical fields generated by the electrode arrays to better focus the stimulation [108].

As part of a European project NEUROS, NIBEC developed a flexible thin film-based stimulation and sensing cuff electrode for FES related application. IrO, Pt and Au electrodes were deposited onto a polyimide substrate. In order to facilitate implantation and ensure good contact between nerve fascicles and electrode surface was self-curling. Polyimide resin with a thermal expansion coefficient differing from that of the polyimide substrate was chosen so that the curing process gives rise to a residual stress and curl in the device. The diameter of the electrode cylinder could be made less than 1 mm.

Diamond like carbon (DLC) encapsulation was deposited onto the device using a plasma enhanced chemical vapour deposition (PECVD) process. Adhesion to the polyimide substrate was found to be satisfactory following the addition of a silane adhesion layer at the interface. [50, 112].

With the aid of mico-fabrication techniques, one can control the area and properties of the electrodes and greatly decrease them in size. However, as electrode area decreases, the interface impedances increase with resultant difficulties in making accurate measurement. The key to success in this case is in the choice of electrode design, material and electrode surface topography.

A similar concept to Chardack's Differential Current Density pacemaker electrode was suggested for use in thin film electrodes. The metal electrode is housed within a hollow chamber (Fig. 3.55). The chamber is filled with electrolyte and has a small aperture to enable electrical contact with tissues [110]. As the metal-electrolyte interface is relatively large, the interface impedance is relatively small. The interface is also protected from mechanical disturbance (similar to the floating

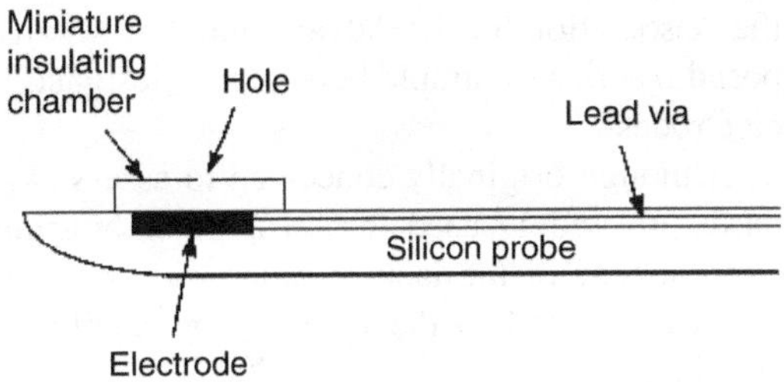

Fig. 3.55 Thin-film "Differential Current Density" electrode after [110]

electrode) and hence should suffer from less artifact. As the small aperture determines the area of contact with the tissue, the effective stimulation or recording area is very small.

Other "3-dimensional" designs with etched "meshes" etc should be assessed for their potentially larger interfacial areas.

Once again, surface roughness is an important factor in decreasing interface impedance and possibly in helping anchor the electrode in position. Rough surfaced electrodes must be used with caution, depending on the application, in case the surface causes damage to the surrounding tissues. Certain materials and the electrode fabrication processes involved may well result in favorable macro, micro and nano surface features. Presently, investigators are studying modifications to the electrode surface using such things as nanotubes. Nanotechnology offers much promise for new sensor devices particularly in the biomedical sector. Not only do individual nanotubes offer the possibility of using them as ultra fine needles for in vivo probing at the cellular level, but surfaces can be created with optimal distributions of clusters of nanotubes to maximize performance.

3.8 Electrode Standards

The Association for the Advancement of Medical Instrumentation (AAMI) produce a range of labeling, electrical and other performance requirements for manufacturers and users to help ensure a acceptable levels or product safety and efficacy. Some of the key electrode-related standards are briefly reviewed below.

3.8.1 Standards for Biosignal Monitoring Electrodes

3.8.1.1 Standards for Disposable ECG electrodes. ANSI/AAMI EC 12 (2000)

Introduction

In an effort to minimize ECG recording problems associated with the performance of electrodes coupled to a standard ECG monitor or electrocardiograph, the Association for the Advancement of Medical Instrumentation (AAMI) has proposed a series of simple bench tests designed to assess pregelled, disposable ECG electrodes.

Although originally conceived to assess disposable ECG electrodes, these standards are widely used to assess other biosignal monitoring electrodes. This is a consequence of the lack of other widely accepted standards for these monitoring applications and to the general applicability of the ECG standards to the other applications.

Although AAMI also lays down stipulations for electrode labeling, adhesion testing testing, etc., only the electrical performance requirements are reviewed here.

AC Impedance

> The average value of 10-Hertz impedance for at least 12 electrode pairs connected gel-to-gel, at a level of impressed current not exceeding 100 microamperes (μA) peak-to-peak, shall not exceed 2 kilohms (kΩ). None of the individual pair impedances shall exceed 3 kΩ.

Low impedance electrodes are desirable to avoid signal attenuation and distortion and to minimize 50/60 Hz interference pickup. High electrode impedances can also give rise to serious burns when the ECG electrodes are used in the presence of electrosurgery or defibrillator discharges [124].

The impedance of the skin's outer layer, the stratum corneum, is many times larger than that of the metal/electrolyte interface and hence the former is of key concern when endeavoring to ensure good electrode performance. The skin preparation technique, the extent of diaphoresis, and the ability of the electrode gel in penetrating and reducing the skin impedance are generally more important than the electrode-electrolyte interface impedance.

The Standards Committee decided that the electrode gel-to-gel impedances should be significantly less than the expected impedance of clean, dry skin to insure a minimal contribution by the electrode itself to the overall impedance [124]. In the UBTL tests carried out on behalf of the Standards Committee, it was found that the mean 10 Hz impedance of a standard pair of ECG electrodes on unabraded skin was of the order of 100 kΩ. The AAMI committee chose 2,000 kΩ as a reasonable limit for 10 Hz gel-to-gel impedance to ensure that the electrodes did not contribute significantly to the overall impedance nor to power dissipation in the presence of defibrillation overload and electrosurgery currents.

As the electrode-gel interface impedance is nonlinear and decreases with applied signal amplitude, the standard stipulates that the level of impressed current must not exceed 0.1 mA peak-to-peak when carrying out the test.

In the UBTL tests it was found that the impedance as measured on abraded skin correlated well (99%) with the impedance measured with the electrodes connected gel-to-gel, whereas the impedance measured with electrodes applied to clean, dry skin correlates very poorly (47%) with the gel-to-gel measurements. Obviously, the bench test simply evaluates the ac impedance performance of the electrode-gel interface and will therefore not accurately predict or represent the clinical performance of an electrode on intact skin. For example, there are cases of electrodes which performed poorly as per the AAMI bench test, yet which proved very satisfactory in vivo. Conversely, some of the best electrodes according to the bench tests performed relatively badly in vivo [49].

DC Offset Voltage

> After a 1-min stabilization period, a pair of electrodes connected gel to gel, shall not exhibit an offset voltage greater than 100 mV.

Ideally, the potentials of both electrodes used to monitor a biosignal should be identical and thus cancel each other out. Slight differences in the gels and metals used, however, result in an offset voltage. The potentials of the skin sites further complicate the recording, especially as these latter potentials (and their amplified difference) tend to be much larger. If the overall electrode-skin potential difference is larger than 300 mV, the amplifier may saturate and the biosignal will not be observed.

The UBTL report studied the correlation between gel-to-gel and electrode-skin offset voltages and found that gel-to-gel offsets were in the order of 2.5 times smaller than those recorded in vivo for the same electrodes on a patient's skin. As the maximum allowable in vivo dc offset should be less than 300 mV, the Committee decided that the limit for gel-to-gel dc offset should therefore be less than 300/2.5 mV – i.e. 100 mV.

Some reviewers of the standard argued that the limit should be reduced to 10 mV as this would help minimize motion artifact problems. The Committee rejected this suggestion, pointing out that there is no clear evidence that links high gel-to-gel offset voltages with motion artifact (largely caused by skin deformation).

Offset Instability and Internal Noise

"After a 1-min. stabilization period, a pair of electrodes connected gel to gel, shall not generate a voltage greater than 150 μV_{p-p} in the passband (first-order frequency response) of the 0.15–100 Hz for a period of 5 min following the stabilization period."

This standard is concerned with the problem of baseline wander which introduces a low frequency component into the monitored biosignal making accurate diagnosis difficult. The American College of Cardiology's Task Force on the Quality of Electrocardiographic Records judged that drift rates less than 400 μV/sec, although not highly rated were "not considered unacceptable".

The UBTL report detailed several experimental limitations which prohibited their detailed study of in vivo dc offset drift. Consequently, no correlational analysis was carried out between dc offset drift measurements made with electrodes applied to human skin and those joined gel-to-gel. They however decided to use the factor of "2.5" they had observed between clinical and bench test result for dc offsets, given that the measurement techniques are fundamentally similar. A limit of 150 μV/sec was therefore arrived at by dividing the 400 μV/sec baseline drift rating by a factor of 2.5. As the test circuit used in the bench test differentiates the offset voltage, the offset instability requirement is specified in μV rather than μV/sec.

The Committee was contacted and asked to decrease the limit from 150 μV to 40 μV p-p in order to be in line with the AAMI standard "Cardiac monitors, heart rate meters and alarms (EC13)". The working group agreed that this requirement could be made more stringent but refused to decrease the limit to 40 μV p–p. This requirement is under study and may well be altered.

This calculation involved in reaching the 150 μV/sec limit implies that skin potential fluctuations are only 2.5 times larger than those of the electrode-gel interface. This is most unlikely and problems arising from drifting electrolyte/skin potentials will depend on skin preparation, electrode design and electrode gel rather on than the electrode-gel interface characteristics per se [136].

Defibrillation Overload Recovery

> Five seconds after each of four capacitor discharges, the absolute value of polarization potential of a pair of electrodes connected gel-to-gel shall not exceed 100 mV.

Also

> During the 30-sec interval following each polarization potential measurement, the rate of change of the residual polarization potential shall be no greater than ± 1 mV/sec.

It is important that a clinician, having defibrillated a patient, be able to see a meaningful ECG within 5–10 s in order to judge the efficacy of the delivered impulse and to decide if another is required. The offset voltage across the electrode-skin interfaces, which drastically increased as a result of the defibrillation impulse, must therefore return to below 300 mV within 5 s following the discharge. Once again, using the 2.5 factor between bench test and in vivo potentials, this requirement translates to a gel-to-gel bench test offset voltage under 100 mV within 5 s of applying an overload of 2 mC (representing the worst possible situation encountered in vivo where the defibrillator paddles are placed in immediate contact with the ECG electrodes). Electrodes made of stainless-steel, for example, tend to acquire offset voltages of several hundred millivolts for minutes and consequently no ECG trace is observable on the monitor [19].

Following the initial 5 s the EGC must not only be visible on the monitor but must also be recognizable and clinically useful. Hence the stipulation that the offset voltage should not drift with time by more than ± 1 mV/sec.

The UBTL results indicate there is good correlation between the results of this bench tests and animal tests, particularly at the higher recovery voltages encountered with non Ag/AgCl electrodes.

Although only a very low percentage of ECG electrodes are in fact subjected to defibrillation impulses in vivo, the AAMI committee decided after some deliberation to insist that all ECG electrodes meet the proposed standard as it is impossible to guarantee that a given electrode would not be used in an emergency defibrillation situation.

Bias Current Tolerance

> The observed dc voltage offset change across a pair of electrodes connected gel-to-gel shall not exceed 100 mV when the electrode pair is subjected to a continuous 200-nanoampere (nA) dc current over the period recommended by the manufacturer for the clinical use of the electrodes. In no case shall this period be less than 8 h.

When a dc current passes through the metal-gel interface of an electrode, the electrode potential deviates from its equilibrium value and the electrode is said to be polarized. If the current is maintained indefinitely the reactants become depleted causing the electrode potential to deviate further, possibly exceeding the limit allowable at the input of the ECG recording device.

Although most modern ECG recorders pass less than 10 nA of bias current through the electrodes, some older models can have bias currents as high as 1000 nA. A number of cardiac monitor manufacturers use dc bias currents to sense high electrode impedances to warn of disconnected leads or poorly affixed electrodes. The standard for cardiac monitors permits input bias currents of up to 200 nA. UBTL therefore adopted the 200 nA limit on the dc input bias current suggested for cardiac monitors for the tests. The ability of an electrode to cope with this value of bias current must therefore be demonstrated by not exceeding the AAMI dc offset requirement of 100 mV over the time period recommended by the manufacturer for the clinical use of the electrodes.

The 200 nA current level is generally well tolerated by Ag/AgCl electrodes. Stainless steel electrodes rapidly fail this test even at 10 nA with major increases in electrode potential.

Discussion

The AAMI standards bench tests are currently the only widely accepted electrode standard tests in use. The tests are simple and inexpensive to set up and have been widely embraced by manufacturers and users for production quality control purposes.

One must bear in mind, however, that these tests evaluate only the electrode-gel interface and that they do not include the more important properties of the gel-skin interface. Assessment of the clinical performance of electrode impedance using the proposed bench tests is only relevant if the skin has been suitably abraded. Skin abrasion is not widely used by the clinical community and hence the relevance of at least some of the standard tests to the clinical situation is open to question.

Especially several decades ago, fulfillment of the AAMI requirements was commonly quoted as a guarantee of the high in vivo electrical performance of an electrode. An electrode with, for example, a dc offset of 1 mV was widely believed by customers to be a much better electrode than one with an offset of 5 mV. This naivety appears to be on the wane, however, and manufacturers and customers are shifting towards low cost electrodes that "score" less highly in the AAMI tests but are "good enough" for a given application.

The author once supplied a leading company with dry metal-loaded polymer electrodes. The company connected the electrodes together and tested them as per the AAMI standards (for "pregelled" electrodes). Perfect electrical performances were measured given that what was effectively being assessed was metal-to-metal contact. DC offsets of 0 mV were obtain, etc. Once the dry electrodes were applied to a patient's skin, a less than favorable result was obtained.

The attitude to adopt therefore when interpreting AAMI standard bench tests results for pregelled, disposable ECG electrodes is that electrodes which meet the AAMI standards have a "tendency" rather than a "certainty" to perform well in vivo. Electrodes which perform better as per the bench tests do not necessarily perform better in vivo. They are a useful set of tests nonetheless.

The ANSI/AAMI standard tests were conceived such that the test apparatus needed can be readily assembled by an electrode manufacturer. However, one can can buy a convenient-to-use, custom-built electrode tester (as per AAMI standards) called the Xtratek electrode tester ET65A (Direct Design Corporation, Lenexa, Kansas, USA) (Fig. 3.56).

Electrocardiograph surface electrode testers also exist for the in vivo testing of the quality of (i) the design ECG electrodes, (ii) the application of the electrodes and (iii) the skin preparation technique used.

The electrode tester generally measures the AC impedance and DC offset of the electrode-patient system. These measurements can be used, for example in "stress testing", to decide if the skin sites have been sufficiently well prepared (i.e. contact impedances are low enough) to proceed with the clinical procedure. They can also be used to detect the presence of loose cables or bad contacts.

Fig. 3.56 Early version of the Xtratek electrode tester ET65A (Direct Design Corporation, Lenexa, Kansas, USA)

3.8.2 *Standards for Stimulation Electrodes*

Although not covered in this review, the following standards exist which stipulate minimum labeling, safety, and performance requirements for the given stimulators. The rationale for the standards is also presented.

- Transcutaneous electrical nerve stimulators ANSI/AAMI NS4
- Implantable spinal cord stimulators ANSI/AAMI NS14
- Implantable peripheral nerve stimulators ANSI/AAMI NS15

3.8.2.1 Standards for Automatic External Defibrillators and Remote-Control Defibrillators. ANSI/AAMI DF 80 (2003)

AC Small Signal Impedance

"The 10 Hz impedance for any of at least 12 electrode pairs connected gel-to-gel, at a level of impressed current not exceeding 100 microamperes (μA) peak-to-peak, shall not exceed 3 kΩ. The impedance at 30 kHz shall be less than 5 Ω". The rationale for this requirement is based on the performance criteria in ANSI/AAMI EC 12 for Disposable ECG electrodes. Interestingly, the permissible gel-to-gel 10 Hz impedance for large area defibrillation pads is higher than that allowed for small area ECG electrodes. The gel-to-gel impedance measured at 30 kHz will be largely that of the gel pads as the interface impedances at this frequency will be almost zero.

AC Large Signal Impedance

"The impedance of an electrode pair connected gel-to-gel, in series with a 50 ohm (Ω) load and measured at the maximum rated energy of the defibrillator shall not exceed 3 Ω." 50 ohm is thought to represent the typical (rather low) in vivo transthoracic impedance between the electrodes. One wants the delivered energy to be dissipated in the patient's chest and not in the electrodes where the wasted energy may give rise to skin burns. The above requirement is therefore thought to provide a reasonable limit on the impedance contributed to the overall impedance by the electrode pair during defibrillation (less than 6%).

Combined Offset Instability and Internal Noise

"A pair of electrodes connected gel-to-gel shall generate, after a 1 min stabilization period, a voltage no greater than 100 μV peak-to-peak in the pass band of 0.5–40 Hz, for a period of 5 min following the stabilization period." The rationale for this requirement is based on the performance criteria in ANSI/AAMI

EC 12 for Disposable ECG electrodes. The frequency range used is more limited in recognition that the cardiac monitor bandwidth is more appropriate in this application.

Defibrillation Recovery

"The potential of a pair of gel-to-gel electrodes in series with a 50 ohm resistor and subjected to 3 shocks at 360 J or maximum energy at 1 min intervals shall not exceed 400 mV at 4 s and 300 mV at 60 s after the last shock delivery". The rationale for this requirement is largely based on the performance criteria in ANSI/AAMI EC 12 for Disposable ECG electrodes. An actual defibrillation impulse is applied instead of that from a simulation circuit. The offset voltage across the simulated electrode-patient load must return to below 400 mV within 4 s following the discharge (slightly different values, 300 mV and 5 s, are used in ANSI/AAMI EC 12). As the patient's chest is represented by the 50 ohm resistor, there is no need for the 2.5 factor used in ANSI/AAMI EC 12 to correlate bench test and vivo results.

DC Offset Voltage

"A pair of electrodes connected gel-to-gel shall, after a 1 min stabilization period, exhibit an offset voltage no greater than 100 mV". The rationale for this requirement is based on the performance criteria in ANSI/AAMI EC 12 for Disposable ECG electrodes.

Universal-Function Electrodes

"With conventional defibrillators, it has been customary to use separate pregelled ECG electrodes for monitoring and defibrillator paddle electrodes for defibrillation. The monitoring electrodes are not capable of effectively delivering a defibrillation shock, and the paddle electrodes have only limited monitoring capability. For recent applications, particularly automatic external defibrillation, it is very desirable to use self-adhesive pregelled disposable combination electrodes that perform well in the dual monitoring and defibrillation functions. These electrodes may also be used for delivery of transcutaneous pacing. ... Hence, combination electrodes may become preferred for defibrillation, and it is appropriate in a standard for defibrillators to consider their use and to outline a few requirements for them"

> If the electrodes are designed and intended for use in multiple modes, ie monitoring, defibrillation, and pacing, ...The electrode shall meet all (of the above) requirements... after 60 minutes of pacing at the maximum current output and maximum pacing rate through a pair of gel-to-gel electrodes in series with a 50 ohm resistor
>
> Because no general performance standards exist for combination pacing/defibrillation/monitoring electrodes, the (above) requirements define the basic minimum controls necessary to ensure safe and reliable operation

3.8.2.2 Standards for Electrosurgical Devices. ANSI/AAMI HF 18 (2001)

Introduction

Although AAMI lays down stipulations for the testing of a range of parameters, only the key electrical performance requirements for the dispersive electrodes are reviewed below.

Maximum Safe Temperature Rise

"The maximum patient tissue temperature rise shall not exceed 6°C when the dispersive electrode carries a current of 700 milliamperes (mA) under the test conditions below, unless the device is labeled in accordance with 4.1.4.2" (i.e. for use on infants). For devices labeled for use on infants, "the maximum patient tissue temperature rise shall not exceed 6°C when the dispersive electrode carries a current of 500 mA under the test conditions" stipulated in the standard. In monopolar electrosurgical procedures the dispersive electrode must be able to reliably conduct the required surgical current without generating a significant rise in skin temperature. It is widely accepted that the maximum safe skin temperature for short-term and long-term exposure is 45°C. As normal resting skin temperature varies between 29°C and 33°C. Electrodes must not generate skin temperature increases approaching 12°C. A 6°C increase in temperature is therefore thought to represent an acceptable upper limit.

"The temperature measurement method must have an overall accuracy of better than 0.5°C and a spatial resolution of at least one sample per square centimetre of the electrode thermal pattern. The thermal pattern must include the area extending 1 cm beyond the geometry of the electrode under test." This degree of special resolution is stipulated as electrosurgical burns may be confined to very small areas and these must be detected. As current tends to flow to the edge of the electrode and spread out further in the skin, the test requires that the surround area of skin is also scanned.

"The electrode under test is to carry a current from an electrosurgical generator of 700 mA_{rms} for 60 s, unless the device is labeled in accordance with 4.1.4.2, in which case the test current may be 500 mA". A current of 700 mA applied for 60 s yields a heating factor of 30 A^2 s. [Heating Factor = I^2t (Ampere^{2}seconds)]. This value is far in excess of the maximum likely current and duration for a TUR (transurethral resection) procedure. A more realistic heating factor is less than 10 and hence the stipulated testing procedure is very conservative.

"These tests must be conducted on human volunteers or on a suitably structured surrogate medium.... When human volunteers are used, the tester must include a variety of body types in the sample group rather than concentrate on a single body type (thin, average, or thick layers of subcutaneous body fat). If surrogate media are used, the tester must demonstrate that the media are electrically and thermally similar to human volunteers." Human volunteer subjects are the reference standard. Current density distribution under an electrode depends on a wide range of factors, including the electrical properties of the skin and underlying tissues. Hence the need to test a given electrode on a wide range of individuals. The use of a surrogate material, even pig skin which is commonly used, will not necessarily replicate with sufficient accuracy the clinical performance of the electrode. If a surrogate medium is used, the tester must demonstrate the equivalence of the test medium to human tissue. It is the Committee's view that no adequate surrogate medium has yet been suggested or used that has all of the properties of human tissue for the purpose of determining electrode performance.

Nessler et al [101] point out that the above experiments are laborious, time consuming and expensive to perform. They have developed a new test device, swaroTEST, which includes a surrogate "electronic skin" which, they claim, simulates the relevant electrical features of human skin and thus can replace the required volunteer experiments Figure 3.57. The "device consists of a 3 dimensional resistor network representing the electric features of the skin and muscle tissue, and a temperature sensing array (one transistor for each cm^2) to measure the resultant temperature increase after a standardized current load (700 mA hf current during 60 s, proposed in the relevant AAMI HF-18 standard)". The authors claim that a comparison of results obtained with their device and those with thermo camera images of volunteer experiments correspond sufficiently well to justify the acceptance of their test device as a surrogate medium.

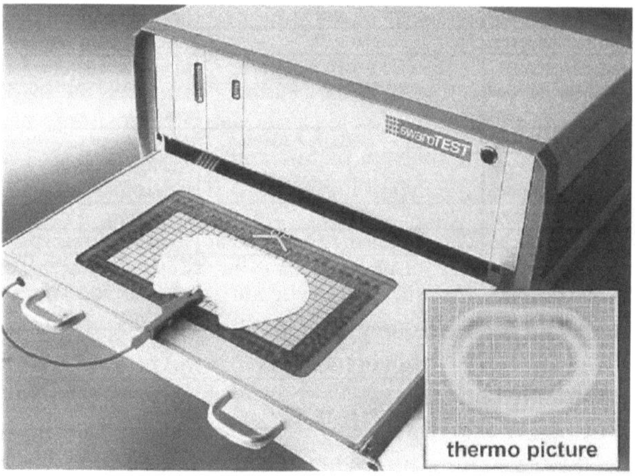

Fig. 3.57 SwaroTEST device with a measuring board "electronic skin" to simulate the electrical properties of human skin. It is hoped that this device will replace the human volunteer experiments required for ANSI/AAMI HF 18

Electrode Contact Impedance

The electrode contact impedance must be low enough that the dispersive electrode represents the preferred current pathway, thus avoiding skin burns at alternative pathways. "For conductive electrodes, the maximum electrode contact impedance shall not exceed 75 Ω over the frequency range of 200 kHz to 5 MHz when measured as described" on a human subject. The frequency range of 200 kHz to 5 MHz encompasses the frequency ranges of existing generators. As electrode-tissue impedance increases as applied current decreases, the committee decided on an impedance measuring current of 200 mA is used as it represents the lower limit of average currents reported for TUR procedures. Under these conditions a maximum contact impedance value of 75 Ω was judged an acceptable for the *conductive* electrodes.

"For capacitive electrodes, the minimum capacitance shall be no less than 4 nanofarads (0.004 microfarads) when measured as described". In this case, electrode contact impedance is measured by placing the *capacitively coupled* dispersive electrode under test on a rigid metal plate larger than the electrode contact area. The test current and frequencies are the same as those specified for conductive electrodes. Their impedance characteristics are described in terms of capacitance as their impedances vary as the inverse of the frequency. The majority of capacitive electrodes that have been found to be clinically acceptable typically have a capacitance value of 4 nanofarads, hence the minimum acceptable capacitance value specified by the Committee.

3.9 Summary

With external biosignal monitoring electrodes, difficult challenges exist in the exciting new area of "Personalised health". Such electrodes must form part of the patient's (or health-conscious citizen's) clothing and must continue to work, day after day, wash after wash, without gelling or preparation of any kind, without suffering from motion artifacts and without causing skin irritation. No mean achievement.

An old monitoring problem still remains to be adequately conquered. A convenient and rapid method of applying many high performance electrodes to the head of a patient for EEG measurement awaits invention. The problem (and that of ECG ambulatory monitoring above) can be side-stepped to some extent by finding new electrode positions (montages or leads) which avoid the most problematic skin sites – hairy head in EEG, muscle and flabby areas in ECG.

For external stimulation, exciting new areas include Public Access Defibrillation. The electrodes and their application to the victim must be almost literally "foolproof", given the seriousness of the possible consequences for all concerned. The electrodes must work after having been stored in the most inhospitable locations and possibly under extreme temperature fluctuations, for example in a car boot (trunk in the US?). In the more "mainstream" areas of Cardiac Pacing and Defibrillation and Electrosurgery, the optimal distribution of current density under the electrodes remains a goal still to be achieved. The solution to this problem offers the hope of decreased electrode areas and the design of truly multi-function pads.

The integration of electrodes into garments for FES and body toning is relatively new area with considerable possibilities.

At present, implant electrodes and associated technologies already offer amazing potential for the deaf, lame and even the blind. The development of multi-microelectrode arrays and waveforms which can help optimally "shape" the electrical fields to facilitate more effective and "natural" stimulation is a thrilling prospect. The interface properties of the micro-electrodes will require further research so as to offset the potentially high interface impedances. Ideas already exploited in, for example, cardiac pacing may prove rewarding when adapted for these areas.

It is hard to overstate the potential of research being undertaken in the area of Brain-machine interface. We live in exciting times.

References

1. Adams G (1785) An essay on electricity, Logographic Press, London
2. Admundson DC, McArthur W, Mosharrafa M (1979) The porous endocardial electrode. PACE 2:40–50
3. Aguiló J (2002) Microprobe multisensor for graft viability monitoring during organ preservation and transplantation. 2nd annual international IEEE-EMB special topic conference on microtechnologies in medicine & biology. Madison, WI, Feb 2002, pp 15–20
4. Aldini G (1803) In: An account of late improvements in Galvanism. Nilkes and Taylor, London
5. Almasi JJ, Schmitt OH (1970) Systemic and random variations of ECG electrode system impedance. Ann NY Acad Sci 170:509–519
6. Almasi JJ, Hart MW, Schmitt OH, Watanabe Y (1972) Bioelectrode voltage offset time profiles and their impact on ECG measurement standards. Can. Med. Biol. Eng. Conf., 4th, Winnipeg, MB, Canada September 6
7. Axelgaard J (2004) Reverse current controlling electrode US patent 2004158305
8. Axisa F, Schmitt PM, Gehin C, Delhomme G, Mc Adams E, Dittmar A (September 2005) Flexible technologies and smart clothing for citizen medicine, home healthcare and disease prevention. IEEE Trans Info Tech Biomed 9(3):325–336
9. Bard AJ, Faulkner LR (1980) Electrochemical methods. Wiley, New York, NY
10. Barker LF (1910) Electrocardiography and phonocardiography: a collective review. Bull Johns Hopkins Hosp 21:358–359
11. Barnes AR, Pardee HEB, White PD et al (1938) Standardization of precordial leads. Am Heart J 15:235–239
12. Barron SL (1950) The development of the electrocardiograph in Great Britain. Br Med J 1: 720–25
13. Bell GH, Knox AC, Small AJ (1939) Electrocardiography electrolytes. Br Heart J 1:229–236
14. Bergveld P (1976) Med Biol Eng Comput 14:479–482
15. Brown L Langer R (1988) Transdermal derlivery of drugs. Ann Rev Med 39:221–229
16. Brummer SB, Turner MJ (1975) Electrical stimulation of the nervous system: the principle of safe charge injection with noble metal electrodes. Bioelectrochem Bioenerget 2:13–25
17. Brummer SB, Robblee LS, Hambrecht FT (1983) Criteria for selecting electrodes for electrical stimulation: theoretical and practical considerations. Ann NY Acad Sci 405:159–171
18. Burch GE, DePasquale NP (1990) A history of electrocardiography with a new introduction by Joel D. Howell, 2nd edn. Jeremy Norman, San Francisco, CA
19. Carim HM (1988) Bioelectrodes. In: Webster, J.G. (ed) Encyclopaedia of medical devices and instrumentation. Wiley, New York, NY, pp 195–226
20. Carim HM, Hawkinson RW (1982) EKG electrode electrolyte-skin AC impedance studies. In: Proceedings of 4th Annual Conference of the IEEE Engineering in Medicine and Biology Society pp 503–504

21. Caruso PM, Pearce JA, DeWitt DP (1979) Temperature and current density distributions at electrosurgical dispersive electrode sites. In: Proc. 7th N. Engl. Bioeng. Conf., Troy, NY, March 22–23, 373–376
22. Chardack WM, Gage AA, Greatbatch W (1961) Correction of complete heart block by a self-contained and subcutaneously implanted pacemaker. J Thorac Cardiovasc Surg 42:418
23. Cheney M, Isaacson D, Newell JC (1999) Electrical impedance tomography. SIAM Rev 41:85–101
24. Chien YW (1982) Transdermal controlled-release drug administration. In: Swarbrick, J. (ed) Novel drug delivery systems. Marcel Dekker Inc., New York, NY, p 149
25. Chien, Y W.(1987) Development of transdermal drug delivery systems. Drug Dev Ind Pharm 13(4&5):589–651
26. Cole KS, Curtis HJ (1938) Transverse electric impedance of squid giant axon. J Gen Physiol 22:3764
27. Crenner F, Angel F, Ringwald C (1989) Ag/AgCl electrode assembly for thin smooth muscle electromyography. Med Biol Eng Comput 27:346–356
28. De Levie R (1965) The influence of surface roughness of solid electrodes on electrochemical measurements. Electrochim Acta 10:113–130
29. de Levie R (1989) On the impedance of electrodes with rough interfaces, J Electroanal Chem 261:1–9
30. Dempsey GJ, McAdams ET, McLaughlin J, Anderson JMcC (1992) NIBEC cardiac mapping harness. 14th Annual International Conference IEEE Eng. In Med. and Biol. Soc., Paris, France, pp 2702–2703, November 1, 1992
31. De Talhouet H, Webster JG (1996) The origin of skin-stretch-caused motion artefacts under electrodes. Physiol Meas 17:81–93
32. Drake KL, Wise KD, J Farraye, Anderson DJ, BeMent SL (1988) Performance of planar multisite microprodes in recording extracellular single-unit intracortical activity. IEEE Trans Biomed Eng 35:719–732
33. Duchenne GBA (1855) In: De l'électrisation localisée et de son application à la physiologie, à la pathologie et à la thérapeutique, Paris
34. Duchenne GBA (1876) In : Mécanisme de la physionomie humaine. J. B. Baillière et fils, Paris
35. Dymond AM (1976) IEEE Trans BME 23:274–280
36. Edelberg R (1971) Electrical properties of the skin. In: Elden HR (ed) A treatise of the skin. Wiley, New York, NY
37. Ellenbogen KA, Wood MA (2002) Cardiac pacing and ICDs. Blackwell Science Malden, Massachusetts: Blackwell Publishing Company, Inc., 2002
38. Ferrari RK (1996) X-ray transmissive transcutaneous stimulating electrode. US Patent 5,571,165
39. Fischler H, Schwan HP (1981) Polarisation impedance of pacemaker electrodes: in vitro simulating practical operation. Med Biol Eng Comput 19:579–588
40. Fricke H (1932) The theory of electrolytic polarization. Phil.Mag 7:310–318
41. Gatzke RD (1974) The electrode: a measurement systems viewpoint, In Miller HA, Harrison DC (eds) Biomedical electrode technology. Academic Press, New York, NY
42. Geddes LA, Baker LE, Moore AG (1969) Optimum electrolytic chloriding of silver electrodes. Med Biol Eng 7: 49–56
43. Geddes LA (1972) Electrodes and the measurement of bioelectric events Wiley, New York, NY
44. Geddes LA, Baker LE (1989) Principles of applied biomedical instrumentation, 3rd edn. Wiley, New York, NY
45. Geddes LA (1995) Historical perspectives 2: the electrocardiograph In: Bronzino JD (ed) The biomedical engineering handbook. Boca Raton: CRC Press, pp 788–798
46. Greatbatch W, Holmes CF (1991, September) History of implantable devices. IEEE Eng Med Biol 38–49

47. Grimnes S (1983) Dielectric breakdown of human skin in vivo. Med Biol Eng Comput 21:379–381
48. Heath R (1989) Tin-stannous chloride electrode element. United States Patent 4852585.
49. Hollander JI (1983) ECG-electrodes. Report No. 83.336, MFI-TNO, Utrecht, The Netherlands
50. Hyland M. McLaughlin J, Zhou DM, McAdams E, (1996) Surface modification of thin film gold electrodes for improved in vivo performance. Analyst (121):705–709
51. Hyman AS (1932) Resuscitation of the stopped heart by intracardial therapy. Arch Intern Med 50:283
52. Hymes AC (1981) Monitoring and stimulating electrode. US Patent 4,274,420, June 23, 1981
53. Janz GJ, Ives DJG (1968) Silver-silver chloride electrodes. Ann NY Acad Sci 148:210–221
54. Jaron D, Briller SA, Schwan HP, Geselowitz DB (1969) Nonlinearity of cardiac pacemaker electrodes. IEE Trans Vol BME 16:132–138
55. Jasper HH, Carmichael L (1935) Electrical potentials from the intact human. Science 81:51–53
56. Jossinet J, McAdams ET (1991) Skin impedance. Innova Tech Biol Med 12(1):21–31
57. Khan A, Greatbatch W (1974) Physiologic electrodes. In:CD Ray (ed) Medical engineering, Year Book Medical Publishers, Chicago, IL
58. Kim Y, Fahy JB, Tupper B (1986) Optimal electrode designs for electrosurgery, defibrillation, and external cardiac pacing. IEEE Trans Biomed Eng 33:845–853
59. Kim Y, Schimpf PH (March 1996) Electrical behavior of defibrillation and pacing electrodes. Proceed IEEE 84(3):446–456
60. Kingma YJ, Lenhart J, Bowes KL, Chambers MM, Durdle NG (1983) Improved Ag/AgCl pressure electrodes. Med Biol Eng Comput 21:351–357
61. Klingman AM (1984) Skin permeability: Dermatologic aspects of transdermal drug delivery. Am Heart J 108(1):200–207
62. Krasteva V, Papazov S (2002) Estimation of current density distribution under electrodes for external defibrillation. BioMedical Engineering Online, 1:7, URL http://www.biomedical-engineering-online.com/content/1/1/7
63. Krug K, Marecki NM (1983) Porous and other medical and pressure sensitive adhesives. Adhes Age 26(12):19–23
64. Lagergren H, Johansson L (1963) Intracardiac stimulation for complete heart block. Acta Chir Scand 125:562–566
65. Lagergren H, Edhag O, Wahlberg I (1976) A low threshold non-dislocating endocardial electrode. J Thorac Cardiovasc Surg 72:259
66. Lawler JC, Davis MJ, Griffith EC (1960) Electrical characteristics of the skin. J Invest Dermatol 1960 May; 34: 301–308
67. Lenarz T, Battmer R-D, Goldring JE, Neuburger J, Kuzma J, Reuter G (2000) New electrode concepts (Modiolus-Hugging Electrodes). Adv Otorhinolaryngol Basel, Karger, 57: 347–353
68. Lewes D (1965) Multipoint electrocardiography without skin preparation. Lancet 2:17 –18
69. Lewin G, Myers GH, Parsonnet V, Zucker IR (1967) A non-polarizing electrode for physiological stimulation. Trans Am Soc Artif Intern Organs 13:345
70. Licht S (1959) History of electrotherapy. In Licht S (ed) Therapeutic electricity and ultraviolet radiation. Elizabeth Licht Pub, New Haven, CT, pp 1–69
71. Lilly JC, Hughes JR, Alvord EC, Galkin TW (1955) Brief, noninjurious electric waveform for stimulation of the brain. Science 121:468–469
72. Linquette-Mailley SC, Hyland M, Mailley P, McLaughlin J, McAdams ET (2002) Electrochemical and structural characterisation of electrodeposited iridium oxide thin film electrodes applied to neurostimulating electrical signal. Mater Sci Eng 21:167–175
73. Low J, Reed A (1994) Electrotherapy explained: principles and practice. Butterworth – Heinmann Ltd, Oxford, England

74. Lymberis A (2004) Research and development of smart wearable health applications: the challenge ahead. In: Lymberis A, de Rossi D (eds)wearable e-health systems for personalised health management, studies in health technology and informatics 108, IOS Press
75. Manley AG (1976) Medical electrode. US Patent 3977392
76. Mannheimer JS, Lampe GN (1987) Clinical transcutaneous electrical nerve stimulation, F.A. Davis, Philadelphia, PA
77. Maritan A, Toigo F (1990) On skewed arc plots of impedance of electrodes with an irreversible electrode process. Electrochim Acta 35:141–145
78. Mastrototaro JJ, Massoud HZ, Pilkington TC, Ideker RE (1992) Rigid and flexible thin-film multielectrode assays for transmural cardiac recording. IEEE Trans Biomed Eng 39: 271–279
79. Maynard EM (2001) Visual Prostheses. Annu Rev Biomed Eng 3:145–68
80. McAdams ET (1989) Effect of surface topography on the electrode-electrolyte interface impedance. Part 1: The high frequency, small signal interface impedance. Surf Topogr 2:107–122
81. McAdams ET, (1989b) Effect of surface topography on the electrode-electrolyte interface impedance, part 2: the low frequency (F<1 Hz), small signal interface impedance. Surf Topogr 2:223–232
82. McAdams ET (1990) Surface biomedical electrode technology. Int Med Device Diagn Ind, Sep./Oct. 1990, pp. 44–48
83. McAdams ET, Jossinet J (1990) Hydrogel electrodes in bio-signal recording In: Proceedings of the 12th annual international conference of the IEEE, engineering in medicine and biology society, Philadelphia, PA, pp 1490–1491
84. McAdams ET, Jossinet J, (1991) DC nonlinearity of the solid electrode-electrolyte interface impedance. Inn Tech Biol Med 12:329–343.
85. McAdams ET, Jossinet J (1991b) The importance of electrode-skin impedance in high resolution electrocardiography. Automedica 13:187–208
86. McAdams ET, Jossinet J, (1992) A physical interpretation of Schwan's limit current of linearity. Ann Biomed Eng 20:307–319.
87. McAdams ET, Henry P, Anderson JMcC, Jossinet J (1992) Optimal electrolytic chloriding of silver ink electrodes for use in electrical impedance tomography. Clin Phys Physiol Mes 13(1):19–23
88. McAdams ET, Andrews P (2003) Biomedical electrodes and biomedical electrodes for electrostimulation. US Patent 2003134545
89. McAdams ET, Jossinet J (1994) A Physical Interpretation of Schwan's Limit Voltage of Linearity. Med Biol Eng Comput March:126–130
90. McAdams ET, McLaughlin JA, Anderson JMcC (1994a) Multi-Electrode Systems for Electrical Impedance Tomography. Physiol Meas 15:A101–A106
91. McAdams ET, McLaughlin JA, Brown BN, McArdle F (1994b) The NIBEC EIT harness chapter 8 In: Holder D (ed) Clinical and physiological applications of electrical impedance tomography. UCL Press, London, 1993, pp 85–92
92. McAdams ET, Lackermeier A, Jossinet J (1994c) AC impedance of the hydrogel-skin interface. In: 16th annual international conference on IEEE engineering in medicine and biology society, Baltimore, MD, USA, pp 870–871
93. McAdams ET, Lackermeier A, Woolfson ET, Moss GP, McCafferty DF (1995) In vivo ac impedance monitoring of percutaneous drug delivery. In: Proceedings of 9th international conference on bioimpedance, Heidelberg, Germany, pp 344–347
94. McAdams ET, Jossinet J, Lackermeier A, Risacher F (1996) "Factors affecting the electrode-gel-skin interface impedance in electrical impedance tomography." Med Biol Eng Comput 34 (6):397–408
95. McLaughlin J, McAdams ET, Anderson JMcC (1994) "Novel dry electrode ECG sensor system". In: 16th annual international conference on IEEE engineering in medicine and biology society, Baltimore, MD, USA, p 804, Nov 1994

96. Mittal T (2005) Pacemakers – a journey through the years. Ind J Thorac Cardiovasc Surg 21:236–249
97. Mond H, Stokes KB (1991) The electrode–tissue interface: the revolutionary role of steroid elution. PACE 15:95–107
98. Mond HG, Stokes KB (1996) The steroid-eluting electrode: a 10-year experience. Pacing Clin Electrophysiol Jul 19(7):1016–20
99. Mortimer JT, Bhadra N (2004) "Peripheral nerve and muscle stimulation" In: Horch KW, Dhillon GS (eds) Neuroprosthetics: theory and practice (Series on bioengineering & biomedical engineering), vol 2. World Scientific, Singapore, pp 638–744
100. Myers GH, Parsonnet V (1969) "Pacemaker electrodes." In: Engineering in the heart and blood. Wiley, New York, NY
101. Nessler N, Reischer W, Salchner M (2003) Electronic skin replaces volunteer experiments. Meas Sci Rev 3(2):71–74
102. Netherly SG, Carim HM (1998) Biomedical electrode with Lossy dielectric properties. US patent 5836942, 1998
103. Oh SY, Leung L, Bommannan D, Guy RH, Potts RO (1993) Effect of current, ionic strength and temperature on the electrical properties of skin. J Controlled Release 27:115–125
104. Olson WH, Schmincke DR, Henley BL (1979) Time and frequency dependence of disposable ECG electrode-skin impedance. Med Instrum 13:269–72
105. Onaral B, Schwan HP (1982) "Linear and non-linear properties of platinum electrode polarization: part 1. frequency dependence at very low frequencies." Med Biol Eng Comput 20:299–306
106. Parsonnet V, Zucker IR, Asa MM (1962) Preliminary investigation of the development of a permanent implantable pacemaker utilizing an intracardiac dipolar electrode. Clin Res 10:391
107. Pearce JA (1980) "The thermal performance of electrosurgical dispersive electrodes." PhD Thesis, Purdue University, West Lafayette, IN
108. Peckham PH, Knutson JS (2005) Functional electrical stimulation for neuromuscular applications. Annu Rev Biomed Eng 7:327–60
109. Prausnitz MR, Bose VG, Langer R, Weaver JC (1993) Electroporation of mammalian skin: a mechanism to enhance transdermal drug delivery. Proc Natl Acad. Sci 90:10504–20508
110. Prohaska OJ, Olcaytug F, Pfundner P, Dragaun H (1986, Feb) "Thin film multiple electrode probes: possibilities and limitations". IEEE Trans Buiomed Eng 33:223–229
111. Reilly JP (1992) Electrical Stimulation and Electropathology. Cambridge University Press, Cambridge
112. Rieger R, Taylor J, Comi E, Donaldson N, Russold M, Mahony CMO, McLaughlin JA, McAdams E, Demosthenous A, Jarvis JC (2004) "Experimental determination of compound A-P direction and propagation velocity from multi-electrode nerve cuffs." Med Biol Eng Comput Phys 26:531–534
113. Robinson AJ, Snyder-Mackler L (1995) Clinical electrophysiology: electrotherapy and electrophysiologic testing. Williams and Wilkins, New York, NY
114. Rosendal T (1945a) Further studies on the conducting properties of human skin to direct and alternating current. Acta Physiol Scand 8:183–202
115. Rosendal T (1945b) Concluding studies on the conducting properties of human skin to alternating current. Acta Physiol Scan 9:39–49
116. Rosell J, Colominas J, Riu P, Pallas-Areny R, Webster JG (1988) Skin impedance from 1 Hz to 1 MHz. IEEE Trans Biomed Eng BME-35: 649–51
117. Rothman S (1956) Electrical behavior. In: Physiology and biochemistry of the skin. University of Chicago Press, Chicago, IL, pp 9–25
118. Rowbottom ME, Susskind C (1984) "Electricity and medicine: history of their interaction." San Francisco Press, Berkeley, CA
119. Rutten WLC (2002) Selective electrical interfaces with the nervous system. Annu Rev Biomed Eng 4:407–52

120. Salter DC (1980) A study of some electrical properties of normal and pathological skin in vivo. D.Phil. Thesis, University of Oxford
121. Sarlandière (1825) "Mémoires sur l'électropuncture considérée comme moyen nouveau de traiter efficacement la goutte, les rhumatismes et les affections nerveuses. Paris 1825
122. Schechter DC (1983) "Exploring the origins of electrical cardiac stimulation." Medtronic, Minneapolis, MN
123. Schmitt OH, Almasi JJ (1970) Electrode impedance and voltage offset as they affect efficacy and accuracy of VCG and ECG measurements. In: Proceedings of X1th international vectorcardiography symposium, New York, NY, USA, pp 245–253
124. Schoenberg AG, Klingler DR, Baker CD, Worth NP, Booth HE, Lyon PC (1979) Final report: development of test methods for disposable ECG electrodes. UBTL Technical Report No. 1605–005, Salt Lake City, UT, 1979
125. Schwan HP (1966) "Alternating current electrode polarisation." Biophysik 3:181–201
126. Schwan HP (1968) "Electrode polarization impedance and measurements in biological materials." Ann NY Acad Sci 148:191–209
127. Searle A, Kirkup L (2000) A direct comparison of wet, dry and insulating bioelectric recording electrodes. Physiol Meas 21:271–283
128. Simpson RW, Berberian JG, Schwan HP (1980) "Nonlinear AC and DC polarization of platinum electrodes." IEE Trans BME-27:166–171
129. Singh S, Singh J (1993) "Transdermal drug delivery by passive diffusion and iontophoresis: a review". Med Res Rev 13(5):569–621
130. Sluyters-Rechbach M, Sluyters JH (1970) "Sine wave methods in the study of electrode processes." In: Bard AJ (ed) Electroanalytical chemistry, vol 4. Marcel Dekker, New York, NY, pp 1–128
131. Stankevich BA (1980, December) "4% of professional liability claims involve electromedicine equipment." Mod Health Care 10(12):74–76
132. Stokes K (1990, June) "Implantable pacing lead technology." IEEE Eng Med Biol 9(2): 43–49
133. Stokes K (1996, March) "Cardiac pacing electrodes." Proc IEEE 84(3):457–467
134. Schwartz AB (2004) Cortical neural prosthetics. Annu Rev Neurosci 27:487–507
135. Szeto AYJ (1988) 'Pain relief from transcutaneous electrical nerve stimulation (TENS).' In: Webster JG (ed) 'Encyclopedia of medical devices and instrumentation.' Wiley, New York, NY, 2203–2220
136. Tam HW, Webster JG (1977) Minimizing motion artifact by skin abrasion. IEEE Trans Biomed Eng BME-24:134–40
137. Timmis G (1990) "The electrobiology and engineering of pacemaker leads." In: Saksena S, Goldschlager N (eds) Electrical therapy for cardiac arrhythmias. W.B. Saunders, London
138. Ungerleider HE (1939) "A new precordial electrode." Am Heart J 18:94
139. Waller AD (1887) A demonstration on man of electromotive changes accompanying the heart's beat. J Physiol 8:229–234
140. Waller AD (1889) On the electromotive changes connected with the beat of the mammalian heart, and of the human heart in particular. Phil Trans R Soc London Ser B 180:169–194
141. Waller AD (1888) Introductory address on the electromotive properties of the human heart. Br Med J 2:751–754
142. Webster JG (1998) Medical instrumentation: application and design. Wiley, New York, NY
143. Weiland JD, Liu W, Humayun MS (2005) Retinal prosthesis. Annu Rev Biomed Eng 7:361–401
144. Weinman J (1965) "Biphasic stimulation and electrical properties of metal electrodes." J Appl Physiol 20:787–790
145. Welch W (1951) Self-retaining electrocardiographic electrode. JAMA 147:1042
146'. Wiley JD, Webster JG (1982) "Analysis and control of the current distribution under circular dispersive electrodes". IEEE Trans Biomed Eng BME-29:381–385
147. Williams DF (1999) The Williams dictionary of biomaterials. ISBN 0-85323-921-5

148. Wolferth CC, Wood FC (1932) The electrocardiographic diagnosis of coronary occlusion by the use of chest leads. Am J Med Sci 183:30–35
149. Yamamoto T, Yamamoto Y (1976) Electrical properties of the epidermal stratum corneum. Med Biol Eng 14:151–158
150. Yamamoto T, Yamamoto Y (1977) 'Analysis for the change of skin impedance.' Med Biol Eng Comput 15:219–227
151. Yamamoto Y, Yamamoto T (1978) Dispersion and correlation of the parameters for skin impedance. Med Biol Eng Comput 16:592–594
152. Zinc R (1991) "Distortion and interference in the measurement of electrical signals from the skin (ECG, EMG, EEG)." Inno Tech Biol Med 12(special issue 1):46–59

Chapter 4
Readout Circuits

R. Firat Yazicioglu

4.1 Introduction

Biopotential signals are routinely monitored in current medical practice for diagnostics of several different disorders. Commonly, patients are connected to a bulky and mains powered instrument, which reduces their mobility and creates discomfort. This limits the acquisition time, prevents the continuous monitoring of patients, and affects the diagnostics of the illness. Therefore, there is a growing demand for low-power and small-size biopotential acquisition systems [1–5].

These biopotential acquisition systems can be divided into two groups. The first one targets the non-invasive monitoring of patients, namely portable biopotential monitoring systems, where as the second one focus on implantable biopotential monitoring systems mainly targeting closed-loop sensing and prevention of disorders. Although both of these fields are very closely related, the requirements in terms of system development are very much different due to different signal characteristics and monitoring environment. In order to be able to achieve the optimum signal quality with minimal system power dissipation, the circuit and system designers should understand the needs and the requirements of these applications prior to the development.

This Chapter focuses on the readout front-end circuit development for both portable and implantable biopotential monitoring systems. First a brief introduction to biopotential signals will be presented. This introduction will include the characteristics of biopotential signals, biopotential electrodes, and aggressors of biopotential signals. Later, main types of instrumentation amplifier architectures will be presented. It is very much important to understand the different properties of these instrumentation amplifiers to be able to make the best choice for a target application. Later, current mode instrumentation amplifiers will be presented, which may

R.F. Yazicioglu (✉)
Interuniversity Microelectronics Center (IMEC), Leuven, Belgium
e-mail: firat@imec.be

H.-J. Yoo, C. van Hoof (eds.), *Bio-Medical CMOS ICs*,
Integrated Circuits and Systems, DOI 10.1007/978-1-4419-6597-4_4,

be the most suitable instrumentation amplifier topology for the applications requiring very high performance instrumentation amplifiers. Finally, different integrated circuits for the extraction of biopotential signals will be presented.

4.2 Biopotential Acquisition

4.2.1 Biopotential Signals

Biopotential signals are generated due to the electrochemical activity of certain class of cells that are components of the nervous, muscular or glandular tissue. Electrically, these cells exhibit a resting potential, and when they are stimulated they generate an action potential. Biopotential signals refer to the actions potentials from a single cell or to the average activity from groups of cells. Figure 4.1 shows the frequency and amplitude characteristics of most commonly recorded biopotential signals [6].

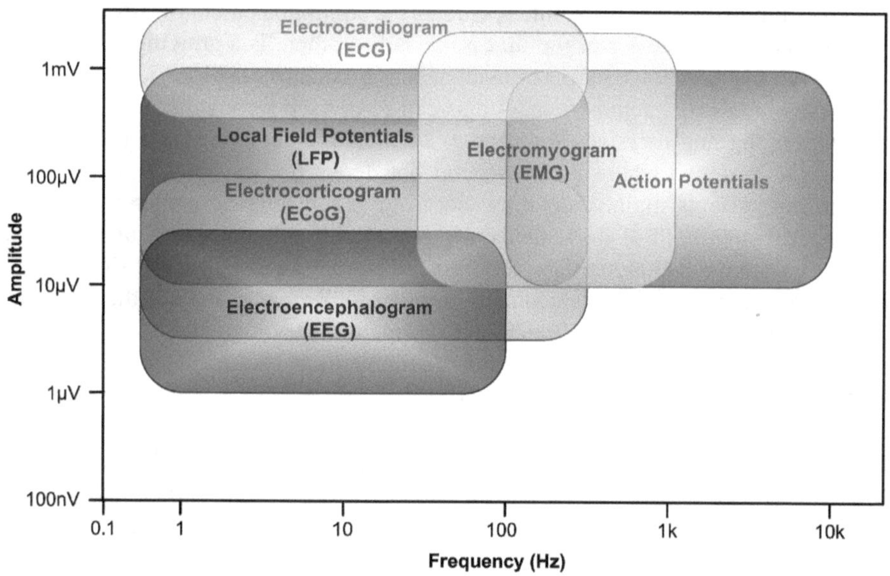

Fig. 4.1 Amplitude and frequency characteristics of common biopotential signals

Electroencephalogram (EEG), Electrocorticogram (ECoG), Local Field Potentials (LFP), refers to the recording of electrical activity of brain created by a group of neurons. The naming indicates the invasiveness of the recording. The EEG recording is the least invasive of all. It uses surface electrodes that are attached to the tissue of the skull, where as, the electrodes for ECoG measurements are placed directly on the surface of the brain, beneath the skull. More invasively, LFP measurement uses neural probes that are inserted inside the brain, picking up average activity from several neurons. On the other hand, the measurement method and invasiveness of Action Potential measurement is very similar to LFP measurement, where the neural probes are inserted to the brain. However, the electrode surface area for Action Potential measurements needs to be much smaller so that the measurement can focus on single unit action potentials rather than the activity of group of neurons. It is interesting to note that as the invasiveness of the measurement increases, the amplitude of the biopotential signals becomes stronger. Also, another interesting point is the fact that as the measurement focuses on the activity of groups of neurons rather than single unit potentials, the frequency of the biopotential signals decreases. This interesting property of the biopotential signals will be further discussed in Section 4.3, where we will investigate how these applications actually define the specifications of the instrumentation amplifier, which in turn, defines the topology to be used.

Finally, the Electrocardiogram (ECG) and Electromyogram (EMG) signals are both referring to the muscle activity. The key difference between these two biopotential signals is the fact that the prior refers to the recording of cardiac muscle activity, where as, the latter refers the recording of the skeletal muscle activity. The measurement of ECG signals is generally performed using surface electrodes on the skin of the chest. Therefore, they can be considered as non-invasive. On the other hand, the measurement of the EMG signals are generally considered as non-invasive, unless needle electrodes are used in order to monitor single motor unit potentials.

Referring again to Fig. 4.1, it can be realized that the common characteristics of biopotential signals are their very low frequency characteristics and extremely small amplitudes (especially in the case of EEG measurements). This possesses strict requirements on the readout circuit design in terms of noise. It should be noted that the design of low-noise CMOS circuits at these frequencies are not straight forward due to the presences of the flicker noise in CMOS technology.

4.2.2 Biopotential Electrodes

In order to be able pick-up biopotential signals from human body a non-zero current should flow between the tissue and the acquisition electronics. It should be noted that this current is carried by ions in the body, whereas it is carried by electrons on the wires connecting the electrodes to the electronic circuit. Therefore, a transducer interface is necessary between the body and the readout circuit that converts the ionic current into electronic current, or vice versa. This interface is called a biopotential electrode.

The basic electrical model of an electrode can be described as shown in Fig. 4.2. C_A and R_A represent the impedance associated with the electrode-tissue interface, and R_S is the resistance of the tissue. Due to the charge imbalance of the tissue and electrode, there is an equivalent charge build-up at the interface, which is represented by the half-cell potential, V_{HC}, also named as polarization voltage. The absolute values of these electrical parameters heavily depend on the electrode type

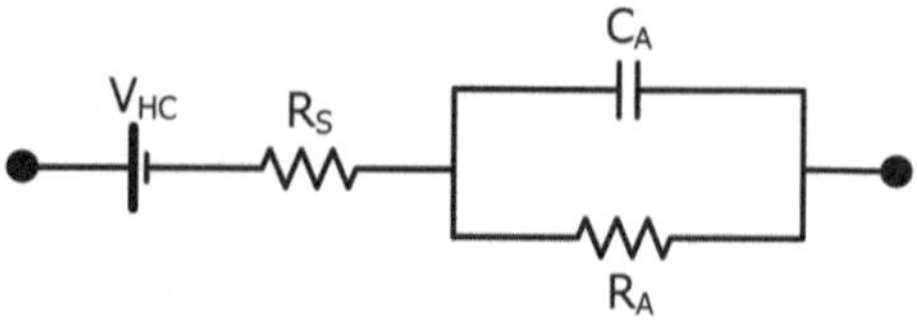

Fig. 4.2 Electrical model of an electrode-tissue interface [6]

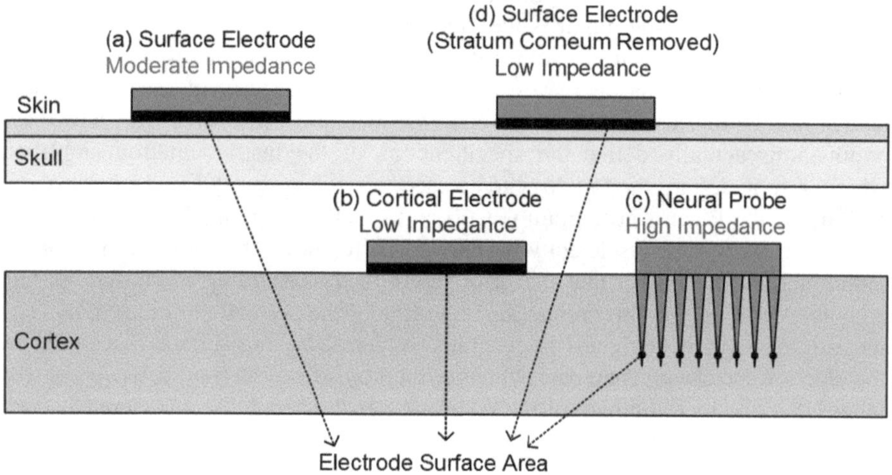

Fig. 4.3 Different types of biopotential electrodes. **a** Surface electrodes standing on the surface of the skin. **b** Cortical electrodes placed beneath the skull on the top of the brain surface. **c** Neuroprobes standing on the surface of the brain, recording electrodes are on the probes penetrating through the brain. **d** Surface electrode sitting on the skin surface, different from the **a** stratum corneum is removed by abrasive material

that is being used, as well as, the invasiveness of the electrode placement. Figure 4.3 shows the different types of electrodes for the extraction of different biopotential signals. Surface electrodes are generally used for EEG, ECG, and EMG signals. In current practice, abrasive skin preparation is used to remove the dead layer of the skin, i.e., stratum corneum, reducing the total impedance of the electrode-tissue interface. Unless this layer is removed, significant increase in total impedance can be expected due to the non-conducting behavior of the stratum corneum layer [6]. On the contrary, the cortical electrodes, which are being used for ECoG and LFP measurements, do not suffer from the presence of this dead skin layer due to the fact that they are already placed beneath the skull. Hence, long term brain activity monitoring with better signal quality may prefer the use of cortical electrodes rather than surface EEG electrodes, due to their superior impedance. Finally, the last electrode type is the neural probes. These electrodes are generally used for monitoring the activity from a single neuron. Due to the dedicated application, the electrode diameter of neural probes is generally limited to 10's of μm's. Hence, the impedance value of these electrodes can be extremely high, in the range of 100 kΩ to 10 MΩ depending on the signal frequency.

Electrode impedance can play an important role during the design of the readout circuit, especially during the selection of the instrumentation amplifier topology. As the electrode surface area decreases its impedance tends to increase. Furthermore, high impedance stratum corneum layer can further increase the impedance of the electrodes. The electrode impedance and the input impedance of the amplifier form a voltage divider. Hence, to prevent the scaling of the biopotential signal before amplification, the input impedance of the amplifier should be maximized. In addition, input bias current of the amplifier must be minimized to prevent the tissue and electrode damage [6].

As a result, it can be concluded that with large electrode surface area and being placed directly on the brain tissue, cortical electrodes can achieve the lowest impedance value compared to rest of the electrodes. However, the invasiveness of these electrodes is the biggest disadvantage. Secondly, surface electrodes may achieve very low impedance as well, only if the stratum corneum layer is removed. However, this is not convenient for long-term monitoring applications, since stratum corneum layer can quickly regenerate itself, requiring the repetition of the abrasive skin preparation. Therefore, the main attention for applications requiring long-term monitoring is the use of surface electrodes without any skin preparation, however, it should be noted in this case the equivalent electrode-tissue impedance is much larger.

On the other hand, neural probes can be considered as separate case, since they are generally targeting the monitoring of activity from a very much localized region of brain. The use of minimal size electrodes results in very high electrode impedance dictating the choice of instrumentation amplifiers with very high input impedance.

4.2.3 Interference Theory

Biopotential acquisition systems are often disturbed by the interference from the power lines. Two main types of interference are called electromagnetic and electrostatic interference. In the case of the electromagnetic interference, the magnetic field created by the alternating mains current cuts the loop enclosed by the human body, the leads of the circuit, and the biopotential amplifier. This induces an electromotive force (EMF), which creates an AC potential at the input of the circuit. The electromagnetic interference can be reduced by decreasing the area of the loop by twisting the cables [7]. Further reduction is possible by using miniaturized portable biomedical acquisition systems that can be placed much closer to the electrodes, which in turn reduces the cable length. An alternative approach is to use active biopotential electrode architectures, where the readout circuit is integrated with the biopotential electrodes itself [8].

Figure 4.4 shows the equivalent circuit for describing the electrostatic interference [9]. The human body is capacitively coupled to the power lines via C_{bp} and also to the earth ground via C_{bg}. In addition to these two capacitances, there exists an isolation capacitance between the earth and the ground of the system battery. As

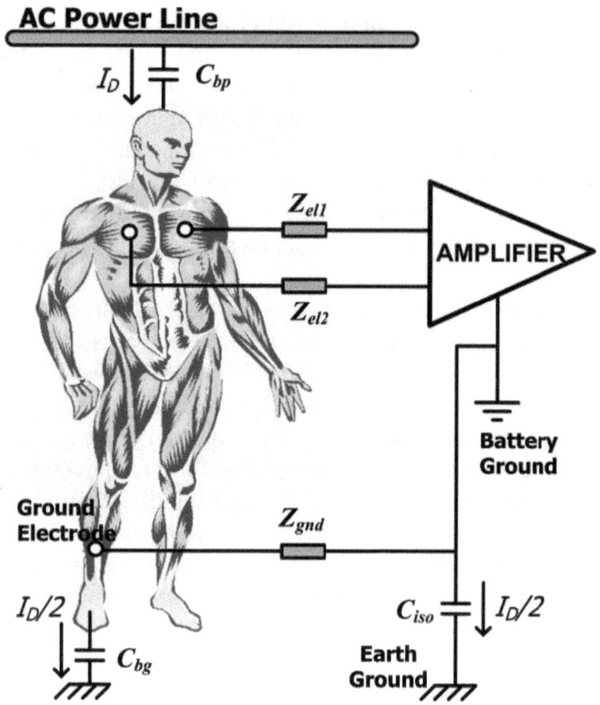

Fig. 4.4 Electrical equivalent circuit of electrostatic interference to human body

a result, the path through the coupling capacitors creates a displacement current, I_D, passing through the human body and splitting equally between the C_{bg} and C_{iso} [10] (C_{bg} and C_{iso} are assumed to have similar capacitance values and R_{gnd} is much smaller than the impedance of C_{iso} and C_{bg} at 50 Hz/60 Hz). Therefore, an AC voltage with magnitude:

$$V_{CM} = \frac{I_D}{2} \times R_{gnd} \tag{4.1}$$

is generated on the human body. Unless there is a mismatch between Z_{el1} and Z_{el2}, this voltage appears as a common-mode input to the amplifier, and can be rejected by the amplifiers CMRR. However, there is always a mismatch between the electrode impedances due to the difference in the nature of the electrode-tissue interface on different locations. Due to this mismatch, a differential error signal is created at the input of the amplifier, whose amplitude can be defined by:

$$\Delta V = V_{CM} \left(\frac{|Z_{el1} - Z_{el2}|}{Z_{in}} \right) \tag{4.2}$$

where Z_{in} stands for the input impedance of the amplifier [7]. As a conclusion, a high CMRR, alone, is not sufficient for an instrumentation amplifier to completely reject the electrostatic interference, very large input impedance must be realized to be able to prevent the conversion of common mode signals to differential mode due to the mismatch of electrode-tissue interfaces.

4.2.4 Noise Considerations

High signal-to-noise ratio is one of the main specifications for instrumentation amplifiers in biomedical signal acquisition. Therefore, before designing any acquisition system or circuit noise sources should be well understood and studied.

Referring to Fig. 4.4, the main noise sources of a biopotential acquisition system are interference, electrode noise, and readout circuit noise. Section 4.2.3 describes the presence of interference noise in biopotential acquisition systems. This noise appears due to electromagnetic and electrostatic interference to the cables and human body, respectively. The prior can be minimized by using shorter cables or active electrode architectures, where as, the latter requires the implementation of high CMRR instrumentation amplifiers. However, as indicated in Section 4.2.3, input impedance of the amplifier also plays a critical role in terms of rejecting the common-mode signals. Hence, an ideal instrumentation amplifier not only should have a large CMRR but also should have large input impedance.

A significant improvement in systems' CMRR (~20 dB) can be realized if a driven-right-leg (DRL) circuit is incorporated in the system [10]. The DRL circuit uses an active ground electrode where the voltage on the ground electrode is derived from the common mode voltage at the inputs of the amplifier. This way displacement current through C_{bg} can be minimized, which sets the patient body to virtual ground in terms of common mode signals. Unfortunately, this ideal case is never fully accomplished due to the finite impedance of the ground electrode and stability requirements of the negative feedback loop. The latter is an important limitation of DRL circuits, leading to large power dissipation.

Another noise source in a biopotential acquisition system is the biopotential electrodes. The impedance of an electrode is a noise source itself, which will be added to the biopotential signals. Hence, the noise of the electrode will be dependent on the surface and the chemical parameters of an electrode. The detailed analysis of these noise sources is not in the scope of this Chapter. However, it is interesting to note that there is a logical relation between the electrode surface area and the biopotential signal that needs to be monitor. As the monitoring gets more invasive, the amplitude of the biopotential signal increases, therefore, electrode area can be reduced while still meeting the noise requirements.

An additional noise source from electrodes is related to motion artifacts. The operation of the biopotential electrodes requires the charge balance between the electrode and the tissue [6]. The half-cell potential represents this charge balance. However, if the electrode moves with respect to the tissue, then the charge distribution at the electrode tissue interface is disturbed, which generates a voltage change.

This voltage change can be orders of magnitude larger than the biopotential signals, which can significantly disturb the measurements.

At last but not the least, another important noise source is the readout circuit itself. While extracting the biopotential signals noise of the CMOS transistors are also added to the signal reducing the SNConsidering R the fact that biopotential signals have extremely low frequencies, two important noise sources needs to be considered, namely, the thermal noise and the flicker (1/*f*) noise. The thermal noise of a CMOS transistor is defined by its transconductance, where as, the flicker noise is a process dependent noise that can be reduced with the increasing gate area [11]. It should be noted that the flicker noise increases with reducing frequency, meaning that, it has more affect in the frequency range of EEG and ECoG signals than the frequency range of action potentials. Hence, readout circuit design for the extraction of biopotential signals generally requires the use of circuit architectures that minimizes and/or eliminates the flicker noise of CMOS transistors. As a result, the circuit designer's main target is to reduce the total integrated noise of the amplifier in the signal bandwidth. This noise should be lower than the smallest signal of interest. Knowing that the thermal noise of CMOS transistor is inversely proportional to its drain-to-source current, and the flicker noise is inversely proportional its gate area, a direct solution to achieve a low noise instrumentation amplifier implies the use of CMOS transistors with increasing power dissipation and large gate area. However, both the increased power and large gate area are not desired in battery power portable/implantable biopotential monitoring applications.

The term, noise efficiency factor (NEF), describes the trade-off between the power dissipation and the noise of an amplifier. This term has been first introduced by [12] in order to compare the power-noise performance of different amplifiers and can be expressed as:

$$\mathrm{NEF} = V_{\mathrm{in,rms}}\sqrt{\frac{2 \times I_{\mathrm{tot}}}{\pi \times V_t \times 4kT \times BW}} \tag{4.3}$$

where BW is the −3 dB bandwidth of the amplifier (assuming that the amplifier has a single dominant pole) and $V_{\mathrm{in,rms}}$ is the total input referred voltage noise of the amplifier. The NEF of a single bipolar transistor having only thermal noise is 1, which is the theoretical limit for any practical circuit. NEF can be used to compare the power-noise performance of different amplifiers. The amplifier with lower NEF can achieve lower power dissipation for a given noise level.

4.3 How Application Affects the Choice of Instrumentation Amplifier Topology

The instrumentation amplifier is the most critical building block of the analog readout front-end in terms of signal quality and clarity. It affects the noise level and the CMRR of the readout front-end, and filters the differential DC electrode offset. Hence, it is generally the most power consuming building block. Therefore, circuit designer's effort generally focuses on implementing a low-power and low-noise instrumentation amplifier with high signal quality. However, it is important to understand the requirements of different applications to be able to implement dedicated instrumentation amplifiers with optimum power dissipation and highest signal quality.

Different measurement types of cortical activity may present a very good example how the requirements may change according to the intended application. Figure 4.5 shows the frequency and amplitude characteristics of different cortical signals. It should be noted that as the measurement goes in-vivo signal amplitudes increases considerably, more than order of magnitude. Hence, it can already be concluded that in-vivo measurement may allow larger total integrated noise compared to the ex-vivo applications such as EEG measurement using surface EEG electrodes. On the other hand, it should be noted that as the electrode area increases the frequency of the signal that is being monitored is reduced. This is mainly due to the fact that the larger the electrode area the more the average activity of several neurons is picked-up. As a results of this discussion, we can concluded that due to their amplitude levels and frequency content, measurement of action potentials may suffer less from the flicker noise of CMOS transistors.

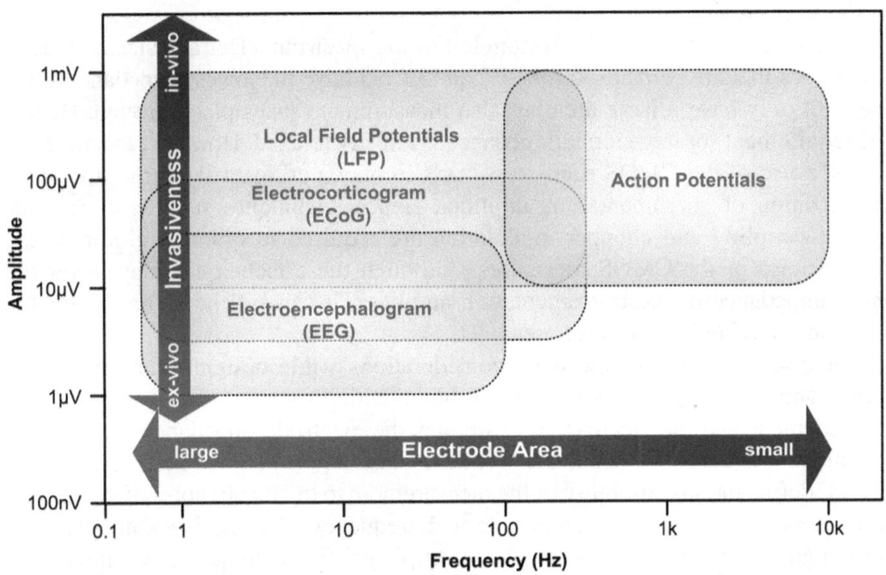

Fig. 4.5 Comparison of different cortical measurements in terms of invasiveness and electrode area. Cortical measurements with high frequency (i.e. action potentials) require the use of electrodes with very small area. This is due to the fact that the signal source is very small and measurement is interested with the signal generated by this small source. On the other hand, as the measurement becomes more invasive the signal amplitudes increases. This is due to the fact that the electrode gets closer to the signal source

On the other hand, the electrode impedance is the combined affect of electrode area and the invasiveness of the measurement. As the measurement gets more invasive, the impedance of the electrode decreases. Already, removal of the stratum corneum layer of the skin can make a considerable difference on the electrode impedance. On the other hand, the electrode area also reduces the electrode impedance. From this perspective, it can be concluded the measurements such as LFP and ECoG has the most advantage since they are both invasive and they can be measured using an electrode with large surface area. Therefore, the design of an instrumentation amplifier for these measurements can have more relaxed specifications in terms of input impedance.

As a result, the specifications of an instrumentation amplifier will mainly be defined by the measurement type of the biopotential signals. For instance, as we go to the top right of the spectrum in Fig. 4.5, the electrode are gets smaller. This means that our instrumentation amplifier should have maximal input impedance. The input impedance of an amplifier is highest when it is limited by parasitic and stray capacitances. This can be achieved when the input signal is connected directly to the gate of a CMOS transistor. If a switching circuit, such as a sampling circuit or a chopper modulator switch, is included before the input of the instrumentation amplifier, the input impedance of the amplifier is negatively affected. This is due to the fact that the switching nature requires more current to be drawn from the electrode in order to charge the sampling capacitor or the stray capacitance at the gate of the amplifier.

Conversely, as we go to the bottom left of the spectrum, electrode size increases. Electrodes that are being used for LFP and ECoG have the lowest impedance since they not only have a large area but also measurement takes place in-vivo. Hence, the requirement for large input impedance is already relaxed. However, this time the flicker noise of the CMOS transistors starts to play a critical role during the noise optimization of instrumentation amplifier. Hence, techniques such as correlated-double-sampling and chopper modulation are required in order to eliminate the flicker noise of the CMOS transistors. Although these techniques may lower the input impedance of the instrumentation amplifier, it can still be sufficient for the electrodes having large surface area.

Table 4.1 summarizes the main considerations while designing an instrumentation amplifier for the extraction of biopotential signals. Invasiveness of the measurement and the electrode area defines the electrode impedance of the measurement. On the other hand, the frequency and amplitude characteristics of the signals define the susceptibility of the measurement to the flicker noise of the CMOS transistors. In addition, the amplitude and frequency characteristics also specify how high should be the CMRR of the amplifier. If the frequency of interest is much larger than the mains frequency, for instance in the case of action potential measurements, then interference free measurements can be achieved with a lower CMRR. Knowing that the requirements are different for each different measurement type of biopotential signals, we can continue with the state-of-the-art instrumentation architectures and discuss their applicability to different types of biopotential measurements.

Table 4.1 Summary of the considerations during the design of instrumentation amplifiers for different biopotential signals. A negative sign (–) indicates a low/small value, where as a positive sign (+) indicates a high/large value

	Measurement Invasiveness	Electrode Area	Electrode Impedance	Susceptibility to 1/f noise	CMRR Requirement
ECG	– –	+ +	– –	+	+
EMG	– –	+	– –	+	+
LFP	+ +	+	+	+	–
ECoG	+	+	– –	+	+
EEG	–	+	–	++	++
Action potentials	+ +	– –	+ +	–	–

4.4 Power Efficient Instrumentation Amplifier Topologies for Biopotential Signal Extraction

A typical biopotential monitoring system may consist of sensors, actuators, front-end, microprocessor, DSP, and radio. The power dissipation of each block has the uttermost importance for minimizing the total power dissipation of the system. It can be realized that commercially available instrumentation amplifier topologies are not meeting the requirements of low-power systems. For instance, an existing high performance instrumentation amplifier [13] consumes 120 μW, while achieving only 60 nV/Hz. Obviously, in a multi-channel system, this will lead to excessive power dissipation, considerably reducing the battery lifetime. Therefore, this Section will describe power efficient instrumentation amplifier topologies suitable for the extraction of biopotential signals.

4.4.1 Limitations of Existing Off-the-shelf Instrumentation Amplifier Topologies

The most commonly employed instrumentation amplifier topology is the three-opamp instrumentation amplifier [14–17], Figure 4.6. It uses three opamp and six resistors in order to realize an instrumentation amplifier whose gain is defined by the ratio of the resistors as:

$$V_{\mathrm{OUT}} = V_{\mathrm{IN}} \left(1 + \frac{2 \times R}{R_G}\right) \frac{R_3}{R_2} \tag{4.4}$$

Hence, the gain of the instrumentation amplifier can be adjusted by only changing the value of a single resistor, R_G. In addition, the fact that the input signal only sees the gates of the input transistors of the opamps, this instrumentation amplifier is ideal for application requiring very high input impedance.

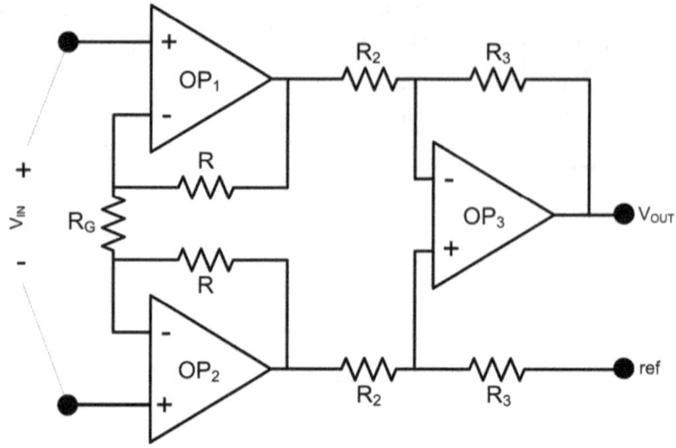

Fig. 4.6 Circuit schematic of a three-opamp instrumentation amplifier

An interesting property of this instrumentation amplifier is the fact that the differential gain of this architecture can be defined by the first stage, while its common mode gain is always one. Hence, the common mode signal at the input of this instrumentation amplifier directly appears at the input of the second stage. This second stage rejects the common mode signals and only amplifies the differential signals. Hence, the ideal common-mode gain of the whole chain is zero setting the CMRR of this amplifier to infinity under ideal conditions, i.e., perfectly matched components. Unfortunately, the actual common-mode gain of the three opamp instrumentation amplifier heavily depends on the matching of the resistors [18], and real-life implementations generally use laser trimming for these resistors in order to reduce the common-mode gain and increase the CMRR. It should also be noted that as the gain of the first stage increases, the CMRR of the instrumentation amplifier is improved due to the fact that the common mode gain of the first stage is independent of its differential gain and always equals to one.

Although, this architecture is ideal for achieving high input impedance and it can also be trimmed to achieve high CMRR, two fundamental limitations prevent its use in low-power and low noise applications. This first one is due to the use three opamps. This not only increases the power dissipation of the architecture but also elevates the total noise. In addition to this, the use of several resistors further elevates the total noise of this architecture. Therefore, it can be concluded that this architecture is not ideal for designs targeting optimum NEF, hence, should be carefully evaluated for applications requiring very low power dissipation and noise. A slight improvement to the architecture in terms of power can be achieved by using two-opamp instrumentation amplifier topology. However, the limitations on the elevated noise due to the resistors and the requirement of matched components still applies to this topology [13, 19].

A second technique for implementing instrumentation amplifiers uses switched-capacitor (SC) architectures [20, 21]. The main advantage of the SC architectures over the three opamp instrumentation amplifiers is their capability of eliminating the flicker noise of the CMOS transistors by incorporating correlated double sampling technique [22]. This may be an attractive feature especially for biopotential signals suffering from the presence of flicker noise. Unfortunately, the main limitation of these amplifiers is the presence of the sampling capacitors in the single path between the electrode and the amplifier. The charging and discharging of this capacitance significantly reduces the input impedance of the amplifier.

Another limiting factor in SC amplifiers is the noise fold-over problem above Nyquist frequency (half of the sampling frequency) [22]. The sample and hold nature of these amplifiers shifts the noise over Nyquist frequency back into the amplification bandwidth. Hence, the noise of the amplifier is elevated above the value that would be achieved using continuous-time amplifiers. Therefore, both the low input impedance of the switched capacitors amplifiers and their elevated noise make this architecture not suitable for biopotential acquisition systems.

As a conclusion, it is not surprising that neither the three opamp nor the switched capacitor instrumentation amplifiers are being considering in the design of next generation biopotential acquisition systems. Instead, the research is focusing on different instrumentation amplifier topologies that can achieve optimum NEF while meeting the strict performance requirements on CMRR, PSRR, and THD.

4.4.2 Instrumentation Amplifiers Utilizing Pseudo Resistors

A new instrumentation amplifier architecture has been introduced in the recent years [23, 24] (Fig. 4.7). Unlike the three opamp instrumentation amplifier, which

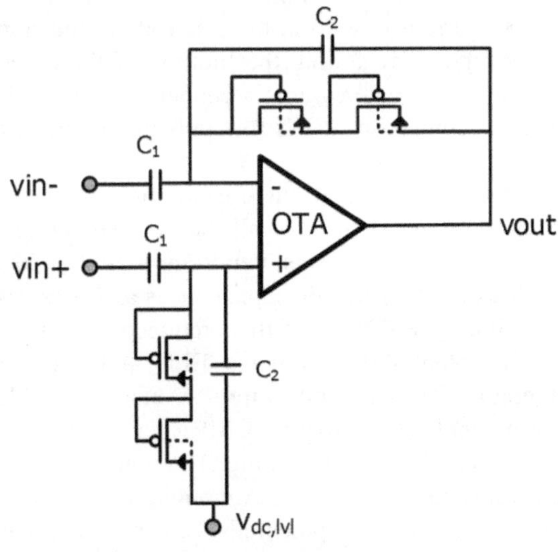

Fig. 4.7 Circuit schematic of an instrumentation amplifier using pseudo resistors

uses resistors in the feedback path of the amplifier, this architecture incorporates capacitors to define the gain of the instrumentation amplifier as:

$$V_{\mathrm{OUT}} = V_{\mathrm{IN}} \left(\frac{C_1}{C_2} \right) \tag{4.5}$$

On the other hand, in order to set the DC patch between the output, reference voltage, and inputs, weakly conducting transistors are employed. These transistors implement resistors in the range of T Ω, hence they are also ideal for setting the high-pass filter cut-off frequency much below 1 Hz range, which is required for most of the biopotential signals in order to block the DC polarization voltage from biopotential electrodes.

In addition, this instrumentation amplifier architecture has also some attractive features in terms of ease of implementation, power dissipation, noise, and input dynamic range. The only active block in the circuit is the OTA, which is a significant advantage in terms of power dissipation, total noise, and ease of design. Therefore, implementing a low noise and low power OTA architecture will be sufficient to minimize the noise and power dissipation of the instrumentation amplifier. A more detailed noise analysis of the complete architecture can be found in [23], which indicates that the parasitic capacitances at the input node of the OTA is also important for the optimization of the total noise of this instrumentation amplifier.

Another advantage of the architecture is the inherent AC coupling. The use of the weakly conducting transistors for setting the DC path of the feedback loop introduces a high-pass filter when combined with C_1. This AC coupling scheme is capable of filtering supply range DC polarization voltage, which makes this architecture ideal for applications that may suffer from large polarization voltages, such as, recording of action potentials using neural probes [24]. In addition, the complete DC isolation of the amplifier inputs from the electrodes eliminates the possibility of DC current flow from the electrode to the amplifier, which can be an important consideration regarding the lifetime of the biopotential electrodes. As a result of its attractive features, this architecture is extensively being used in several biopotential acquisition ASICs for applications such as action potentials, LFP, and ECG monitoring [25–27].

On the other hand, there exist some fundamental limitations of this amplifier, which prevent its use for all the biopotential signal monitoring applications. The first and may be the most important limitation its dependency to element matching. Unless, all the parasitic capacitances and passives of this architecture are perfectly matching, the CMRR of this architecture is much lower than what is required for the biopotential signals such as EEG, and ECoG. The highest CMRR reported in the literature for this architecture is around 84 dB [23], where most of the EEG monitoring applications require CMRR values in excess of 110 dB. Another, limitation is the high flicker noise of the OTA that will elevate the total integrated noise of the instrumentation amplifier. Although, this architecture may be designed for applications, such as action potentials, LFP, and ECG signals, with negligible flicker noise contribution, designs targeting EEG and ECoG applications may significantly suffer from flicker noise, due to the fact that the total integrated noise requirement of such applications in lower than 1 μV_{rms}. One possible solution to overcome flicker noise problem is to introduce chopper stabilization to this architecture, where the input chopper modulator can be placed at the virtual ground of the architecture [28]. This way the flicker noise of the OTA can be modulated to higher frequencies and low-pass filtered. However, the designer should be aware of the fact that the CMRR of such implementations will still be limited by the matching of the capacitors and parasitic capacitances. In addition, the parasitic input capacitance of the OTA has an increasing affect over the noise.

As conclusion, it can be stated that this architecture is very attractive for applications where large input impedance and very low power are the key requirements. As it will be discussed further in this Chapter, the total noise of this type of amplifiers are generally in the range of 2–5 μV_{rms}, which is suitable for LFP applications, however one can discuss the convenience of this architecture for other biopotential applications such as EEG, ECoG, EEG, and LFP due to the low frequency characteristics of these signals. In any case, application specific circuits using this architecture exist for ECG and EEG signal monitoring applications [23, 26].

4.4.3 Introduction to Chopper Modulation

CMOS amplifiers occasionally suffer from passive component mismatches and flicker noise problem. The prior reduces the CMRR of the amplifier, where as, the latter increases the total noise especially for applications requiring the extraction and amplification of very low frequency signals. The monitoring of biopotential signals falls exactly into this category.

The operation principle of the chopper modulation is described in Fig. 4.8 [29]. In addition to the amplifier, the architecture consists of a modulator and a demodulator connected to the input and output of the amplifier, respectively. The modulator and demodulator uses square wave signals, m(t), with a frequency $f_{chop} = 1/T_{chop}$. The modulation of the signal by the input modulator shifts the frequency spectrum of the input signal, X(s), to the odd harmonics of f_{chop}. Then, the modulated input signal is amplified by the amplifier with transfer function A(f), and demodulated with m(t). This shifts the modulated spectrum back to its original location, leaving harmonics at the odd multiples of f_{chop}. These harmonics can simply be filtered by using a low-pass filter.

Using this modulation and demodulation technique aggressors such as flicker noise and mismatch related non-idealities can be eliminated. Figure 4.8 describes the principle of how the chopper modulation technique can be useful for reducing the flicker noise and increasing the CMRR of the core amplifier. The input referred noise and the offset of the core amplifier are indicated by v_n and v_{off}. While the input signal is modulated and demodulated by the input and output modulators, respectively, the aggressors are only modulated by the output modulator. Hence, the

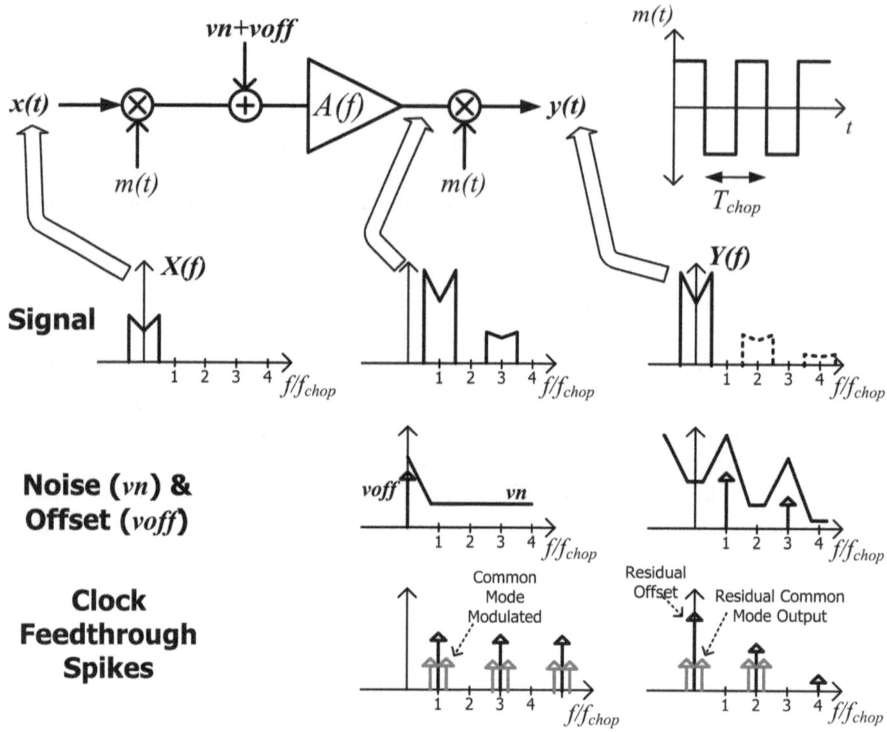

Fig. 4.8 Circuit schematic of an instrumentation amplifier using pseudo resistors

aggressors and the signal of interest are clearly isolated in frequency domain at the output. Therefore, aggressors can be discarded by only using a low-pass filter.

In this Chapter, we will not get into the details of deriving the equations related with the chopper amplifiers. Instead, we are going to summarize the results of the papers which are very extensively discussing the noise, distortion, and gain performance of chopper modulated amplifiers [22, 30–32].

The equivalent input referred noise of a chopper modulated amplifier is equivalent to the input referred noise, v_n, of the core amplifier modulated by the input modulator. Hence, the double-sided input-referred voltage noise power spectral density (PSD), $S_{in}(f)$, of the amplifier can be expressed as:

$$S_{\text{in}}(f) = \left(\frac{2}{\pi}\right)^2 \sum_{n=-\infty,\text{odd}}^{+\infty} S_{\text{vn}}\left(f - n \times f_{\text{chop}}\right) \tag{4.6}$$

After necessary manipulations and assumptions the total input referred noise of a chopper modulated amplifier can be approximated as [22]:

$$S_{\text{in,total}}(f) = S_0 \left(1 + 0.8525 \frac{f_{\text{corner},1/f}}{f_{\text{chop}}}\right) \text{ for } \left|\frac{f}{f_{\text{chop}}}\right| < 0.5 \ \& \ \frac{f_{\text{corner},1/f}}{f_{\text{chop}}} >> 1 \tag{4.7}$$

where S_0 represents the input referred thermal noise spectrum of the core amplifier, $f_{\text{corner},1/f}$ is the corner frequency where power of flicker noise equals to the power of the thermal noise. As results, if the chopping frequency is selected large enough from the flicker noise corner frequency, then the total input referred noise of a chopper modulated amplifier is equivalent to the total input referred noise of the core amplifier without any flicker noise. This property of chopper modulated amplifiers can be particularly interesting for applications requiring very low flicker noise such as EEG and ECoG measurements.

Another advantage of amplifiers using chopper modulation is the elimination of component mismatch related non-idealities. The offset and the CMRR of the core amplifier fall into this category. Similar to the flicker noise, these non-idealities are also modulated by the output modulator and hence separated from the signal of interest in frequency domain. However, it should be noted that the performance of chopper modulation on eliminating the offset of the core amplifier is limited with circuit non-idealities, i.e. non-ideal implementation of the modulators.

The modulator only consists of four cross-coupled CMOS switches. These switches inject a finite amount of charge to input of the amplifier correlated with the modulation clock, m(t). The voltage generated due to this charge injection is also demodulated by the output choppers and results in a finite output offset voltage that is proportional with the input capacitance of the core amplifier, size of the chopper switches, modulation frequency, and source resistance [33]. Designers should note that under the conditions where the electrode impedance is very large, the chopper modulation can introduce significant offset to the amplifier.

Fortunately, several techniques exist in the literature to cope with this problem. Reference [34] uses a bandpass filter between the input and the output choppers. However, matching of the bandpass filter center frequency with the chopping frequency limits the efficiency of this technique. Reference [35] uses nested choppers, where in addition to the input and output modulators, another pair of slow chopping modulators are used to modulate the output spikes of the fast modulator. Although, this technique can be very efficient for slow signals, some biopotential signals have too large bandwidth for this solution. Another technique is proposed by [36]. It uses a SC notch filter with synchronous integration after the output modulator to filter both the chopping ripple and the modulated amplifier offset, however results in excess quiescent current and complexity in the signal path.

Another non-ideality of the chopper modulated amplifiers is the signal distortion. This appears due to the finite bandwidth of the core amplifier [37] and is due to the fact that the core amplifier actually filters some of the harmonics of the modulated input signal due to its finite bandwidth. The main consequence of this distortion is an equivalent reduction in the gain of the amplifier as represented in the formula below:

$$A_{\text{Gain,Chopped}} = A_{\text{Gain}} \times \left(1 - \frac{4\tau}{T}\right) \tag{4.8}$$

where $1/\tau = 2\pi f_c$ and f_c is the −3 dB of the amplifier. It should be noted that techniques that are effective for reducing the chopping ripples are also effective for reducing the amplifier distortion due to finite bandwidth.

4.4.4 Chopper Modulating Amplifiers for Biopotential Signal Extraction

Chopper modulating amplifiers has several key advantages for the applications including the extraction of biopotential signals. The most significant advantage is regarding the elimination of the flicker noise as well as the removal of the mismatch related non-idealities, i.e., reduced common mode gain. On the other hand, some attention has to be paid on the input impedance of the chopper modulated amplifiers since input signal is modulated to higher frequencies. Hence, any stray capacitance at the input of the instrumentation amplifier can reduce the input impedance of the instrumentation amplifier.

Different chopper modulated amplifier topologies exist in the literature dedicated to the applications requiring the monitoring of the biopotential signals. The main challenge in these amplifiers is the elimination of the polarization voltage of the biopotential electrodes, as well as, the realization of an optimum power-noise performance.

One popular implementation technique is the use of capacitive feedback amplifiers with chopper modulation. Figure 4.9 shows one example of such instrumentation amplifier architecture [37]. Voltage gain of the instrumentation amplifier is

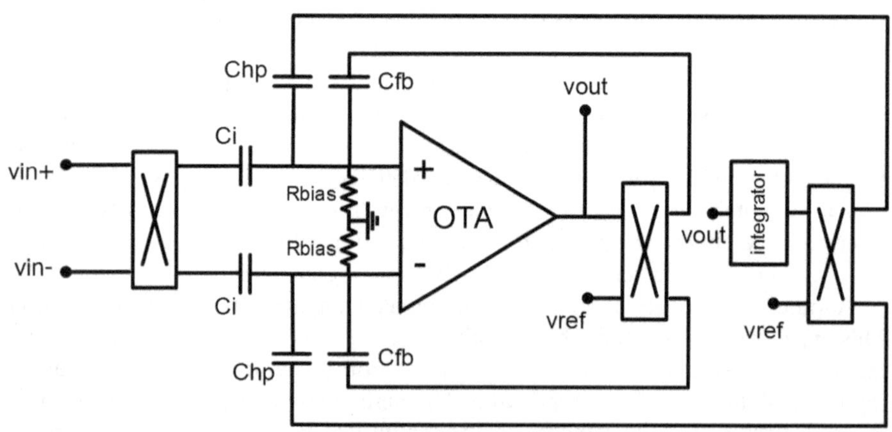

Fig. 4.9 Chopper modulated instrumentation amplifier of [33]. The DC servo is based on voltage feedback

defined by the ration of C_i to the feedback capacitor C_{fb}. Therefore, the voltage gain of this instrumentation amplifier can be precisely controlled by changing the ratio of the capacitors. On the other hand, the polarization voltage of the electrodes can be filtered by the high pass filter realized by the servo-loop incorporating an integrator and coupling capacitors, C_{hp}. Resistors, Rbias, sets the DC voltage at the input of the core amplifier. It should be noted that due to the complete AC coupling of the common mode input voltage through C_i and R_{bias}, this architecture exhibits rail-to-rail input common mode range. In addition the use of a single core amplifier relaxes the power requirements of the topology and improves the power-noise performance of this instrumentation amplifier. On the other hand, one limitation of the topology is the limited headroom for the polarization voltage of the biopotential electrodes. The ratio between C_{hp} and C_{i} sets this headroom and unfortunately, increasing this headroom by would result in increased total noise [37]. Another point that needs to be considered while using this instrumentation amplifier is the input impedance. Due to the capacitive path seen from the input to the reference voltage v_{ref}, the total input impedance is reduced considerably. This may be a limiting factor for this topology to be used for wide range of biopotential applications. For instance, as shown in Table 4.1, some applications of biopotential monitoring use electrodes with high impedance requiring the use of instrumentation amplifier with very high input impedance. An example realization of this instrumentation amplifier realizes more than 100 dB CMRR, 100 nV/Hz input referred noise density and higher than 8 MΩ input impedance, while consuming only 1 μA from 1.8 V supply.

Shifting the input modulator to the virtual ground of the amplifier may significantly improve the input impedance by preventing the modulation of the input voltage before the input of the amplifier. This is actually the proposal of implementation presented in [28]. This ways the high frequency path to the ground can be eliminated, significantly improving the input impedance of the amplifier. On the other hand, the output demodulator, integrated inside the OTA, modulates the flicker noise of the transistors, which can be rejected by passing the output voltage through a low-pass filter. Then the input modulator only serves the purpose of keeping the opamp in negative feedback configuration. Although, this configuration is advantages in terms of the input impedance of the amplifier, one disadvantage is appearing regarding the affect of chopper modulation on eliminating the mismatch of the components. Since the input signal is not modulated around the passives of the instrumentation amplifier, then the mismatch of the capacitors and parasitic capacitances are susceptible to process related mismatches. This can significantly degrade the CMRR of the instrumentation amplifier. Indeed, the measurement results of [28] indicate that the flicker noise is suppressed, however, only 60 dB CMRR is achieved. Although, this may limit the use of this amplifier to applications requiring moderately low CMRR, on the other hand, the advantage of reduced flicker noise and having rail-to-rail input range is attractive especially for biopotential signals such as ECoG and LFPs. Another important advantage of this amplifier is its capability of filtering large polarization voltages from the biopotential electrodes.

A third type of instrumentation amplifier focusing on the use of chopper modulation with a purpose of achieving high input impedance, low-noise and high CMRR

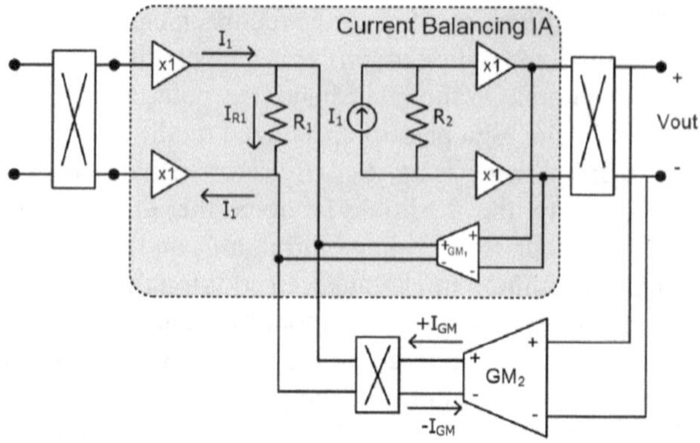

Fig. 4.11 Chopper modulated instrumentation amplifier of [38]. The DC servo is based on current

is presented in Fig. 4.11. The main elements of the current type implementation are a current balancing instrumentation amplifier and a DC servo, which can be implemented with a transconductance stage with LPF characteristics [38]. The current feedback instrumentation amplifier uses a transconductance stage, where the input voltage is converted into current on a resistor R1, and a transimpedance stage, where the current created in the first stage is converted into voltage on a second resistor. Therefore, the gain is simply defined by the ratio of two resistors. Unlike the implementation of Fig. 4.9, the input signal only sees the input parasitic capacitance of the buffers, therefore, the input impedance can be much larger. On the other hand, the noise-efficiency factor of this topology also has a dependency on the available headroom for the polarization voltage of the biopotential electrodes. The headroom can only be increased by increasing the available output current from the GM stage, reducing the noise-efficient factor of the amplifier.

4.4.5 Summary and Comparison of Topologies

Up to know, we have discussed different types of instrumentation amplifiers and stated their advantages and disadvantages. From this discussion, it is obvious for the reader that there is not an ideal instrumentation amplifier that could be used for extracting all the biopotential signals, but different instrumentation amplifiers suits better for dedicated applications. In order to be able to match the instrumentation amplifiers with the applications, it is important to understand the relative advantages of different topologies; where the requirements of each application are presented in Table 4.1.

The first comparison is the based on the noise-efficiency factor and the total noise performance of different instrumentation amplifiers. Figure 4.12 shows the noise-efficiency factor of different amplifiers from the literature. The amplifiers are grouped according to their topologies. Each color represents a different instrumentation amplifier topology. In addition, the total noise of the different amplifiers is also represented in the graph by the diameter of the circle. Hence, it is interesting

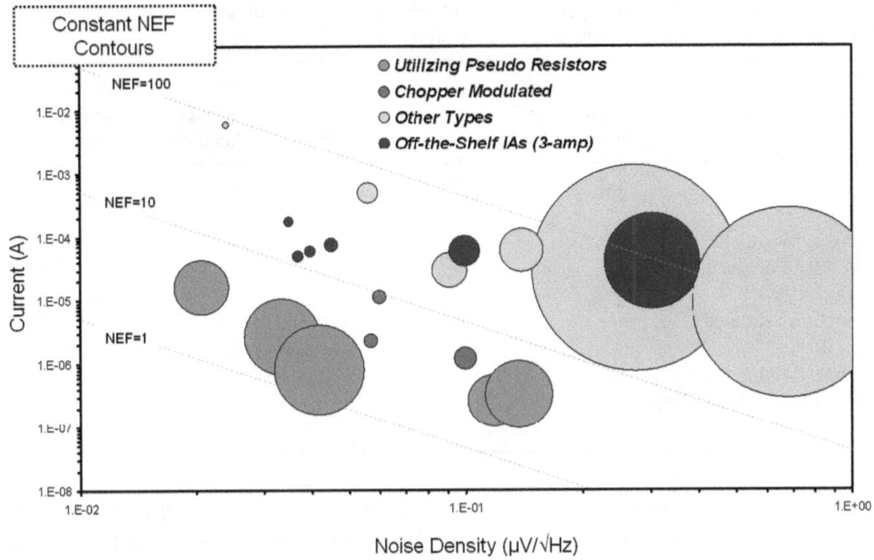

Fig. 4.12 Comparison of Noise-Efficiency-Factor for different biopotential amplifiers. The diameter of the circuit indicates the total noise of the instrumentation amplifier

to see from the graph that two different topologies can achieve low noise levels, which is required for biopotential signals such as EEG. The first group involves off-the-shelf instrumentation amplifiers. These amplifiers' total integrated noise significantly vary, however, if we have a look at the NEF of such instrumentation amplifiers, they are all located above the NEF $= 10$ line, indicating poor power efficiency compared to other architectures. On the other hand, amplifiers using chopper modulation appears to be very attractive both in terms of total integrated noise and overall NEF values. This means that these amplifiers are capable of achieving very good power efficiency together with very low total integrated noise. This indicates a clear advantage for chopper modulated amplifiers for applications requiring very low total integrated noise. On the other hand, the reader should keep in mind that chopper modulation may reduce the differential input impedance of the instrumentation preventing the use of chopper modulated instrumentation for all the biopotential applications.

Therefore, it can be concluded that the application defines the type of instrumentation amplifier that will be incorporated in the system. In order to be more descriptive in this statement, Table 4.2 shows the comparison of different instrumentation amplifiers in terms of their various relevant properties regarding the extraction of biopotential signals.

The main advantage of amplifiers using pseudo resistors are their very large input impedance and polarization voltage headroom, which makes this kind of amplifiers very attractive for application requiring the use of very small electrodes, such as action potentials. On the other hand, the presence of $1/f$ noise prevents the use of these amplifiers for applications requiring very low noise.

Table 4.2 Comparison of different instrumentation amplifier topologies. A negative sign (–) indicates a low/small value, where as a positive sign (+) indicates a high/large value

	Input impedance	1/*f* Noise	CMRR	Polarization voltage headroom	Noise efficiency factor
Pseudo-Resistors (Fig. 4.7)	++	++	– –	++	– –
Chopper Modulated – Voltage Feedback (Fig. 4.9)	– –	– –	+	–	– –
Chopper Modulated (Fig. 4.10)	++	– –	– –	++	+
Chopper Modulated – Current Feedback (Fig. 4.11)	++	– –	++	–	– –

On the other hand, instrumentation amplifiers combining chopper modulation with a voltage based DC servo removes the problem of 1/*f* noise at the cost of significantly lowered differential input impedance. However, some applications requiring low noise, such as the monitoring of local filed potentials or ECoG, do not require as large input impedance as neither EEG nor action potentials. Hence, these types of amplifiers are very much fitting to the applications involving the monitoring of LFP and ECoG. Even some modifications can be made to these instrumentation amplifiers for increasing their differential input impedance by shifting the modulator from input to the virtual ground, Fig. 4.10. However, then the CMRR of these amplifiers are significantly reduced as explained in Section 4.4.4.

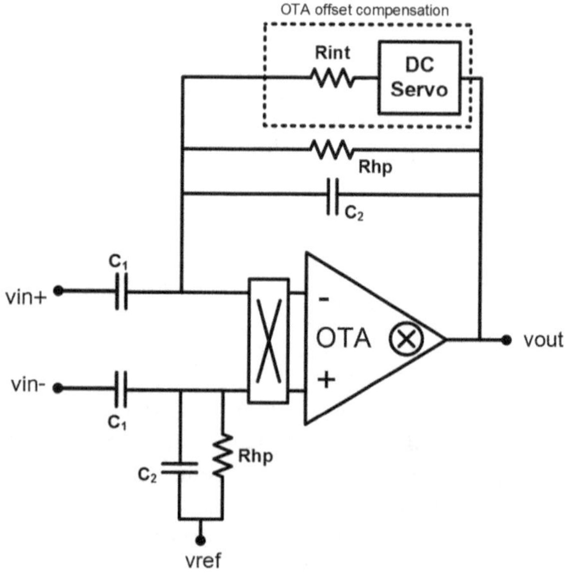

Fig. 4.10 Chopper modulated instrumentation amplifier of [28]

Finally, the use of a current based DC servo may significantly improve the input impedance of chopper modulated instrumentation amplifiers enabling applications requiring minimal total input referred noise and high differential input impedance. One of these applications is the monitoring of EEG signals, where the requirements are the most strict in terms of instrumentation amplifier performance. On the other hand, it is already explained that the use of current based DC servo may reduce the NEF of the instrumentation amplifier, if very large DC polarization voltage headroom is addressed in the design.

4.5 Current Mode Instrumentation Amplifiers

It has been described in the previous section that chopper modulated amplifiers with current based DC servo can be an attractive solution for applications such as EEG signal monitoring since the requirements for this application is low-power dissipation, low-noise, and high input impedance. This Section will introduce the current mode instrumentation amplifiers and explain the implementation of chopper modulated current mode instrumentation amplifiers convenient for EEG monitoring applications.

4.5.1 Open-Loop Current Mode Instrumentation Amplifiers

A current mode instrumentation amplifier is simply a combination of a transconductance stage and a transimpedance stage, where prior is used as an input stage and latter as an output stage Fig. 4.13. This decouples the input stage from the output stage, leading to the decoupling of output voltage swing and input voltage swing. Also, it is interesting to note that both the input transconductance stage and the output transimpedance stages are operating in an open-loop mode, significantly reducing the stability concerns of the implementation. This open-loop nature also leads to a gain-bandwidth product independent of the instrumentation amplifier gain. Another big advantage of this topology is the fact that the CMRR does not depend on the matching of the resistors unlike conventional instrumentation amplifiers as shown in Fig. 4.6. However, it should be noted that this time the CMRR is limited by the matching of CMOS transistors. Besides, flicker noise is still present in this architecture significantly increasing the low frequency noise.

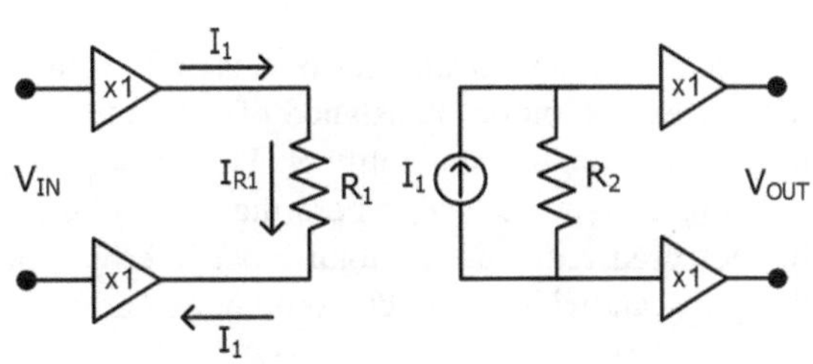

Fig. 4.13 Simplified schematic of a current mode instrumentation amplifier

The implementation of current mode amplifiers reaches back to 1990s. That time the drawbacks of the 3-opamp architecture (mainly the gain-bandwidth product limitation due to the resistances in the feedback loop) lead to the implementation of alternative topologies. One of the first implementations was based on supply current sensing [39]. The transconductance stage was using the supply current sensing to detect the current passing through R_1 and copying it to the output transimpedance stage. This way the gain of the instrumentation could be defined as the ratio of two resistors.

4.5.2 Closed-Loop Current Mode Instrumentation Amplifiers (Current Balancing/Feedback Instrumentation Amplifiers)

The fact that the current-mode instrumentation amplifiers are working in an open-loop mode brings some concerns about the gain accuracy gain and requires the use of input buffers with very low output impedance. An improved approach to current-mode instrumentation amplifier is called a current balancing instrumentation amplifier [40], which is also one of the first integrated instrumentation amplifier architectures.

A typical current balancing instrumentation amplifier is shown in Fig. 4.14. The main difference compared to the conventional current mode instrumentation amplifier is the addition of a feedback loop that supplies the current through R_1 instead of the input buffers. This way the current from the input buffers are balanced by the feedback loop, and the input buffers are only buffering the input voltage to the terminals of R_1 but not supplying any current.

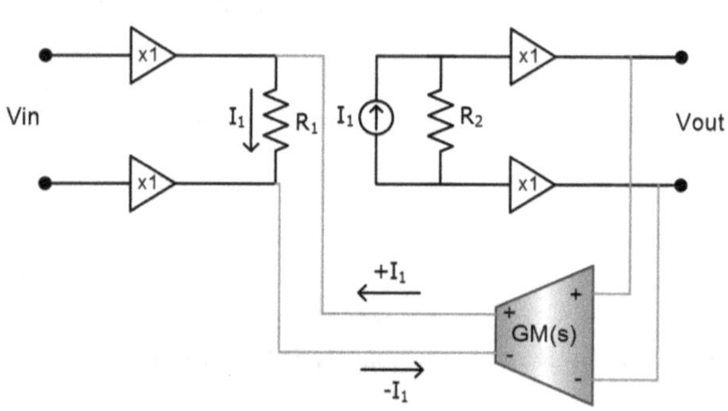

Fig. 4.14 Simplified schematic of a current balancing (feedback) instrumentation amplifier

One of the key advantages of this architecture is the fact the value of R_1 is not limited by the output resistance of the input stage, unlike the open-loop current mode instrumentation amplifiers. This may be particularly helpful for applications targeting low-noise [26], where the total transconductance of the input stage can be increased reducing the total input referred noise of the amplifier and boosting the gain-bandwidth product. Another key advantage of this architecture is the fact that the input stage is not affected by the large input signals, which require large currents over R_1. Since this current is supplied by the feedback loop, even under the presence of large differential input signal, the CMRR of this amplifier can be very large.

Several current feedback instrumentation amplifiers exist in the literature [12, 38, 34, 41]. All these amplifiers are focusing on the implementation of current balancing architecture towards achieving high performance instrumentation amplifiers. However, not necessarily all of them are targeting optimum power efficiency. It was [12] who has first introduced the definition of NEF and emphasizes the importance of power efficiency in an instrumentation amplifier. Since then low-power instrumentation design uses NEF as an indicator to show how good the power efficiency of an instrumentation amplifier is [23]. This particular target was the main driver for the development of instrumentation amplifiers dedicated to portable and implantable applications during the recent years.

Due to all of its advantages over other topologies, the current balancing architecture can be an interesting choice towards the realization of high-performance and low-power instrumentation amplifiers. Figure 4.15 shows an example of such instrumentation amplifier proposed by [38]. In this figure, the input buffers of Fig. 4.14 are represented by transistors M_1 and M_2, and the feedback transconductance is represented by transistors M_3 and M_4. Resistors R_1 and R_2 implements the gain resistors and their ratio defines the gain of the instrumentation amplifier. An important property of this instrumentation amplifier is the fact that the parallel branches are minimized compared to the previously published instrumentation amplifiers. This significantly reduces the total power of the instrumentation amplifier, while realizing all the properties of current balancing architecture. For instance, the current through M_1 and M_2 are not disturbed even tough a large differential voltage is applied to the instrumentation amplifier. This is due to the fact that the current through R_1 is supplied by the feedback transconductance, implemented by transistors M_3 and M_4.

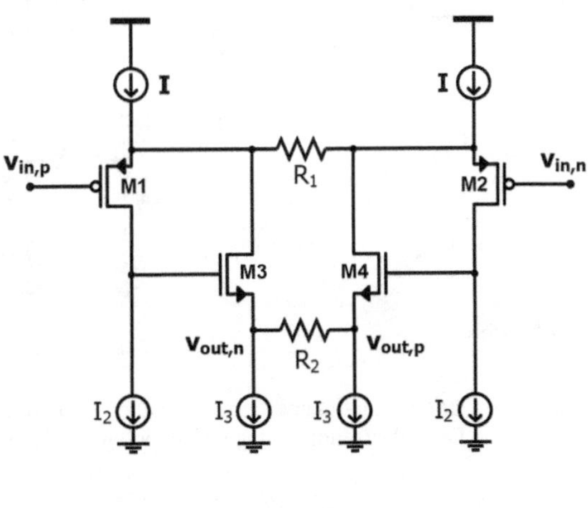

Fig. 4.15 A low-power current balancing instrumentation amplifier architecture [38]

4.5.3 Chopper Modulated Current Balancing Instrumentation Amplifiers

As the name implies, current balancing instrumentation amplifier has the advantage of balanced input stage transistors. This improves the CMRR of the instrumentation under large differential input signals. However, process related mismatches still affects the maximum achievable CMRR of the instrumentation amplifier. Besides, the flicker noise of CMOS transistor may dominate the total in-band noise of the instrumentation amplifier, especially for biomedical applications, where the signals are appearing at very low frequencies. Hence, the application of chopper modulation to current balancing instrumentation amplifiers is inevitable in order to eliminate the flicker noise at low frequencies and also in order to remove the performance limitations due to component mismatches, such as CMRR limitation.

A challenge arises when applying chopper modulation to instrumentation amplifiers addressing biopotential monitoring applications. This is due to the polarization voltage, existing at the electrode-electrolyte interface and needs to be rejected by amplifier to prevent saturation. This challenge can be tackled by incorporating high-pass filter characteristics together with chopper modulation. This has been already introduced in Fig. 4.11, where a DC servo has been included in the feedback path of the current balancing instrumentation amplifier. Figure 4.16 shows the detailed implementation of the chopper modulated current balancing instrumentation amplifier. The transconductance, GM_1 of Fig. 4.11, has been realized with transistors M_3 and M_4, where as the transconductance, GM_2 of Fig. 4.11, has been implemented using a transconductance amplifier and a transconductance stage as highlighted with the blue color. Therefore, while the transconductance stage inside the current balancing instrumentation amplifier is balancing the AC current through R_1, the DC

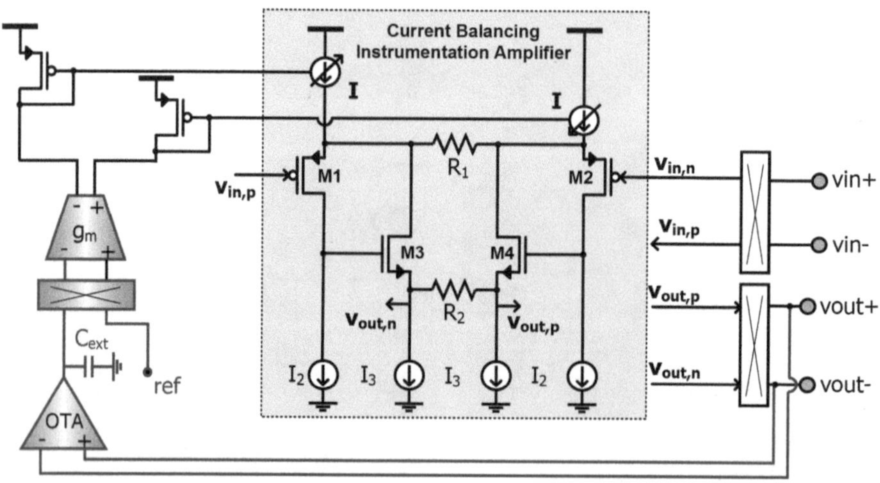

Fig. 4.16 Chopper modulated current balancing instrumentation amplifier with a DC servo for filtering the DC polarization voltage of biopotential electrodes [38]

component of the current is balanced by the DC servo-loop. Since, the internal feedback loop of the current balancing instrumentation amplifier is not involved in the DC component of the current through R_1, the current passing through R_2 has only the AC component of the current passing through R_1. Therefore, DC component of the input voltage is not amplified, leading to a chopper modulated instrumentation amplifier with high-pass filter properties. It can be calculated that the high-pass filter cut-off frequency can be realized as:

$$f_{\mathrm{HPF}} = gm \times R_2 \left(\frac{gm_{\mathrm{OTA}}}{2\pi C_{\mathrm{ext}}} \right) \tag{4.9}$$

which clearly indicates that by selecting a proper C_{ext} value, the high-pass cut-off frequency of the instrumentation can be freely adjusted.

4.6 Examples of ICs for Biopotential Acquisition

The development of integrated circuits for biopotential acquisition, mainly involves the realization of high performance and low-power building blocks. This arises from the fact that the biopotential signals are extremely weak and correlated with various aggressors such as flicker noise and interference. During the recent years several researchers were targeting the development of low-power integrated circuits for the extraction of different biopotential signals [42]. These integrated circuits were sometimes dedicated to a specific biopotential signals and other times targeting the extraction of several different biopotential signals using the same integrated circuit architecture [27, 38]. Application drivers are mainly the monitoring of neural activity, portable ECG and EEG monitoring devices, and implantable ECoG and/or LPF monitoring.

One of the first complete integrated circuit for the acquisition of biopotential signals have been presented during the recent years [25]. This system was involving 100 recording channels, analog-to-digital converters and telemetry system for complete extraction of action potentials. Later, the improvement to the circuit has been realized by involving the part of signal feature extraction in the ASIC leading to further increase in the smartness of the system [43]. These miniaturized systems enable the monitoring of small animals with interesting neural properties, correlating their movement with activity of neurons [36]. The front-end instrumentation amplifiers of this topology use the architecture shown in Fig. 4.7. As discussed through-out the text this choice was due to the requirement for filtering supply level polarization voltage, as well as, for the need of very high input impedance.

Another interesting ASIC architecture is targeting applications for implantable brain monitoring in the area of deep-brain stimulation [37]. ASIC uses the instrumentation amplifier architecture in Fig. 4.9 to be able to achieve very low flicker noise and still be able to filter limited polarization voltage of 15 mV from DBS electrodes. The large DBS electrodes were characterized to have mean value of 0.1 mV with a standard deviation of 2 mV, which states that the design range comfortably includes the maximum possible offset voltage from electrodes. Hence, while achieving very good power efficiency, this ASIC can achieve very good performance specifications. Another important property of this ASIC is the introduction of analog signal processing. The use of chopper modulation with synchronous demodulation enables the band-power extraction of biopotential signals [44]. This way algorithms relying on band-power fluctuation can be implemented in digital domain without the requirement for the band-power extraction processes. As a result the overall algorithm power can be minimized.

A similar ASIC has been proposed by [28], where the target is the detection of epileptic seizures through non-invasive surface electrodes on the skull. The aim is again to monitor the activity of EEG signals in specific frequency bands and decide on the presence of epileptic seizures. However, this time the band-power monitoring has not been performed in the analog domain but digital band-pass filters are employed to monitor the signal activity in a specific frequency band. As an instrumentation amplifier, the architecture shown in Fig. 4.10 has been employed. Therefore, both the input impedance and the offset filter capability of the architecture have been boosted compared to the instrumentation amplifier of Fig. 4.9. On the other hand, the noise efficiency factor and the CMRR are reduced. The prior is due to the increased affect of parasitic capacitance at the virtual ground and latter is due to the mismatch of the passives.

Another complete ASIC architecture targets the portable monitoring of biopotential signals [35], Fig. 4.17. It uses a modified version of the instrumentation

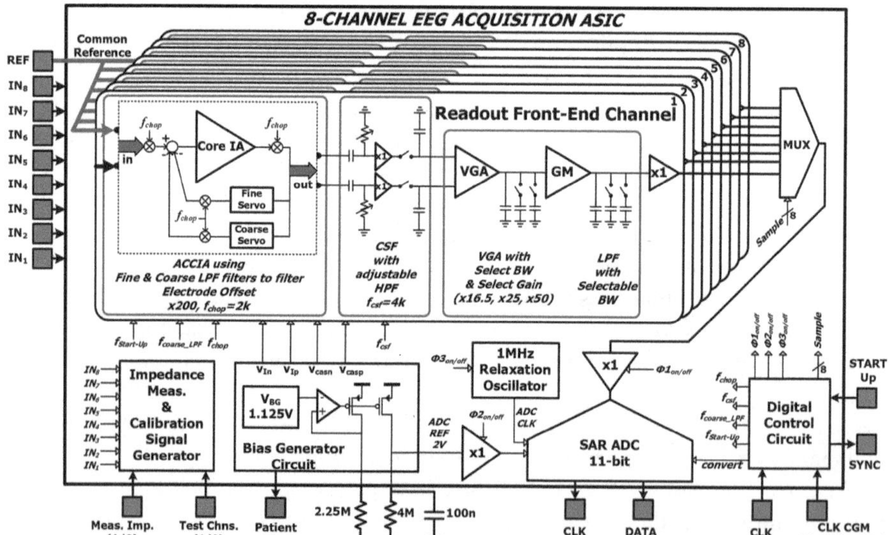

Fig. 4.17 Architecture of the ASIC presented in [EEGASIC] for the portable acquisition of EEG signals

amplifier presented in Fig. 4.16. Hence, the ASIC was targeting very low flicker noise together with very high input impedance and CMRR. The polarization voltage filtering capability of the ASIC was designed as 50 mV according to the polarization voltage measurements from several Ag/AgCl electrodes [38]. In addition to extracting high quality EEG signals, this ASIC also considers the testability and usability. For the prior, the ASIC includes a calibration signal generator, where a differential square wave is connected to the inputs of all channels. This way the operation of the channels, ADC, and all related blocks can be controlled. On the other hand, impedance measurement tries to improve the usability of the system. In common practice, the connectivity of electrodes with the tissue needs to be checked to ensure the signal quality. The impedance measurement mode provides an automated means for measuring the impedance of each electrode.

At last but not the least, several designs target the low-voltage operation of biopotential acquisition system [26–28]. For this reason, low-voltage instrumentation amplifiers are developed, together with low-voltage ADC's. This way the analog instrumentation can be incorporated on the same technology as the low-voltage digital signal processing circuits reducing the overall system power. However, it should be noted that the only topologies that could achieve such low-voltage operation are the architectures of Figs. 4.7 and 4.10. The reader should be aware of the fact that these instrumentation amplifiers are very low-power but their performance is not as high as other chopper modulated instrumentation amplifiers.

4.7 Conclusion

The design of CMOS readout circuits for the acquisition of biopotential signals is significantly affect by the application and monitoring environment. The choice of instrumentation amplifier, which is the most critical building block in terms of signal clarity, heavily depends on the application, the electrode type, and the invasiveness of the measurement.

In this Chapter, we have presented the details of these dependencies, i.e., how the application affects the choice of the instrumentation amplifier selection. The reader should realize from this chapter the fact that the knowledge of different instrumentation amplifier architectures is necessary for designing an acquisition circuit for biopotential applications. Only then the delicate balance between the signal quality and the power dissipation can be optimized by the designer.

References

1. Park S, and Jayaraman S (2003) Enhancing the quality of life through wearable technology. IEEE Eng Med Biol Mag. 22(3):41–48, May–June 2003
2. Gyselinckx B., van Hoof C., Ryckaert J., Yazicioglu R., Fiorini P., Leonov V (2005) Human++: autonomous wireless sensors for body area networks. In: custom integrated circuits conference. Proceedings of the IEEE 2005, 18–21 Sept. 2005. pp 13–19

3. Mundt C., Montgomery K., Udoh U., Barker V, Thonier G, Tellier A., Ricks R, Darling R, Cagle Y., Cabrol N., Ruoss S., Swain J, Hines J, Kovacs G (2005) A multiparameter wearable physiologic monitoring system for space and terrestrial applications. IEEE Trans vol. 9(3):382–391, Sept. 2005
4. Paradiso R, Giannicola L, Taccini N (2005) A wearable health care system based on knitted integrated sensors. IEEE Trans Inform Technol Biomed.3:337–344
5. Gyselinckx B, Vullers R, Hoof C, Ryckaert J, Yazicioglu R Fiorini P, Leonov V (2006) Human++: Emerging technology for body area networks., In: Very large scale integration, 2006 IFIP International Conference, Oct. 2006, pp 175–180
6. Webster JG (1992) Medical instrumentation: application and design, 2nd edn. Boston (Mass.), Houghton Mifflin
7. Huhta JC, Webster JG (1973) 60-Hz interference in electrocardiography. IEEE Trans Bio-Med Eng BME-20(2):91–101, March 1973
8. Van Rijn AC, Kuiper AP, Dankers TE, Grimbergen CA (1996) Low-cost active electrodes improves the resolution in biopotential recordings. IEEE EMBC
9. Metting van Rijn AC, Peper A, Grimbergen (1990) High-quality recording of bioelectric events; part 1, interference reduction, theory, and practice, Med Biol Eng Comput 28:389–397
10. Winter BB, Webster JG (1983) Driven-right-leg circuit design. IEEE Trans Bio-Med Eng. BME-30(1):62–66, Jan. 1983
11. Razavi B (2001) Desing of analog CMOS integrated circuits. McGraw-Hill Science/Engineering Math; 1 Edition (August 15, 2000)
12. Steyaert M, Sansen W (1987) A micropower low-noise monolithic instrumentation amplifier for medical purposes., IEEE J Solid-State Circuit 22(6):1163–1168, Dec 1987
13. Burr-Brown (1997) INA122: single supply, micropower instrumentation amplifier, online, Oct. 1997
14. Burke M, Gleeson D (2000) A micropower dry-electrode ECG preamplifier, IEEE T, Bio-Med Eng. 47(2):155–162, Feb. 2000
15. Pallas-Areny R, Webster J (1993) AC instrumentation amplifier for bioimpedance measurements, IEEE Trans Bio-Med Eng. 40(8):830–833, Aug. 1993
16. Spinelli E, Martinez N, Mayosky M, Pallas-Areny R (2004) A novel fully differential biopotential amplifier with DC suppression. IEEE Trans Bio-Med Eng. 51(8):1444–1448, Aug 2004
17. Spinelli E, Pallas-Areny R, Mayosky M 2003 AC-coupled front-end for biopotential measurements. IEEE Trans Bio-Med Eng. 50(3):391–395, March 2003
18. Huijsing JH (2001) Operational amplifiers: theory and design. Kluwer Academic, Springer; 1 edition (December 2000)
19. Mancini R Don't fall in love with one type of instrumentation amp available online: http://www.edn.com/contents/images/217678.pdf
20. Van Peteghem P, Verbauwhede I, Sansen W (1985) Micropower high performance SC building block for integrated low-level signal processing. IEEE J Solid-State Circuit 20(4):837–844, Aug 1985
21. Degrauwe M, Vittoz E, Verbauwhede I (1985) A micropower CMOS instrumentation amplifier. EEE J Solid-State Circuits 20(3):805-807, Jun 1985
22. Enz C, Temes G (1996) Circuit techniques for reducing the effects of opamp imperfections: autozeroing, correlated double sampling, and chopper stabilization. Proc IEEE, 84(11): 1584–1614, Nov 1996
23. Harrison R, Charles C (2003) A low-power low-noise CMOS amplifier for neural recording applications. IEEE J Solid State Circuits 38(6):958–965, June 2003
24. Olsson R, Gulari A, Wise K (2003) A fully-integrated bandpass amplifier for extracellular neural recording. In: Neural Engineering, 2003. Conference Proceedings. First International IEEE EMBS Conference, 20–22 March 2003, pp. 165–168
25. Harrison RR, Watkins PT, Kier RJ, Lovejoy RO, Black DJ, Greger B, Solzbacher F (2007) A Low-Power Integrated Circuit for a Wireless 100-Electrode Neural Recording System. IEEE J Solid-State Circuits 42(1):123–133, Jan. 2007

26. Wu H, Xu PA (2006) 1 V 2.3 μW Biomedical Signal Acquisition IC. In: Proceedings of the 2006 IEEE International Solid-State Circuit Conference, Feb. 5–9, pp. 119–128, 2006
27. Zou X, Xu X, Yao L, Lian Y (2009) A 1-V 450-nW Fully Integrated Programmable Biomedical Sensor Interface Chip. IEEE J Solid-State Circuits 44(4):1067–1077, April 2009.
28. Verma N, Shoeb A, Guttag J, and Chandrakasan A (2009) A Micro-power EEG Acquisition SoC with Integrated Seizure Detection Processor for Continuous Patient Monitoring. 2009 Symposium on VLSI Circuits, June 2009.
29. Menolfi C, Huang Q (1999) A fully integrated, untrimmed CMOS instrumentation amplifier with submicrovolt offset. IEEE J Solid State Circuits 34(3):415–420, March 1999
30. Enz C, Vittoz E, Krummenacher F (1987) A CMOS chopper amplifier. IEEE J Solid-State Circuits 22(3):335–342, June 1987
31. Menolfi C, Huang Q (1997) A low-noise CMOS instrumentation amplifier for thermoelectric infrared detectors. IEEE J Solid State Circuits 32(7)968–976, July 1997.
32. Menolfi C (2000) Low noise CMOS chopper instrumentation amplifiers for thermoelectric microsensors. Ph.D. dissertation, Swiss Federal Institute of Technology, ETH
33. Menolfi C and Huang Q (1997), A low-noise CMOS instrumentation amplifier for thermoelectric infrared detectors. IEEE J Solid State Circuits 32(7)968–976, July 1997
34. Eatock F 1973 A monolithic instrumentation amplifier with low input current. Solid-state circuits conference. Digest of technical papers. 1973 IEEE International, XVI: 148–149, Feb 1973
35. Yazicioglu RF, Merken P, Puers B, Van Hoof C (2008) A 200 μW Eight-channel EEG acquisition ASIC for ambulatory EEG systems. IEEE J Solid State Circuits 43(12):3025–3038, Dec 2008
36. Fotowat H, Harrison RR, and Gabbiani F (2009) Measuring neural correlates of insect escape behaviors using a miniature telemetry system. In: Proceedings of the 35th annual northeast bioengineering conference, Cambridge, MA
37. Denison T, Consoer K, Kelly A, Hachenburg A, Santa W (2007) 2.2 μW 97 nV/Hz, chopper stabilized instrumentation amplifier for EEG detection in chronic implants. In: Solid-state circuits, 2007 IEEE International Conference Digest of Technical Papers, 2007, pp. 162–163
38. Yazicioglu RF, Merken P, Puers R, Van Hoof C (2007) A 60 μW 60 nV/Hz readout front-end for portable biopotential acquisition systems. IEEE J Solid-State Circuits 42(5):1100–1110, May 2007
39. Toumazou C, Lidgey FJ, and Makris CA (1989) Current-mode instrumentation amplifier, IEE Colloquium on Current Mode Analogue Circuits, pp. 8/1-8/5, Feb
40. Krabbe H (1971) A high-performance monolithic instrumentation amplifier. In: Solid-State Circuits Conference. Digest of Technical Papers. 1971 IEEE International XIV:186–187, Feb 1971
41. Martins R, Selberherr S, Vaz F (1998) A CMOS IC for portable EEG acquisition systems. IEEE T Instrum Meas 47(5):1191–1196, Oct 1998
42. Jochum T, Denison T, Wolf P (2009) Integrated circuit amplifiers for multi-electrode intracortical recording. J Neural Eng 6
43. Chae M, Liu W, Yang Z, Chen T, Kim J, Sivaprakasam M, and Yuce MR (2008), A 128-channel 6 mW Wireless Neural Recording IC with On-the-fly Spike Sorting and UWB Transmitter. IEEE Int Solid-State Circuits Conf (ISSCC'08), Feb 2008
44. A.-T. Avestruz, Santa W, Carlson D, Jensen R, Stanslaski S, Helfenstine A, and Denison T (2008) A 5 μW/Channel spectral analysis IC for chronic bidirectional brain–machine interfaces. IEEE J Solid-State Circuits 43(12)3006–3024, Dec 2008

Chapter 5
Low-Power ADCs for Bio-Medical Applications

J. Craninckx and G. Van der Plas

In this chapter, recent innovations reducing power consumption in A/D converters will be discussed. Indeed, in many applications the function performing a conversion from the analog continuous-time domain to the discrete-time digital domain takes a large proportion of the power consumption. Especially for biomedical systems an aggressive reduction in power consumption of all blocks including A/D converters opens up a window for higher performance and more versatile solutions.

A/D converters find multiple applications in biomedical systems. They are indispensible in the signal processing chain where sensor signals are amplified and sometimes filtered before they are converted to the digital domain for further processing and interpretation. They are also used in low-power wireless communication circuits, systems which enable to build complex applications that use (multiple) sensor data from inside and around the body. In this case the radio signals are typically down-converted and filtered before being converted to the digital domain for signal processing. In both applications, power savings with increased sample rate and/or resolution are instrumental for improving biomedical systems.

In this chapter two new A/D converter architectures will be introduced: the Charge-Sharing Successive Approximation ADC, in Section 5.2, and the Comparator-based Asynchronous Binary Search ADC in Section 5.3. Both architectures in their respective performance ranges have pushed the boundaries of achievable energy efficiency. Before discussing these architectures in detail, in the next section the specifications of A/D converters are introduced, including the figure of merit energy per conversion step E_Q expressing the energy efficiency of converters.

J. Craninckx (✉)
Interuniversity Microelectronics Center (IMEC), Leuven, Belgium
e-mail: jan.craninckx@imec.be

H.-J. Yoo, C. van Hoof (eds.), *Bio-Medical CMOS ICs*,
Integrated Circuits and Systems, DOI 10.1007/978-1-4419-6597-4_5,

5.1 ADC Specifications

An A/D converter must translate a continuous time, continuous value analog signal to a discrete time, discrete value signal. It must therefore sample and quantize the signal. While any techniques to accomplish this can be easily generalized to deal with discrete time input signals, they can not be generalized to yield continuous time output signals. We also assume that the quantization step should be constant, although other relationships (e.g. logarithmic) are possible and sometimes desired.

In this section the ADC specifications used commonly throughout this chapter will be defined. The most basic ADC specifications relate to the properties of an ideal sampler and quantizer. These will be introduced first, and then the common spectral properties of the ADC output will be explained. Finally ADC specifications detailing practical ADC implementations will be defined.

5.1.1 Ideal ADC Specifications

An ideal sampler is described by its sampling period T_S: the difference between the times at which the input is sampled. This also corresponds to the time between evaluation of the discrete time output signal and is more commonly expressed as a sampling frequency or speed $f_S = 1/T_S$, in samples per second.

An ideal quantizer can be completely characterized by its resolution and input range. The resolution is usually expressed as a number of bits $N = \log_2(m)$, with m being the number of discrete output values. The input range $[in_{\min}; in_{\max}]$ describes the lower and upper bounds of the input signal for which the quantizer has constant quantization steps. This is illustrated in Fig. 5.1 for a 3-bit resolution. Each vertical transition in this input output curve will be called a quantizer threshold while the distance between these thresholds is called a least significant bit $LSB = \dfrac{in_{\max} - in_{\min}}{m}$.

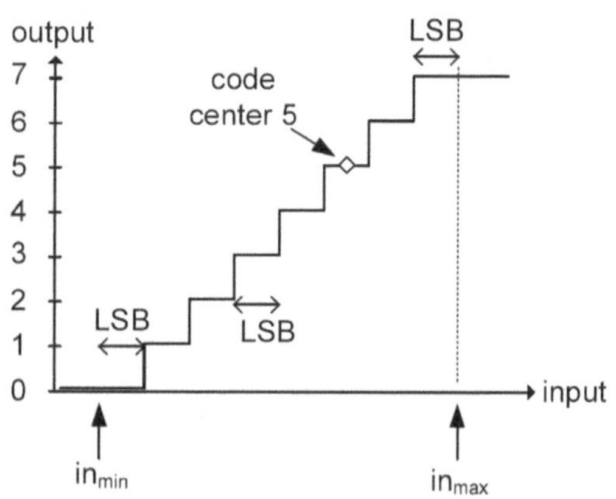

Fig. 5.1 Input output curve of an ideal 3-bit quantizer

The analog input value in the center of two consecutive thresholds is called the code center, the area between them is referred to as a code of a certain output value.

5.1.2 Practical ADC Specifications

Practical ADCs are never ideal and must be described by a larger set of metrics than ideal ADCs. Some of these will detail how the conversion of a certain ADC differs from that of an ideal ADC as depicted in Fig. 5.1, while others are related to the spectral properties of the ADC input and output signal and are frequency-dependent.

Two specifications related to the conversion of a practical ADC will be defined: the integral non-linearity (INL) and the differential non-linearity (DNL). Figure 5.2 shows a section of the input/output curve of a non-ideal ADC, with the dashed line representing the desired (ideal) case. The vector of differences between each code center of a practical ADC and the corresponding code center of the ideal case is defined to be the INL. The DNL vector is defined as the vector containing the difference between the width of each code and 1 LSB, or in other words: the difference between two consecutive thresholds minus one LSB.

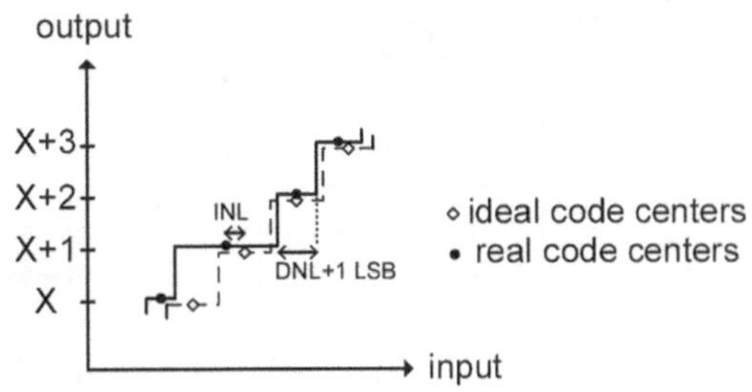

Fig. 5.2 Input output curves of an ideal quantizer (*dashed*) and a non-ideal quantizer (*solid*)

Since INL and DNL define the static linearity of an ADC they are related to the low frequency SFDR and SNDR in an ADC. In [1] it is shown that the low frequency SNDR degrades by less than 3 dB if the INL is within ±0.5 LSB. Additionally, if the INL is within ±0.5 LSB a converter is monotonic, meaning that the static output will not decrease when the input is increased. It should however be noted that not all monotonic converters will have an INL bound within ±0.5.

Considering the ADC's spectral performance, a sinewave with a certain frequency applied to the ADC input should translate to the output with identical spectral properties. The output power spectral density (PSD) will however have a certain power which can be attributed to the input signal, but will also contain some additional and unintended power components. This additional power can be attributed to two types. The first type is the noise energy which is uncorrelated with the input signal and represents the quantization noise, i.e. the noise energy caused by the effect of "rounding" continuous-amplitude analog signals to discrete-level digital numbers. The second type is distortion energy, which is correlated with the input signal. It can be located at specific frequencies with respect to the input frequency, depending on the exact source of the distortion. Based on these distinctions some spectral properties of ADC outputs can be defined: the signal to noise and distortion ratio (SNDR) and spurious-free dynamic range (SFDR), both of which are a function of the input frequency. Based on how the SNDR varies as a function of the frequency, the effective resolution bandwidth (ERBW) is defined.

The SNDR is simply defined as the ratio between the signal power and the sum of all other power in the ADC output, typically expressed in decibels (dB). It can be shown [1] that if a sinewave is applied to an ideal N-bit ADC, the maximum SNDR is given by $6.02 \cdot N + 1.76\,dB$. Based on this relationship, the effective number of bits (ENOB) is derived from the SNDR. It is defined as $ENOB = \frac{SNDR-1.76}{6.02}$. The SFDR is defined as the ratio between the maximum signal power density to the maximum non-signal, or spurious, power density, as shown in Fig. 5.3. It can be easily seen that the SNDR and the SFDR are not independent. Since the SNDR includes the distortion specified by the SFDR, it can not be larger than the SFDR.

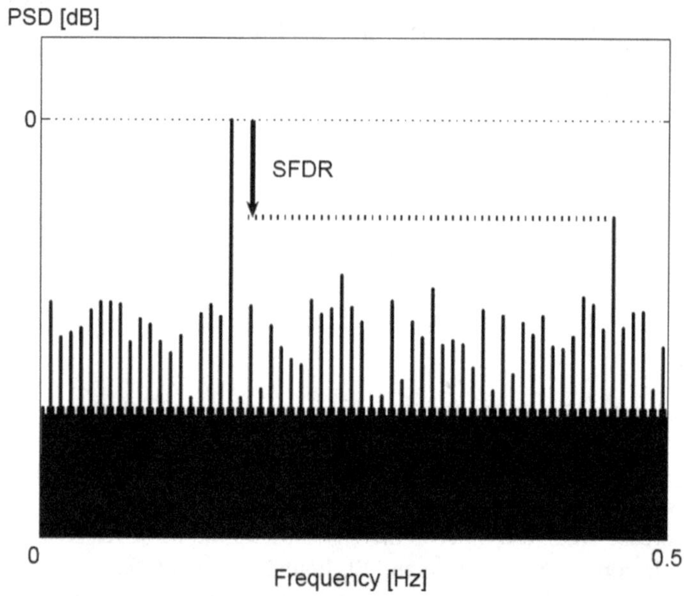

Fig. 5.3 Power spectral density with SFDR marked

Both the SNDR and the SFDR are defined for each given input frequency, and typically these degrade as the input frequency increases. A final spectral property for ADCs is the effective resolution bandwidth (ERBW), which is the frequency at which the SNDR of an ADC is degraded by 3 dB, or the ENOB by 0.5 bits with respect to their values at very low frequencies. If an ADCs ERBW is equal to or larger than its Nyquist frequency, *fs/2*, an ADC is called a "Nyquist converter".

5.1.3 ADC Implementation Issues

The most obvious practical issue is the cost or area of an ADC implementation in a given CMOS technology: a smaller area indicates a lower cost. A second implementation issue for ADCs is the input impedance. The ADC input must always be driven by a preceding stage in the larger system. All else being equal, an ADC with a high input impedance is usually preferred. This high impedance will reduce the power required for this preceding stage. A final practical issue is the power consumption of the ADC. Given a tight energy budget, minimizing the power consumption is often the critical issue in ADC design.

A derived performance indication related to the power consumption P is the energy per conversion step $E_Q = \frac{P}{2^{\mathrm{ENOB}} 2 f_{\mathrm{in}}}$ with f_{in} equal to the Nyquist frequency or the ERBW, whichever is less. This energy of conversion step is expressed as a number of Joules per conversion step and is often used as a figure of merit (FoM) to compare different implementations.

Some discussion is possible about the suitability of the E_Q as a figure of merit. Figures of merit are used to compare different designs, and a good figure of merit can be recognized in the fact that it should remain roughly constant for many points in the design space. In this case, the expected figure of merit should be roughly constant from one combination of sampling rate and ENOB to another.

For example, using E_Q as a figure of merit, one would expect that increasing the ENOB by one would double the power consumption. The physical limits, however, usually dictate that the power consumption increases by a factor much closer to 4 than to 2. It is therefore somewhat surprising that in the state of the art overview [2] E_Q was empirically shown to be more or less constant throughout a wide range of resolutions, sampling frequencies and architectures. Since then, the energy per conversion step of ADCs has improved vastly, but the conclusion remains valid. Today, state of the art efficiencies are in the order of tens of femtojoules per conversion step [3–7, 25].

In this work, a number of design techniques and example prototypes will be introduced that bring the E_Q down to femto-joule range. It should also be clear that given the extensive number of ADC specifications, a figure of merit taking into account only three of these can not exhaustively compare with different implementations as many ADC specifications such as the required area, the input range or input impedance are not taken into account at all. These issues will be covered in the next sections.

5.2 Charge-Sharing Successive Approximation ADCs

When it comes to achieving low-power consumption in A/D conversion, the successive approximation (SAR) principle appears to be a very attractive candidate. For an N-bit ADC, only N operations are needed to determine the output word, a process that can be done with little overhead.

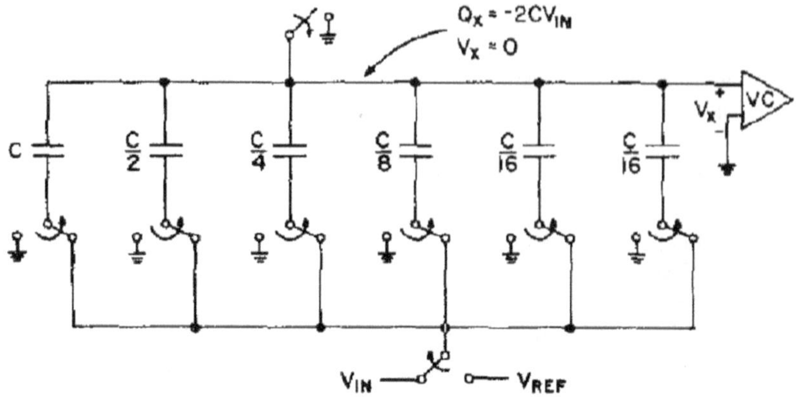

Fig. 5.4 Conceptual 5-bit charge-redistribution SAR ADC [24]

The use of the SAR architecture has been widely adopted since the original introduction of the charge-redistribution ADC in 1975 [24]. In this technique, illustrated Fig. 5.4, the input voltage is first sampled on an array of binary scaled capacitors. Then, by sequentially switching the bottom plates of the capacitor to the reference voltage or to ground, this input voltage can be compared with different levels and in a binary-search the exact input level is determined. During this process, charge is indeed "redistributed" between the two sides of a voltage divider constructed between the capacitors. A full conversion cycle thus consists of 1+ N events (1 input sampling and N comparisons).

This conceptual ADC schematic is deceivingly simple: there is only one active element (a comparator) and it further contains just some passive capacitors and a set of switches. An easy conclusion would be that a really low-power consumption can be obtained. Some aspects not shown in the conceptual circuit however limit this ideal picture. All of them are related to the fact that the total conversion takes 1+ N events, so only a limited amount of time can be attributed to each action.

First, an input voltage buffer is needed in order to make sure that during the (short) input sampling phase the voltage on the (large) capacitor array can settle accurately to the actual value. Second, for the same reason also an accurate and high-speed reference buffer is needed to make sure that a stable reference voltage can be maintained during the charge-redistribution process. Finally the clock generation must be taken into account. Since the SAR algorithm runs at a speed higher than the actual conversion rate, a high-frequency clock is needed. Of course, depending on the actual number of bits or conversion rate needed in a specific application, these effects become more or less important. But they are not always taken into consideration when power consumption numbers are reported, so care must be taken.

In the remainder of this section, a novel charge-sharing SAR architecture will be presented that solves the charge-redistribution problems, and will hence lead to a major breakthrough in the realization of low-power ADCs.

5.2.1 Basic Operation Principle

The SAR ADC topology proposed does not need any high-speed buffers. Instead of operating in the voltage-domain, the useful signal is represented by *charge*. Its operation is explained in the following steps ([3, 8]).

5.2.1.1 Input Sampling

An accurate charge proportional to the input signal must be created, and the most obvious way is of course to sample the input voltage on a capacitor. However, to avoid the problems of high-speed settling, a full conversion period must be available for this. However, if we manage to avoid that the input must be sampled on the binary capacitor array, but instead can be sampled on a separate capacitor, a time-interleaved Sample & Hold (S/H) can be used.

The time-interleaved sampling circuit, shown in Fig. 5.5, processes a $0.8V_{pp}$ single-ended input signal with 0.5 V common mode voltage V_{CM}. The circuit is based on bootstrapped NMOS switches at the input (BM1) [9], offering a low, signal-independent on-resistance, and pass-gates at the output (TG1). When phase Φ_2 is active, the input signal is tracked on capacitors $C_{TL1} + C_{TH1}$ while the charge previously sampled on $C_{TL2} + C_{TH2}$ is already available for conversion. So after sampling, the capacitors C_T hold a charge proportional to the input:

$$Q_{in} = C_T \cdot V_{in} \tag{5.1}$$

An extension to the basic operation is the fact that, after the MSB will have been determined and the differential voltage on the nodes $V_{QP,N}$ becomes small enough to

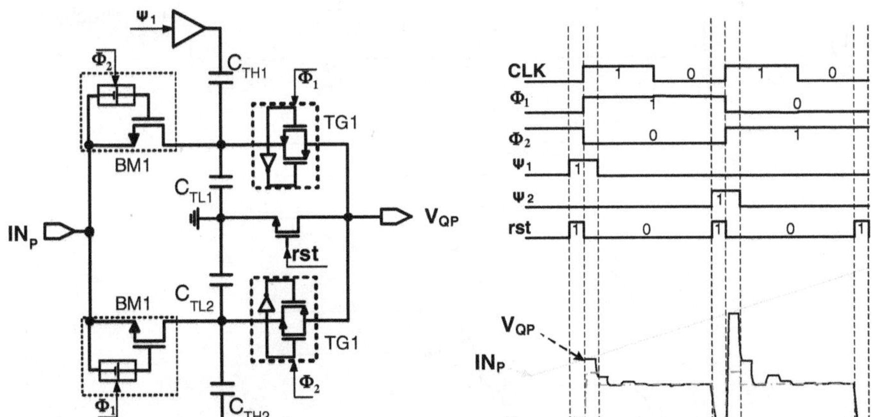

Fig. 5.5 Single ended schematic of the time-interleaved S/H and relevant waveforms

avoid clipping, the common mode voltage V_{CM} is lowered. To create this CM shift, the connection of the capacitors C_{TH} is switched from V_{DD} to ground by the signal $\Psi_{1,2}$. With this lower common mode, the on-resistance of NMOS switches that will be connected to the sampling nodes will significantly decrease, an effect that will show its benefit once the core of the SAR operation is explained.

5.2.1.2 Successive Approximation, MSB

Now the A/D conversion can start. The most significant bit is determined first by using the comparator to check if the differential voltage on C_T is positive or negative. Depending on that decision, a signal proportional to half of the input range must be subtracted or added to the input signal.

In order to avoid tough specs on the voltage settling time with opamps, we will not add a certain voltage to the signal, but instead add a charge. This operation can be done by passive charge sharing and happens very fast without taking any power. As shown in Fig. 5.6, this charge is taken from a capacitor of a certain size that has been pre-charged previously to a fixed reference voltage.

If the comparator input voltage is positive ($V_{QP} > V_{QN}$), the switches controlled by c_{Op} are closed, after which the charges on the three capacitors (C_{TP}, C_{TN} and

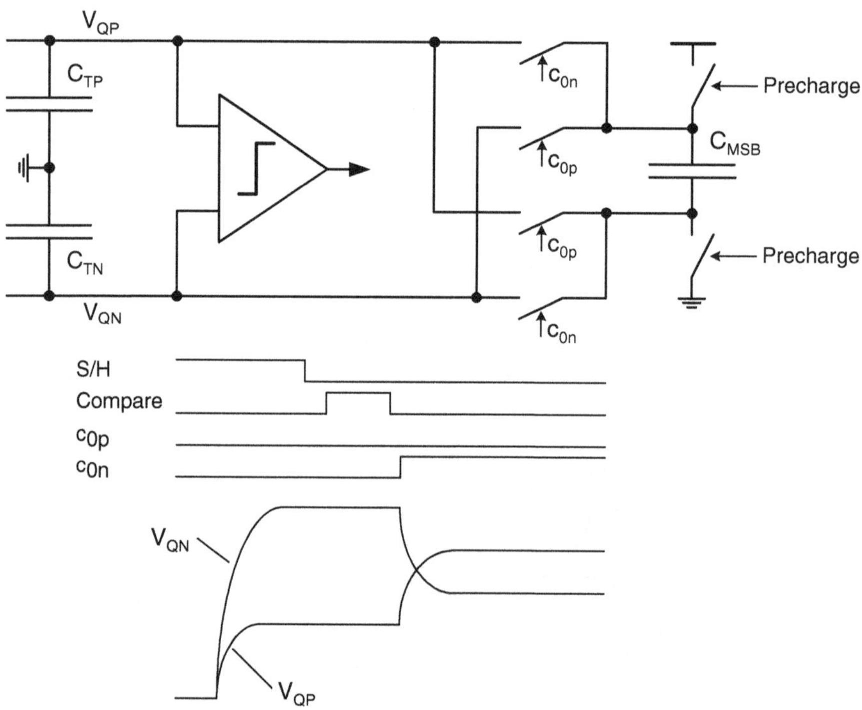

Fig. 5.6 Comparison and charge-sharing action for MSB detection

C_{MSB}) equalizes. The voltage V_{QN} will rise and V_{QP} will fall. Likewise, if the comparator input voltage is negative ($V_{QP} < V_{QN}$) as in the figure, the switches controlled by c_{0n} are closed, after which voltage V_{QN} will fall and V_{QP} will rise. The total charge on the capacitor set is now given by

$$Q_0 = C_S \cdot V_{in} + S_0 \cdot C_0 \cdot V_{ref} \tag{5.2}$$

where S_0 represents the output of the comparator (+1 for positive, −1 for negative signal). The time needed for this charge-sharing process is of course proportional to the time constant given by the product of the switches on-resistance and the capacitor size. Since NMOS switches are used for this, the on-resistance is improved by lowering the common-mode voltage as described earlier, and hence the ADC speed increases.

It is crucial to note here that the reference voltage itself is not loaded by this action. During the pre-charge phase, that reference voltage was sampled on the capacitor C_{MSB}, and it is that reference *charge* that is now used to provide the feedback DAC action required in the SAR ADC. The reference is thus independent on the input voltage, and the constraints posed on the reference buffer are almost negligible. Even if an error would have been made during the reference sampling, because of its independency on the input, it would only result in a gain error of the ADC, which is not important.

5.2.1.3 Successive Approximation, MSB-1

The following bit is determined in the same way, but now a precharged capacitor of size $C_{MSB\text{-}1} = C_{MSB}/2$ will be used, since in the binary search for the correct digital output code, the range is now reduced by a factor 2. The sign of the voltage on the capacitor set formed by C_S and C_{MSB} represents the sign of the current signal Q_0, so the comparator can be used again to determine it.

Depending on the comparator output, the switches c_{1p} or c_{1n} are closed, and the following charge sharing action between C_{SP}, C_{SN}, C_{MSB} and the newly connected $C_{MSB\text{-}1}$ will cause the voltages V_{QP} and V_{QN} to rise or fall.

5.2.1.4 Successive Approximation, Remaining Bits

Intuitively one can see that the SAR algorithm at each step uses these pre-charged capacitors to add or subtract a binary scaled-down charge to the initial charge (that represented the sampled input voltage) until the results converges to zero. If too much charge was added during a certain step, the next comparison returns the opposite sign, and in the next step the charge will be subtracted. The actual value of the voltage on the nodes (V_{Qp}, V_{Qn}) is not needed, just the sign is used to determine whether the next binary scaled down capacitor (and hence, charge) must be connected positively or negatively.

5.2.1.5 First Block Diagram

The block diagram of this initial charge-sharing SAR ADC architecture is shown in Fig. 5.7. As already explained, it consists of

- a passive Sample&Hold with capacitor and switches
- a binary scaled array of unit capacitors that are pre-charged to the reference and afterwards connected positively or negatively to the sampling capacitor, depending on the outcome of the comparator
- a comparator that returns the sign of the differential voltage on the sampling capacitors (VQp-VQn)
- a control block that implements the SAR algorithm, i.e.
  - generate the control signals for the S&H switches
  - generate the signal precharge
  - go through a loop that for every bit of the ADC
    - activate the comparator
    - interprete the result and close one of the switches c_p or c_n
  - output the digital code that represents the digitized value of the input voltage.

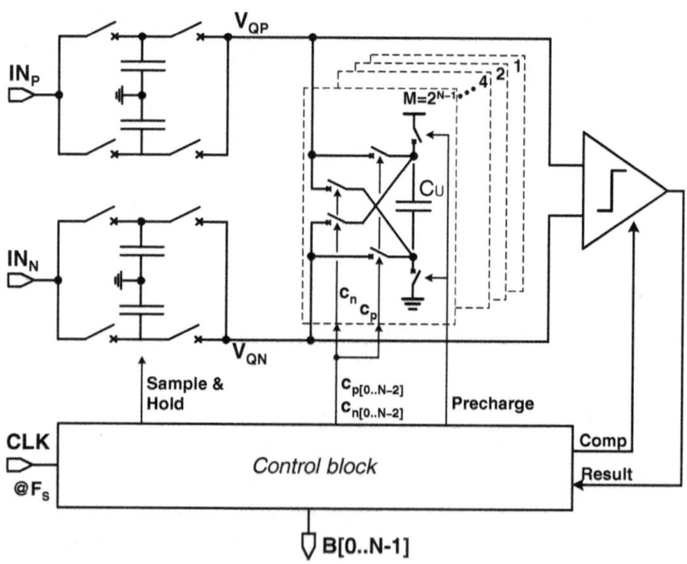

Fig. 5.7 Block diagram of initial charge-sharing SAR architecture

5.2.2 Asynchronous Operation

The synchronous operation of the control block described above would need a high-frequency clock, which has to be generated externally and hence also results in a power consumption penalty. Moreover, the maximum speed possible with the circuit is not exploited, as this way of working may require the control block to wait until the falling edge of the clock to close one of the switches c_p or c_n, although the comparator result S is already available earlier. All blocks (comparator speed, settling time for charge sharing) must be designed fast enough to certainly finish within the available clock period.

An asynchronous operation must be implemented that removes the need for an extra high-frequency clock, and allows A/D conversion at the highest possible speed. The timing of this asynchronous controller is fairly simple, as a straightforward sequential list of actions must be taken during the binary search algorithm. It is further aided by the use of a comparator that also provides a "valid" signal, as shown in Fig. 5.8

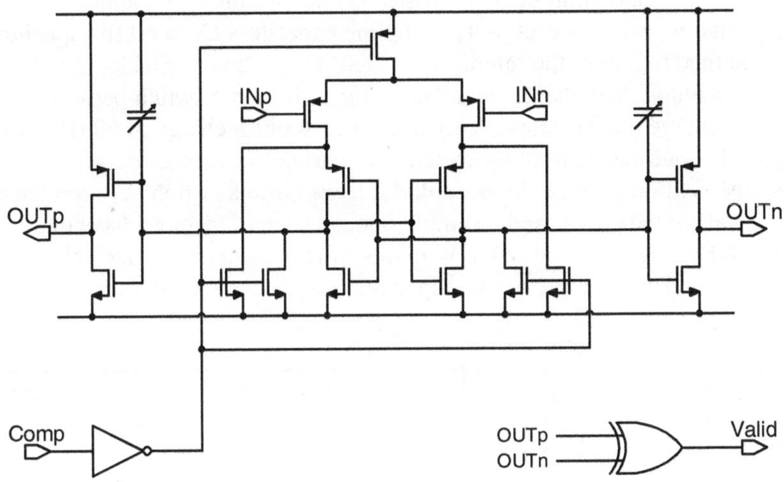

Fig. 5.8 Comparator for asynchronous operation

The comparator is based on [10]. The "valid" signal goes high after the two cross-coupled inverters leave their metastable operating point, meaning that one of the two sides has gone high. This means the comparison results are ready.

This comparator circuit also has the nice feature, a fully dynamic implementation that does not consume any power when inactive, and thus provides for the whole ADC the feature that its power consumption scales linearly with the sampling frequency

5.2.3 Binary Scaled Capacitor Array

The linearity (INL/DNL) performance of a SAR ADC is determined by the INL/DNL performance of the feedback DAC, and in our case by the matching of the capacitors in the binary scaled reference array. From [11] we know that for 99.7% yield the units of a 9 bit DAC need a standard deviation less than 0.7%, which is the key number in determining the size of the reference capacitor array. With some margin, the total size is set at 2 pF, such that $C_{MSB} = 1$ pF, $C_{MSB\text{-}1} = 0.5$ pF, etc.

The resulting LSB capacitor size now equals about 8 fF, which is obviously too small to be used. The parasitics from the connections will be too large w.r.t. the actual capacitance, and since the units can be positively or negatively connected to the sampling capacitor, any difference or mismatch in these parasitics deteriorates the INL/DNL behavior.

An alternative for the most significant bit is certainly to use a bigger unit, e.g. 8 times bigger. The capacitor controlled by $c_{0p,n}$ can now consist of 16 units of 60 Ff, which is a value that can be practically used. The following one has 8 units, and so on, until the one controlled by $c_{4p,n}$ which has 1 unit.

For the next charge sharing, a 30 fF unit could be used, but this will not match correctly with a "half" unit of 60 fF. Instead, since we only care about the amount of charge that we will connect to the sampling capacitors C_T, a 60 fF capacitor can also be charged to half of the reference voltage. This is done by taking 2 units, keep one of them empty and charge the other to V_{REF}. If then a switch between them is closed, the charge redistributes evenly and on each one a charge of 60 fF $\times$ $V_{REF}/2$ remains. This one can be used for the charge sharing by switches $c_{5p,n}$.

Also the next bits can be done similarly. If we close a switch between the other unit with half of the charge and an empty one, on each of them we have one quarter of charge for the next bit, and so on. With this structure a practical size unit capacitor can be used, and an example of this capacitor array is depicted in Fig. 5.9.

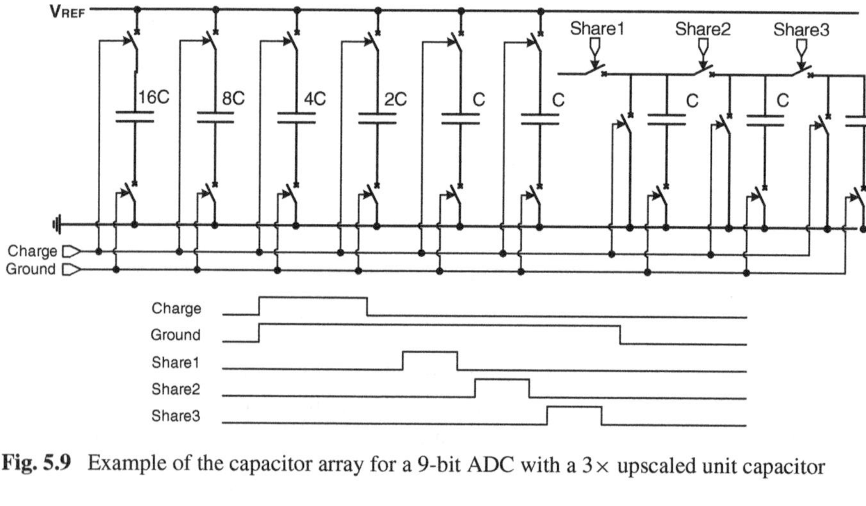

Fig. 5.9 Example of the capacitor array for a 9-bit ADC with a 3× upscaled unit capacitor

The effect of parasitics in the capacitor array must of course be evaluated carefully. In first order, the parasitics to the substrate of the unit capacitors do not pose a problem. They behave just as useful capacitors, i.e. they have a certain charge on them which will be connected to the positive or negative sampling capacitor and this way helps the DAC to perform the SAR ADC operation. Some constraints do apply however.

First of all, the parasitics on both sides of the capacitor must be balanced. If not, connecting the units positively or negatively to the sampling capacitor will have a different result. Therefore, the use of MIM capacitors that use lateral capacitance between a large number of closely space metal fingers is necessary. Capacitor structures that use vertical capacitance have obviously different parasitics on the top and the bottom plate. Mismatch of the parasitics will also result in missing the INL spec of the ADC, but since often no process data is available to estimate these mismatches, the only solution is to keep the parasitics small. An important part of the parasitics is caused by the switching transistors as well. Their drain-bulk and source-bulk junction constitute a non-linear capacitance. Their gate capacitance varies from very large to very small when the transistor switches from on to off. At first sight, this could be a performance-degrading effect, but to a first degree this non-linearity is not important. The signal is represented by charge, and the fact that this charge is present on a nonlinear capacitance is of no importance.

Besides matching, also sampling noise that determines the minimum input capacitance of the ADC, and hence also the size of the reference array because these two are proportional. Each time a sampling switch closes, a noise power (given by the integrated noise voltage kT/C) remains on the capacitor. However, with capacitor sizes in the picoFarad range as dictated by matching requirements, kT/C noise is negligible.

5.2.4 Comparator Noise

Although also not obvious from the basic architecture of Fig. 5.7, care must be taken when designing the comparator. During the SAR algorithm, sometimes the input signal to the comparator is very small, and when this value becomes comparable to the inherent noise of the comparator, an error can occur. This effect was the reason why the first prototype of the charge-sharing SAR that was implemented [3] had the worse measured performance than originally estimated. Reducing comparator noise by increasing its power consumption quickly has a detrimental effect on the efficiency, since for a 6 dB ($2\times$) noise reduction, a quadratic increase ($4\times$) in power is needed.

To resolve this issue, a noise robust design approach was developed for fully dynamic SAR ADCs by leveraging redundancy in the search algorithm. The strategy behind the proposed correction technique is the fact that during the SAR operation, at most two out of N comparisons are critical, i.e. the one when the signal is right below the threshold and the one when it is right above. All other comparisons will be done on a relatively large input signal, and hence can use a low-power but noisy comparator.

A low-noise (higher-power) comparator is only needed for the critical decisions, but of course it is unknown when they will appear. However, one of them will certainly be the last one: an error in this decision can be avoided by using the comparator in its low-noise/high-power state. The other one can be any of the previous (N − 1) comparisons, and avoiding it by always employing a low-noise comparator is not power efficient. As shown in Fig. 5.10 for a 5-b example, the SAR algorithm uses the comparator in its high-noise state during the first (N − 1) iterations, thus allowing errors in these cycles. However, if σ_H, the input comparator RMS noise in that mode, is less than one half of the LSB value, only one error can be made. The ADC then switches into its low-noise mode (with comparator input noise $\sigma_L \approx \sigma_H/2$) to avoid errors for the Nth comparison, and an extra $(N + 1)^{th}$ iteration is added to correct for the error possibly made in the first phase. As shown in Fig. 5.10(a), if the last 2 comparisons give different results, no error was made and no action has to be taken. On the other hand, in case the last two bits are equal, then a digital addition or subtraction needs to be performed on the final N-bit result.

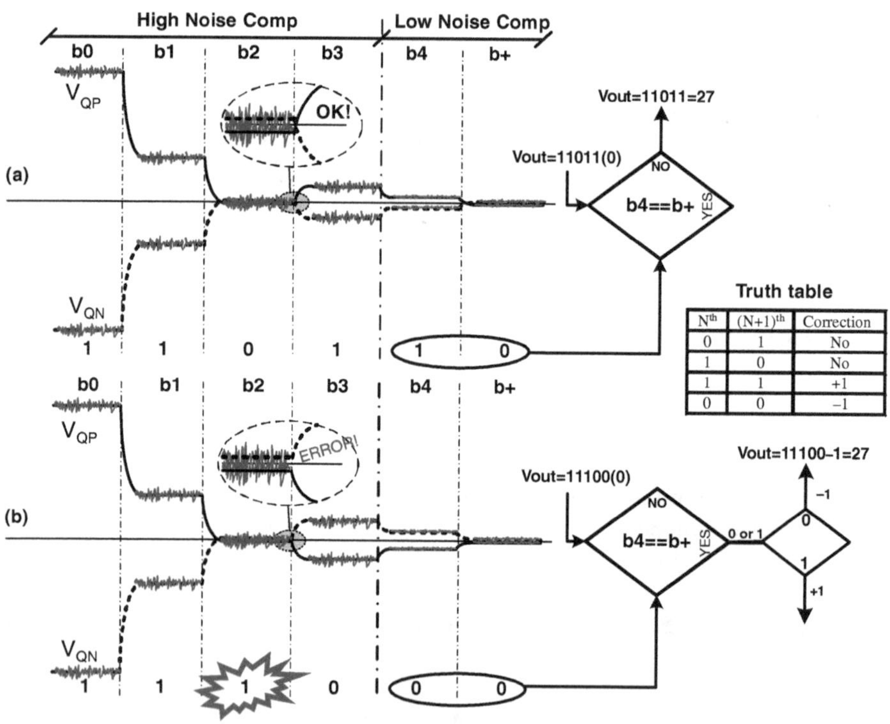

Fig. 5.10 Correction algorithm for comparator noise (5b example) in case no error is made (**a**) and in case of error (**b**)

Being pipelined to the SAR conversion, the simple digital adder needed to correct does not work at the internal SAR frequency (N times the sampling frequency), thus limiting its power consumption. Importantly, the correction is effective not only for thermal noise, but also for other error sources, including static non-linearities, as far as they are not bigger than 1 LSB.

5.2.4.1 Comparator Offset

In comparison with classical charge-redistribution SARs, in the charge-sharing SAR offset of the comparator does have an effect on the INL/DNL performance. The reason for this is that the signal is represented by charge and the offset in the comparator has always a certain voltage. During the successive approximation process, the capacitance size is changing and hence the relationship between charge and voltage is not fixed. To make the comparator offset small enough not to have an effect, the same technique as in [10] was again used, as already indicated by the varicaps shown in Fig. 5.8. At startup or at regular time intervals, the two inputs of the ADC must be shorted and the correct digital value that results in equal probability of a positive/negative comparator decision must be searched.

5.2.5 Implementation

The ADC prototype that was designed to show the performance of the proposed techniques was implemented in a 90 nm 1 V 1P9M digital CMOS process [8]. The die photo is shown in Fig. 5.11. Only regular transistors and MIM-caps are used in the whole design, making it ideally suited for implementation in digital CMOS.

Figure 5.12(a) shows the static INL/DNL performance when the correction is active. The peaks in DNL and INL are due to incomplete settling during the common mode switching, a small design mistake that could easily be corrected. The actual error caused by this effect is even worse (uncorrected INL/DNL is worse than $\pm$1LSB), but as already stated the correction algorithm does not only detect errors caused by noise, but also this static non-linearity. The peak DNL and INL are 0.7/$-$0.45 and 0.56/$-$0.65, respectively.

As shown in Fig. 5.12(b), when the input signal is sampled at 40 Ms/s, the measured ENOB is 8.56 (53.3 dB SNDR) at low frequencies, mainly limited by static distortion, and lowers to 8.23 at Nyquist. The effective resolution bandwidth extends up to 32 MHz. At 40Ms/s the ADC consumes 820 μA from a 1 V supply voltage of which 290 μA is drawn by the asynchronous controller, and 530 μA is shared between the S/H, the pre-charging phase of the capacitor array, and the flexible comparator. Because of the dynamic architecture, power consumption scales linearly with the sampling frequency. The resulting FoM is only 54 fJ per conversion step, an absolute record and much lower than comparable SAR ADCs.

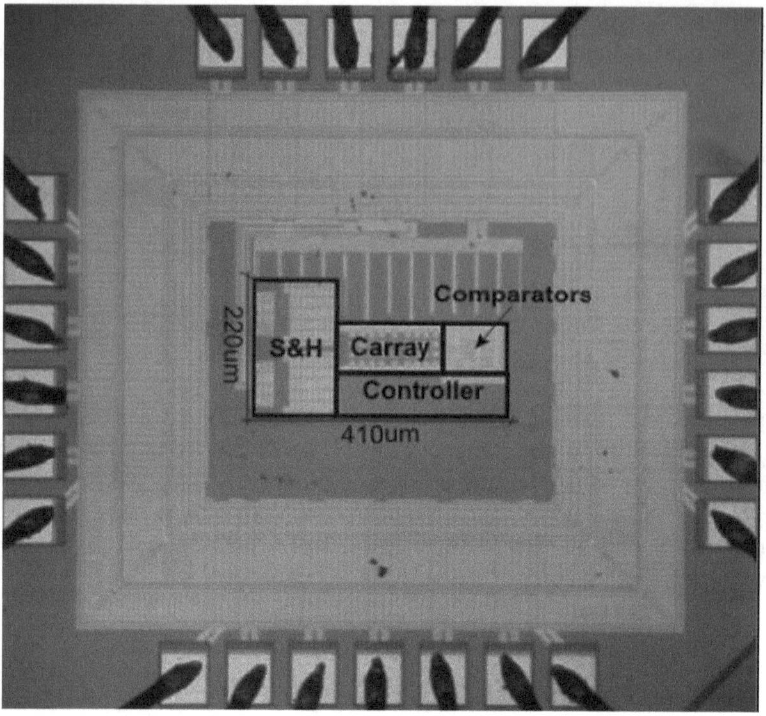

Fig. 5.11 9-bit 40 MS/s Charge-sharing SAR ADC photograph

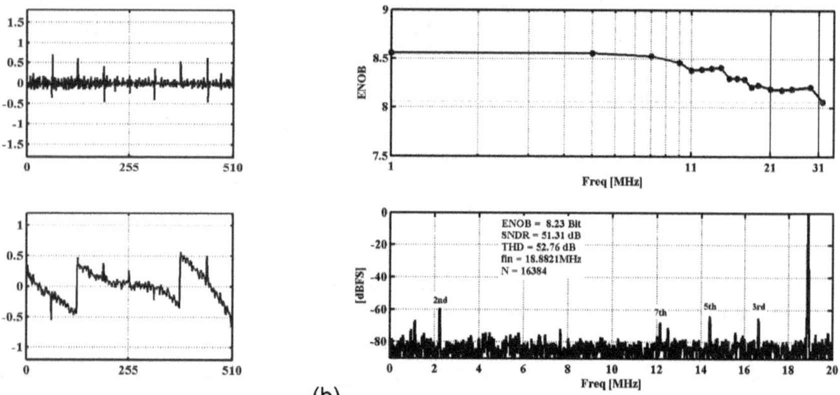

(a) (b)

Fig. 5.12 (a) Measured INL/DNL plot; (b) ENOB vs. frequency and near-Nyquist FFT at 40 MS/s

5.3 Comparator-Based Asynchronous Binary Search ADCs

For moderate resolution (5–7-b), A/D converters have traditionally been implemented with flash-type architectures, enhanced with folding and/or averaging techniques to reduce power consumption [12, 13]. These designs are typically not extremely power efficient, since they still use a large number of comparators to determine the quantized value of the input signal. More recently highly interleaved SAR ([14,15]) has been proposed to reach speeds of a few 100 Ms/s at much improved power efficiency since the number of comparisons can be reduced compared to flash converters. By carefully optimizing the logic and aggressively going for maximal speed in [16] the interleaving factor of a 600 MS/s 6-b SAR converter has been reduced to two in 0.13 μm CMOS, reducing system complexity and size considerably. This resulted in a power efficiency of 220 fJ/conversion step. Pipeline converters are also attractive candidates, intrinsically they offer a higher speed capability and by utilizing the zero crossing based switched capacitor technique a power efficiency of 380 fJ/conversion step has been obtained in [17]. In conclusion, the architectures and circuits proposed up to now for 7-b, 150 MS/s converters achieve a power efficiency of approximately 200 fJ/conversion step in nanometer technologies (130 nm to 65 nm).

The new architecture proposed in this chapter results in a power efficiency of 10 fJ/conversion step for a 150 MS/s 7-b converter in 90 nm digital CMOS technology [18]. The data converter is implemented using a passive track-and-hold followed by a two-step conversion process. One bit is determined using a comparator that steers a charge feedback D/A converter to bring the input signal into the range of the sub-A/D converter. This circuit is a 1-b implementation of a SAR A/D converter with comparator-based controller (SAR-CC). The remaining 6 bits are determined using a comparator-based asynchronous binary search (CABS) sub-A/D converter. In this architecture only the required comparators are clocked, reducing the power consumption. This low-power consumption has further been reduced by employing dynamic circuits only (power consumption is proportional to activity), and relying extensively on calibration to remove matching constraints and designing the circuits for thermal noise limitation.

This section is structured as follows. Next the operating principle of the new A/D converter technique is explained. In the following section the implemented architecture is described in detail. In the next section the experimental results are reported and discussed.

5.3.1 Operating Principle

The operating principle of the CABS converter is based on binary search, the same principle that is used in a successive approximation A/D converter. But instead of approximating the input signal, in the CABS architecture, comparators with built-in thresholds are used to bracket the input signal (shown in Fig. 5.13). Similar to a flash converter, the sampled input signal is applied to all comparators, but unlike flash converters, not all comparators are clocked. Instead the comparators are connected in a binary tree, in which the root comparator compares the input signal with zero and based on its decision asynchronously triggers one of its children, comparators with threshold $\frac{1}{2}$ and $-\frac{1}{2}$. If the input signal is greater than zero the comparator with threshold $\frac{1}{2}$ is triggered, if smaller than zero, the comparator with threshold $-\frac{1}{2}$ is triggered. This second comparator in turn triggers one of its children in the

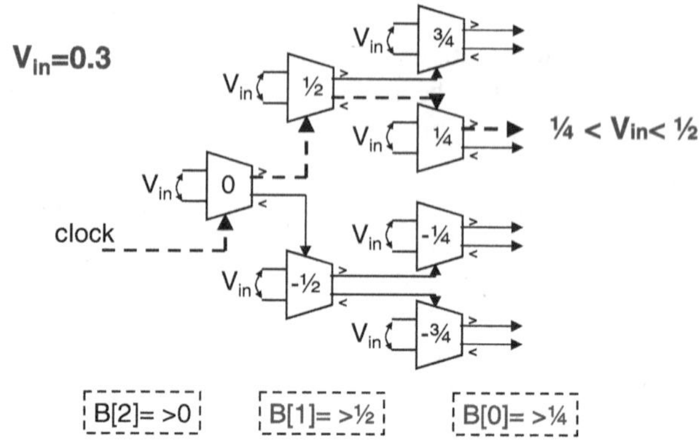

Fig. 5.13 Operating principle of a comparator-based asynchronous binary search (CABS) A/D converter, shown for a 3-bit A/D converter with 0.3 input on a −1 to +1 range (dashed lines indicate active path)

third layer, closing in on the input signal. Based on the outputs of the activated comparators an unsigned binary code is derived: a logic 1 is encoded for "greater than" and a logic 0 for "smaller than" (Fig. 5.13). In this manner the input value is converted into an unsigned binary representation.

In Fig. 5.13 the operating principle is illustrated for 3 bits, but it can be extended to more bits. The number of comparators in the tree is for an n-bit converter equal to $2^n - 1$ as is the case in a (full) flash converter, but only n comparators are triggered. So the power consumption is reduced to the power required to do a minimal binary search, hence much smaller than the power of a parallel search implemented in a standard flash converter. The quantization starts on the rising clock edge: successively, layer by layer, one comparator decides. A falling clock edge resets all the activated comparators following the same path through the tree that was followed during quantization. As a result this architecture does not need explicit controller circuitry.

Although this architecture reduces power consumption in an A/D converter, it does have the disadvantage of exponential complexity. This can be avoided by implementing a SAR A/D converter in which the comparators also implement the controller function (SAR-CC). The operating principle of this architecture is shown in Fig. 5.14. Once again the sampled input signal is applied to all comparators in the architecture in parallel. But instead of a tree, a chain of comparators is implemented of which each controls a feedback D/A converter that modifies the sampled input signal. The algorithm implemented on this architecture is also successive approximation. Normally this algorithm is implemented in a synchronous or asynchronous controller, in the presented approach the comparators serve as state machine of the algorithm: if a comparator is reset, both "greater than" and "smaller than" outputs are zero, if it is activated one of them is logic 1 and its feedback D/A converter modifies the input signal. The output of a comparator however can not immediately trigger the next one in the chain, since the D/A converter feedback signal needs to settle. Hence an appropriate delay block with delay τ is inserted, such that the next

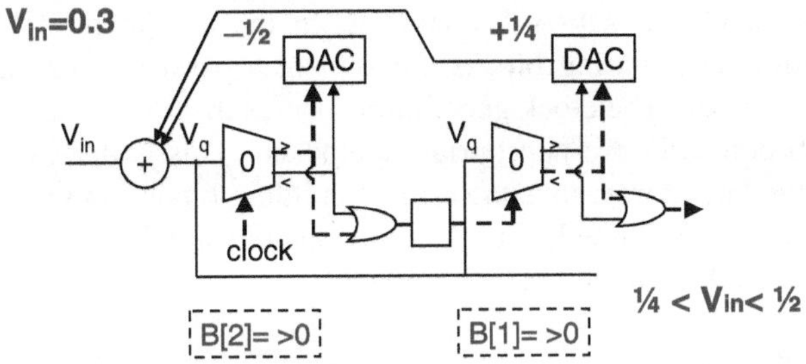

Fig. 5.14 Operating principle of a SAR A/D converter using a comparator-based controller, after sampling the input every bit is determined by a comparator which controls a feedback D/A converter (*dashed* lines indicate active path)

comparator is only triggered when the feedback signal of the D/A converter is stable. Also since this architecture is edge triggered, a rising clock edge at the start of the chain starts the quantization, and a falling clock edge resets the structure.

In the prototype the SAR-CC is used as a 1-b front-end for a CABS converter. This combines advantages of both techniques: it avoids the exponential complexity of the CABS converter in terms of the number of bits albeit at a higher energy consumption (feedback D/A-converter extra) and slower speed (delay block). In the next subsection, the architecture chosen for the prototype will be described in detail.

5.3.2 Implemented Two-Step 1-b Coarse 6-b Fine Architecture

To demonstrate both new techniques a prototype A/D converter has been implemented (Fig. 5.15). It is a 7-b two-step architecture in which a 1-b coarse quantization is performed with a SAR-CC converter, and 6-b fine quantization is

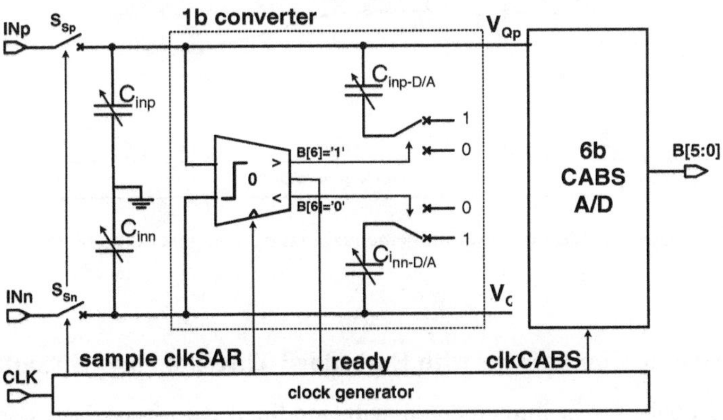

Fig. 5.15 Implemented architecture, a 2-step 1-b SAR comparator-based coarse converter and a 6-b CABS fine converter

performed with a CABS sub-A/D converter. In the architectural block diagram (Fig. 5.15) additional building blocks are shown: a passive track-and-hold and a clock generation block. The clock generation ensures that the track-and-hold, SAR-CC and CABS converter are appropriately clocked. This is discussed in the next paragraph. In the later paragraphs the critical building blocks and the calibration of the converter will be discussed. This section concludes with a power consumption breakdown of the 7-b architecture.

5.3.2.1 Clock Generation and A/D-Converter Timing

The timing signals of the A/D converter are all derived from a single-ended 50% duty cycle clock input (Fig. 5.16). This clock input is first buffered to drive a track-mode high, hold-mode low signal ("sample" signal in Fig. 5.16). A delayed and inverted version of the "sample" signal is used to start the level triggered 1-b coarse converter ("clkSAR"). The SAR-CC internally directly controls the D/A converter and returns (asynchronously) a "ready" signal to the clock generator. The D/A converter feedback signal modifies the sampled value and brings it into the input range of the 6-b CABS converter. The delay block in the clock generator delays the start of the level triggered 6-b CABS converter until the input signal has settled. The 6-b CABS then quantizes the reduced range input signal. When the "sample" signal goes high, the tracking mode is resumed, and all sub-A/D converters are reset. The reset phase overlaps with part of the tracking phase.

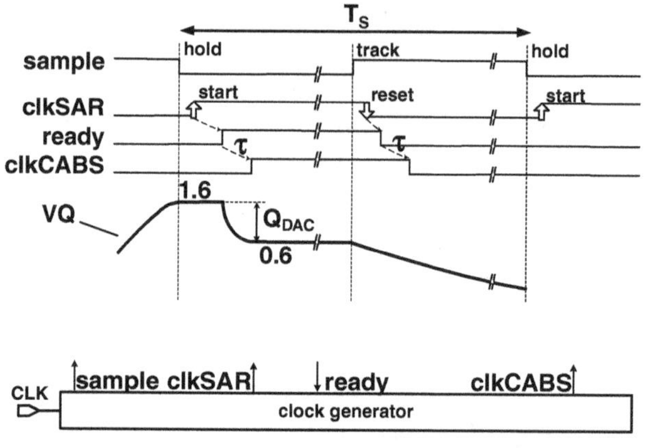

Fig. 5.16 Timing of the A/D converter implemented in the clock generation block

5.3.2.2 Dynamic Comparator with Embedded Threshold and Encoding

The CABS converter is implemented with modified Strongarm [19] comparators with built-in offset, shown in Fig. 5.17. When the comparator "comp" input is made high, the input pair converts its input voltage into a current that turns on

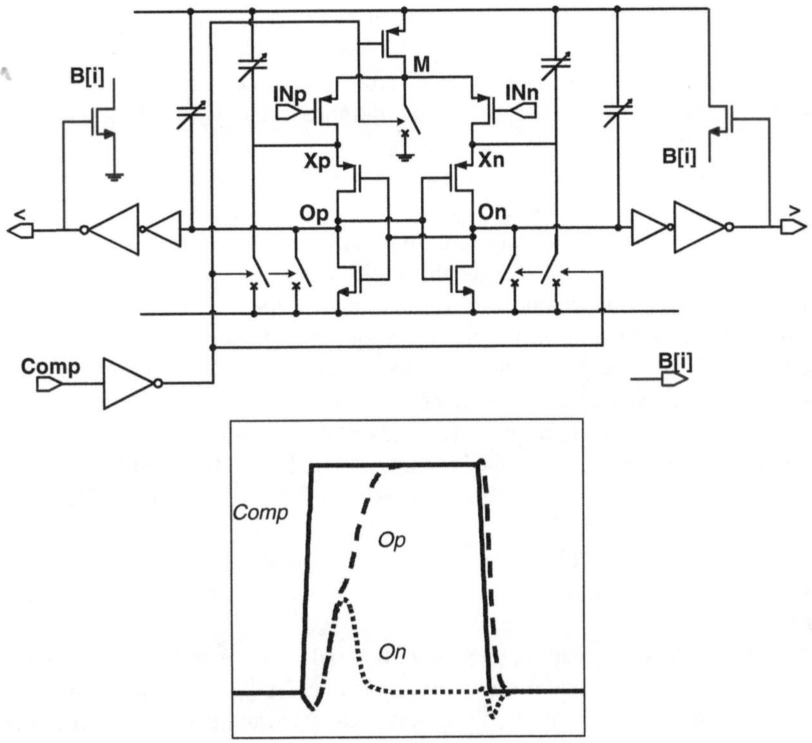

Fig. 5.17 Modified Strongarm comparator with encoder function that is used in the CABS converter and the waveforms on "comp" and "O" nodes

the cross-coupled inverter pair. For low input signals and a balanced realization, the cross-coupled inverter pair at first goes to a metastable region after which any imbalance caused by the input signal is amplified to a full scale signal. The time constant on this node determines how fast this happens. When the "comp" input is low, this comparator is reset. In this phase the comparator presents a capacitance to the inputs which depends on the voltage on the common node ("M" in Fig. 5.17). The input signal of the A/D converter is sampled on this capacitance. To avoid any non-linear behavior the "M" node is grounded (instead of left floating as in the original Strongarm), pushing the input devices in the strong accumulation region to linearize input capacitance. A further modification is the introduction of imbalance in the input pair [10]. This imbalance causes an input threshold shift (built-in offset).

To reduce the power consumption, the comparator has been sized as small as possible, limited by thermal noise performance [20]. The reduced transistor sizes cause mismatch which is compensated by calibration. This is implemented with the digitally controlled variable capacitors added on the internal nodes ("X" and "O") of the comparator.

In the prototypes there is a binary scaled bank on the "X" node, and a single varicap on the output node "O". Both nodes "O" and "X" have been used to maximize flexibility at run-time. The input referred imbalance has been analyzed using transient simulations, on node "X" a range of approximately 60 mV is available, on node "O" an additional 45 mV is available.

For both the relative width of the input pair (imbalance in input pair) and relative capacitance imbalance on "X" and "O" nodes of the comparator, the overdrive of the input devices in first order determines the threshold shift of the comparator: a larger overdrive yields a larger threshold shift. The input referred imbalance (through overdrive) thus depends on common mode input voltage (the device above the input differential pair is in the linear region when the comparator decides) and further depends on the supply voltage and temperature of the circuit. All these uncertainties are compensated for using the provided calibration. The sensitivity towards these will be further discussed in the experimental results section.

To reduce their time constant the regeneration nodes are buffered from their load by two inverters. Each of the buffered outputs of the comparator drives a child in the CABS tree. To avoid any accidental propagation of the comparator output signal during the metastable phase, the input threshold of the first inverter in the buffers has been made high. This has been done by altering the NMOS/PMOS drive strengths.

The encoder function is implemented with two devices, a pull-up device to create a logic 1 on the bit-line when the comparator decides "greater than", and a pull-down device to create a logic 0 on the same bit-line when the comparator decides "smaller than". Since only one comparator on each layer is activated, only one output goes high, and one device drives the bit-line. When the comparator is reset, the capacitance of the bit-line retains its state and the digital code can be clocked in by D-flipflops at the clock edge starting the next conversion. The A/D converter latency is one clock period.

These comparators are structured in a binary tree and clock signals propagate as shown in Fig. 5.18. The control signal coming from the clock generator starts the conversion at the root of the tree, and it propagates along one path (signals shown on timing diagram) till one of the leaves of the tree. For each conversion of the CABS six comparators are activated of which at most two comparators have an input signal close to their threshold. With a large effective input voltage comparators decide quickly, while the comparators with an input value close to their threshold decide slower. On average it takes 330 ps to propagate an edge on the "comp" pin to one of its outputs, limiting the total conversion time to approximately 2 ns. The reset time is 100 ps per comparator, or 600 ps for the converter. The maximal operating speed (with a 50% duty cycle clock) of the sub A/D converter (CABS) is limited by the quantization phase to a clock period of 4 ns or 250 MS/s.

5.3.2.3 Passive Track-and-Hold

The passive track-and-hold samples the input signal and provides a stable signal to the quantizer blocks. This sampled value is stored on a capacitor consisting of the

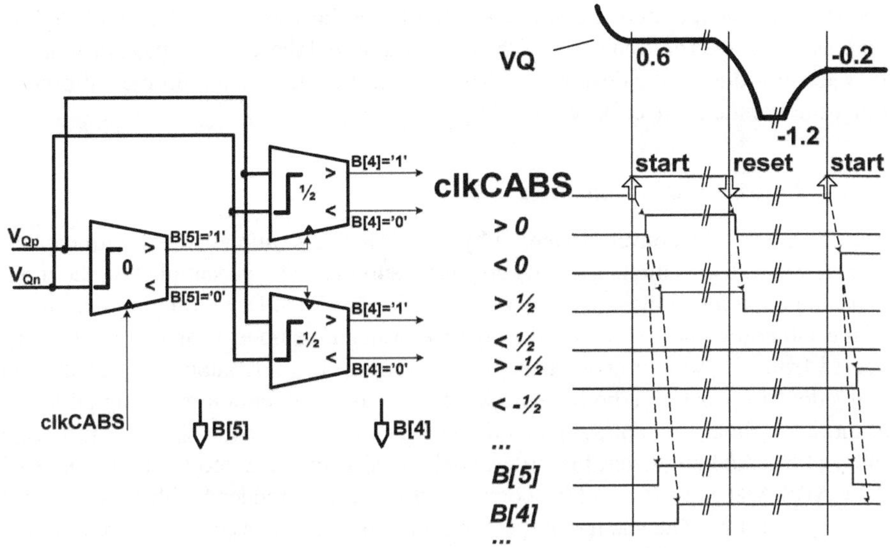

Fig. 5.18 CABS converter and timing diagram

following parts (in order of importance): the input capacitance of the comparators in the 1-b SAR-CC and 6-b CABS converters (MOS gates, where the operating point is in the accumulation region), feedback capacitance of the 1-b D/A converter, an explicit tuning capacitance (PMOS devices, controlled from the shorted Drain-Source), parasitic routing capacitance and track-and-hold switch capacitance. In total about 250 fF (estimate based on parasitic extraction followed by spice simulation) is present on a single ended input. Since this capacitance has been linearized as much as possible, the major contributor to distortion is then the non-linear switch resistance. No bootstrapping has been used to keep the V_{gs} constant. The switch resistance has been sized to ensure sufficient dynamic performance (ERBW exceeding Nyquist with margin). Simulations have been performed to verify that a switch 4.5 μm wide and with minimal length gave a 3rd order distortion component of −48 dBc at 250 MHz at full scale input.

During the conversion, the charge on the sampling capacitance is changed by charge injection from the CABS comparators. This, however, does not compromise the correct operation of the converter. The first effect of this charge injection is a change of the common mode level of the sampled signal. In simulation it has been verified that this effect is limited, and the input capacitance of six comparators is small compared to the sample capacitance. Secondly, the comparator operations depend on the input signal level. Except for the root comparator, the input pairs of these comparators are imbalanced, which creates a differential charge that is injected into the sample capacitance. This effect, however, is also present during calibration and is therefore compensated for by calibrating the comparator thresholds. The last effect is that the input signal charge is redistributed on the input capacitance of the activated comparators, causing a small differential signal loss. This will slightly increase the input referred noise of the comparator. All these effects have been taken into account during the design and extra margin has been added in the calibration ranges to handle these effects.

5.3.2.4 Feedback D/A Converter

The feedback D/A converter steered by the SAR-CC is implemented with capacitive charge redistribution. The D/A converter subtracts charge sampled on the input capacitance in a single-ended way as shown in Fig. 5.15. Depending on the output of the 1-b comparator either on the positive or negative node a capacitor is disconnected from the supply and grounded resulting in this way subtracting charge on that side. This operation both changes differential and common mode signal levels at the same time. By subtracting a charge which corresponds to ¼ of the full scale input of the A/D converter, the differential signal is brought into the input range of the CABS A/D converter and the common mode level is reduced with 1/8 of the full scale input range. The change of the common mode improves the regeneration time and speed of the CABS comparators.

This charge redistribution works as described if there is no pure floating direct capacitance between the differential input nodes. Otherwise the switching of the feedback capacitor on one side influences the charge on both differential input nodes (through the floating direct capacitor). This results in reduced differential signal change and would require a larger feedback capacitor to compensate for. In the layout and the design, precaution has been taken to minimize this direct coupling capacitance. It was verified with parasitic extraction that this coupling capacitance is only a few femtofarad. Furthermore the grounding of the "M" node in the comparators is also important here. Would the common mode node "M" be left floating, there would be additional unwanted parasitic coupling through this path (Fig. 5.17).

This single-ended subtraction is inspired by the passive charge sharing technique of [3]. During the SAR-CC quantization process, the only action is shorting a capacitance to ground, which is passive and fast. Charging this capacitance to a reference voltage occurs during the reset/tracking phase of the converter, where more time is available for settling.

The D/A converter is implemented using MIM capacitors. Since the actual value of the sampling capacitance is not known accurately at design time, the feedback capacitor is implemented in units of 8 fF which at runtime are enabled following a calibration procedure. Since the sampling capacitor is approximately 250 fF and the reference voltage is 1 V (supply), 8 fF corresponds to a voltage step of 8/250 $\times$ 1 V = 30 mV, or 5 LSB of the A/D converter. Additional calibration capability has been added to the D/A converter on the input nodes. These capacitors don't take part in the feedback action, simplifying their implementation. We chose PMOS based varactors to modify the input capacitance. These offer sufficient granularity to do a fine calibration so that a 1.5 fF capacitance value results in a 1 mV calibration step. The non-linear capacitive behavior of the PMOS varactors has been verified and does not adversely affect the A/D converter performance. During calibration first the coarse MIM setting is determined and then the fine setting of the PMOS varactors is determined.

In the current prototype, the reference is taken from the supply, which is low-noise because of the low-power consumption of the A/D converter and thus it does not compromise the performance. However, long term changes affect the step of the feedback D/A converter. A 1.5% reduction in supply voltage (15 mV) linearly translates in a 1.5% reduction of feedback step. Since the feedback is ¼ of the full range, this corresponds (for a 7-b converter) to an error of 0.5LSB (0.4% of full scale).

5.3.2.5 Calibration

The A/D converter presented relies on calibration to achieve the improved power efficiency. In this paragraph the method used to calibrate the various parts of the A/D converter will be discussed in detail. The calibration method used is a binary search (in the digital domain), explained in detail in [21]. In the implemented prototypes no calibration circuitry has been included on-chip. In the test setup this circuitry has been implemented with measurement equipment and the calibration algorithm is run in matlab on a PC.

There are two calibrations to be performed in the two-step 7-b A/D converter. First of all the D/A converter has to be calibrated, and secondly the thresholds of all comparators (both the 1-b coarse and the 6-b fine) have to be calibrated. For a comparator that has to be calibrated for a certain threshold, first the wanted threshold is applied at the input. A binary search algorithm then determines the correct calibration setting. The capacitance on the "O" node is controlled by the MSB of the calibration value, the binary weighted bank on the "X" node by the LSBs of the calibration value.

For the feedback D/A converter a similar principle is used. First the threshold of the root comparator of the 6-b fine A/D converter is calibrated (to 0), then the input corresponding to the threshold detected by the root comparator (which is both the ¼ and ¾ thresholds of the full range) is applied to the 7-b two-step converter. For the ¾ threshold, one side of the feedback D/A converter is calibrated, in two steps, first coarse setting, then fine setting. This process is repeated for the ¼ threshold to calibrate the other side of the feedback D/A converter.

The number of samples to calibrate the two-step A/D converter can be determined by applying the analysis of [21]:

$$N_{\mathrm{A/D,cal,samples}} = (N_{\mathrm{comp}} + N_{\mathrm{D/A}}) \times (M + 1) \times N_{\mathrm{average,samples}} \tag{5.3}$$

where the total amount of samples required for calibration ($N_{\mathrm{A/D,cal,samples}}$) is determined by multiplying the number of settings to calibrate (N_{comp}+ $N_{\mathrm{D/A}}$), with the number of binary search steps (M +1, where M is the number of bits in the calibration) and the number of samples required to determine without error the direction to continue the search in ($N_{\mathrm{average,samples}}$). This number of samples depends on the comparator noise and the minimal step of the calibration. For the 7-b converter the comparator noise is 0.25LSB, and the calibration step is 0.2LSB. Based on Table 1 in [21] this gives $N_{\text{average,samples}} = 640$. The total number of samples ($N_{\text{A/D,cal,samples}}$) then equals 340,000. Running at 150MS/s the time needed to collect this amount of samples is 2.3 ms.

5.3.2.6 Power Breakdown

In simulation the power breakdown of the 7-b A/D converter has been determined. The circuit is fully dynamic, and has a (measured) power consumption of 0.89 μW per MHz of clock rate. The dominant contributors to power consumption (67%) are the comparator-buffer circuits, seven of which are clocked in the converter for each sample. This corresponds to about 10% of overall power consumption per bit. The feedback D/A converter takes up 18%, and is the second contributor. The encoder takes up about 8% of the power. The encoder power is mainly due to the charging of the bit-lines, and hence activity dependent, 8% is an average value. 7% of the power is taken up by the clock generator including the buffer driving the passive track-and-hold. This analysis indicates that indeed most of the power is taken up by the core function of the A/D converter: comparing the input signal versus a threshold.

5.3.3 Experimental Results

In this section the experimental results of the two prototype chips will be discussed. First the layout will be described and relevant aspects will be highlighted. Next the measurement setup will be described and the static and dynamic measurements of the two prototypes will be presented. This section concludes with an analysis of the energy efficiency of the converters.

5.3.3.1 Layout Implementation

The chip micrographs of the 7-b two-step and 6-b CABS converters are shown in Fig. 5.19. The prototypes are implemented in a 90 nm digital 1P9M CMOS process with a nominal supply voltage of 1 V. In the designs only CMOS devices are used, with regular and low V_{th}. No RF process options such as thick metal layer or RF models have been used. Both prototypes use the same basic layout template. The 32 LSB comparators are placed in the column on the right, the MSB downto LSB + 1 comparators (32 in the case of 7-b, 31 in the case of 6-b CABS) have been placed in the column on the left. For the two-step 7-b converter, the 1-b SAR-CC comparator has been placed at the bottom of the left column. The 6-b CABS comparators in the left column have been placed as in the tree: the root comparator central, the ½ comparators halfway the two halves of the column and so on. In this way the tree has been folded into one column. Between the two comparator columns the encoder bitlines have been inserted. These lines are routed on a high metal layer to reduce parasitic capacitance and hence power consumption. The differential input

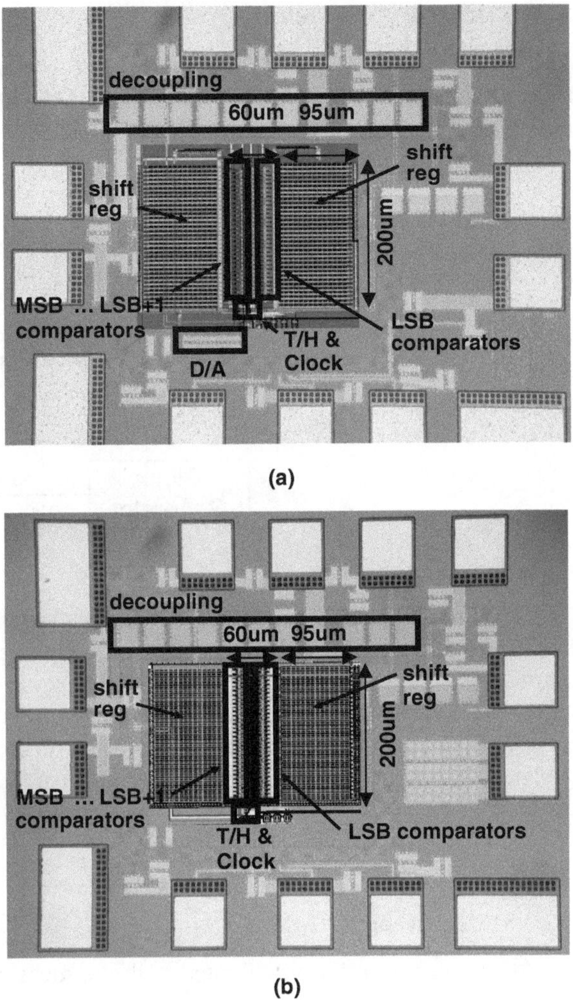

Fig. 5.19 Chip micrograph with super-imposed layout view and sub-blocks highlighted, 7-b two-step chip (**a**) and 6-b CABS converter (**b**)

signal is distributed to both columns. In order to reduce direct coupling capacitance a grounded shield wire has been inserted in between the wires. Below the comparator columns the track-and-hold switches and the clock buffering are placed. For the 7-b two-step converter, the feedback D/A converter (implemented with MOM) is placed at the bottom of the structure. The remainder of the A/D converter area is taken up by a shift register that stores the calibration bits. This shift register has not been optimized in terms of area efficiency. In total (including shift register, but excluding decoupling capacitance) the core area of the 7-b converter is $220 \times 250\ \mu\text{m}$.

5.3.3.2 Measurement Setup

The measurement setup used to calibrate and characterize the A/D converters is shown in Fig. 5.20. The input and clock signals are applied through DC/RF sources and a clock generator respectively. These are controlled from a host PC through a GPIB interface.

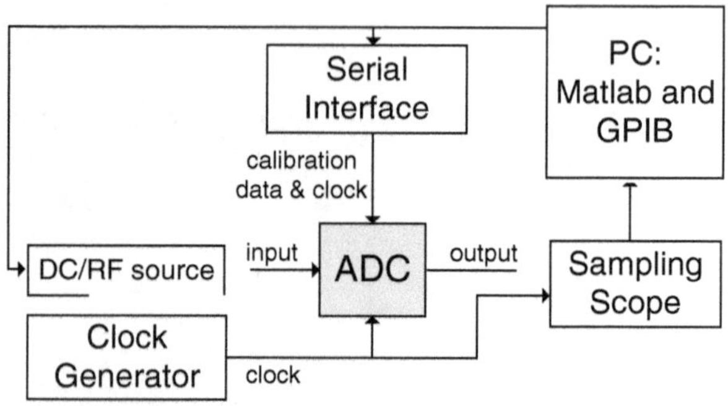

Fig. 5.20 Measurement setup used to calibrate and characterize the A/D converters

The output of the A/D converters is brought off-chip in real-time using 50 Ω wide bandwidth drivers. These driver circuits are implemented on-chip. The seven bits output from the A/D converter are clocked into seven D-flipflops and then buffered to the 50 Ω drivers. The power consumption of these output circuits is approximately 30mW. The output waveforms are captured and processed by a real-time scope.

The calibration algorithm is running on the host PC in matlab. The input sources are steered by the algorithm and the output signal is analyzed. The calibration bits are downloaded to the chip using a two-wire serial interface.

5.3.3.3 Measurement Results

The prototypes have been calibrated with the measurement setup as described in previous paragraph, following the procedure explained in section 3.0. The sampling frequency at which calibration has been performed is 33 MS/s, and the converter has subsequently been measured using this calibration setting. The static linearity after calibration is plotted in Fig. 5.21. For the 6-b CABS converter an INL and DNL of 0.27LSB and 0.57LSB, respectively, is obtained, and the LSB size is 6 mV. For the 7-b two-step converter an INL and DNL of 0.48LSB and 0.93LSB, respectively, are obtained with the same LSB size. In the case of the 7-b converter only 64 of the 127 thresholds ($2^n - 1$) are calibrated directly, and the 63 remaining thresholds are fixed by the operation of the feedback D/A-converter. As can be seen from the linearity plots, the thresholds at the extremes have an increased DNL. These thresholds have not been used to calibrate the A/D converter and are fixed by the step of the D/A converter. This non-linear behavior of the D/A conversion is however limited to less than 1LSB DNL. An overall non-linearity in the INL of the 7-b converter is visible (shape of the curve is second order like), which is most likely due to the effect of

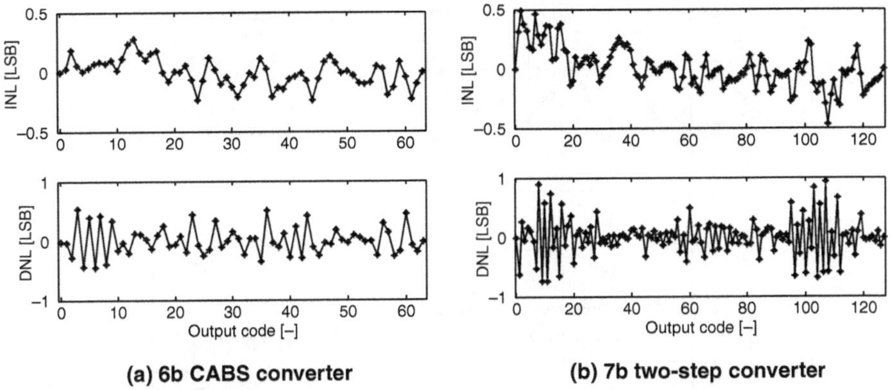

Fig. 5.21 INL and DNL of 6-b CABS converter (**a**) and 7-b two-step converter (**b**)

non-linear capacitances on which the signal is sampled. The measures taken during design (pushing input pair into accumulation, use of MOS varactors with sufficient linearity) to mitigate this effect are sufficient to bring the INL of the converter down to 0.5LSB.

Next the converters have been characterized for dynamic performance. In Fig. 5.22 the SNDR of the 7-b converter versus clock frequency has been plotted for two input signals, a low input signal frequency (10 MHz) and the Nyquist signal frequency. SNDR is 41 dB for the 10 MHz input signal and 40 dB for the Nyquist signal at frequencies up to 150 MS/s. At 200 MS/s SNDR is reduced by 5 dB. 200 MS/s corresponds to a clock period of 5 ns, of which 2.5 ns are available for quantization. This is clearly insufficient to resolve the last bit, explaining the loss of 5 dB in SNDR. The 7-b converter has next been characterized running at 150 MS/s: the SNDR and SFDR versus input signal frequency are plotted in Fig. 5.23. The ERBW is 270 MHz, which is well above the Nyquist frequency of 75 MHz. The SFDR stays above 54 dBc in the first Nyquist band, dropping to about 46 dBc at 250 MHz

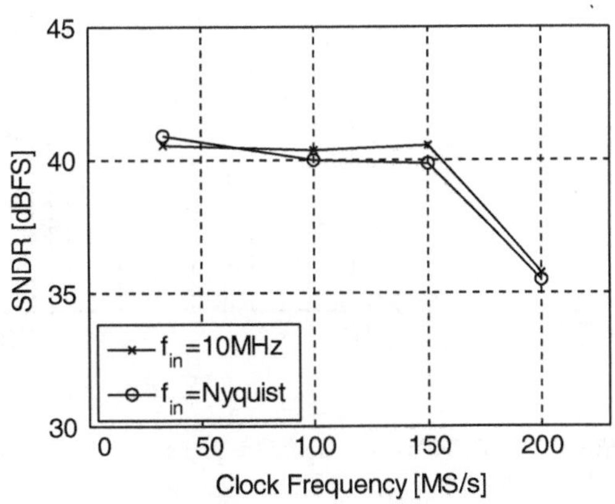

Fig. 5.22 SNDR of the 7-b converter for a full-scale low frequency input signal (10 MHz) and Nyquist input signal versus sampling rate

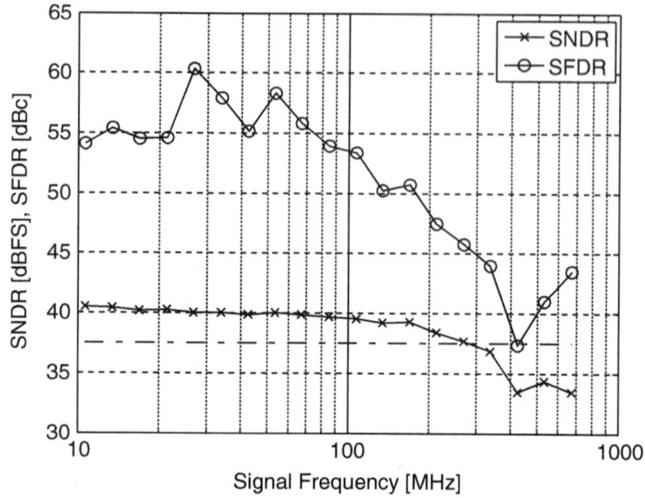

Fig. 5.23 SNDR and SFDR of the 7-b two-step converter for a 150 MS/s sample rate, versus input signal frequency, ERBW is 270 MHz

input signal frequency. This corresponds closely to the 48 dBc predicted by track-and-hold simulations, and confirms that this block limits the input bandwidth of the converter.

The stand-alone 6-b CABS converter, which is the sub-A/D converter of the 7-b, has similarly been characterized for dynamic performance. In Fig. 5.24 the SNDR of the CABS converter running at 3 different clock frequencies is shown for input

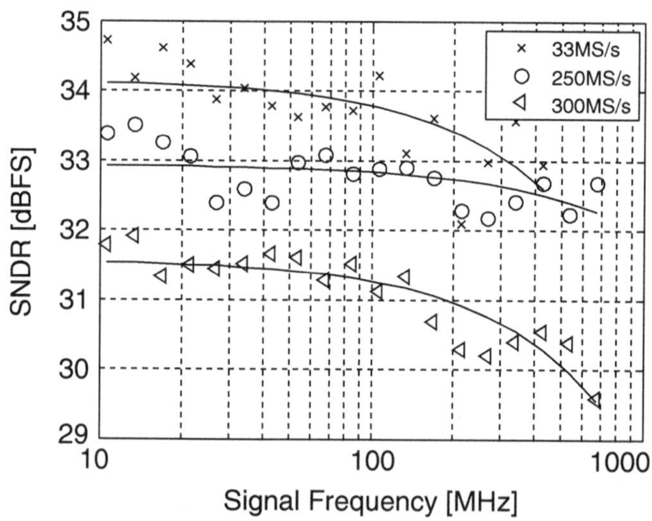

Fig. 5.24 SNDR versus signal frequency for 3 different clock frequencies of the stand-alone 6-b CABS converter

signals with frequencies up to a few 100 MHz. At low clock frequencies the SNDR peaks at 34 dB, at 250 MS/s the SNDR has dropped approximately 1 dB. At 300 MS/s the SNDR at low frequencies is 31 dB, which is 5-b ENOB. This is consistent with the conversion time that is required. As explained before, 2 ns are required (which corresponds to 4 ns clock cycle in 50% duty cycle system) to convert 6 bits, when the A/D converter runs faster than 250 MS/s there is insufficient time to resolve the last bit. The ERBW of the 6-b CABS A/D converter extends to 2.8 GHz due to the reduced input range increasing the input bandwidth of the track-and-hold significantly. The measured power consumption of the 6-b CABS converter is 0.56 μW/MHz.

5.3.3.4 Sensitivity to Environmental Parameters

As already indicated during the description of the architecture, the comparator, the central block in the architecture is sensitive to supply voltage and temperature variations. The feedback D/A converter is sensitive to reference voltage variations and since the supply pin is used as reference, it is sensitive to supply variations. During the measurement the supply voltage was varied (after calibration) and the ENOB at low input frequencies was varied. In a band of ±15 mV around the nominal supply voltage the deterioration in ENOB is limited to 0.2 bit. When the supply voltage is reduced, the feedback D/A converter has a smaller step and the built-in thresholds shift to 0. The full range of the A/D converter is reduced. The opposite happens when increasing the supply voltage. With a supply voltage change of up to ±30 mV the ENOB dropped to 5.8-b (0.6-b below the maximum ENOB of 6.4). The shifts of thresholds and feedback D/A converter are no longer sufficiently aligned and the linearity of the converter is compromised. Simulations indicate that for the built-in threshold there is a similar shift in built-in threshold for a temperature change of ±15 degrees as for a ±15 mV supply change. Similar sensitivities and effects are reported in [22]. Both these effects can be compensated for by either recalibrating the A/D converter when environmental constraints change (supply/temperature). Alternatively, a sufficiently stable supply/reference (<10 mV change) can be created and the temperature variation can be handled by a lookup table approach. Temperature dependent calibration values are then stored in a table and by monitoring the temperature range in which the chip functions [23], correct values can be loaded.

5.3.3.5 Energy Efficiency

The power efficiency of an A/D converter is expressed by the energy per conversion step figure of merit (see Eq. 5.1). For the 7-b two-step converter the figure of merit is 10.3 fJ/conversion step running at sampling frequencies between 33 MS/s and 150 MS/s, and for the 6-b CABS converter the figure of merit is 15 fJ/conversion step running at sampling frequencies between 33 Ms/s and 250 MS/s. For the calculation the average power is taken (over full scale input signals at different frequencies) and the *ENOB* is taken at DC since the value at Nyquist is above *ENOB*-0.5 (ERBW is higher than $F_S/2$). However it is not only on the supply pin that power is delivered to or taken from the A/D converter. In the next paragraphs we will consider the analog input, the clock input and the digital output pins. For the digital input and output pins an equivalent energy per conversion step can be calculated as follows, starting from P_{SW}, the switching power:

$$P_{\text{sw}} = aCV_{\text{DD}}^2 F_S \tag{5.4}$$

where a is the switching activity, C is the switched capacitance, F_S is the sampling frequency and V_{DD} is the supply voltage (switching swing). The switching power P_{SW} is converted to energy per conversion step E_{SW} as follows:

$$E_{\text{sw}} = \frac{P_{\text{sw}}}{2^{\text{ENOB}} F_S} = \frac{aCV_{\text{DD}}^2}{2^{\text{ENOB}}} \tag{5.5}$$

where *ENOB* is the effective number of bits. For the clock input pin the switching activity is 1, while the capacitance is approximately 8 fF, resulting in an E_{SW} of 0.1 fJ/conversion step. For the digital code outputs, the switching activity is at worst 0.5 (0–1 toggling on each output), the load is approximately 10 fF which is the input capacitance of the D-flipflops resampling the A/D converter output data. This results in (for seven outputs) approximately 0.4 fJ/conversion step.

As a last entry point, the power on the analog input pins is taken into account. Every sample clock charge is taken in by the converter. The charge added on the input capacitor depends on the change of the input signal. The worst case (largest change of voltage) is at Nyquist condition ($f_{in} = f_S/2$), for a full scale input signal. In each sample period when one of the differential input capacitances is charged (while the other input is discharged), a charge of (per sample clock) is given as:

$$Q_{\text{signal,FS,Nyquist}} = C_{\text{in,se}} \frac{V_{\text{signal,FS}}}{2} \tag{5.6}$$

where $C_{in,se}$ is the single ended input capacitance (250 fF), $V_{signal,FS}$ is the full scale input signal (peak-to-peak differential). If this charge is delivered from the supply (1 V), an energy per conversion step is derived as follows:

$$E_{\text{signal,FS,Nyquist}} = \frac{Q_{\text{signal,FS,Nyquist}} V_{\text{DD}}}{2^{\text{ENOB}}} = C_{\text{in,diff}} \frac{V_{\text{signal,FS}} V_{DD}}{2^{\text{ENOB}}} \tag{5.7}$$

where the differential input cap $C_{\text{in,diff}}$ has been put in and a factor of two has been removed. An energy transfer of 1.1 fJ/conversion step is obtained.

In conclusion, the energy budget of the A/D converter is dominated by the energy taken from the supply pin. The energy taken from the input and delivered to the output pins is about an order of magnitude lower, and the energy on the clock pin is negligible.

5.3.4 Summary

In this section two new A/D converter techniques have been presented. The SAR-CC technique implements the SAR controller on the comparators which are also used to do the quantization. By removing the explicit controller circuits and only by using comparators, D/A converter and delay elements the power consumption is reduced. The CABS technique implements a new approach to successive approximation with a comparator-based asynchronous binary search. The signal is bracketed in a binary tree of asynchronously clocked comparators with built-in thresholds. This eliminates the D/A converter from a traditional SAR A/D-converter to reduce the power consumption even further. Both of these techniques have been merged in a 7-b two-step prototype using 1 bit of SAR-CC and 6 bits of CABS conversion. The 7-b prototype consumes 133 μW from a 1 V supply voltage at 150 MS/s, achieving 6.4 ENOB and an energy efficiency of 10.3 fJ/conversion step.

5.4 Conclusions

To solve the power bottleneck that ADCs often pose in the practical implementation of battery-powered or energy-harvesting biomedical systems, research towards ultra-low-power ADC architectures is crucial. In this chapter, two techniques have been presented that fulfill this need.

Charge-sharing SAR ADCs offered sampling speeds of several ten's of MHz and resolutions in the 8–10bit range, at an efficiency level of 50 fJ per conversion-step, a factor 5–10X lower that classical charge-redistribution SAR ADCs. A typical application example of such a system could be a brain monitoring system where 64 neural probe signals are time-multiplexed and digitized by a single ADC.

For slightly lower resolutions (up to 7-bit) but for an even higher efficiency of only 10 fJ per conversion-step, the comparator-based asynchronous binary search ADC architecture was presented. This technique combines the simplicity of flash ADCs with the efficiency of the successive approximation algorithm, and easily achieves high sampling rates up to 150 MHz.

We hope that these innovations will result in an important reduction of the power consumption of biomedical systems, and enable new architectural trade-offs where the advances in ADC performance are exploited to achieve a new optimum in the mixed analog/digital design space.

References

1. van de Plassche R (2003) CMOS Integrated Analog-to-Digital and Digital-to-Analog Converters, 2nd edn. Kluwer, Dordrecht
2. Walden RH (1999) Analog-to-digital converter survey and analysis. J Select Areas Commun 17(4):539–550, April 1999

3. Craninckx J, Van der Plas G (2007) A 65 fJ/Conversion-Step 0-to-50 Ms/s 0-to-0.7 mW 9b Charge sharing SAR ADC in 90 nm Digital CMOS. ISSCC Dig Tech Papers, pp. 246–247, Feb 2007
4. van Elzakker M, van Tuijl E,.Geraedts P, Schinkel D, Klumperink E, Nauta B (2008) A 1.9 μW 4.4fj/conversion-step 10b 1 ms/s charge-redistribution ADC. In: Proceedings of Digest of Technical Papers. IEEE International Solid-State Circuits Conference ISSCC 2008, p 244–610, 3–7 Feb 2008
5. Van der Plas G, Verbruggen B (2008) A 150 MS/s 133 μW 7 bit ADC in 90 nm digital CMOS. IEEE J Solid-State Circuits, 43(12):2631–2640, Dec 2008
6. Agnes A, Bonizzoni E, Malcovati P, Maloberti F (2008) A 9.4-ENOB 1 V 3.8 μW 100 kS/s SAR ADC with time-domain comparator. In: Proceedings of Digest of Technical Papers. IEEE International Solid-State Circuits Conference ISSCC 2008, pages 246–610, 3–7 Feb 2008
7. Boulemnakher M, Andre E, Roux J, Paillardet F (2008) A 1.2 V 4.5 mW 10b 100 MS/s pipeline ADC in a 65 nm CMOS. In: Proceedings Digest of Technical Papers. IEEE International Solid-State Circuits Conference ISSCC 2008, pages 250–611, 3–7 Feb 2008
8. Giannini V, Nuzzo P, Chironi V, Baschirotto A, Van der Plas G, Craninckx J (2008) A 820 μW 9b 40 MS/s Noise Tolerant Dynamic SAR ADC in 90 nm Digital CMOS. ISSCC Digest. of Technical Papers, pp. 238-239, Feb 2008
9. Abo M, Gray P (1999) A 1.5-V, 10-bit, 14.3-MS/s CMOS pipeline analog-to-digital converter. IEEE J Solid State Circuits 34(5):599-606, May 1999
10. Van der Plas G, Decoutere S, Donnay S (2006) A 0.16 pF/conversion-step 2.5 mW 1.25 GS/s 4b ADC in a 90 nm digital CMOS process. ISSCC Digest of Technical Papers, pp. 566–567, Feb 2006
11. Van den Bosch A (2004) Static and dynamic performance limitations for high speed D/A converters. ISBN 9781402077616, Springer
12. Scholtens P, Vertregt M (2002) A 6-b 1.6-Gsamples Flash ADC in 0.18 μm CMOS using averaging termination. IEEE JSSC 37(12):1599–1609, Dec 2002
13. Lin J, Haroun B (2002) An embedded 0.8 V/480 μW 6B/22 MHz flash ADC in 0.13-μm Digital CMOS process using a nonlinear double interpolation technique. IEEE JSSC 37(12)1610–1617, Dec 2002
14. Draxelmayr D (2004) A 6b 600 MHz 10 mW ADC Array in Digital 90 nm CMOS. IEEE digest of ISSCC 2004, paper 14.7
15. Ginsburg BP, Chandrakasan AP (2008) Highly Interleaved 5b 250 MS/s ADC with redundant channels in 65 nm CMOS. ISSCC Digest of Technical Papers, pp. 240–241, Feb. 2008
16. Chen S, Brodersen R (2006) A 6-bit 600-MS/s 5.3-mW asynchronous ADC in 0.13-μm CMOS. IEEE J Solid-State Circuits 41(12)2669–2680, Dec 2006
17. Brooks L, Lee H-S (2007) A zero-crossing-based 8-bit 200 MS/s pipelined ADC. IEEE JSSC 42(12):1896–1906, Dec 2007
18. Van der Plas G, Verbruggen B (2008) A 150 MS/s 133 μW 7b ADC in 90 nm digital CMOS Using a comparator-based asynchronous binary-search sub-ADC. IEEE Digest of ISSCC 2008, paper 12.3
19. Kobayashi T et al (1993) A current-controlled latch sense amplifier and a static power-saving input buffer for low-power architecture. IEEE JSSC 28(4):523–527, April 1993
20. Nuzzo P et al. (2008) Noise analysis of regenerative comparators for reconfigurable ADC architectures. IEEE Trans Circuits Sys I: Fundam Theory Appl 55(6): 1441–1454, July 2008
21. Nuzzo P et al. (2006) Efficient calibration through statistical behavioral modeling of a high-speed low-power ADC. Proceedings of PRIME, pp. 297–300, Jun 2006
22. Daly D, Chandrakasan A (2008) A 6b 0.2-to-0.9 V highly digital flash ADC with comparator redundancy. ISSCC Digest of Technical Papers, pp. 554–555, Feb 2008
23. Petrescu V et al. (2006) A signal-integrity self-test concept for debugging nanometer CMOS ICs. ISSCC Digest of Technical Papers, pp. 544–545, Feb 2006
24. McCreary J, Gray P (1975) All-MOS charge redistribution Analog-to-Digital conversion techniques—Part I. IEEE J Solid-State Circuits 10(6):371–379, Dec 1975
25. Verbruggen B et al. (2008) A 2.2 mW 5b 1.75 GS/s Folding Flash ADC in 90 nm Digital CMOS. ISSCC Digest of Technical Papers, pp. 252–253, Feb 2008

Chapter 6
Low Power Bio-Medical DSP

Hyejung Kim and Hoi-Jun Yoo

6.1 Introduction

Many micro-watt power processors have been proposed to improve the processing efficiency for the possible application to Bio Signal Processing [1–5]. Figure 6.1 denotes the energy of recent low power (energy) processors, indicating the trend of the processor's energy efficiency. The first group is the general purposed processor [1–3, 5]. They have developed for low power operation. Yet, they still require the long operating time, which is the important factor of the energy consumption. Thus, the application specific processor rather than general purpose processor has been developed [4]. Even though it consumes more power than the general purposed processors, the operating time can be reduced remarkably due to the dedicated hardware and instructions. Thus, if the application is clearly defined such as the Bio Signal Processing, it becomes very attractive to improve the energy efficiency.

Recently, with the increase of the interests in the healthcare, the need for the ambulatory arrhythmia monitoring system has been rising exponentially. The monitoring system records ECG signal continuously in ambulatory condition for a sizable time like several hours. The system transmits the record data to the user or the healthcare center like hospital when the alert ECG signal is detected or the recording period is finished. In order to monitor and analyze the ECG signal, the functions operated at the clinical instrument such as signal sensing and the classification should be integrated into the light-weight, ambulatory monitoring system.

The most important requirements for the ambulatory monitoring system are ultra low energy operation for the long battery life time and a small footprint for wearability. In general, since the highest energy consuming parts are the memory transaction blocks and the wireless communication blocks than the processing block, the data processing as much as possible before transmission is the most efficient method to reduce the total system energy consumption.

H. Kim (✉)
Ultra Low Power and Extreme Electronics Group IMEC, Kapeldreef 75, B-3001 Leuven, Belgium
e-mail: hyejung@imec.be

H.-J. Yoo, C. van Hoof (eds.), *Bio-Medical CMOS ICs*,
Integrated Circuits and Systems, DOI 10.1007/978-1-4419-6597-4_6,

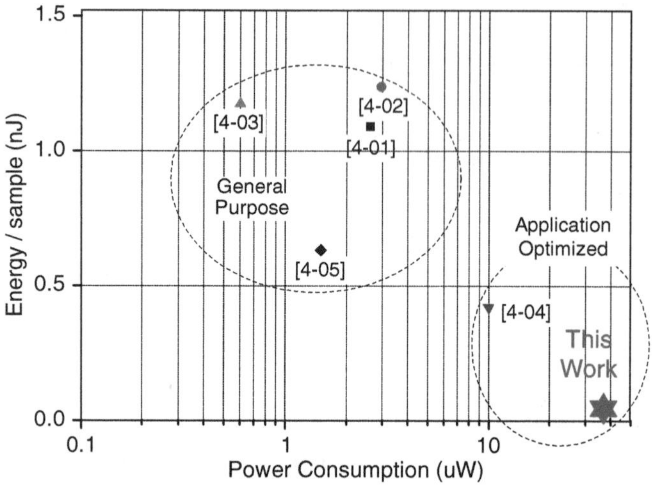

Fig. 6.1 Energy consumption trend for recently reported low energy processor

In this chapter, a Bio Signal Processor, especially ECG signal processor (ESP) will be investigated. Especially, low energy ECG signal processor design will be addressed to improve the required signal processing for the ECG monitoring system under very low energy budget. The energy reduction schemes will be explained in terms of the architectural improvements, the QLV based pre-processing, data reduction, the arrhythmia detection and the voltage scaling.

6.2 ECG Signal Processor Design

6.2.1 Algorithm Overview

The ECG signal processor executes mainly four functions: filtering, compression, ECG classification and encryption. Figure 6.2 shows the flow diagram of the proposed ECG signal processing algorithm. The ECG sensing data is digitized and transmitted to the ESP module. At first, the filtering unit is applied to reduce the noises such as baseline wander, power line interference, and high frequency noise. After filtering in the preprocessing stage, the Quad Level Vector (QLV), which indicates the ECG waveform delineation and its information level, is generated for the next processing. The QLV supports the both flows to achieve better performance with low computational complexity. The main processing stage consists of the compression flow and the classification flow. The compression flow combines the lossy and the lossless algorithm, which are the skeleton, delta coding and the Huffman coding. The classification flow extracts the features and analyzes whether the current heartbeat has the abnormality. Finally, the compressed data and the analysis results are encrypted for protecting the user privacy and authentication and stored to the data memory.

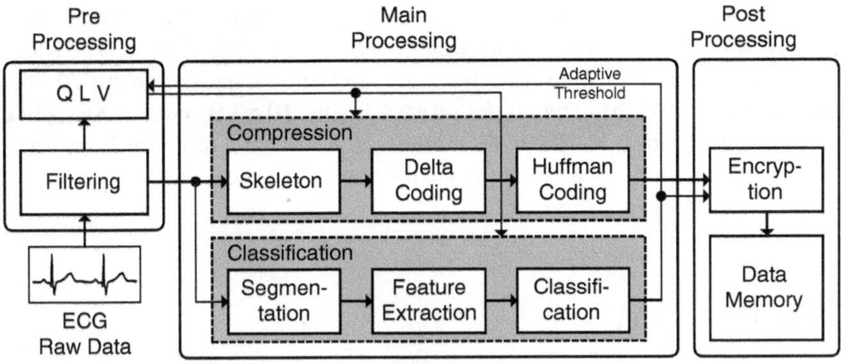

Fig. 6.2 Flow diagram of the proposed ECG signal processing algorithm

6.2.2 Hardware Implementation

The ESP consists of three heterogeneous processors with specific functionalities as shown in Fig. 6.3. Since the compression and the encryption takes thousands of cycles to generate on general RISC processor, that cycle limits to reduce the operating frequency and the voltage. To optimize to these functions—filtering, skeleton, Huffman coding, encryption, the dedicated hardware accelerator should be designed and integrated. Pre-processor and the post-processor consist of the specific function-based accelerator to meet the performance requirement. And the RISC architecture

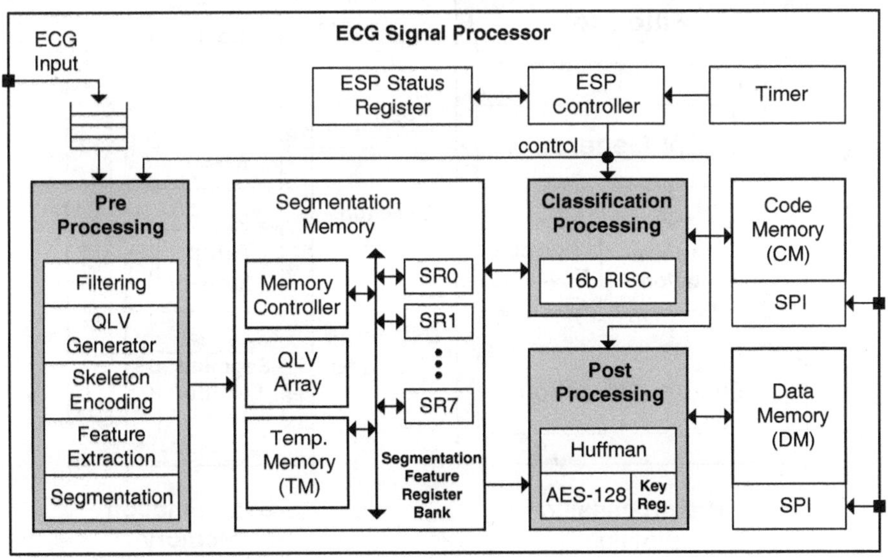

Fig. 6.3 Top architecture of ECG signal processor

is adapted for the classification stage to enhance programmability to apply the various classification algorithms and to overcome the variability of users. The sensing data is incoming through the sensor interface, and SPI (Serial Peripheral Interface) blocks are integrated for external data transmission. 10.5 kB internal SRAMs are integrated, which are 2 kB for code memory, 0.5 kB for temporary memory and 8 kB for data memory. The system controller including the status register and the timer performs the system control.

6.3 Pre Processing

The aim of the pre-processing is the efficient reduction of the raw data while maintaining the crucial information for further processing. Moreover, since the pre-processor should process ECG input stream, the throughput is the most important factor for the pre-processing. It is designed by fully pipelined, which consists of 4 stages with 5 functions as shown in Fig. 6.4. The pipeline flow provides 1-cycle/sample throughput. The main functions consist of filtering, QLV generation, segmentation, skeleton encoding, and the feature extraction. The results of each stage are stored to the segmentation memory.

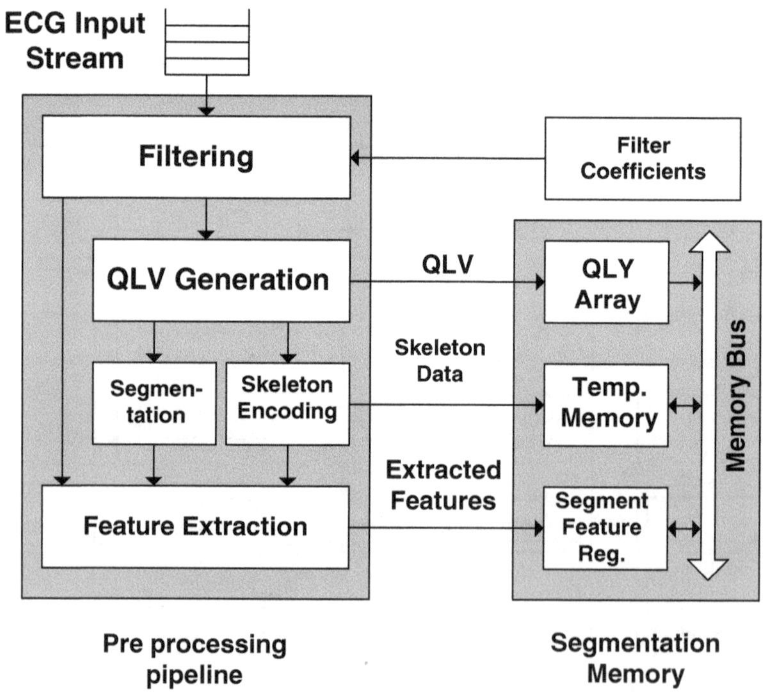

Fig. 6.4 Block diagram of pre processing pipeline

6.3.1 Filtering

The filter can be classified into two major groups of a finite impulse response (FIR) filter and an infinite impulse response (IIR) filter. A FIR filter has a unit impulse response that has a limited number of terms, as opposed to an IIR filter which produces an infinite number of output terms when a unit impulse is applied to its input. The FIR filters are generally realized nonrecursively, which means that there is no feedback involved in computation of the output data. The output of the filter depends only on the present and past inputs. In this chapter, the typical FIR filter is implemented for the low pass filter (LPF). The FIR filter consists of 8 taps and the coefficient registers. The order of the filter can be constructed up to value of 8 with programmable coefficients as shown in Fig. 6.5.

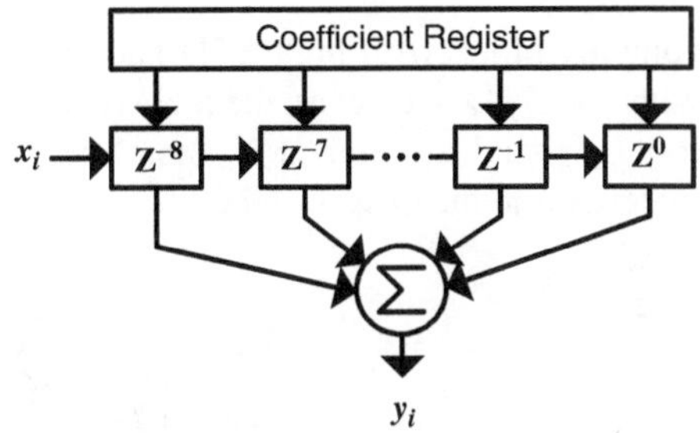

Fig. 6.5 Block diagram of FIR filter

6.3.2 Feature Extraction

Figure 6.6 shows the PQRST wave forms of ECG signal. The significant features of ECG, such as the R point, RR interval, amplitude of R point, average RR interval, QRS duration and existence of QRS [6], should be extracted for the next classification stage. The significant features are shown in Fig. 6.6. And one heartbeat signal begins from the P wave and finishes at the next P wave of the following heartbeat. Moreover, the heartbeat can be divided into a crucial part and a plain part [7]. The QRS complex wave is the most important part of the cardiology system to determine arrhythmia [8]. The P and T wave also have a high level of information and

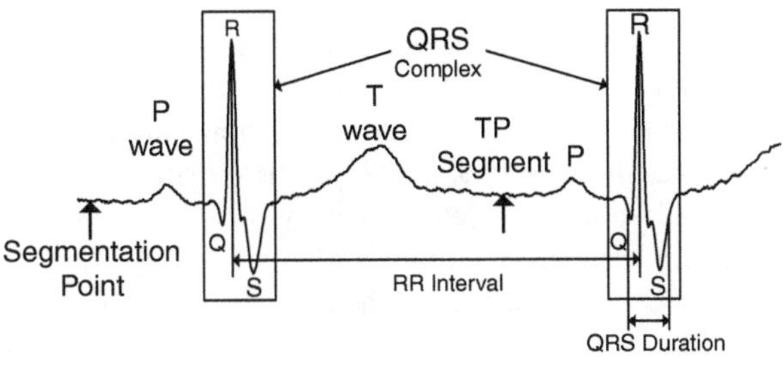

Fig. 6.6 Significant features of one beat ECG signal

the remaining plane parts of TP segment contain less information. Therefore, in this work, the ECG signal is classified into four different levels to preserve as much property of the information as possible. Afterward, the number of bits is assigned differently according to the level. For example, more bits are assigned to the highest level block, and fewer bits are assigned to the lower level block.

In this way, the ECG data is divided into smaller blocks and every block is encoded in real time as an independent entity. In this work, the unit block size is selected in 0.04 second which is the half duration of the QRS complex duration. It is suitable period to detect the change the ECG signal precisely. After block division, the QLV of the block is calculated. For normal ECG signals, the QRS complex part can be regarded as a typical representative signal with high standard deviation ($STD = \sqrt{\sum_{I=0}^{N_B-1} (x_i - \overline{x})^2/N_B}$) in comparison with the plain part [7]. The complex block with high STD has more crucial information than the plain block with low STD. However, the STD requires the complex calculations such as square root ($\sqrt{x}$) and squaring (x^2). Therefore, the mean deviation (MD) value is proposed to determine the QLV instead of the STD. The MD is defined as:

$$MD = \frac{\sum_{i=0}^{N_B-1} |x_i - \overline{x}|}{N_B} \tag{6.1}$$

where, x_i is the sampled data, $\overline{x}$ is the mean value of single block, and N_B is the block size. The MD requires the only absolute operation, thus it leads to lower computation complexity than STD with the almost same results [9].

Afterward, each block is decomposed into four compression levels by comparing the MD value with the three threshold values (TH_0, TH_1, TH_2) as given by (6.2), the proposed skeleton equation.

$$QLV(CR_{\text{block}}) = \begin{cases} 0(8\alpha : 1) & \text{if } MD < TH_0 \\ 1(4\alpha : 1) & \text{if } TH_0 \leq MD < TH_1 \\ 2(2\alpha : 1) & \text{if } TH_1 \leq MD < TH_2 \\ 3(\alpha : 1) & \text{if } MD \geq TH_2 \end{cases} \tag{6.2}$$

In order to obtain the accurate QLV, it is important to choose properly the three threshold values (TH_0, TH_1, TH_2). Since the amplitude of ECG signal varies with the environmental conditions such as noise injection and body movements, the QLV threshold should be adaptable to deal with those variations in a real time. The threshold values are determined by the maximum MD value ($\mathrm{MD}_{\max}$) of the previous 8 heartbeats, then the results is applied to the preprocessing by the feedback path for QLV adjustment of the next heartbeat. The threshold values are determined as (6.3).

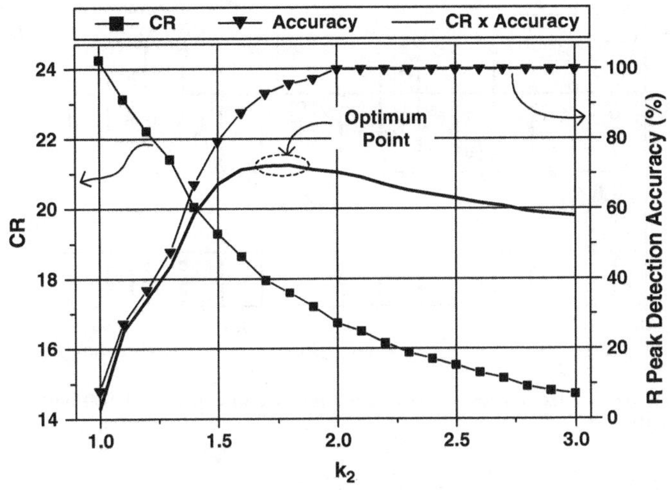

Fig. 6.7 Optimum point selection of the threshold coefficients (k_2)

$$TH_l = \frac{1}{k_l} \cdot \frac{\sum_{i=0}^{7} MD_{\max,i}}{8}, \quad l = 0, 1, 2 \tag{6.3}$$

where the threshold coefficients, k_l, are the programmable coefficient. Figure 6.7 shows the effect of the compression ratio (CR) and the R peak detection accuracy according to the value of the coefficient, k_2. The CR increase with the decrease of the k_2. However, the lower k_2 value is very susceptible to the noise interference or the amplitude variation, and its detection accuracy is very low. On the contrary, if the coefficient values go up, the CR decreases while the accuracy improves. Therefore, there exists the optimum threshold values between the noise robustness and the accuracy, and the optimum point for the k_2 is between 1.6 and 2.0 according to the Fig. 6.7.

The accuracy of R peak detection is crucial for the reliable analysis in this flow, because the R peak contains the primary data for arrhythmia analysis like RR interval [10]. During the R peak detection operation, the QLV helps to reduce the peak searching cost. In case of the conventional system [10], the searching window would be 1 second same as the one heartbeat duration. In this proposed system, searching for QLV array is performed first. Then the only selected searching window, 40–80 ms, is applied to find the real peak. Although this searching method has 1% memory capacity overhead for the QLV array, the number of memory access time is reduced by 90%. Figure 6.8 shows the successful R peak detection results using MIT/BIH record 100 with serious noise injection (SNR = −10 dB) and the selected search window. By using the QLV, only 8% search window is enough for the investigation compared to the entire range as shown in Fig. 6.8(b). Only one fault negative result is detected as R peak due to the steep noise denoted by *x* near the 3800 point of

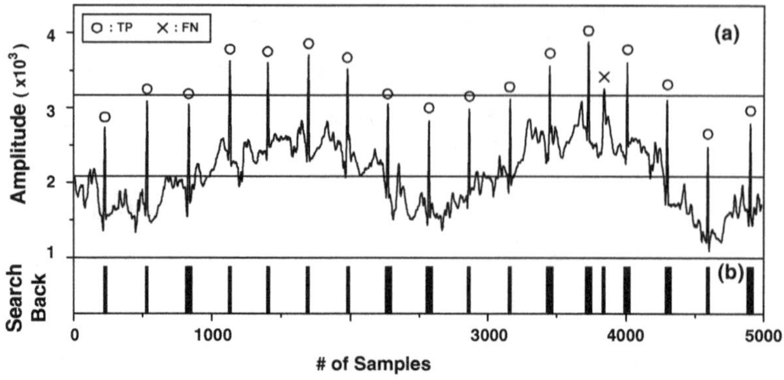

Fig. 6.8 (**a**) R peak detection results. (**b**) Searching window in record 100 with noise injection (SNR = −10 dB)

Fig. 6.8(a), but more preprocessing like filtering and the post-processing can correct the fault detection.

The performance of the classification can be represented by the sensitivity (Se) and the positive predictivity (+*P*). The Se and +*P* are defined as:

$$\text{Sensitivity(Se)} = \frac{\text{TP}}{\text{TP} + \text{FN}} \tag{6.4}$$

$$\text{Positive Predictivity}(+P) = \frac{\text{TP}}{\text{TP} + \text{FP}} \tag{6.5}$$

False positive (FP) is the number of false beat detection, true positive (TP) is total number of correct R peak detection by the algorithm, and false negative (FN) is the number of the failures to detect the true beat. The proposed method has good sensitivity and positive predictivity, Se = 100% and +*P* = 100%.

6.3.3 ECG Skeleton

Many ECG signal compression algorithms were introduced, and they can be classified into two major groups, the lossless and the lossy algorithms [11]. The lossless algorithms such as LZW [12] and Huffman [13] do not show sizable quantization error, while the compression ratio is generally smaller than that of the lossy algorithm. The compression ratio is typically between 2:1 and 4:1. The lossy algorithm has a comparatively higher compression ratio, typically between 10:1 and 20:1, while it has a possibility to lose the significant information. The lossy algorithm can be classified further into two categories: The direct signal compression and the transformation compression. The direct compression techniques are based on the extraction of a subset of significant samples, such as the FAN [14], CORTES [15], AZTEC [16], and Turning Point [17] algorithms. The transformation techniques retain the coefficients of its particular features, and the signal reconstruction can be achieved by an inverse transformation process. Wavelet transform [8–9,

18–20], Fourier transform [21–22], and the Karhunen-Loeve transform [23] have been introduced for the transformation compression techniques. In contrast to the direct compression techniques, the transformation techniques require heavy arithmetic calculation and large temporary memory capacity due to their large scale frame based transformation operation. For the lossy compression techniques, the reduction of the reconstruction error rate is also important issue, because the error may distort diagnostic information. Moreover, the processing cost is a critical factor in the design of the Holter system. The processing cost is composed of the encoding delay time, computational complexity and the memory capacity. Thus, the trade-off should be made between the compression ratio, the reconstruction error and the processing cost according to the target applications.

In this work, the compression flow consists of three steps: skeleton, delta coding, and Huffman coding. The first step of skeleton is constructed with essential sample to reduce not only the transmission bandwidth, but also the on-chip memory capacity and number of the memory access during the processing. The main idea of the proposed skeleton algorithm is that the number of bits is assigned differently according to the information level by QLV [9]. In other words, more bits are assigned to the highest level block like QRS complex, and fewer bits are assigned to the lower level block like TP segment.

Figure 6.9 shows the skeleton algorithm. The input consists of the block-wise discrete signal $\{x_i(m), i = 1, 2, \ldots n, m = 1, 2, \ldots N_B\}$. And let $N_B^l = N_B/2^{3-l}$, then the output of each block is the set $y\_l = \left(y_{l,1}, y_{l,2}, \ldots, y_{l,N_B^l}\right)^T$ at levels $l = 0, 1, 2, 3$. The final output $\{y_i(m_l), i = 1, 2, \ldots n, m_l = 1, 2, \ldots N_B^l\}$ is determined by the MD value corresponding to QLV in (4). If the compression ratio of the block (CR_{block}) of the 3rd level is $\alpha : 1$, those of the 2nd, 1st, and 0th level are $2\alpha : 1, 4\alpha : 1$, and $8\alpha : 1$, respectively.

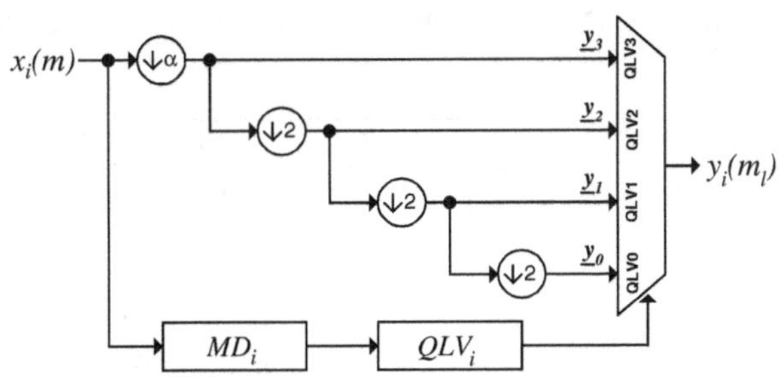

Fig. 6.9 Skeleton algorithm

The ECG data from MIT/BIH [24] is used to verify the efficiency of the proposed algorithm. The sampling rate and the resolution of the signal are 360 samples/s and 12 bits, respectively. Figure 6.10 shows the skeleton steps by the part of the record 231 data. Since the skeleton is the lossy compression algorithm, the goal of the skeleton is reduction of the error rate while maintaining the high compression ratio (CR). The essential samples are extracted according to the QLV values, and the output format of the skeleton consists of the signal amplitude and the sampling interval

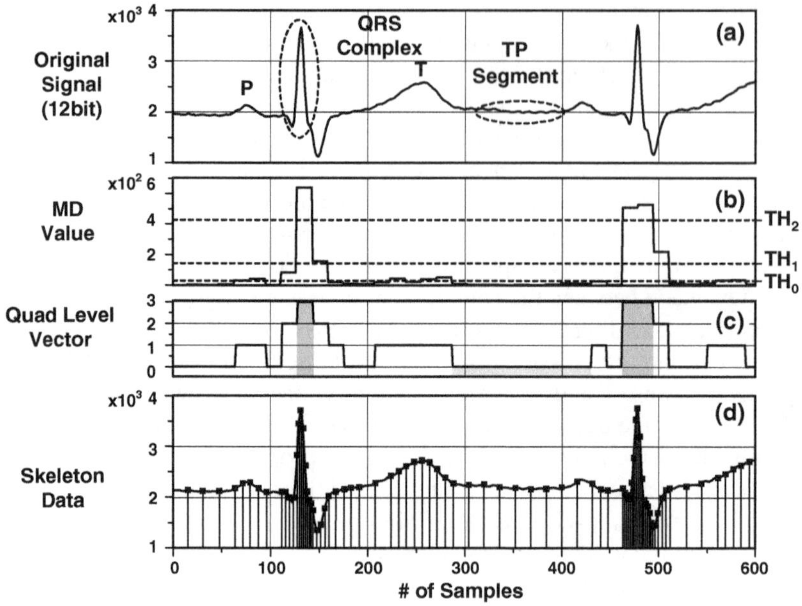

Fig. 6.10 Evaluated result of skeleton with MIT/BIH record 231 **a** Original ECG signal **b** MD values **c** Quad level vector **d** Skeleton results

for the later decoding operation. When decoding the skeleton data, the linear interpolation method is used for the smooth reconstructed waveform with small error rate. Figure 6.11 shows the original signal, encoded signal, and the reconstructed signal as the example of the ECG skeleton method.

The Fig. 6.12(a, b) shows the original ECG signal and the reconstructed results. The result shows that high quality signal is reconstructed with small error rate. Even

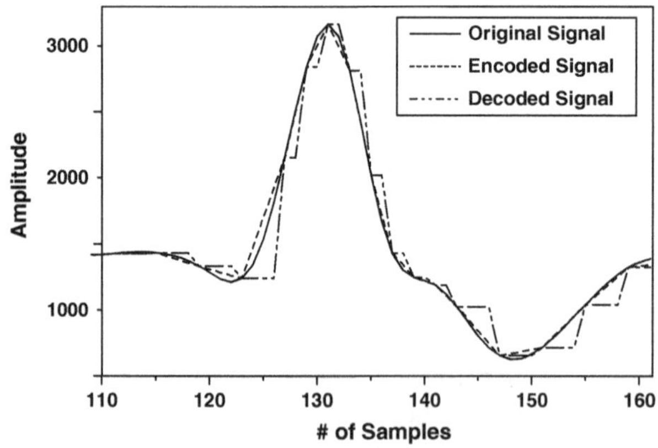

Fig. 6.11 Encoding and decoding results of skeleton

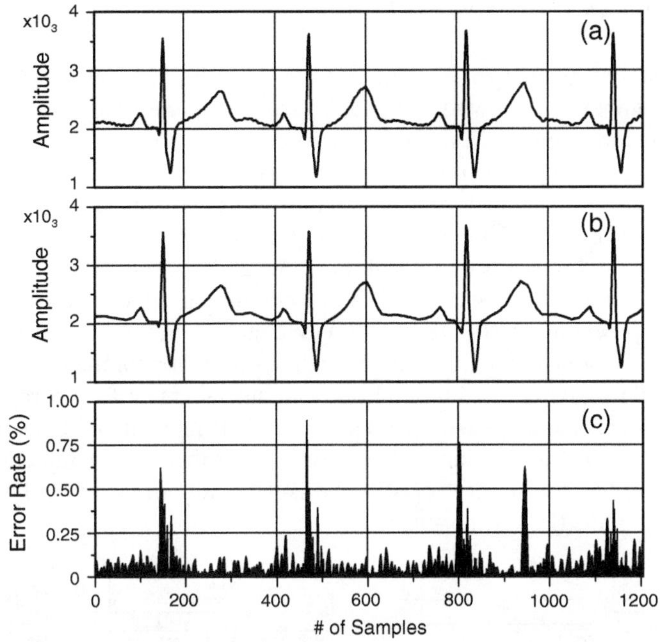

Fig. 6.12 Evaluation results of skeleton. **a** Original ECG signal of record 231. **b** Reconstructed ECG signals. **c** Reconstructed error rate between original and reconstructed signal

though the maximum peak error is 0.85%, the most of samples shows <0.1% error as shown in Fig. 6.12(c).

The coding performances can be evaluated by encoding rate, compression ratio (CR) and percentage root mean square difference (PRD). The PRD is usually used to quantify the performance quality of the compression algorithm [25]. The PRD indicates the error between the original ECG samples and the reconstructed data, and is defined as:

$$\mathrm{PRD}\,(\%) = \sqrt{\frac{\sum_{i=1}^{n} (x_i - \tilde{x}_i)^2}{\sum_{i=1}^{n} x_i^2}} \times 100 \tag{6.6}$$

where n is the number of samples, x_i and $\tilde{x}_i$ are the original data and the reconstructed data, respectively.

The CR and PRD have the close relationship in the lossy compression algorithm. In general, the CR goes higher with the higher lossy level, while the error rate goes up. The final goal of the proposed compression algorithm is to keep the PRD value smaller than that of the conventional methods [7–8, 11–23] while maintaining the similar CR.

6.3.4 Segmentation Memory

The segmentation memory is implemented to keep the information of previous 8 heartbeats, temporary. It includes the QLV array, temporary memory, and the segmentation feature register files. The three processors share the processing data the results through the segmentation memory. Therefore the shared memory architecture should be implemented, and the memory management unit is necessary to protect the congestions. The priority scheduling technique is used, which the pre processor has the highest priority, and the RISC and post processor have the next, and the lowest priority, respectively. If all processing units try to access at the same time, the pre processing task is treated as a first, and the post processing task should wait until the other tasks are finished. Since the information for the 8 previous heartbeats are used to diagnose the current heartbeat, the 8 segment-feature register banks (SR0–SR7) stores the extracted features such as R-R interval and QRS duration as shown in Fig. 6.13. The classification processor refers them. The temporary memory stores the recent ECG data. By applying the skeleton algorithm, about 10 heartbeats can be stored in the 0.5 kB capacity.

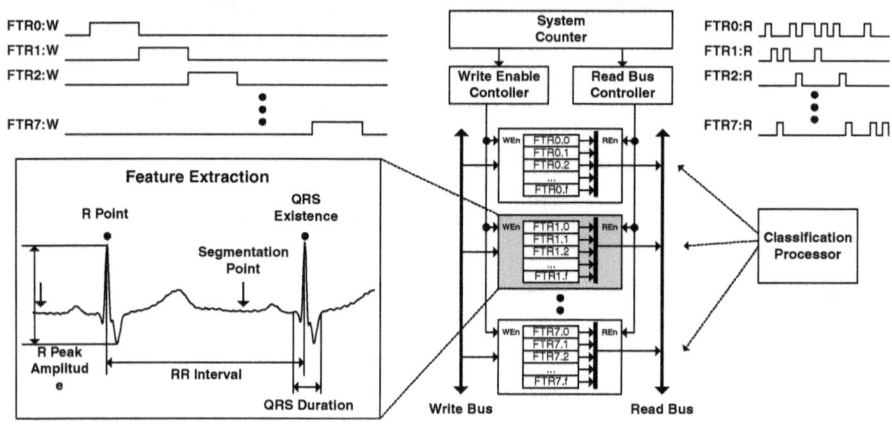

Fig. 6.13 Segmentation register

6.4 Classification Processing

6.4.1 ECG Classification Algorithm

After the pre-processing, the classification algorithm checks whether the current heartbeat has abnormal arrhythmia. Figure 6.14 shows the overall flow diagram for the arrhythmia detection the alert mode operation. When the extracted features meet the specific condition, the current heartbeat is classified as the disorder heartbeat. Otherwise, the heartbeat is regarded as normal. If the abnormal heartbeat is

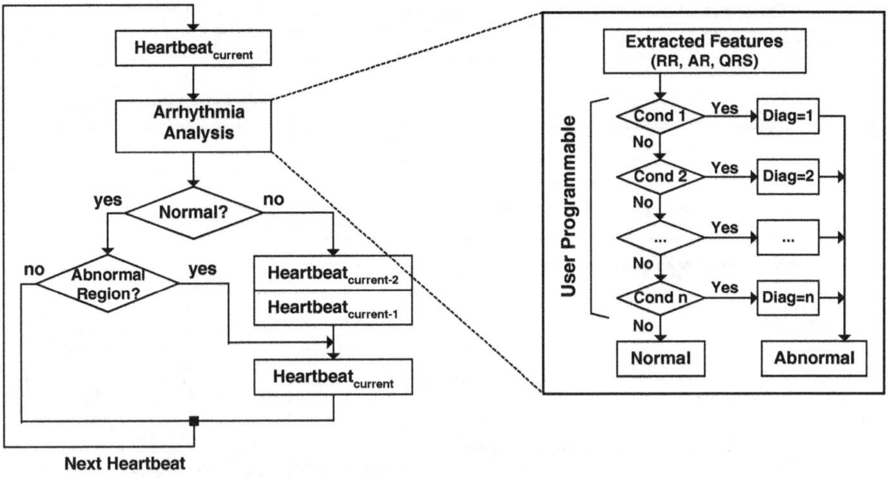

Fig. 6.14 Flow diagram of the ECG classification algorithm

detected, the region including the 2 seconds before and after the detected abnormal heartbeat is decided as abnormal region. Afterward, the abnormal region is sent to the post processing stage. Otherwise, the next heartbeat is treated without any further operation.

The nine major disorder symptoms are chosen, such as bradcardia, tachycardia, asystole, skipped beat, R-on-T, bigeminy, trigeminy, PVC, and APB, and each symptom can be characterized by the simple numerical calculation [26]. Table 6.1 summarizes the mentioned arrhythmia conditions.

6.4.2 Micro Architecture of RISC

Figure 6.15 shows the micro architecture of the RISC for the classification stage. The processor is designed in 3 pipeline RISC architecture because the 3 stages pipeline is known as the optimum for low power consumption and transistor utilization [27]. The pipeline consists of fetch, decode and execution stage. The fetch block fetches the instructions from the code memory in the first stage. The control block decodes the fetched instructions to execute in the second stage. The last stage executes ALU operations, memory access, and write-back to the register file. The general instructions of RISC output the operation result in 3 stages, but some special instructions require multi-cycle. Since the both operations of read and write the register file occurs in the same stage, the data hazard can be eliminated. The branch is performed with 2-cycle penalty.

The RISC sleeps in ordinary times by clock gating until the events occur. The two start modes exist: reset and wakeup mode. The reset mode is performed by reset signal and begins zero address for system initialization. The wakeup mode

Table 6.1 Selected 9 Arrhythmia symptoms and numerical conditions [26]

Arrhythmia	Description	Conditions	ECG Shape
Normal			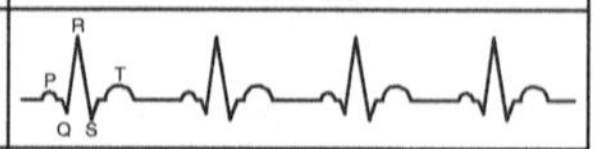
Bradycardia	Heart rate of under 50 beat/min.	RRt > 1.5s ARt > 1.2s	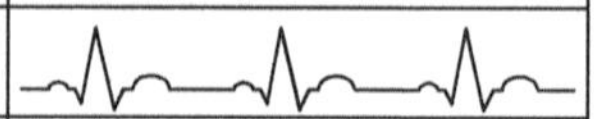
Tachycardia	Heart rate greater than 100 beat/min.	ARt < 0.5s	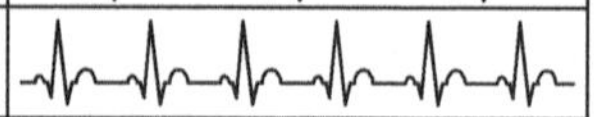
Asystole	State of no cardiac electrical activity.	No QRS > 1.6s	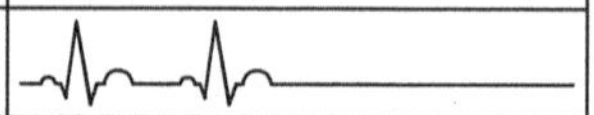
Skipped Beat	Skip one heart beat.	RRt > 1.9ARt-1	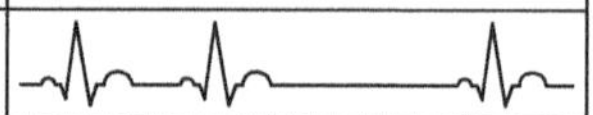
R-on-T	QRS in the electrocardiogram interrupting the T wave of the preceding beat.	RRt < 0.33ARt-1	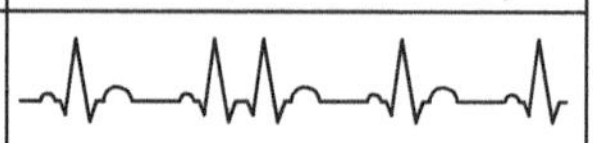
Bigeminy	Abnormal heart beats occur every other concurrent beat.	RRt-3 < 0.9 ARt-4 RRt-1 < 0.9 ARt-4 RRt-3 + RRt-2 = 2ARt-4 RRt-1 + RRt = 2ARt-4	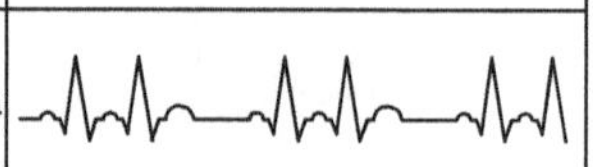
Trigeminy	A cardiac arrhythmia in which the beats are grouped in trios.	RRt-2 < 0.9 ARt-3 RRt-1 < 0.9 ARt-3 RRt-2 + RRt-1 + RRt = 2ARt-3	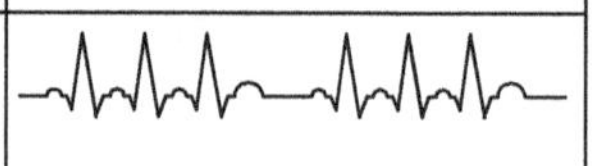
PVC (Premature Ventricular Contraction)	Contractions of the lower chambers of the heart, the ventricles, which occur earlier than usual, because of abnormal electrical activity of the ventricles.	RRt-1 < 0.9ARt-1 RRt-1 + RRt = 2ARt-1 Wider QRS, Opposite T	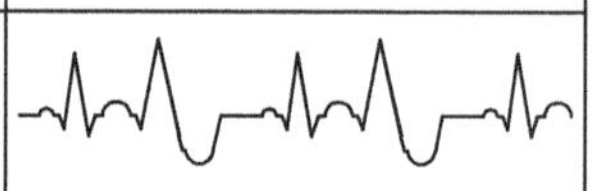
APB (Atrial Premature Beat)	As P waves are small and rather shapeless the difference in an APB is usually subtle	RRt-1 < 0.9ARt-1 RRt-1 + RRt = 2ARt-1	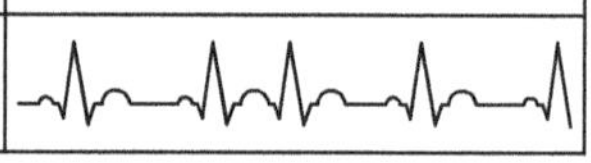

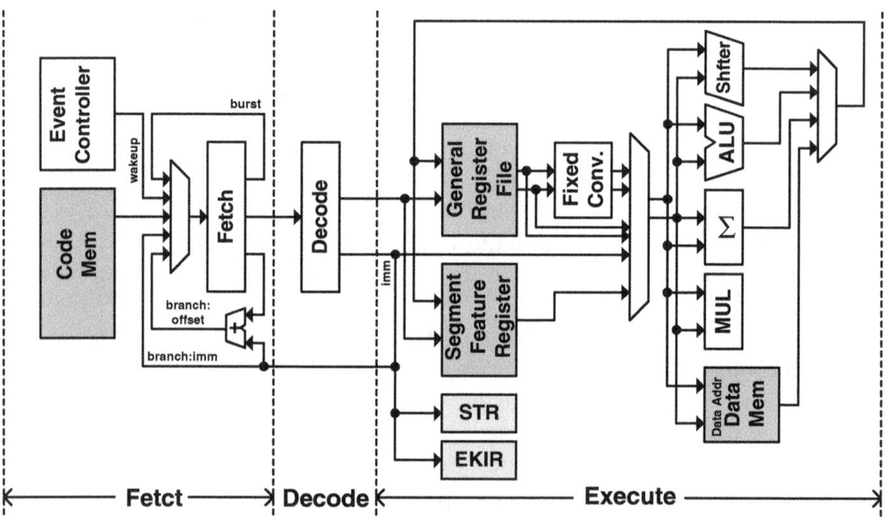

Fig. 6.15 The micro architecture of RISC for classification processor

is performed when external wakeup occurs. If the wakeup occurs, the wakeup controller sends an active clock with wakeup signal and stored PC value. Then the RISC wakes up to operate from specified PC value. After all the arranged program codes are processed, the RISC sends a sleep command to wakeup controller. When receiving the sleep command, the wakeup controller gates the RISC clock to go to sleep mode.

The RISC has 32-byte general purpose register file (GPR) with 16 entries, and it can access the external registers such as segmentation feature register (FTR), status register (STR), encryption key register (EKIR), and the temporary memory (TM). The GPR is used for storing the temporary results during the classification program execution. The FTR has 256-byte register with 128 entries, which are used for storing the extracted features for the previous 8 heartbeat segments. The STR has 16-byte register with 8 entries, which are used for storing the system status configuration such as number of the unit block and the transmission mode. The EKIR used for programming and holding the input key for the AES-128 block.

The datapath of the RISC consists of two's complement integer ALU, shifter, 16×16 multiplier, and SUM block. The classification algorithm requires the average calculations, which requires many operating cycles. In order to reduce the operating cycle, SUM unit is implemented, which performs the summation and the average calculation from the multi-elements from the FTR. In addition, since the programmable fixed point number system is adopted for the floating point number calculation, the point alignment unit is integrated before the datapath units.

During the ALU processing, the power consumption for the code memory access occupies more than half. Moreover, since the memory consumes the largest factor of leakage current, efficient designs is necessary to reduce memory storage requirements. To resolve this request, the well-defined compact 16-bit instruction encoding is proposed. The ISA is summarized in Fig. 6.16, which consists of 6 major categories. Special instructions are prepared for the special function such as feature extraction and the system control. The ISA consists of the single operand format, which the second operand is used as a destination register. The MOV instructions transfer data efficiently between registers. Branch operation can access both immediate point and point with offset. The burst mode provides 128b data movement

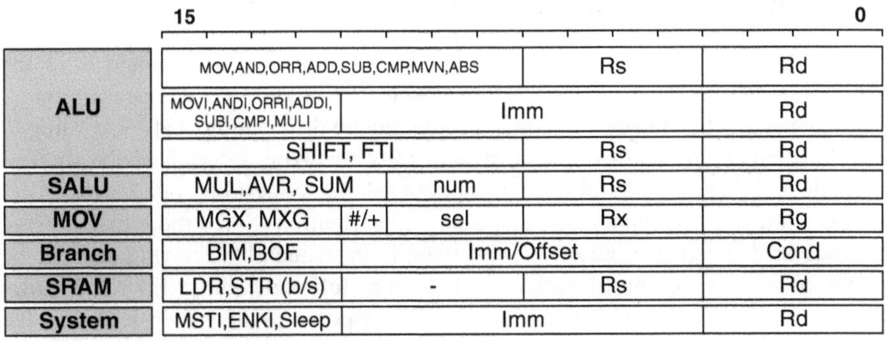

Fig. 6.16 16-bit compact ISA

with a single instruction. System instruction is for the sleep signal generation and the system status configuration. The 4-bit condition field (C/N/Z/V) determines the circumstances under which branch is to be executed.

6.5 Post-processor

The aim of the post-processing is packing data format for efficient memory access, and the encryption to protect the privacy. The delta and the Huffman coding are applied for the packing and compression the output data, and AES-128 is applied for the encryption as shown in Fig. 6.17. The overall latency for post-processing is achieved as 6.56 cycle/sample. It is higher than the pre and classification processing, however, the processing bandwidth is already reduced by the factor of eight by the skeleton operation, its latency is enough to meet the target performance. The power consumption of each block can be controlled by clock gating. If the compression enable signal (COMPEn) or the AES enable signal (AESEn) is disabled, the data is bypass the disabled operation. After this post-processing, the data format is changed into the 16-bit-wise word oriented format in this post-processing stage for the efficient memory access.

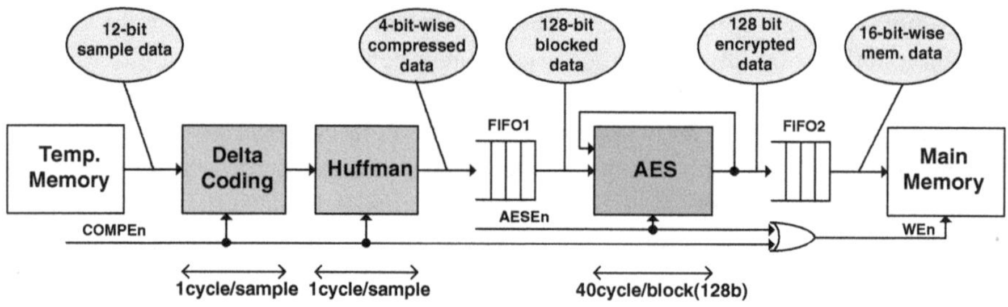

Fig. 6.17 Flow diagram for post processing

6.5.1 *Huffman Coding*

The delta coding and the lossless compression algorithm are adapted after the skeleton method. The Huffman coding is selected because it provides minimum encoding cost when the original data has the unique distribution [13]. According to the Huffman coding scheme, the most frequently occurring values have the shortest bit code length, and the rarely occurring data has the longest bit code length. After the skeleton step, the input data has Gaussian distribution, which more than 50% of the data are located near the zero. Thus, these high frequently occurring data can be transformed with the short length of code by the Huffman coding. Table 6.2 shows the modified 4-bit-wise Huffman coding table proposed in this paper. It divides the entire range into the 4 groups according to the input value to reduce the length and the number of the prefix code bits. Its output result consists of the prefix code and

Table 6.2 Modified 4-bit-wise Huffman coding table

Group	Value	Number of bits: Prefix code	Number of bits: Encoded data	Number of bits: Entire code	Prefix code (in bits)	Comments
0	[−1, 1]	1	3	4	0	0, 1, −1, EOB, Change of QLV
1	[−31, 31]	2	6	8	10	
2	[−255, 255]	3	9	12	110	
3	[−4095, 4095]	3	13	16	111	

the encoded data. The prefix code is selected by the data probability distribution and indicates the group of the input range. The group 0 is reserved specially to notice the end of the block (EOB) and the information of the QLV, while the other groups show the encoded data. The modified Huffman coding method transforms the sample oriented format into the 4-bit-word oriented stream. So, it can obtain the unified data format for the efficient memory access, although the variable sample resolution is provided. When decoding the Huffman code, the bit stream is decoded into the 4-bit-word. The first bit is picked up and compared with the Huffman table, and the original value is reconstructed from the remaining encoded data. The average compression ratio of the Huffman coding is approximately 2:1 without the compression error rate.

Figure 6.18 shows the detailed block diagram of modified Huffman coding. At first, the Huffman coding compresses the input data with the variable length coding. In other words, the modified Huffman coding transforms the sample oriented data format into the 4-bit-word oriented stream. 4-bit FIFO register aligns the encoded output again to the 16-bit-word oriented output to provide the efficient memory access. The 128 bit-block output from AES accelerator is sent to the main memory in 16-bit-wise word format.

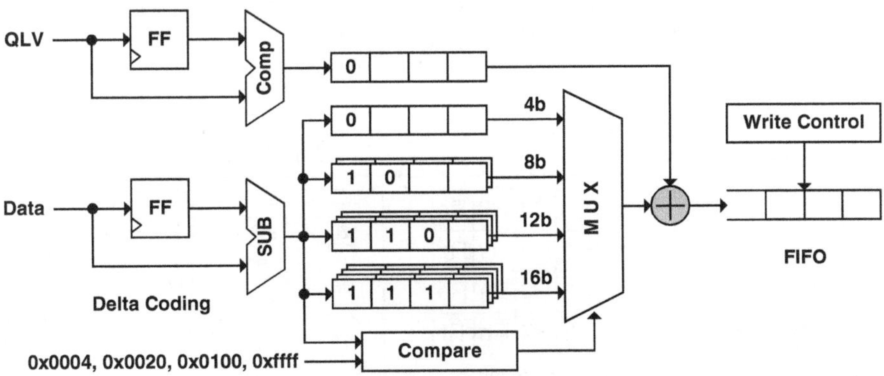

Fig. 6.18 Implementation of modified 4-bit-wise Huffman coding table

6.5.2 AES-128

AES-128 algorithm for data encryption is selected because it is widely used [10]. Most of the previous researches were focused on getting high throughput. However, the throughput is not a main issue in this work because the data rates are not high for this application. We focused to a low energy operation by reducing the registers and the datapath complexity. AES-128 performs the round function iteratively 10 times with 128 bit cipher key [11]. The substitution, shift row, mix column and key addition are operated in each round. Figure 6.19 shows the block diagram of AES datapath. The datapath consists of two paths of key generation and the data encryption. The 10 round keys are pre-computed before encryption, stored in SRAM, and accessed when they are necessary. 16 encryption register are integrated for round operation. Contrary the shift row and the key addition operations are simple, the substitution and the mix column operations have heavy calculation.

The byte substitution step is a nonlinear operation that substitutes each byte of the round data independently according to a substitution table. Because the largest portion of the area and the overall encryption rate are contributed by substitution step, the number of the substitution table is important. In the mix column step, the modular matrix multiplication can be divided into two steps of shifting and addition as (6.7).

$$\begin{bmatrix} b3 \\ b2 \\ b1 \\ b0 \end{bmatrix} = \begin{bmatrix} 2\,3\,1\,1 \\ 1\,2\,3\,1 \\ 1\,1\,2\,3 \\ 3\,1\,1\,2 \end{bmatrix} \begin{bmatrix} a3 \\ a2 \\ a1 \\ a0 \end{bmatrix} = \begin{bmatrix} 2\,2\,0\,0 \\ 0\,2\,2\,0 \\ 0\,0\,2\,2 \\ 2\,0\,0\,2 \end{bmatrix} \begin{bmatrix} a3 \\ a2 \\ a1 \\ a0 \end{bmatrix} + \begin{bmatrix} 0\,1\,1\,1 \\ 1\,0\,1\,1 \\ 1\,1\,0\,1 \\ 1\,1\,1\,0 \end{bmatrix} \begin{bmatrix} a3 \\ a2 \\ a1 \\ a0 \end{bmatrix} \quad (6.7)$$

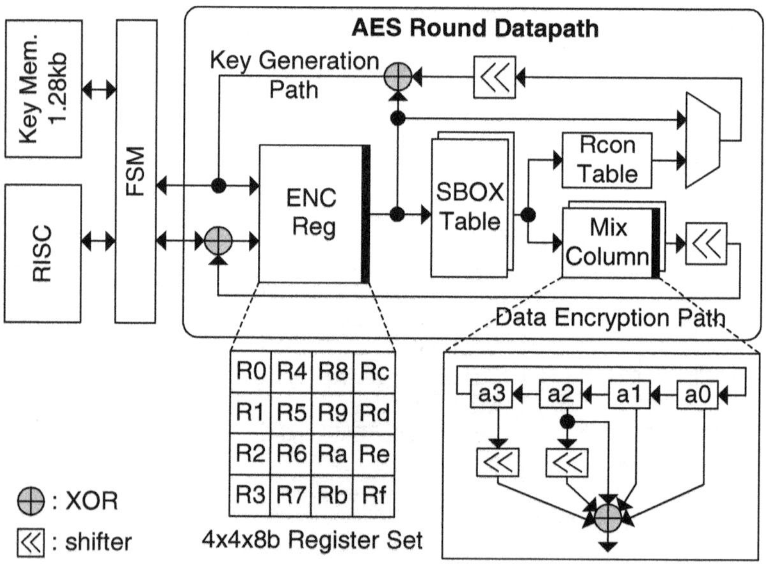

Fig. 6.19 Block diagram of AES datapath

6.6 Low Energy Techniques

6.6.1 Heterogeneous Processor Integration

The operating energy consumption is given by product of the power and the operating time Eq. 6.8. In this work, the energy consumption reduction is achieved by reducing the supply voltage (V), and the operating time (T_{op}).

$$\text{Energy}_{op} = CV^2T_{op} \tag{6.8}$$

As a first, let's reduce the T_{op} by applying three-heterogeneous processor architecture. The three application specific processor is integrated to improve the parallelism. The three processors and the segmentation memory take large area occupation (C) more than 12 times. Since the dedicated hardware and the instructions are implemented to accelerate the significant functions, the operating time is reduced by 1 cycle/sample. It is 420 times reduced compared to the conventional work. Although the area is increased, the energy consumption can be reduced by 31 times as shown in Table 6.3.

Table 6.3 Comparison of energy consumption

	Conventaional [VLSI 2009]	This Work
Type	Single General Purpose Processor	Three-Hetero Processor
		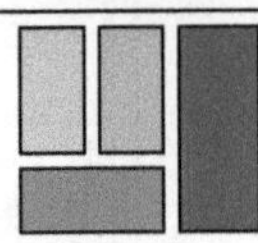
Area	C	12 C
T_{op}	420T	T
Energy	370 CV^2T	12 CV^2T

6.6.2 Low Supply Voltage Operation

The second step is supply voltage reduction. The operating time reduction causes the low power consumption. In addition, the operating time reduction enables the voltage scaling which provides the low energy consumption by power of two. The sampling frequency of the biopotential monitoring system is low enough, i.e. up to 1 kHz, the real-time throughput performance is guaranteed as shown in Fig. 6.20(a). By reducing the voltage from 1.8 to 0.6 V, the energy consumption is reduced by 89% as shown in Fig. 6.20(b).

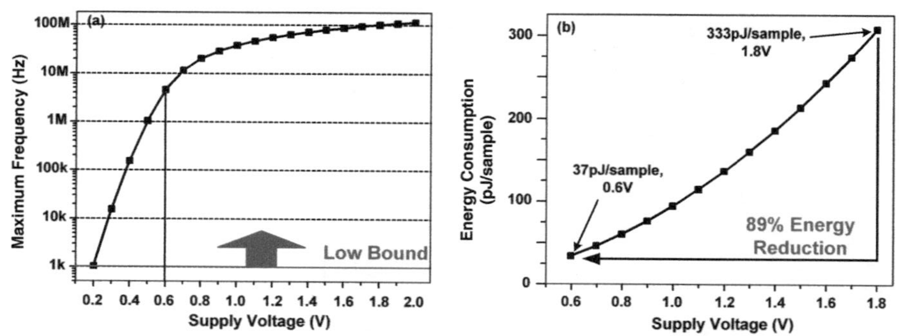

Fig. 6.20 **a** Maximum operating frequency reduction, and **b** Energy consumption reduction by voltage scaling

6.6.3 Segmentation-Based Pipelined Operation

To increase the throughput, the three processors are segmentation-based pipelined. However, since the duration of the heartbeat and their workloads are not same all the time, it is difficult to know when the processors should start or sleep. In this work, the segmentation-based wakeup method is proposed for the efficient relation between the processors. When the processor finishes the work for the single segment, it signals to the next stage to wake it up as shown in Fig. 6.21(a). The arrows represent the wakeup signal for the next processor. The pre-processor executes the input heartbeat segmentation one by one. The classification processor sleeps until the pre-processor signals the wakeup for the n-th heartbeat processing. When the classification operation for the n-th has been processed, it signals the wakeup to the post-processor to work, then the classification processors return to sleep mode. The post-processor is also in sleep mode until the wakeup signal is arrived. Unlike other stages, the post-processor can execute out of order, and treat the multi-segmentation operation. The classification processor decides whether the post-processor operates. What kind of and how many segments should be operated at the post-processor are also decided by the classification processor.

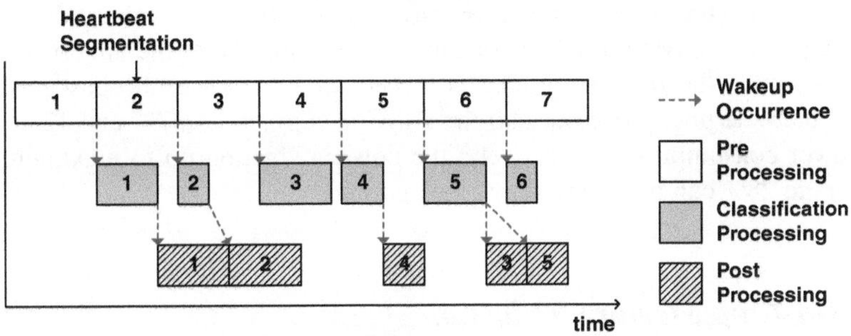

(a) The Timing Diagram of Segmented-based Pipeline Operation

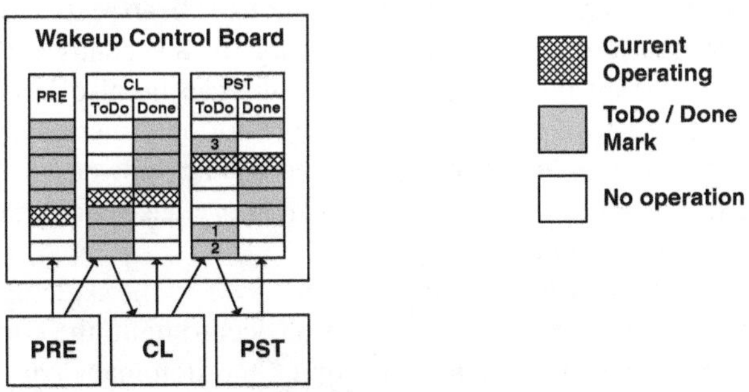

(b) The Wakeup Control Board

Fig. 6.21 Segmented-based pipelined operation and the wakeup control board

The wakeup control board is designed to perform the efficient handling the wakeup event. Figure 6.21(b) shows the block diagram of the proposed wakeup control board. The board contains the currently operating segment, finish, and reservation information for each processor. The post-processing board also contains the order information for multi, and out of order operation. The control board has 8 event registers for processing the 8 previous heartbeat segments. The wakeup control board sends the wakeup signal when the previous stage processor has done (Done = 1), and the processor is reserve to run (ToDo = 1). Each processor sends the Done signal to the board just before return to sleep.

6.6.4 Clock Gating

Because the classification and post processors don't running all the time, the power management is necessary when they don't run to reduce the operating power consumption. In this work, the power management is achieved by clock gating technique. The wakeup control board handles not only wakeup signal, but also clock signal. Each processing unit can be individually enabled or disabled according to the necessity by the clock gating method as shown in Fig. 6.22(a). When the new segmentation is incoming, the clock controller enables the clock signals (CLK_CL, CLK_PST) with the wakeup signal. Figure 6.22(b) shows the timing diagram of the partial clock activation in case of mode2 operation. Only the abnormal heartbeat is detected, the post processing is waked up. Otherwise, the post processor sleeps. Since the post-processor and the RISC occupy about 37% and 12% of the total power consumption, respectively, the power reduction up to maximum 49% and average 28% can be achieved by clock gating.

6.6.5 On-Chip Memory Reduction

A large memory capacity is necessary to holding the temporal data during the classification operation. In general, the previous 8 heartbeats are necessary to decide the arrhythmia [26]. (i.e. 6 kB is necessary for 8 seconds with 500 sample/sec) In addition, the main memory for final results can be also large, if all the raw data is stored. A large memory capacity brings not only increment of the area occupation, but also increment of the access and leakage power consumption as shown in Fig. 6.23. Therefore, if the number of memory access is same, the large capacity memory consumes much more power.

The data memory bandwidth can be reduced by the skeleton method. Figure 6.24 shows the effect of the skeleton operation according to the compression ratio (CR). In low CR region, the memory occupies the dominant part. On the contrary, in high CR region, the energy consumption for the computation is dominant. So there should be the tradeoff between the memory and the computation overhead. According to the graph in Fig. 6.24, the optimum point is in between 7 and 10.

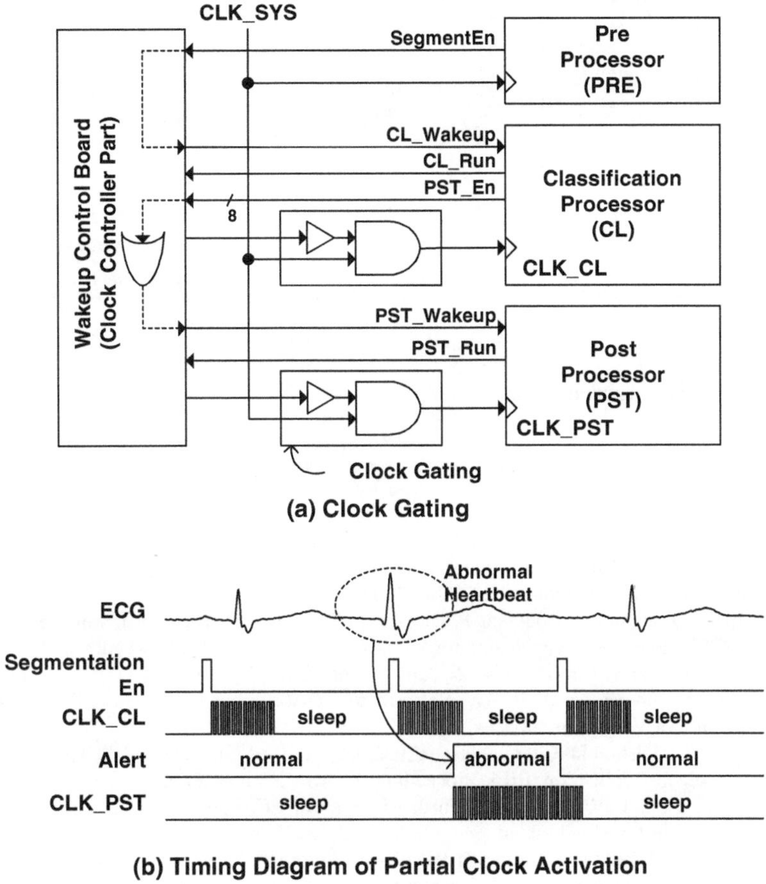

Fig. 6.22 Segmentation based clock gating management

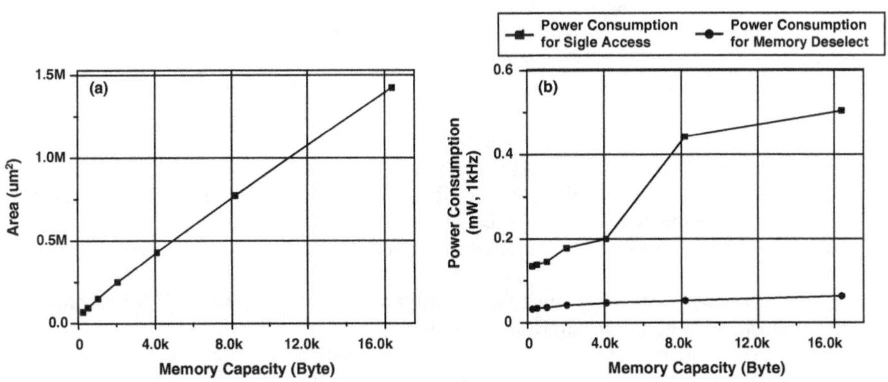

Fig. 6.23 **a** Area occupation. **b** Access and deselect power consumption with variable memory capacity

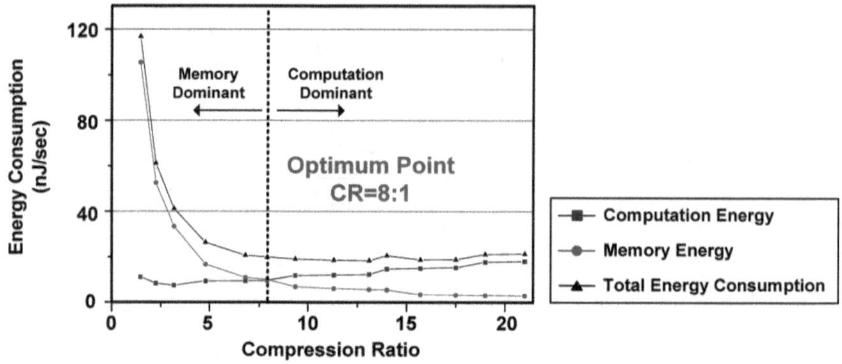

Fig. 6.24 Energy consumption comparison

References

1. Zhai B, Nazhandali L, Olson J, Reeves A, Minuth M, Helfand R, Pant S, Blaauw D, Austin T (2006) A 2.60pJ/Inst subthreshold sensor processor for optimal energy efficiency. IEEE Proceedings of Symposium of VLSI, Jun 2006
2. De Nil M, Yseboodt L, Bouwens F, Hulzink J, Berekovic M, Huisken J, van Meerbergen J (2007) Ultra low power asip design for wireless sensor node. IEEE Proceedings of ICECS
3. Seok M, Hanson S, Lin Y-S, Foo Z, Kim D, Lee Y, Liu N, Sylvester D, Blaauw D (2008) The Phoenix processor: A 30 pW platform for sensor applications. IEEE Proceedings of Symposium of VLSI, Jun 2008
4. Ickes N, Finchelstien D, Chandrakasan AP (2008) A 10-pJ/instruction, 4-MIPS Micropower DSP for sensor application. IEEE Proceedings of ASSCC, Nov 2008
5. Jocke SC,. Bolus1 JF, Wooters SN, Jurik AD, Weaver AC, Blalock TN, Calhoun BH (2009) A 2.6-μW Sub-threshold mixed-signal ECG SoC. IEEE Proceedings of VLSI, Jun 2009
6. de Chazal P, Palreddy S, Tompkins WJ (2004) Automatic classification of hearbeats using ECG morphology and heartbeat interval features. IEEE Trans Biomed Eng 51(7):1196–1206, Jul 2004
7. Kim BS, Yoo SK, Lee MH (2006) Wavelet-based low-delay ecg compression algorithm for continuous ECG transmission. IEEE Trans Infor Tech Biomed 10(1), Jan 2006
8. Istepanian RSH, Petrosian AA (2000) Optimal Zonal Wavelet-based ECG Data Compression for mobile telecardiology system. IEEE Trans Infor Tech Biomed 4(3), Sep 2000
9. Kim H, Kim Y, Yoo HJ (2008) A low cost quadratic level ecg compression algorithm and its hardware optimization for body sensor network system. IEEE Proceedings of EMBC, Aug 2008
10. Arzeno NM, Deng Z-De, Poon C-S (2008) Analysis of first-derivative based qrs detection algorithms. IEEE Trans Biomed Eng 55(2):478–484, Feb 2008
11. Zigel Y, Cohen A, Katz A (2000) The Weighted diagnostic distortion (WDD) Measure for ECG signal compression. IEEE Trans Biomed Eng 47(11), Nov 2000
12. Welch TA (1984) A technique for high-performance data compression Computer 17(6):8–19, Jun 1984
13. Health informatics. Standard communication protocol. Computer-assisted electrocardiography. British-Adopted European Standard BS EN 1064:2005
14. Dipersio DA, Barr RC (1985) Evaluation of the fan method of adaptvie sampling on human electrocardiograms. Med Bio Eng Comp 401–410, Sep1985
15. Abenstein JP, Tompkins WJ (1982) A new data reduction algorithm for real time ECG analysis. IEEE Trans Biomed Eng 29(1):43–48, Apr1982

16. Cox JR, Nolle FM, Fozzard HA, Oliver GC (1968) AZTEC, A preprocessing program for real time ecg rhythm analysis. IEEE Trans Biomed Eng 15(4):128–129, Apr 1968
17. Mueller WC (1978) Arrhythmia detection program for an ambulatory ecg monitor. Biomed Sci Instrument 14:81–85
18. Hilton ML (1997) Wavelet and wavelet packet compression of electrocardiograms. IEEE Trans Biomed Eng 44(5) May1997
19. Djohan A, et al. (1995) ECG compression using discrete symmetric wavelet transform. IEEE Proceedings of EMBC
20. Tai S-C, Sun C-C, Yan W-C (2005) A 2-D ECG Compression method based on wavelet transform and modified SPIHT. IEEE Trans Biomed Eng 52(6):999–1008, Jun 2005
21. Manikandan MS, et al. (2005) ECG signal compression using discrete sinc interpolation. IEEE Proceedings of ICISIP, Dec 2005
22. Manikandan MS, Dandapat S (2005) ECG signal compression using discrete sinc interpolation. IEEE Proceedings of ICISIP, Dec 2005
23. Olmos S, MillAn M, Garcia J, Laguna P (1996) ECG data compression with the Karhunen-Loeve transform. Comput Cardiol 8–11:253–256, Sep 1996
24. http://www.physionet.org/physiobank/database/mitdb/
25. Fira CM, Goras L (2008) An ECG signals compression method and its validation using NNs. IEEE Trans Biomed Eng 55(4):1319–1326, Apr 2008
26. Reddy DC (2005) Biomedical signal processing—principles and techniques. McGraw Hill. ISBN 007-124774-2
27. Nazhandali L, et al. (2005) A second-generation sensor network processor with application-driven memory optimization and out-of-order execution. IEEE Proceedings of CASES, pp. 249–256, Sep 2005

Part II
Bio-Medical Wireless Communication

Chapter 7
Short Distance Wireless Communications

Robert Puers and Jef Thoné

7.1 Introduction

The era of biotelemetry started in June 1957, with the publication of the first biomedical telemetry device by Mackay. A miniature swallowable RF transmitter was described, sending out pressure and temperature information from within the gastro-intestinal (GI) tract [63]. It formed the base of an ever emerging research field, and has enabled measurements of physiological parameters that otherwise would have been impossible. A major advantage of biomedical telemetry is the lack of restraining parameters for the monitored subject, allowing monitoring during virtually any activity. Subjects for monitoring range from trees to humans, including flying, swimming or underground animals. The biotelemetry transmitter may be wearable, inserted through body openings, surgically implanted or swallowed. Power for the monitoring unit can be provided with batteries, induced through a wireless link, or harvested from the environment it is used in.

The fact that telemetry devices should not interfere with the physical functions to be monitored, drives the need for extreme miniaturization, especially in the case of implantable or swallowable devices. This often leads to custom designed integrated circuits (IC), allowing most functionality to be integrated on a small silicon die or a hybrid substrate. In case of battery powered devices, low power consumption is important for increasing the lifetime, although this is less of a problem for wirelessly powered devices. It is advisable though to keep the power consumption low, to limit heat generation of the implanted device. Radiated power, in the case of radiofrequency (RF) telemetry, or coupled power, in the case of inductive powering, is subject to governmental regulations. These regulations are often based on the guidelines issued by the International Commission on Non-Ionizing Radiation Protection (ICNIRP), to provide protection against adverse health effects produced by electromagnetic field exposure [44].

R. Puers (✉)
Katholieke Universiteit Leuven, Kasteelpark Arenberg 10, 3001 Leuven, Belgium
e-mail: robert.puers@esat.kuleuven.be

H.-J. Yoo, C. van Hoof (eds.), *Bio-Medical CMOS ICs*,
Integrated Circuits and Systems, DOI 10.1007/978-1-4419-6597-4_7,

Table 7.1 Comparison of different wireless standards

		WLAN [99]	Bluetooth [12][11]	UWB [110]	Zigbee [120]	MICS [66]
Frequency	MHz	2400, 5000	2400	3100–10700	868, 915, 2400	402–405
Bandwidth	MHz	22–40	1	500	0.6	0.3
data rate	Mbps	11–200	1–3	480	0.02–0.250	–
Range	m	100	10	10	100	–
Output power	dBm	15–20	1–20	–	<0	–16
Supplied power	mW	750–2000	100	400	30	–

Industrial wireless standards like WLAN, Bluetooth or Zigbee offer a complete and rather complex hardware and software stack, leaving little design effort to have a working but power-hungry solution, as represented in Table 7.1. However, the low power consumption requirement pushes the biotelemetry designer to find simplified approaches for handling wireless communication. This often comes down to a completely stripped transmitter and receiver system, leaving only the bare necessities: the physical and datalink layer. The next section treats several possible methods for transferring information from the telemetry unit to the receiver.

7.2 Biomedical Telemetry Methods

Based on the physical connection between the biomedical implant transmitter and the external receiver, wireless communication can be divided into three classes: wave propagation, electrical conduction and near-field coupling, see Fig. 7.1.

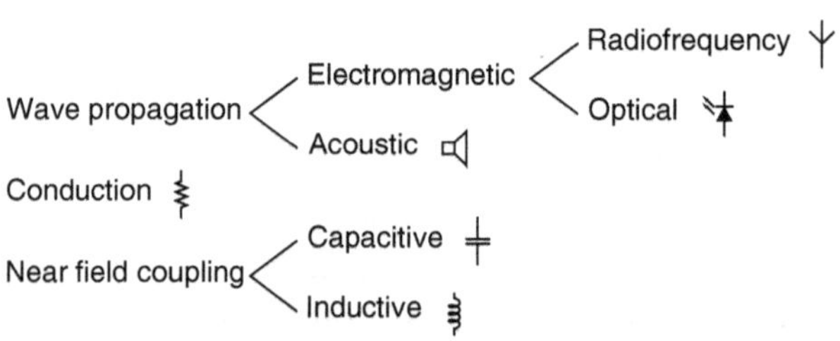

Fig. 7.1 Classification of the communication links based on the physical connection between transmitter and receiver

7.2.1 Wave Propagation

A large part of the biomedical telemetry systems uses wave propagation for transmitting information, through the air or through the body. Wave propagation can be divided into two subclasses: electromagnetic (EM) and acoustic wave propagation.

7.2.1.1 EM Wave Propagation

Radio frequency (RF) and optical links both rely on propagation of electromagnetic waves (Fig. 7.2). Radio frequencies are defined between 30 kHz and 300 GHz, whereas light ranges from far infrared (10^{12} Hz) to ultraviolet (10^{17}Hz).

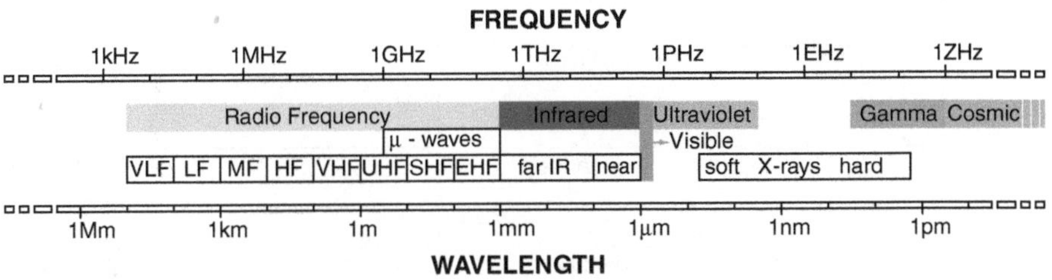

Fig. 7.2 The electromagnetic spectrum (HALLIDAY and RESNICK, 1977)

RF Wave Propagation

The distance a wave of a certain frequency travels through vacuum at the speed of light, during one period, is what we call the wavelength λ. In formula: $\lambda = c/f$, with c the speed of light (300 000 km/s) and f the frequency. This can be interpreted as follows: when a wave travels its wavelength (or a multiple), its amplitude and phase in that point are the same as its starting point. Very low frequencies have an extremely large wavelength, where very high frequencies have a very small wavelength.

Any pair of wires, conducting power from a source to a load, can be modeled as a combined distributed series resistance, series inductance and shunt capacitance. For a lossless line, a characteristic impedance Z_0 can be calculated from the inductance L and capacitance C: $Z_0 = \sqrt{L/C}$. When the length of the conductors is much smaller than the wavelength of the voltage or current passing through, we can safely assume the voltage and current are the same over the complete conductor length, as depicted in the left case of Fig. 7.3. In this case, we can ignore the shunt capacitance and series inductance. When the wire length is in the same order of magnitude as the wavelength of the conducted wave, for high frequencies and/or long conductors, the voltage and current are depending on time and the location in the conductor, and the source to load matching, as represented in the right case of Fig. 7.3. The input wave will travel along the conductor at the speed of light.

When the impedance of the source is not the same as the line impedance Z_0, or the line impedance does not match the load impedance, a part of the power will be reflected back to the source, resulting in an inefficient power transfer between source and load. To improve the power transfer, we have to adapt the source and the load to the line impedance by a technique called matching. This can be performed by discrete components (inductors and capacitors) or by adding pieces of transmission line, of predefined lengths. When short-circuiting one end of the conductors, there is a complete mismatch between the source (limited resistance) and the load (zero resistance) impedance, leading to a complete reflection of the power to the source.

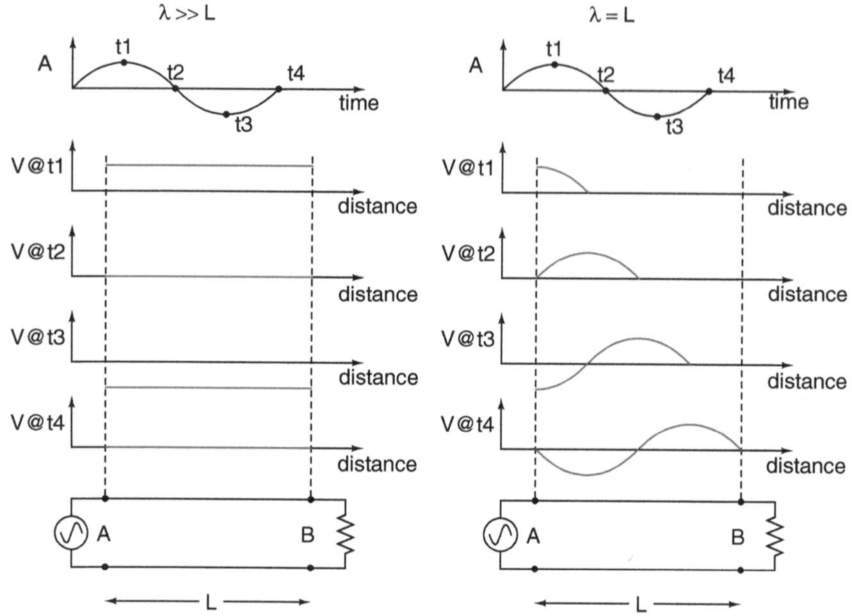

Fig. 7.3 Voltage on a conductor, for different wavelengths vs. conductor length, assuming a perfect impedance match

The voltage wave coming from the source sees a zero resistance at the conductor end. As no power is transferred to the load ($V \times I = 0$), the reflected voltage can only go back to where it came from.

The reflected wave combines with the source wave to a standing wave (in the case of a sinusoidal wave) between the conductors. The standing wave ratio (SWR) is the ratio between the maximal and minimal amplitude of the standing wave pattern, and is often used to quantify the matching between source and load. For a perfectly matched source-line-load condition, the SWR equals 1, while for a total mismatch, where all power is reflected, the SWR goes to infinite. The generation of a standing wave in the case of a short-circuit is depicted in Fig. 7.4. When two parallel conductors carry power from a source to a load, with an opposite current direction and close to each other, their generated magnetic fields largely cancel out. The voltage between the conductors generates an electric field between them, largely confined in a small surrounding volume. There is thus a very limited conversion of conducted to radiated power, what is actually what we want in a wired connection.

When gradually enlarging the length of the short circuit, as depicted on the left in Fig. 7.5, the magnetic fields are cancelling out less and less, and the electric fields are not confined anymore to a small volume. A part of the electric field and magnetic field escapes from the conductor, propagating at the speed of light from the conductor end as a self-sustaining electromagnetic (EM) wave, as depicted on the bottom left pictures of Fig. 7.5. When the short circuit has a length of $\lambda/2$, the electric field between the two conductor ends is maximal and the magnetic field is

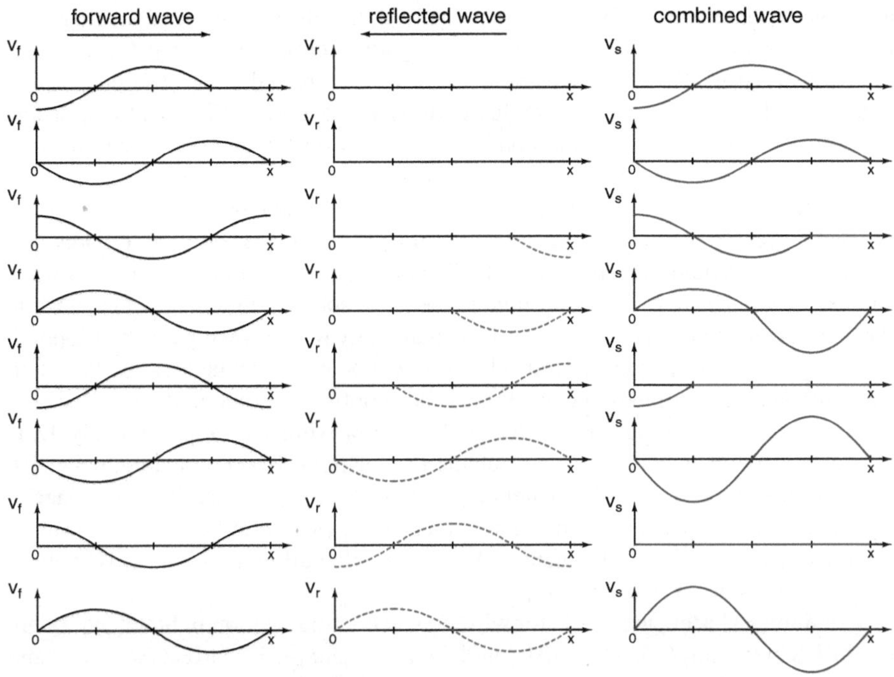

Fig. 7.4 Voltage wave travelling along a conductor of length $x = \lambda$, having a short-circuit at its end. This leads to 100% reflection of the incoming voltage, combining to a standing wave pattern

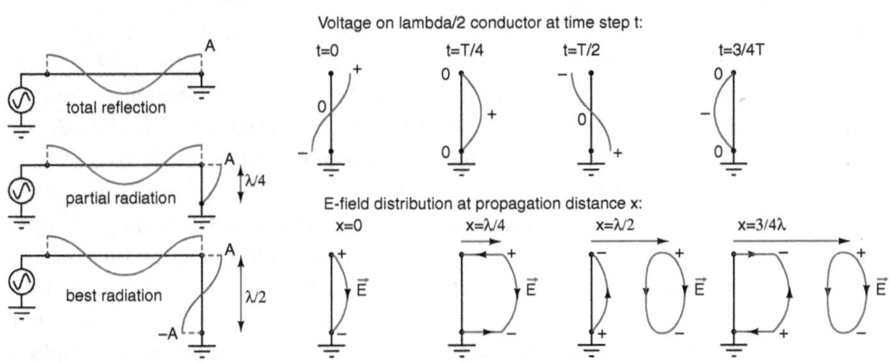

Fig. 7.5 Forward voltage distribution along a short-circuit of different lengths (*left*) (the reflection have been left out for clarity). Note that the maximal voltage difference between the 'short-circuit' ends is maximal for a conductor length of $\lambda/2$. *Top right* shows the voltage along the $\lambda/2$ conductor for different timesteps, while the *bottom right* represents the electric field generation and propagation. The magnetic field is generated in a similar way, also propagating away from the conductor at the speed of light

no longer counteracting. Every half wave period, the electric field and the magnetic field are inversed and maximal, while propagating through the air at the speed of light—away from the $\lambda/2$ conductor. We basically created a $\lambda/2$ antenna, which is a very efficient converter of conducted to radiated power. Electrically, it seems as if the load impedance at the conductor ends is no longer a short circuit anymore, as current and voltage are transformed into EM waves, forming a complex impedance. The real part of this impedance is what is called the radiation resistance. By matching, we try to improve the power transfer from the source to the load, by equalizing the radiation resistance to the line and source resistance. A radiated wave can exist by itself: once it has been generated, the source can be taken away, while the wave is still propagating. There is a clear analogy with throwing a rock in a pond, where the waves propagate isotropically from the point where the rock hit the water. The propagation continues and eventually fades out—even though the source is not present anymore. The field generated within the proximity of approximately 1λ is also called the near-field, the field outside the far-field. Communication using RF EM waves is therefore called far-field communication, as it generally takes place at a distance $>> \lambda$. Near-field communication is also possible, although this is usually not done with a standard RF antenna. A more detailed differentiation between near- and far-field is made in section 7.2.4.

Common EU accepted bands for wireless RF communication in biotelemetry are the MICS band [66] (402–405 MHz) and the ISM band (433.05 MHz, 868 MHz and 2.4 GHz) [46]. Common examples of near and far field communication are treated later in this text.

Optical Wave Propagation

Optical wave communication uses (visible or IR) light as a carrier. Although these are also EM waves, the difference with RF waves mainly lies in the way they are generated. For RF waves near or far-field antennas are required, whereas for optical communication light emitting and sensitive devices are used, like photodiodes or phototransistors. The generation of optical waves is achieved by LEDs or lasers.

Data is usually transmitted by mere on-off keying (OOK) of the light source, propagating the light through a transparent medium. The major advantage of optical data transmission is the extreme robustness against electromagnetic interference (EMI). For close range line-of sight applications, the advantage is the extremely high available bandwidth, allowing high data rates, up to 100 Mbps and more. Although the skin between the transmitter and receiver partly reflects and disperses the optical carrier, causing a huge attenuation, it has been used for high speed transcutaneous communication. Typical applications include hybrid inductively powered - IR communication systems for neural monitoring, as described by Guillory [36], Song [91] and Callewaert [17].

7.2.1.2 Acoustic Wave Propagation

Ultrasound links use acoustical waves at frequencies ranging from 10 kHz to 10 MHz. The modulated carrier drives an ultrasound transducer, generating an acoustic wavefront. Sound is a small pressure disturbance, superimposed on the ambient pressure, propagating by the back and forth exchange of potential and kinetic energy of adjacent medium sections. For this reason a physical medium is required (air, water), as opposed to EM waves, that also propagate in vacuum.

Ultrasound transmission is very attractive for marine telemetry, as acoustic waves propagate much better in water than RF waves, which are absorbed by the relative high electrical conductivity of surrounding water [87]. A carefully designed transducer can avoid reflection of ultrasound energy at the transducer-medium interface. Horn loading of the transmitter and receiver transducers, a technique adopted from loudspeaker engineering, can improve their matching with the medium characteristics. It is also possible to enhance the transducer-to-medium transition by the interposition of intermediate layers. Coupling gel is applied between the skin and the transducer medical ultrasonography for this very same reason.

Typical applications are underwater telemetry and sea animal tracking devices. An acoustic telemetry system for divers has been described in [87] and [47], ranging over 100 m. An ultrasonic tracking device for monitoring crab foraging with a range up to 500 m has been reported in [111]. Liew [58] obtained ranges of 1–2 km in seawater, while tracking green turtles with commercial ultrasonic temperature depth and swim speed transmitters.

7.2.2 Conduction

A modulated current or voltage is transferred through an electrical connection between transmitter (TX) and receiver (RX), using the body as a conductor. Although not common yet in daily-life telemetry, there is quite some research being done on body-area-networks (BAN), using the body as communication channel. A digital wideband human body transceiver has been described in [90], consuming a mere 0.2 mW at a data rate of 2 Mbps. The general idea is to connect multiple physiological sensors (EEG, ECG, blood pressure, body temperature, ...) in a human body channel network, all communicating over the human body.

Currently, the Miro wireless endoscopic capsule is under development, using the body to conduct endoscopic images to an external receiver. The idea has been patented in [53], and test results have been published in [8].

The main potential of this technology lies in its low power consumption, and the possibility to transmit non-modulated information across the channel, just using the baseband information, considerably simplifying the interface electronics. A disadvantage is the lack of reliable models of the human body, making it difficult to do estimations on the link quality based on mathematical models.

7.2.3 Near-Field Coupling

Data and power is transferred from the transmitter to the receiver, using a non-radiating magnetic or electric field.

7.2.3.1 Capacitive Links

A capacitive link consists of two electrode plates, one at the transmitter and one at the receiver side, to provide energy and data transfer. To yield a substantial output level, large capacitors and high carrier frequencies should be used [82]. Although never before applied in biotelemetry applications,a recent publication [89] shows its feasibility to provide an efficient power and data link. Basically, two capacitors are used to transfer power and data, with one plate of the capacitor subcutaneous, the other one external. By supplying an AC carrier of several MHz to the external plates close to the skin, the capacitors form a small impedance, enabling the transfer of power and data-through power carrier modulation- to the connected implant.

7.2.3.2 Inductive Links

Magnetic or inductive near-field links are the most widely used communication technique for biomedical implants. They offer the advantage of simultaneously supplying energy and providing a carrier for bi-directional communication. Induction telemetry however has a fairly small range, compared to RF, ultrasound or optical telemetry. The technique is based on magnetic coupling between a -often subcutaneously implanted- remote and a primary coil, forming a loosely coupled transformer. This is generally known as transcutaneous energy transfer system (TET) or skin transformer. The primary coil is driven by an alternating current to provide an alternating magnetic field, with a frequency in the range of 1 kHz–30 MHz. The remote and the primary coil are tuned to the same resonance frequency using a series or parallel capacitor, leading to higher link efficiencies [80, 55, 27, 82]. The remote-coil voltage is rectified and stabilized to a regulated DC supply to power the rest of the remote electronics.

Inductive power links were originally developed in the sixties and the seventies, for powering artificial heart-assist devices and auditory prostheses. The heart-assist systems demanded continuous power transfers up to 50 W. The coil coupling factor for this early applications (ratio between captured flux and generated magnetic flux) was typically 10–20 %. There is, however, a trend to apply inductive links also in less favourable conditions, with coupling factors dropping below 1%, and varying over about a decade in a very unpredictable matter [42]. Some representative examples include:

1. Wireless endoscopic capsules [57, 19]
2. Cochlear implants [119]
3. Multichannel functional electrical stimulation (FES) of paralyzed muscles where each stimulation channel consists of a complete micro-implant, locally stimulating the muscle [60, 72]
4. 125 and 134.2 kHz RFID transponders for item management and husbandry control (ISO 11785:1996; ISO 14223-1:2003; ISO/IEC 18000-2:2004) and [112]
5. 13.56 MHz proximity ID cards, credit cards, and biometric passports (ISO/IEC 14443-2:2001; ISO/IEC 18000-3:2008)
6. Small telemonitoring implants [70]

Data is mostly transferred from the primary to the remote coil by amplitude modulation of the power carrier. Demodulation at the remote coil is achieved by envelope detection of the carrier amplitude. The other way around, to transmit data from the remote implant to the primary coil, the load current of the primary coil is modulated, by intermittently switching a load at the remote coil. Another possibility is intermittent detuning of the remote coil, causing a similar effect (load modulation) on the primary coil. A block diagram of a typical inductive link for biomedical telemetry applications is depicted in Fig. 7.6.

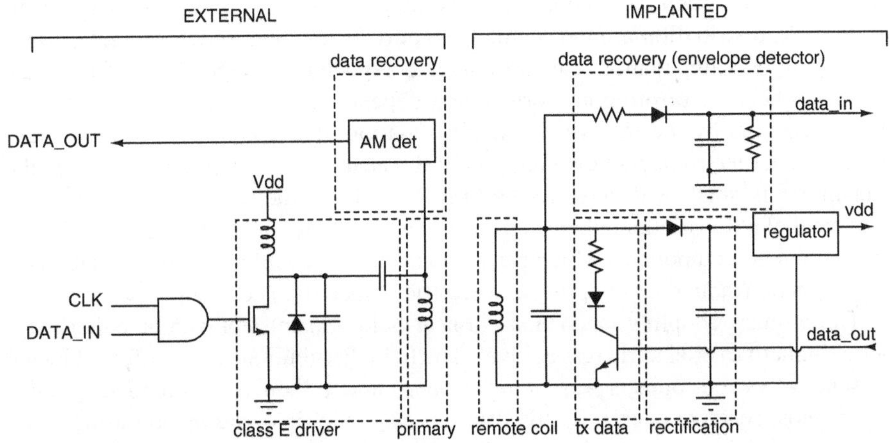

Fig. 7.6 Generalized circuit diagram of a class E inductive link system, with two-way communication. The uplink uses OOK (100% ASK), the downlink ASK by load modulation

7.2.4 Near-Field versus Far-Field

The field generated around a transmitting antenna can be divided into three regions: reactive near-field, radiating near-field (Fresnel) and radiating far-field (Fraunhofer) [7].

In the reactive near-field, the energy is stored and not radiated. For antenna sizes in the order of a wavelength, the boundary of the reactive near-field is commonly approximated at a distance $R_{\text{ReNF}} = 0.62\sqrt{D^3/\lambda}$ from the antenna, with D the largest dimension of the antenna. For an electrically small antenna, with a size well below one wavelength, this region is defined by a distance $R_{ReNF} = \lambda/2\pi$. For inductive links, we want as much energy concentrated in the reactive near-field, and as little as possible radiated. These fields decay as $1/r^3$, with r the distance to the antenna.

In the radiating near-field, radiating fields predominate, and the angular field distribution is depending on the distance from the antenna. For antennas $<< \lambda$, as used a lot in biotelemetry, the radiating near-field region is small or may not exist. The inner boundary for this region is the outer boundary of the reactive near-field region, the outer boundary is at a distance $R_{\text{RaNF}} = 2D^2/\lambda$. These fields decay as $1/r^2$, with r being the distance from the antenna.

In the the far-field, electric and magnetic fields propagate perpendicular to each other and the direction of propagation, and outward as an electromagnetic wave. The angular field distribution is independent of the distance from the antenna. The fields are related to each other via the free-space impedance and decay as $1/r$, with r being the distance from the antenna. The far-field starts at the outer boundary of the radiating near-field: $R_{RaNF} = 2D^2/\lambda$.

When the receiver antenna is located in the near-field of the transmitter antenna, the coupling between the antennas affects the impedance of both antennas as well as the field distribution around them. The equivalent antenna performance parameters (gain and impedance) can no longer be specified independently from each other and become position and orientation-dependent. In general, to calculate the power received by the receiver in such a situation, one needs to perform a three-dimensional electromagnetic simulation of the near-field problem, except when the transmitter is small and does not perturb the field of the receiving antenna. The near-field of a transmitter antenna can have several tangential and radial electric and magnetic field components which can all contribute to coupling. Two ultimate cases are magnetic (inductive) coupling and electric (capacitive) coupling.

The primary coupling mechanism in near-field transmission can be either magnetic (inductive) or electric (capacitive). The field distribution can be affected by the presence of various objects present in the environment. Inductively coupled systems, where most reactive energy is stored in the magnetic field, are mostly affected by objects with high magnetic permeability. The magnetic permeability of biological tissue is practically equal to the magnetic permeability of air. Capacitively coupled systems on the other hand, where most reactive energy is stored in the electric field, are affected by objects with high dielectric permittivity and loss. Since the body has high relative permittivity, inductive coupling is much more efficient for this kind of applications.

7.3 Modulation Methods

A single frequency carrier does not contain any information. In order to add information to a carrier, it needs to be modulated by a source signal. Depending on the type of source signal, we distinguish between analog and digital modulation. Both modulation classes contain similar modulation techniques, that will affect the amplitude, phase or frequency (or a combination) of the carrier. The source signal can be processed prior to modulation by pulse modulation encoding, to improve the signal's noise immunity. Some of the presented pulse modulation techniques applied to analog source signals are also valid for digital source signals. Baseband pulse modulation forms the natural transition from analog to digital modulation.

For biomedical telemetry systems generally the rule holds: the simpler the better. Simplicity reduces circuit complexity and current consumption while increasing circuit robustness. When only little power is available at the (implanted) transmitter, circuit complexity and signal processing can be shifted towards the receiver side.

7.3.1 Analog Modulation

Analog modulation uses an analog baseband signal to directly modulate the amplitude (AM), frequency (FM) or phase (PM) of the carrier. The baseband signal is the amplified and/or filtered sensor signal, e.g. an ECG signal.

7.3.1.1 AM

In amplitude modulation, the analog baseband signal modulates the amplitude of the carrier, as depicted in Fig. 7.7. The amount of modulation is expressed as the ratio between the baseband and carrier amplitude, the modulation depth, between 0 and 100%. Modulation over 100% will cause artifacts in the reception, making the

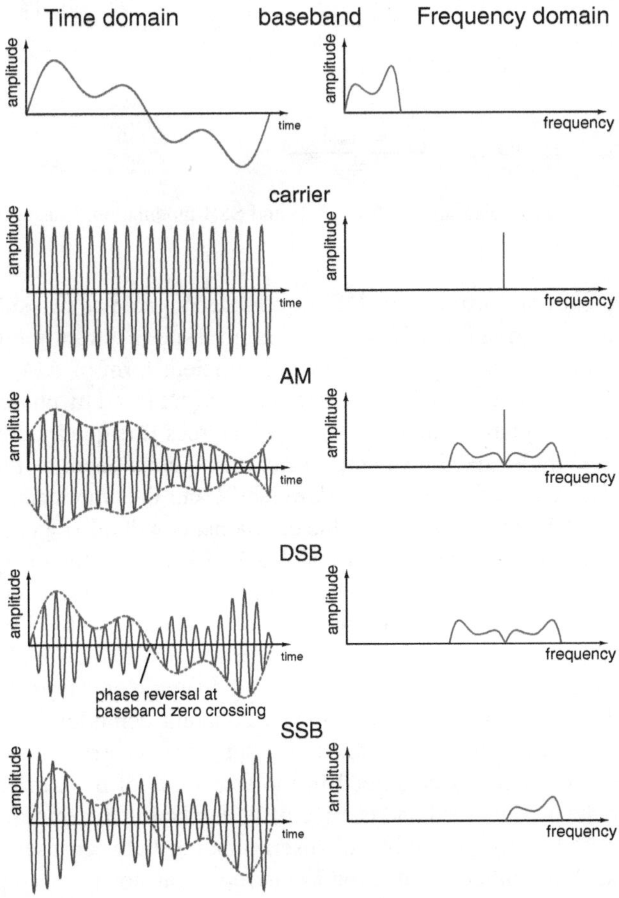

Fig. 7.7 The time and frequency domain representation of AM, DSB-SC and SSB modulation. Note that the AM spectrum occupies two times the baseband spectrum, except for SSB modulation

baseband information (partially) unrecoverable. The larger the modulation depth, the less the influence of noise coupling into the transmission channel.

The simplest form of AM modulates the amplitude of the carrier, still leaving a reasonable amount of power in the information-less carrier. The carrier can thus be suppressed without any loss of information. This is achieved by double-side-band-suppressed-carrier (DSB-SC) AM modulation, the middle circuit in Fig. 7.8. DSB-SC uses a double-balanced modulator or Gilbert-cell as modulator element. This can be seen as a double-pole-double-throw switch, inverting the baseband signal at the rate of the carrier. This can be seen in Fig. 7.7 as in inversion of the carrier when the source signal crosses zero.

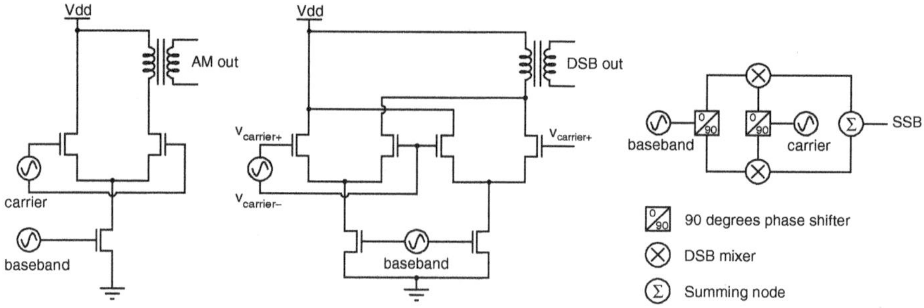

Fig. 7.8 Simplified circuit diagrams of AM, DSB and SSB modulation. Biasing and filtering is not included

As the frequency spectrum of the DSB modulated signal has two sidebands with exactly the same information content, there is no issue in filtering out one of the sidebands, leading to a much more bandwidth efficient form of AM, called single side band (SSB) modulation. SSB modulation can be achieved by phase-shifting the source signal (the in-phase component I) by 90 degrees (the quadrature component Q). The I component is mixed up by the carrier, the Q component by the 90 degrees phase shifted carrier. When adding the up-mixed I and Q component, a SSB spectrum will result. Adding or subtracting the components will allow generation of the left or right sideband. Where for AM and DSB modulation, the envelope clearly follows the source signal, this is no longer the case for SSB modulation.

Figure 7.8 shows the system topologies for generating AM, DSB and SSB modulated carriers. It is clear that the modulator for AM is the simplest to implement, as well as the easiest to demodulate, using a simple envelope detector. DSB and SSB modulators require a more complex coherent demodulator, that needs a phase-locked loop (PLL) to synchronize to the suppressed carrier.

In AM the information is contained in the amplitude of the carrier wave. Any interference in the modulation band is effectively deforming the amplitude, and thus information, without any possibility of discriminating between useful signal and unwanted noise. It is neither possible for AM to represent slowly varying or absolute source signals, as amplitude fading because of transmitter movements or changes in atmospheric conditions provide similar effects on the carrier.

Because of its weak noise immunity, pure analog AM modulation is seldom used in biomedical telemetry. The patent of Barker [9] describes an implantable blood pressure sensor, transmitting its output as a pure AM modulated carrier. AM modulation in combination with baseband modulation (e.g. subcarrier modulation) was used a lot in early biotelemetry applications. Nowadays AM is used a lot in its digital form amplitude shift keying (ASK),with only two possible carrier amplitudes, making the transmission much less sensitive to noise. Using analog baseband modulation, the baseband can be converted into a frequency modulated signal before being AM modulated. Digital baseband modulation will convert the baseband to a discrete amplitude signal, with the baseband information translated to the time domain, in the form of pulses. Both analog and digital baseband modulation will render the AM modulated signal much more noise resistant, by spreading the baseband information over a larger bandwidth.

7.3.1.2 FM and PM

FM is achieved by modulation of the carrier phase or frequency with the source signal. For analog source signals, it is difficult to see the difference between phase and frequency modulation: phase modulation will change the phase of the carrier according to the continuous time-varying source signal, while the change in carrier phase over time results in a carrier frequency change.

Early generation narrow-band PM/FM used a phase shift network and a balanced modulator [5, 93], as depicted in Fig. 7.9. By combining a carrier I with 90 degrees phase shifted version Q, a phase shift can be obtained resulting from the vector sum of I and Q. If I and Q are equal in amplitude, a maximal phaseshift of ±45 degrees can be obtained. Apart from the PM, the vector summation will result in an amplitude modulation, which can be avoided by placing a limiting circuit after the vector summation. The maximal frequency swing δf of the carrier is equal to the maximal frequency of the modulating source signal. The modulation index for PM is defined as ratio of the maximal phase deviation and the input swing. If a larger frequency swing is required, e.g. for better noise performance, the circuit is followed by frequency multipliers. This will multiply the carrier frequency as well as the frequency swing. The need for frequency multipliers for influencing the modulation index has made analog PM an extinct modulation technique, as opposed to digital phase modulation.

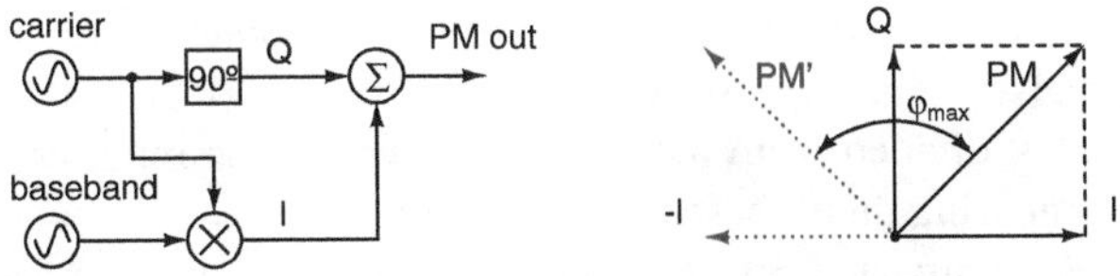

Fig. 7.9 Block diagram of vector generated narrow-band PM, with the vector diagram showing the resulting PM output

It is possible to transform the FM modulator from the circuit in Fig. 7.9, by making the modulating source signal inversely proportional to its frequency, i.e. integration of the source signal. The disadvantage of a small modulation index M remains, for FM defined as the ratio of the FM frequency swing and the source bandwidth. The higher the modulation index, the higher the bandwidth occupation of the modulated signal, and the better the S/N ratio.

A second, more common method will directly modulate the frequency with the source signal. A voltage controlled oscillator (VCO) generates a carrier at a certain frequency, depending on its control input. By modulating the control input of the VCO, the generated frequency is modulated, as represented in Fig. 7.10. Note that the FM output frequency changes with the derivative of the source signal, which is somewhat counter-intuitive. Changes in amplitude because of fading or interfering noise in the transmitted band are less problematic because the information of the baseband signal is contained in the frequency, as opposed to the amplitude in AM. The noise needs an amplitude high enough before it is recognized as a change in frequency at the receiver.

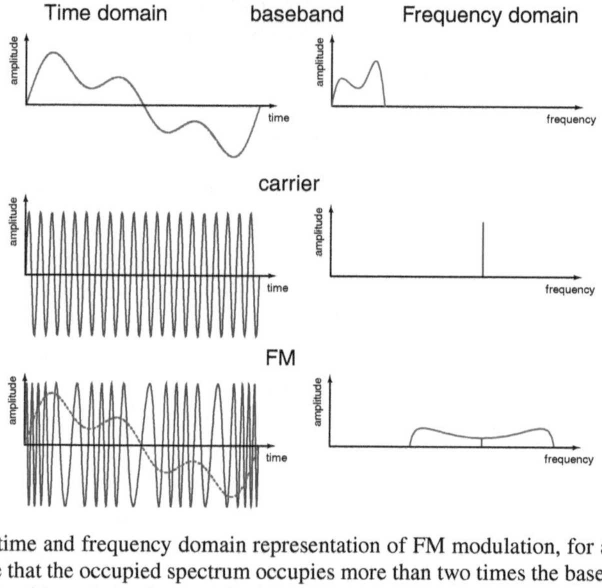

Fig. 7.10 The time and frequency domain representation of FM modulation, for a large modulation index. Note that the occupied spectrum occupies more than two times the baseband spectrum, and there is no intuitive relation between both

The VCO frequency is generally stabilized using a crystal oscillating at its fundamental frequency, 3th, 5th or even 7th overtone. Another more versatile option to generate a precise frequency uses a phase-locked-loop (PLL) [48]. A PLL compares the frequency divided from the VCO output with the input crystal-reference frequency, by comparing their phases. The output of the phase comparator is filtered with a loop filter, tuning the VCO frequency according to the phase difference. The feedback loop is closed by the frequency divider. A PLL allows generation of very precise reference frequencies, making it easier to tune and demodulate at the receiver, as well as to meet regulatory frequency standards. However, for basic biotelemetry applications requiring low power consumption, this is sometimes left out.

FM modulation has been extensively used in biotelemetry in its purest form, using a single transistor Colpitts [22] or Hartley [40] oscillator. The oscillation frequency is defined by the resonance frequency of the tank circuit inductor(s) and capacitor(s), as depicted in Fig. 7.11. In a Colpitts oscillator the feedback is provided through a capacitive path, where a Hartley oscillator provides feedback through an inductive path.

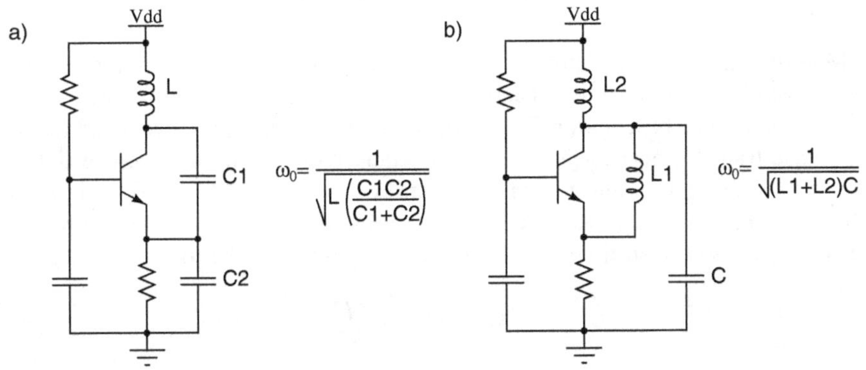

Fig. 7.11 The Colpitts (**a**) capacitive feedback and Hartley (**b**) inductive feedback LC oscillator circuit. For frequencies of operation below 10 MHz, the parasitic capacitances can be neglected, and the oscillation frequencies are given by ω_0

The oscillator circuit will start oscillating when the Barkhausen criterion is met, i.e. the phase of the loop gain is zero and the loop gain magnitude is unity [83]. By using a resonant LC circuit, with capacitive or inductive feedback, it is guaranteed that the Barkhausen criterion is only met for a single frequency, ensuring a single output frequency. The transistor transconductance provides gain to start up and sustain the oscillations, while the feedback circuit will ensure a loopgain of 1.

Figure 7.13 shows different approaches for controlling the frequency of a Colpitts oscillator, although similar methods can be applied for Hartley VCOs. The coil of the LC tank is often doubling as an antenna, as the resonant circuit makes sure the coil current is maximized under all tuning conditions, leading to an efficiently generated magnetic field.

The most common method uses a variable capacitor (varicap) connected in parallel to the LC tank circuit, whose capacitance is modulated by changing its DC bias. The capacitance of the varicap then adds to the resonant tank capacitance, directly influencing the frequency of oscillation. In most cases this is a special reversely biased diode, where the voltage bias modulates the width of the depletion layer. Changing the depletion layer width—by modulating the voltage—changes the capacitance across the diode accordingly. A typical bias voltage–capacitance curve for a BB156 varicap diode is depicted in Fig. 7.12.

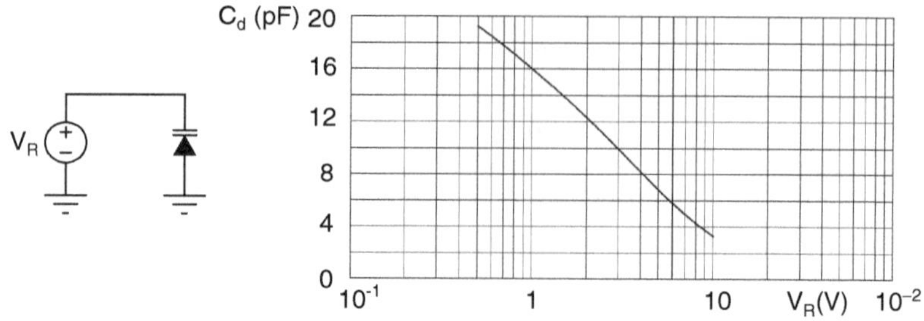

Fig. 7.12 A BB156 varicap capacitance versus bias voltage curve

Modulation of the transistor base current offers another possibility for influencing the oscillation frequency. An active transistor has an amount of minority-carrier charge stored in its base region [83], depending on the collector current and the forward base-transit time $\tau_f.\tau_f$ is a device constant, representing the average time an electron spends in crossing the base region. For small signals, a base-charging or diffusion capacitance C_{de} can be calculated, proportional to the collector current I_c, the forward base-transit time and inversely proportional to the thermal voltage V_T.

$$C_{de} = \tau_f g_m = \tau_f \frac{I_c}{V_T} \tag{7.1}$$

To a lesser amount, a changing base current modulates the base-emitter depletion width, effectively modulating the base-emitter junction capacitance. Changing the base current modulates the diffusion capacitance and base-emitter capacitance, resulting in a change in resonance frequency of the tank circuit.

The bandwidth occupied by an FM modulated signal can be estimated using Carson's rule [18]: 2 times the sum of the baseband bandwidth BW_{BB} and the maximal FM frequency deviation Δ_{FM}. It is worth noting that Carson's paper was actually written to show the superiority of AM modulation—concerning bandwidth occupation. It was only until 1936 that Armstrong [5] showed a practical- and superior application for FM modulation, enabling stabler and cleaner radio broadcasts.

The first generation biomedical FM transmitters used mechanical modulation of the coil, resulting in a change in resonance frequency. The first paper on wireless telemetry from within the body dates 1957 by Mackay [63]. It describes a self-quenching Hartley-like oscillator, with a quenching rate depending on the body temperature. A pressure-sensitive membrane changes the position of an iron rod inside the LC tank inductor, modulating the oscillation frequency with the measured bladder pressure. Similar work was published in the following years by Connell [23], Nagumo [71] and Watson [107]. At the time, a grid-dip receiver was used to detect the emitted frequency. This is a frequency-scanning power meter, showing the measured power vs. frequency on an oscilloscope [62]. Later FM telemetry transmitters modulate the frequency, by acting on the tank circuit capacitance as depicted in Fig. 7.13.

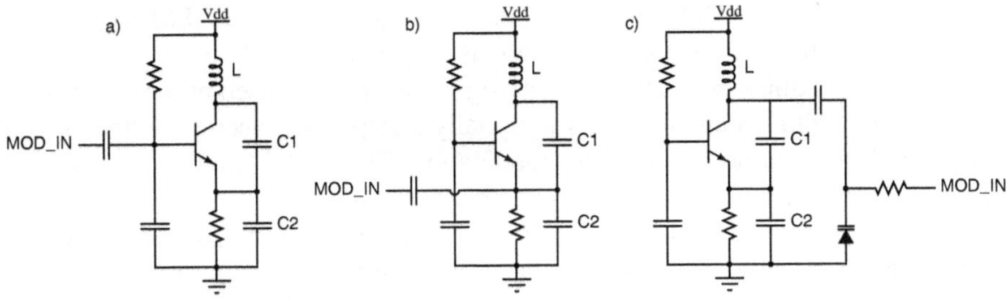

Fig. 7.13 Different methods for implementing frequency modulation on a common-base Colpitts VCO: (**a**) and (**b**) base current modulation, (**c**) variable capacitor modulation. The tank circuit is formed by L, C1, C2, and the diffusion capacitance (**a** and **b**) or varicap (**c**)

7.3.1.3 Discussion on Analog Modulation Methods

FM transmission has proven to be superior to AM concerning noise immunity, at the cost of increased bandwidth occupation. Pure AM modulation causes a degrading S/N ratio or, in the best case, leaves the S/N ratio untouched, in the case of DSB–SC [41]. There is no option to trade bandwidth for an improved S/N ratio, as opposed to FM. For FM modulation, an increase of the modulation index causes an increase in the transmission bandwidth, leading to a quadratic improvement in S/N ratio. FM thus offers a method for exchanging increased transmission bandwidth for better noise performance [41]. Although AM demodulation is slightly easier to implement—a simple envelope detector will do the job—FM modulation has been a popular choice in early biotelemetry applications. A qualitative comparison between AM and FM is shown in Table 7.2. Currently pure analog modulation, using an analog source signal, is not used anymore in biotelemetry applications, as discussed later in Section 7.3.5.

Table 7.2 Comparison of analog AM vs. FM

	FM (PM)	AM
Noise immunity	high	low
Noise trade-off	BW $vs.$ $(S/N)^2$	none
Modulator complexity	low	low
Demodulator complexity	high	low
Baseband frequency minimum	DC	>DC
Bandwidth requirement	$2 \times (\mathrm{BW_{BB}} + \Delta_{\mathrm{FM}})$	$2 \times BW_{BB}$
Linearity requirement	Medium	High

7.3.2 Analog Pulse Modulation Encoding

Direct FM or AM modulation by the analog measurement source has several drawbacks. For AM modulation, the disadvantage is clear: all information is contained in the amplitude, and will be effectively deformed by interference. FM modulation is less sensitive to noise, as the information is contained in the frequency.

The signal to noise ratio of the source signal can be increased at the expense of a larger bandwidth occupation. This is actually a step in between pure analog and pure digital modulation, called pulse modulation. The analog value is represented as a periodic pulse train–the carrier–and a characteristic feature of each pulse (e.g. the phase, amplitude or timing) is varied in a continuous manner, according to that value. The pulse modulated signal remains analog, but the transmission takes place at discrete times [41]. As the pulse modulation conversion implies sampling of the original signal, it has to comply to the Shannon theorem [84]. The pulse frequency has to be at least twice the highest source signal frequency, to avoid overlapping of the aliases.

Several pulse modulation methods exist, classified here as the analog methods: pulse-amplitude (PAM), pulse-position (PPM), pulse-width (PWM) or pulse-duration (PDM) pulse-frequency modulation (PFM). The digital case of pulse modulation is pulse-code-modulation (PCM), where the analog signal is first quantized with a fixed sampling frequency to a code, and then converted to a pulsetrain. An overview of the different pulse modulation methods is depicted in Fig. 7.14.

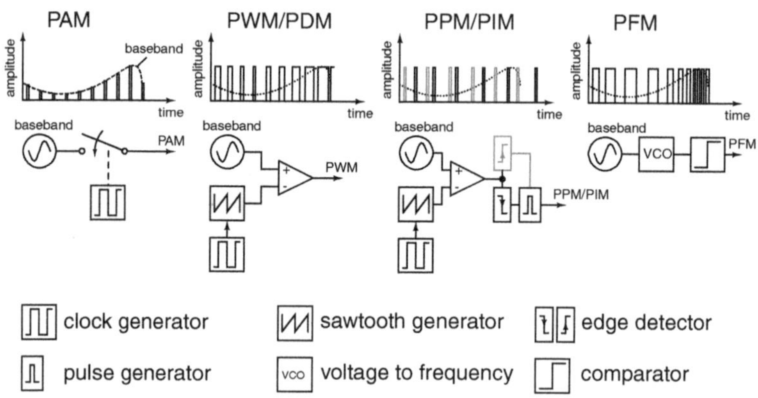

Fig. 7.14 Different pulse modulation encoding techniques

7.3.2.1 Pulse Amplitude Modulation (PAM)

Pulse amplitude modulation is achieved by sampling the original signal at a sampling frequency f_s, and then holding that value for a constant time. Holding the value limits the required bandwidth for the PAM signal, as the bandwidth is inversely proportional to the pulsewidth. Bandwidth limits of the transmitting channel will cause distortion of the original PAM signal, making the reception circuitry more complex. The noise performance of a PAM modulator can never be better than normal baseband transmission. PAM can be used when time-division multiplexing is used, to transmit more channels at the same time.

7.3.2.2 Pulse Width/Duration Modulation (PWM or PDM)

A PWM signal is generated by comparing a sawtooth waveform to a momentary analog input value. This results in a fixed frequency pulse equal to the period of the sawtooth, with a duty-cycle proportional to the sampled value. The receiver can synchronize to the rising edges of the PWM signal. The noise benefit is obtained by representing the sampled value by a time ratio. Large duty-cycles result in a needless waste of power, especially in the case the modulator is followed by an AM modulator. When only transmitting the edge transitions of the PWM signal, we end up with a more efficient pulse position modulation.

A 15-channel PDM-FM modulated telemeter for biomedical monitoring (ECG, temperature, blood pressure, ...) is described in [49], using analog to PWM conversion for each channel, while multiplexing the PWM waveform to a FM transmitter. Each measurement cycle a synchronization pulse is applied to the transmitter, to synchronize the receiver circuit. A 15-channel neural recording interface using PWM time-division-multiplexed FM was described by [115], transmitting up to 320 k-samples/s to a 75-MHz receiver with 8.4 bits of resolution.

7.3.2.3 Pulse Position Modulation (PPM)

PPM is generated by triggering a pulse generator to the falling edge of a PWM signal. PPM as represented in Fig. 7.14 consists of the black pulses: only the falling edges are transmitted. To ease demodulation, pulse interval modulation (PIM) was invented. PIM also sends out the rising edges (the gray pulses) of the PWM signal, considerably relaxing the synchronization requirements for the receiver. PPM exhibits a similar noise performance as FM modulation: the S/N ratio increases with a factor depending on the square of the transmission bandwidth divided by the baseband bandwidth [41].

A telemetry system for long-term monitoring of the intravascular blood pressure of unrestrained animals is described in [20]. The blood pressure sensor signal is first amplified, sampled, converted to a PWM signal and finally converted to a PIM signal. This signal is supplied to a FM transmitter. A bladder pressure telemetry system was conceived by Puers, converting the pressure readout to a PWM signal, convert it to a PIM signal, and transmit it using a FM VCO [104, 6]. An inductively powered intra-ocular pressure-monitoring system was described in [81]. The eye pressure is monitored with a capacitive sensor, converted to a voltage and finally to a PPM modulated waveform. The data is modulated on the power carrier by load modulation, and demodulated at the powering station.

7.3.2.4 Pulse Frequency Modulation (PFM)

PFM is in a way comparable to FM: the frequency is modulated depending on the input signal. The difference lies in the amplitude: the FM signal contains as little harmonics as possible, being ideally a sinewave with modulated frequency, where the PFM signal has only two levels, being harmonically rich. PFM uses variable sampling, depending on the output frequency. The disadvantage is that the lower PFM frequency should at least be twice the input bandwidth [84], and the upper PFM frequency is considerably oversampling the input. This leads to a wider bandwidth occupation than PPM, PWM or PAM. When supplying the PFM waveform to a FM VCO, it is comparable to FM-FM subcarrier modulation. PFM modulation is used for ultra-sensitive CMOS photodetector sensor systems, for detecting low-level bioluminescence in e.g. bacterial cells [85, 13], and for retinal prosthesis photodetectors [73]. Rather than direct modulation of a VCO, the frequency will be integrated using a counter, prior to sending a digital word using digital shift keying.

7.3.2.5 Analog Multiple Channel Modulation Methods

Simultaneous transmission of several measurement channels through the same medium can be done in several ways, as represented in Fig. 7.15.

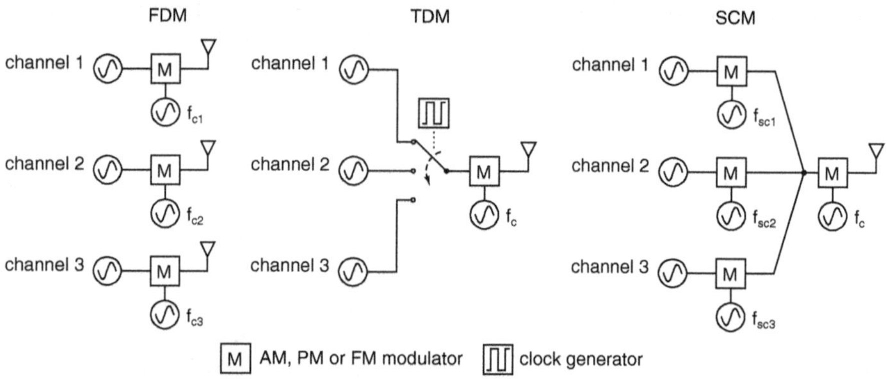

Fig. 7.15 Different analog multiple channel modulation methods: frequency division multiplexing (FDM), time division multiplexing (TDM) and subcarrier modulation (SCM)

Frequency Division Multiplexing

For each physiological parameter, a separate modulator is used, with a different carrier frequency, as depicted on the left picture of Fig. 7.15. For small battery-powered telemetry units this is problematic, as each channel requires its own power-hungry RF transmitter. The telemetry book by Mackay [62] describes a rhesus monkey monitoring setup in 1962, where four separate transmitters at different frequencies were implanted, to measure temperature, ECG, activity and blood pressure. As time multiplexing or subcarrier modulation have obvious advantages, little application and literature are available describing analog frequency multiplexing.

Time Division Multiplexing

The same carrier is shared over time between different channels, by multiplexing the several sensor signals to the modulator, in different time slots, as depicted on the middle picture of Fig. 7.15. In most cases, the analog source signal is first encoded using pulse modulation encoding, increasing the S/N ratio of the source signal, as explained in Sect. 7.3.2. Time multiplexing is also used for multi-channel digital modulation, using pulse-code-modulated baseband streams, see Sect. 7.3.3.1.

A 1963 technical report by Hambrecht [39] discusses in high detail the design of a four-channel time-multiplexed EEG telemetry system. The four EEG channels are first amplified, converted to a pulse-width modulated (PWM) signal and time-multiplexed to modulate a single Colpitts FM VCO. A four-channel FM system is described in [86], measuring two ECGs, respiration rate, and pulmonary ventilation of human subjects during exercise. The four measurement channels are first PWM converted and time-multiplexed to a Colpitts FM modulator.

Subcarrier Modulation

Each measurement channel is first AM or FM modulated on an individual subcarrier, summed and AM or FM modulated on the same carrier. The main advantage is the requirement of a single RF oscillator, considerably limiting the power consumption. The best known example of subcarrier modulation is stereo FM radio broadcast, where the stereo difference signal is DSB-SC modulated to 38 kHz, added to the stereo sum signal and FM modulated between 87.5 and 108 MHz. Although in principle all combinations of subcarrier and carrier modulation are possible, most publications describe FM/FM modulation, because of its superior noise immunity.

A six-channel FM/FM telemetry system for monitoring paralyzed patients was published by Robrock in 1967 [79]. Each of the six monitored physiological parameters controls the frequency of a separate relaxation oscillator at different low frequency subcarriers, between 5 and 70 kHz. These carriers are added and FM modulated with a single Collpits VCO to 110–130 MHz. Another example of FM/FM subcarrier modulation describes a four-channel telemetry system for surgical monitoring of the EEG, EKG and blood pressure monitoring [54]. Subcarrier modulation is also described in a cardiovascular monitoring system for swines [76], with the difference that this design uses ultrasound wave propagation for information transmission.

7.3.3 Digital Pulse Modulation Encoding

7.3.3.1 Pulse Code Modulation (PCM)

Pulse code modulation is the natural extension of analog pulse modulation. Noise, crosstalk and distortion between analog pulses will disturb the amplitude and position of the transmitted pulse, causing non-recoverable errors in the transmitted information. With PCM, the analog signal is first sampled, then digitally encoded before transmission. This encoding step is called quantizing, which will represent each sampled value by a discrete level nearest to its original value. Quantizing is typically performed by a analog-to-digital converter (ADC). The higher the number of quantization steps, the higher the S/N ratio of the PCM system. A nice introduction on PCM is given in a paper by Oliver and Shannon dating 1948 [74]. The biggest advantage of PCM modulation lies in the fact that it can be regenerated infinitely, if the noise or distortion does not cause wrongly interpreted bits. Here again, the S/N ratio is increased at the expense of occupied bandwidth. PCM lends itself ideally for TDM.

Most sampled and quantized biomedical sensor signals show a large correlation between adjacent sample values, allowing the use of compression algorithms, limiting the amount of data to be transmitted. More details on compression methods are given in Sect. 7.4. Adding error correction to the quantized values allows detection and correction of erroneously received bits, cf. Sect. 7.5, decreasing the bit-error-rate (BER) even more.

The waveforms associated with the sampling and quantization process for PCM generation are depicted in Fig. 7.16. In this example the quantized data is mapped to a serial 3-bit binary non-return to zero (NRZ) code. Encoding baseband binary data on a wired connection is called line encoding, of which a few common examples are treated in the next section.

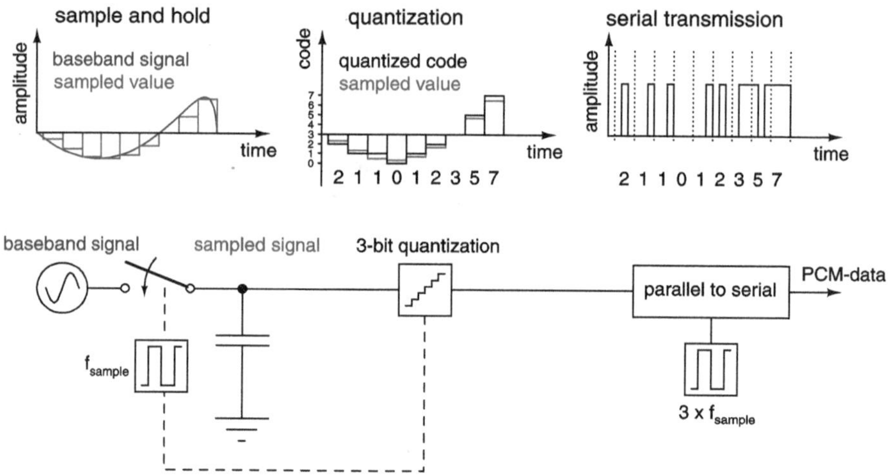

Fig. 7.16 Pulse code modulation of an analog input signal, 3-bit quantized and mapped to NRZ digital signal

7.3.3.2 Line Encoding

A quantized signal to be transmitted can be represented by different bases. As most ADCs deliver directly a binary code or base-2, this is the most common way to represent the signal. For serial data transmissions, the quantized signal is converted to a serial bit stream, where each bit is represented by a 0 or 1. Several methods exist for representing a 0 or a 1 in a binary datastream, each with their specific characteristics. An overview of the treated line codes is depicted in Fig. 7.17.

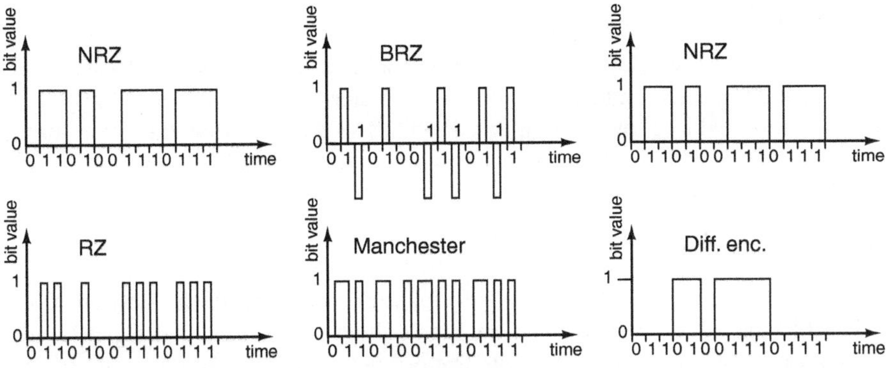

Fig. 7.17 Different line encoding schemes for binary data

Unipolar Non-return-to-Zero (NRZ)

The most common representation is non-return-to-zero (NRZ), where a 0 is represented by a low voltage, a 1 by a high voltage for one bitperiod. These voltage levels can be used for direct modulation of an amplitude-, phase- or frequency-shift keying (ASK, PSK or FSK) transmitter. The disadvantage for this encoding scheme is the difficulty for retrieving the clock from the data stream at the receiver, although not impossible [96]. Especially when sending long streams of zeros or ones, the receiver can lose track of the starting edges of each bit, leading to erroneous reception. Another disadvantage, but rather for wireline than for wireless transmission, is the average DC value of a NRZ stream, requiring DC coupling to the receiver, so transformer coupling is not possible. The DC content of NRZ coding leads to a waste of power, as it does not contribute to the information content of the bit stream.

Unipolar Return-to-Zero (RZ)

In this coding scheme, a 1 is represented as a high pulse of half a bitperiod, a 0 is represented by a low voltage. An advantage is the presence of the bit clock frequency component in the signal [41], which eases clock or data recovery from the original stream. It is however more prone to bit errors than NRZ, and occupies a larger bandwidth.

Bipolar Return-to-Zero (BRZ)

Each bitvalue 1 is represented as a pulse of half a bitperiod, alternating between a positive and a negative pulse. No pulse is used to transmit a zero. The advantages of this encoding scheme is the lack of a DC component. This code is also called alternate-mark-inversion (AMI), and is not commonly used in telemetry applications. The requirement for a dual power supply makes this approach less interesting.

Manchester

To ease clock recovery at the receiver, it is useful to encode the bit clock into the serial bitstream. This can be done using Manchester or split-phase encoding, where a 0 is represented as a low-to-high transition, a 1 as a high-to-low transition (or the other way around). The advantage is the self-contained clock at the cost of a higher bandwidth occupation.

Differential Encoding

By using differential encoding, the information is encoded in the bit transitions, rather than the values. A transition (low-high or high-low) is used to designate a zero bit, no transition designates a one bit. The original values can be decoded by comparing adjacent bits. Differential encoding requires a reference bit before initiating the encoding process, otherwise the received bit stream may be received inverted.

7.3.4 Digital Modulation

For analog signals, all amplitude levels at any timestep are defined and represent information. In between analog and digital, pulse modulation techniques are used to increase the S/N ratio of an analog signal, discretizing the amplitude and converting the analog voltage level to an analog continuous-time value. One step further, when analog signals are discretized in time and in amplitude, a digital signal results. A one-bit binary signal has just two signal levels that contain information, a zero or a one, valid for one bitperiod. Intuitively, this gives the digital signal a much higher noise immunity: noise spikes on the digital signal can be as high as half the signal amplitude before resulting in an erroneous interpretation at the receiver. As a digital signal is valid for a complete bitperiod, the noise immunity can be even increased by averaging the digital signal sampled at different moments during its bitperiod. Noise immunity is traded for wider bandwidth usage, by discretizing the source signal in amplitude and time.

When applying digital or pulse modulated signals as source for modulation, the increased noise immunity is also translates to the modulated signal. Digital and analog modulation methods basically use the same techniques: AM, FM and PM. The difference lies in the source signal, being analog, analog pulse-modulated or digital. Digital AM is called amplitude shift keying (ASK), digital FM frequency shift keying (FSK) and digital PM phase shift keying (PSK). Several subforms of digital modulation exist, most of them being a combination of ASK,FSK or PSK. An overview is given in Fig. 7.18. Possible circuit configurations for generating

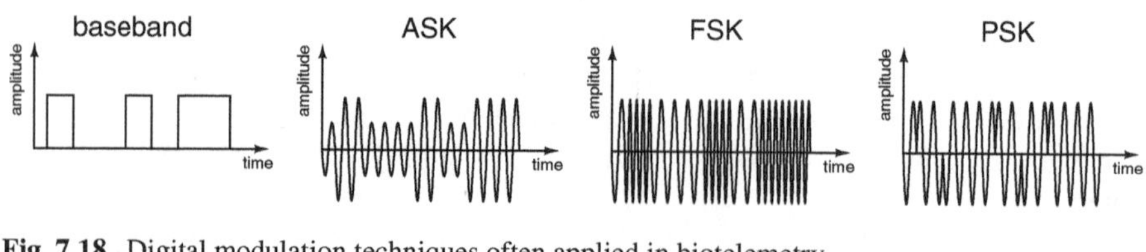

Fig. 7.18 Digital modulation techniques often applied in biotelemetry

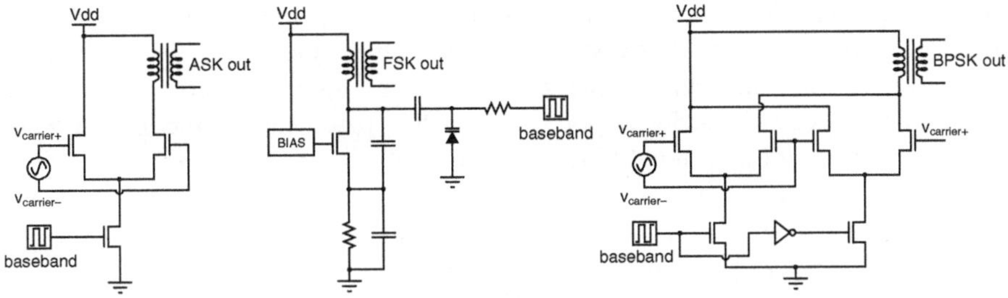

Fig. 7.19 Simplified circuit diagrams for generating ASK (in this case OOK), FSK and BPSK for biotelemetry applications

low-complexity ASK, FSK and BPSK are presented in Fig. 7.19. Several examples of digitally modulated telemetry in biomedical applications is presented in Section 7.7.

7.3.4.1 ASK

An ASK modulated carrier is generated by modulation of an AM transmitter with a digital or pulse modulated input signal. At 100% modulation depth, the transmitter output is actually turned on and off, known as on-off keying (OOK). Especially for low duty cycle source signals, the transmitter can run at a higher peak power and hence a larger range for the same average power consumption. This is an attractive solution for battery-powered telemetry as described in [59].

ASK modulation is used abundantly in inductively powered implants, to provide a power and bi-directional data communication. The tuned coupled coils of an inductively powered implant require a constant frequency to operate in the most efficient way. A modulation depth well below 100% is used for uplink transmission (external coil to implant), to limit the bandwidth occupation of the power carrier, as well as to ensure continuous power transfer to the implant. ASK does not have the noise problem of analog AM, as the input data has been pulse modulated or pulse-code modulated, increasing the S/N ratio at the cost of bandwidth.

7.3.4.2 FSK

When a digital or pulse modulated source modulates a VCO input, a FSK modulated signal is generated. Increasing the frequency deviation will increase the S/N ratio, at the expense of increased bandwidth. The frequency deviation can be controlled by changing the voltage on the varicap, or the current injected in the base. In systems requiring precise control of the spectral output, a PLL is required for controlling the center frequency.

7.3.4.3 PSK

In PSK, the phase of the carrier is modulated by 180 degrees, depending on the digital or pulse modulated source. This can be implemented easily by the use of a Gilbert cell or double balanced modulator, where the phase of the carrier can be inverted by controlling the current in two transistors. This modulation type shows a lot of similarity to AM–DSB SC modulation, where the phase of the carrier is inverted each time the source signal crosses 0. For PSK modulation, the absolute phase of the received carrier is not known, when receiving the modulated wave incoherently. Therefore often differential PSK (DPSK) is used to transmit the data, where the information is contained in the phase transitions, rather than the absolute carrier phase. For DPSK zero will be transmitted as no transition, where a one will be transmitted as a phase inversion of the carrier.

7.3.4.4 Digital Multiple Channel Transmission

Transmission of multiple measurement channels uses similar techniques as in analog multiple channel transmission 3.2.5. We have more degrees of freedom however on how this is exactly implemented.

Time Division Multiplexing (TDM)

For binary digital modulation methods (FSK,ASK and BPSK), time division multiplexing is the easiest to implement, multiplexing each digitized measurement sequentially. A digital multiplexer will switch sequentially through all measurement channels, mostly preceding each channel or series of channels by a start header. This allows synchronization at the receiver to the different measurement channels. TDM can also be implemented on higher constellation modulation schemes. The principle of digital TDM is represented on the top left block diagram of Fig. 7.21.

Multiple Bit Modulation

For binary shift keying methods, the bit rate is equal to the symbol rate, because each symbol consists of one bit. If we are able to map more bits to a symbol, more data can be transmitted for the same symbol rate or bandwidth. In fact, only the noise conditions will place a limit on how many bits can be transmitted in parallel [84].

Any digital modulation method can be represented on a constellation diagram, showing how the amplitude and phase (or real and imaginary component) of the modulated carrier will behave, depending on the selected symbol. The distance between 2 constellation points is a measure for the S/N ratio of a modulation scheme: the closer to each other, the lower the S/N ratio. Somehow this seems intuitive: noise (with a real and complex component) will act on the constellation points by smearing them out. The larger the noise, the larger the smearing area. At a certain noise level, two points on the constellation diagram may start to overlap, making it difficult–not impossible–to retrieve the original bit/nibble value, as depicted in Fig. 7.20, ASK + noise.

It is possible to map as many symbols as we want on a constellation diagram, each having it's unique I and Q coordinate, under the condition that the channel noise still makes it possible to discern between two adjacent constellation points.

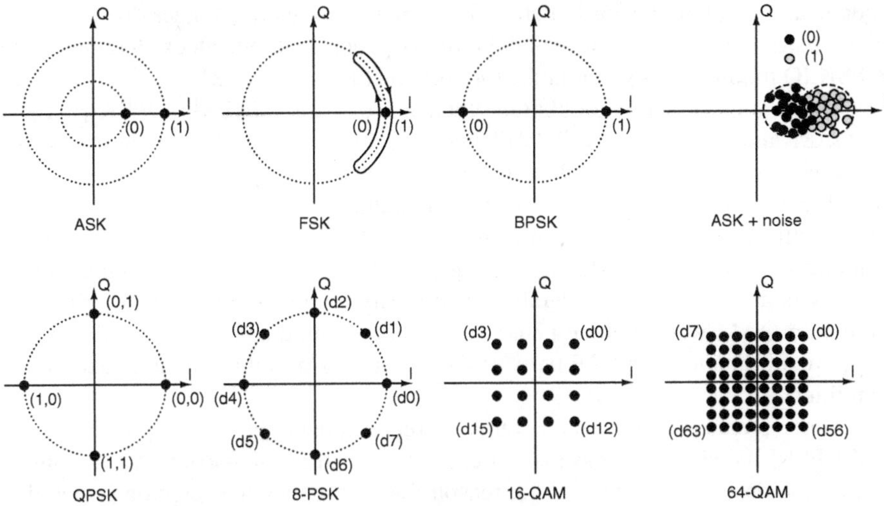

Fig. 7.20 Constellation diagrams of ASK, FSK, PSK, QPSK, 8-PSK, 16-QAM and 64-QAM. Note that the absolute phase of ASK, FSK and PSK is not defined, allowing non-coherent reception of these modulated waves. The constellation diagram at the receiver for ASK modulation is also shown (*top right*), where the constellation points are scattered by the channel and receiver noise

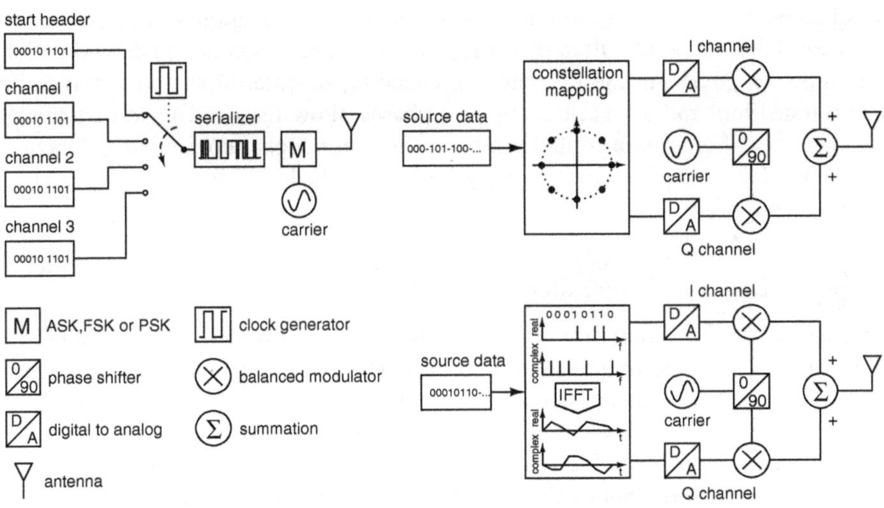

Fig. 7.21 Generalized block diagrams of time division multiplexing (*top left*), constellation IQ modulation (*top right*) and OFDM IQ modulation (*bottom right*)

The more symbols we can map on the diagram, the more bits can be sent in parallel for the same symbol rate, making the modulation scheme more bandwidth efficient.

Using two digital to analog converters (DAC) and two balanced modulators, for the in-phase and quadrature (I and Q) path, it is possible to map each possible input code to any point on the constellation diagram. IQ modulation comes down to combined phase and/or amplitude modulation. A general system block diagram of an 8-PSK IQ modulator is given in the top right picture of Fig. 7.21.

Common multiple bit modulation schemes are quadrature PSK (QPSK), 8-PSK, 16-QAM and 64-QAM. QPSK will map 4 symbols (or 2 bit) on the unity circle, each symbol offset with 90 degrees. For 8PSK, 8 symbols (3 bit) are mapped onto the unity circle of the constellation diagram, each offset with 45 degrees. 16-QAM will map 16 symbols (4 bit) in a square fashion, where 64-QAM maps 64 symbols (6 bit) on the constellation diagram. n-PSK is a constant amplitude modulation scheme, relaxing considerably the linearity requirements of the transmission system. n-QAM will combine phase and amplitude modulation, posing strong linearity requirements on all RF parts of the system (balanced modulators and power amplifier).

Some typical constellation diagrams are represented in Fig. 7.20. For ASK, FSK and BPSK, the absolute phase is not exactly defined in the carrier, this is pointed out by the dotted lines. This is the reason that non-coherent reception is possible for these modulation schemes: there is no need for exact carrier phase or frequency tracking by a PLL. Note that FSK cannot be exactly represented on the constellation diagram: the constant amplitude point will move on the constellation circle to one direction for a zero and the other for a one, with a phase velocity depending on the frequency swing. For multiple bit constellations like QPSK or 16-QAM, the absolute phase is important to know, as it will define the mapping of each word to each constellation point. Often it is still possible to use non-coherent demodulation for BPSK or QPSK, using a differential encoding scheme. In such a scheme, the phase transitions rather than the absolute phases allow the receiver to distinguish between the different transmitted symbols. Currently, higher constellation schemes are not used for low power biotelemetry, because of the low power and low data rate requirements.

Frequency Division Multiplexing

Frequency multiplexing for multiple bit transmission is solved with an elegant solution called orthogonal frequency multiplexing (OFDM). A fixed number of n bits, taken from n-measurement channels, or time-multiplexed channels of each n-bits wide, is mapped to its inverse discrete Fourier transform (IDFT). The IDFT will convert the n input bits which are mapped to n frequency bins, to a continuous time symbol. The continuous time symbol, consisting of a real and complex part, can be used to modulate an I/Q modulator. At the receiver, the discrete Fourier transform (DFT) is taken from the received symbol, recovering the initial bitstream. A basic OFDM transmitter requires a digital signal processing core (DSP) for the IDFT, DACs and again an IQ modulator. If the channel properties are known (as eg. for WLAN or ADSL), OFDM can be used to switch on/off certain frequency bins that are distorted by the channel, decreasing the BER. For complexity reasons and hence increased power consumption, this is currently not used for low power biotelemetry.

7.3.5 Analog or Digital Modulation?

There is a clear trade-off when comparing analog to digital modulation as in Table 7.3. A highly improved S/N ratio and increased flexibility of digital modulation has put pure analog biotelemetry aside. Although there is a penalty of higher bandwidth occupation and higher circuit complexity, the ever decreasing linewidths for integrated circuits (IC) allow a decrease in power consumption and size, that cannot be competed by pure analog solutions. Modern biotelemetry applications always use digital transmission methods, which does not mean we should forget about the "ancient" methods. They offer a lot of design insight, performing complex tasks with simple circuits, and form the base of all biotelemetry electronics that are designed nowadays.

Table 7.3 Analog vs. digital modulation

	Analog modulation	Digital modulation
Noise immunity	Low	High
Bandwidth occupation	Low	High
Improvements S/N	FM modulation	PWM, PPM, PFM, PCM, error correction codes
Multiple channels	FDM, TDM, SC MOD	TDM, (OFDM, multiple-bit constellations)
Modulator complexity	Low	High, ADC is required
Demodulator complexity	Low	High

7.3.6 Data Rates

When examining Table 7.4, respecting the Shannon sampling theorem, the maximal sampling frequency for biopotentials or biophysical measurements is about 20 kHz, while for more common measurement signals like ECG and EEG, this is only 400 Hz. When sampling the amplified signal quantized to 8-bit resolution, the

Table 7.4 Different human biopotential and biophysical measurement signal properties [15]

	Ampl [mV]	BW [Hz]	Data rate, 8 bit/sample [kbps]
Cell action potentials	0.010–100	100–2000	32
Electroneurogram (ENG)	0.005–10	100–1000	16
Electroretinogram (ERG)	0.0005–1	0.2–200	3.2
Electro-oculogram (EOG)	0.01–5	DC–100	1.6
Electro-encephalogram (EEG)	0.002–0.200	0.5–100	1.6
Electro-cardiogram (ECG)	1–10	0.05–100	1.6
Electromyography (EMG)	0.001–2	5–10000	160
Blood pressure	–	10	0.08
Breathing rhythmm	–	10	0.08
Movement	–	10	0.08

maximal bitrate will hence be 160 kbps, for a single channel. For multiple channels, the bit rate increases proportionally. For example: a 10 channel EEG measurement, sampled at 400 Hz and quantized with a resolution of 8 bit, requires a data rate of 32 kbps. In general: the larger the sampled bandwidth or the more channels that need to be transmitted, the higher the output data rate. Most biomedical signals show a high correlation between sequential samples, allowing compression thus a decrease in data rate.

Generally, a data rate well below 200 kbps will do for most biomedical measurements. For long range RF links (>1 m) this can be solved using an off-the-shelf MICS or ISM transmitter. For short range near-field combined power/data links, the data is transmitted through the power transmission channel as depicted earlier in Fig. 6, which does not require a dedicated transmitter. Any signal encoding, like line codes or synchronization, has to be handled by the processor itself. There are exceptions to the low data rates generally required for biomedical data transmissions, as shown in the overview of Table 7.5. In neural signal processing, a large array of low-bandwidth measurement channels has to be transmitted simultaneously, from the cortex to outside the brain. Without any data compression, data rates up to 50 Mbps are no exception. Cochlear implants use an electrode array to stimulate the hair cells in the cochlea. About 10–20 electrodes require parallel stimulation with processed audio data picked up by a microphone. The stimulation data is sent in realtime through a near-field link, requiring approximately 1 Mbps. Swallowable capsules for gastro-intestinal investigation requires 2–3 Mbps to sent compressed RGB images with a resolution of 640 × 480 pixels at 15–20 frames per second (fps).

Especially for multiple channel and high data rate application, data compression is increasingly gaining importance. The data rate can be reduced at the cost of increased circuit complexity and hence power consumption, where it is important to find the right trade-off.

Table 7.5 High data rate biomedical telemetry applications

	Raw data rate [Mbps]
Wireless endoscopic capsule VGA [97]	2–3
Cochlear implant [119]	1
Retinal prosthesis [106]	1–2
Neural monitoring implant [52]	2–90

7.4 Compression

Data compression aims at removing redundancy from digitized sensor data, transmitting only the relevant information. An obvious example of data compression can be found in wireless capsule endoscopy: the more the images are compressed, the more images can be transmitted per second for the same data rate. For multi-channel telemetry applications, like multi-channel EEG or neural monitoring, it makes sense to compress the measurements to keep the available data within transmittable limits.

There are two main types of compression: lossy and loss-less. Loss-less compression allows decompression to the original signal completely to its original form, without any loss of information. Lossy compression will remove a part of the original signal, which is of little importance for the interpreter, but it is impossible to completely recover all original information.

Loss-less compression is mostly based on clever re-organization of the signal data, like Huffman-coding [43] or differential encoding, where only the difference between two samples is encoded. Maximal loss-less compression ratios are in the range of 2–5, depending on the original signal redundancy.

Lossy compression algorithms often filter out the high frequency content of signal, effectively ending up with less data after encoding. Combined with clever encoding, there is basically no limit in compression ratio. It is a trade-off between required signal quality and the amount of transmitted signal samples. For images, compression ratios of 20–30 are no exception, while still maintaining acceptable quality.

We are able to reduce the amount of generated data by two basic principles: quantization and recoding. Recoding will create an optimized way of representing the original data set with less data, being applied abundantly in lossless compression methods. Quantization will limit the amount of possible discrete representations (the resolution) of the sampled sensor signal, immediately influencing the output data rate. By applying quantization and recoding in an intelligent way, very sophisticated and performing compression schemes can be created. Compression will only work efficient on data sets with a limited symbol set, containing non-random and correlated data. A data set comprising only of random non-correlated data is very difficult to compress, as no redundancy can be removed.

Quite a few biotelemetry systems rely on transmitting continuous streams of data across the communication channel, e.g. neural probes or swallowable wireless endoscopic capsules. It is important to limit the amount of data to relax the data rate requirements, to fit an application into existing wireless standards like MICS and ISM. Generally, the trade-off between complexity (power) and data rate reduction has to be made: including complex compression will decrease the data rate at the cost of power consumed by the compressor, while a decrease in data rate will save power consumed by the transmitter.

7.4.1 Loss-Less Compression Algorithms

Several methods exist to represent a data set in a much more compact way, without losing its exact information content. The amount of compression obtained depends largely on the structure of the dataset: the more randomness, the lower the maximal compression that can be achieved.

Run-length-encoding (RLE) represents sequences of the same symbols, e.g. "aaaaaaaa", by a single symbol and its occurrence, e.g. (a,8). This is primarily useful for repetitive data sets, e.g. a black and white images. An example of RLE on a bitstream is depicted in Fig. 7.22.

In variable length coding (VLC), like Huffman encoding, the probability of the input data set is calculated (in a first run or adaptively during runtime), and each input symbol is encoded with a corresponding variable length code. Symbols with a

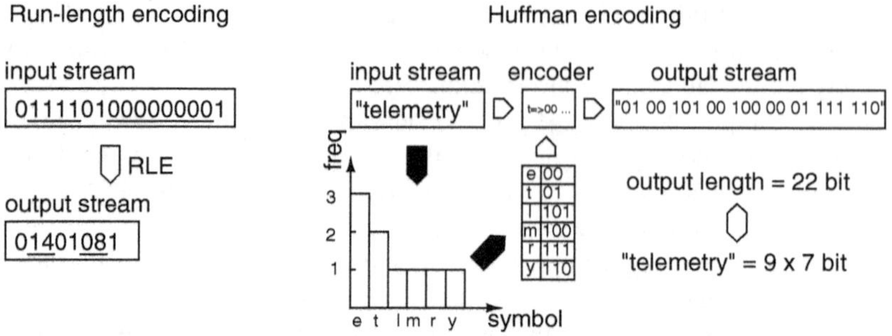

Fig. 7.22 Block level representation of compression through RLE (left) and Huffman VLC encoding (right) of the word 'telemetry'

high probability are encoded with a short bit sequence, symbols with a low probability with a longer bit sequence. An example of Huffman encoding is visualized in Fig. 7.22.

Dictionary based compression will store frequent symbols (or symbol sequences) in a dictionary, replacing the symbol by a pointer to the dictionary. They are based on the fact that typical data sets contain patterns and repetitions. The Lempel Ziv algorithm is based on this method. A combination of dictionary based compression and Huffman encoding is used in the famous zip compression algorithm.

If lossless compression of correlated data is required (like audio and images), often just the differences are transmitted, as they are small. Lossless compression methods are generally used in cases where no data may be lost, as in the storage of program files or text documents. Loss-less compression algorithms are hardly applied in biotelemetry applications, as not all measured information is relevant for the receiver.

7.4.2 Lossy Compression Algorithms

For most data sets, it is possible to permanently remove redundant data, because of its irrelevance for the reader/receiver. This leads to higher compression ratios than generally possible with loss-less compression, at the cost of quality reduction, that is however hardly noticeable at the receiver. Common examples are image and audio compression, where sequential data tends to be correlated. By transforming a sequence of image or audio data to another domain, it is often easier to remove redundant data. A frequency transform (e.g. cosine transform) converts the transient data to the frequency domain, although this step itself does not provide any compression. Quantization is performed on certain data points of the frequency domain data, actually limiting the possible symbols of the data set. High frequency data can often be omitted without noticeable quality problems for the receiver. For image data, the high frequency content is quantized more than the low frequency data. An example for compression of wireless endoscopic images, using a discrete

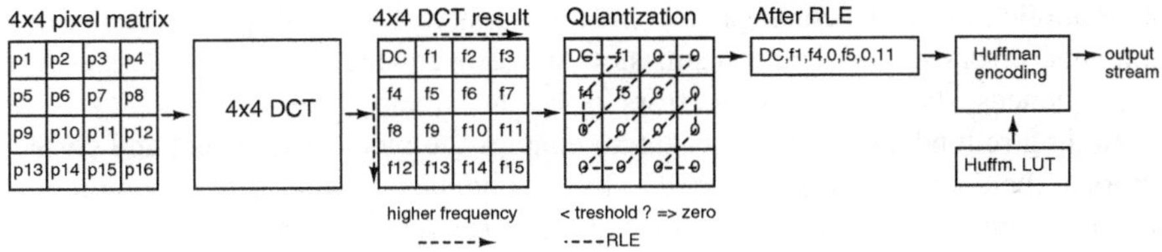

Fig. 7.23 Block level representation of lossy image compression on a 4 × 4 pixel block. First the pixels are converted to the frequency domain by a DCT, quantized to get as much 0 s as possible, RLE and Huffman VLC encoded, finally ending up with the output stream

cosine transform (DCT) is described in [100]. The transformed data is then quantized, Huffman encoded and VLC coded, as depicted in Fig. 7.23. For audio data, a psycho-acoustic model is used to determine which frequency data can be quantized more or less in order to be unnoticeable by the listener.

Color image data is often first converted from the red, green blue (RGB) domain to the luminance and chrominance (YUV) domain. Most of the image information is contained in the luminance Y component, for which the human eye is most sensitive. The U and V components contain only information about the color, and can often be downsampled without any problems for the viewer. Considerable compression can be obtained by quantizing the Y,U and V component, and differentially storing all quantized data. This compression principle has been applied by [101], for wireless endoscopic imaging applications, achieving compression ratios up to a factor 20 without noticeable image quality loss.

Another example of lossy compression can be found in neural probe processing electronics [75, 78]. If m neural probe voltages need to be quantized by a n-bit ADC, a datarate of $m \times n \times f_{\text{sampling}}$ is required, leading to unacceptably high data rates. As mainly the action potential is of interest to the examiner, just 1 bit is enough to show whether the examined neuron has fired. This considerably decreases the required data rate, at the cost of a limited information content. Again, quantizing limits the symbol set to be transmitted. Often it is possible to switch between neuron firing detection of all channels, or full unquantized read-out of a single neuron channel.

7.5 Error Correction

Contrary to compression, where redundancy is removed from the sensor data to enable lower data rates, error correction adds redundancy to the data stream, at the cost of higher required data rates for the same throughput. In pulse modulation or digital modulation, bandwidth is exchanged for a higher S/N ratio, leading to more robust data transmission. In numerous applications however, the bit-error-rate (BER) is still not acceptable for the application. This applies especially for life-sustaining implants, like pacemakers, ventricular assisting devices (VAD), implantable defibrillators, . . . It is unacceptable during wireless monitoring or reprogramming of these devices to have errors that may lead to lethal situations. For other applications, like cochlear implants, wireless endoscopic capsules and retinal implants, errors in the data transmission are also unwanted, but with less severe consequences, showing up as bit errors in the datastream.

Adding redundancy to the datastream enables the receiver to detect and correct errors in the data stream, as opposed to the redundancy present in an uncompressed sensor signal, eventually ending up with an error-free transmission. A daily life example of error correction can be found in CDs, that are playing noise-free music even when moderately scratched. For transmit-only systems, it is necessary to add enough error correction information to detect and correct the errors. The error correction coding is then called forward error correction (FEC). For two way communication systems, it is often enough to detect the errors, and ask to the transmitter to retransmit the faulty data. This is called automatic repeat request (ARQ). We can distinguish two types of FEC codes: block codes and cyclic codes. The principle of both code types is depicted in Fig. 7.24.

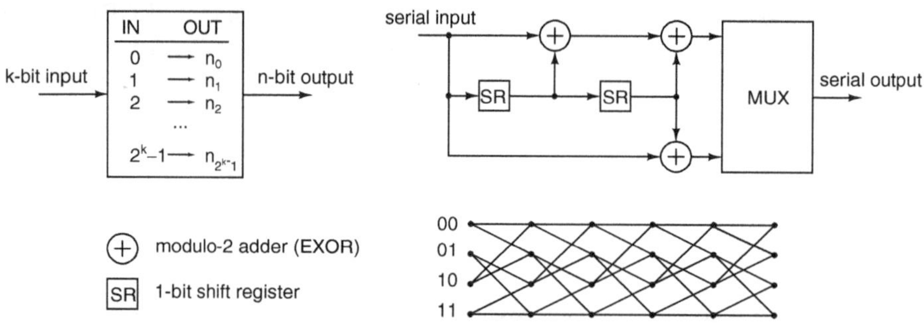

Fig. 7.24 A general representation of the most common FEC codes: block codes (*left*) and convolutional codes (*right*), along with its trellis diagram (*bottom right*)

7.5.1 Block Codes

A (n, k) block code translates each frame of k input bits to n output bits. The code rate R is the ratio of the input and output bits per frame, k/n. The data transmission is made more robust, at the cost of slowed down transmission speed with a a factor $1/R$. For a block encoder, the output bits only depend on the current input bits, i.e. it does not contain a memory of previous input frames.

Block codes are relatively easy to implement: at the transmitter side, a look-up table (LUT) will translate each input bit sequence to an output bit sequence. This allows generation of 2^k valid out of 2^n possible codes, in the case of binary symbols. At the receiver, the LUT is known, and the received codes are compared with the 2^k possible valid codes. A block code is systematic: a certain input sequence will always lead to the same output sequence. When errors occur in the received bit sequence, an algorithm can be implemented to find the valid sequence resembling the closest the faulty sequence. A huge research community is occupied to find the most optimal block codes, that have the best chance for detecting and correcting the errors in a data stream.

A (15, 11) block code error correction system for implantable bladder pressure measurement systems was presented in [4]. The 11-bit word is expanded by 4-parity bits, calculated on the 11 bit word, allowing a single-bit error detection and correction.

7.5.2 *Convolutional Codes*

Convolutional codes use the value of previous input bits to calculate the output bit stream. A (n, k, m) convolutional encoder produces n output bits for k input bits, using m shift registers or memory bits. The example in Fig. 7.24 shows a $(2,1,2)$ convolutional encoder: each input bit produces two output bits, and two memory bits are used to calculate each output bit. This slows down the transmission speed with a factor n/k. A convolutional code is non-systematic: a certain input bit sequence does not necessarily generate the same output sequence, because of the memory involved. This generally leads to a better BER performance compared to systematic block codes of the same length.

A big advantage of a convolutional coder is the relatively little hardware involved in generating an encoded data stream, which is ideal for low-power consuming applications, like biotelemetry. On the other hand, a quite complex—thus power consuming-decoding algorithm is required at the receiver. Often the Viterbi algorithm is used for decoding a convolutional encoded data stream [103, 33]. As a convolutional encoder generates output sequences depending on the input bits and previous input bits, only certain output sequences are valid. The possible sequences can be graphically represented in a trellis diagram (like the structure for climbing plants), as depicted in Fig. 7.24. By following the received sequences in a trellis diagram, invalid bit sequences can be identified, and corrected by tracing back to the last correct received sequence. A detailed explanation of the functioning of a trellis decoding scheme can be found in [92].

7.6 Carrier Frequency Selection for RF Links

The correct selection of the carrier frequency for far-field communication is a tedious trade-off, depending on a lot of parameters: tissue absorption, bandwidth requirements (which is directly depending on the data rate), governmental regulations, biocompatibility regulations, miniaturization requirements, ...

7.6.1 *Tissue Absorption vs. Antenna Size*

A study from Johnson and Guy [50] showed that an EM wave travelling through the human body is partially absorbed, and converted to heat. This wave attenuation increases exponentially with a rising frequency. This can also be expressed as the penetration depth, which is the distance the wave travels until 90% of its surface

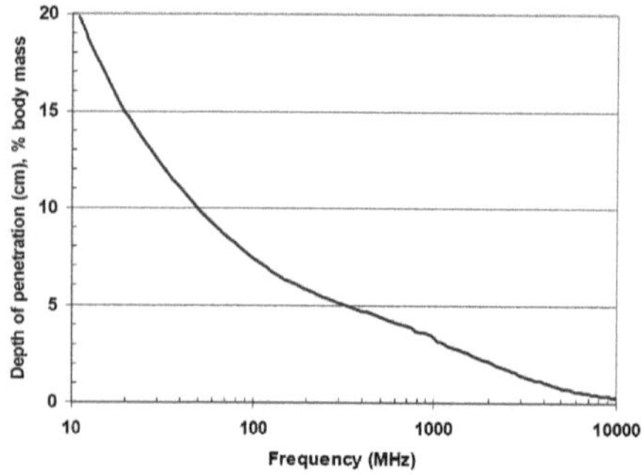

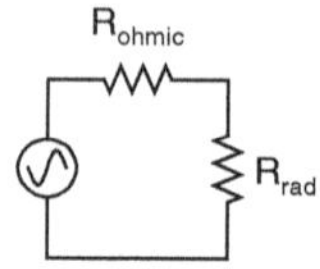

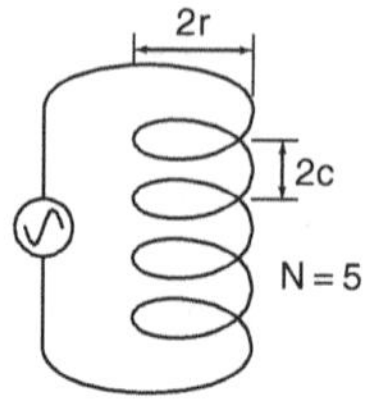

Fig. 7.25 The penetration depth of an RF wave entering muscle-equivalent material [45] (*left*). Simplified circuit representation of an antenna (*middle*). Five-turn coil antenna (*right*)

power density value has been absorbed by the body. The penetration depth versus frequency is depicted in Fig. 7.25.

This fact suggests to select the carrier frequency as low as possible, to limit the absorption of RF energy by the body, decreasing the required transmitted power. Lowering the carrier frequency however, requires a large antenna size to have an efficient radiator, as explained in Sect. 7.2.1.1. For small implantable transmitters, only little space is available for implementing antennas, leading to inherently inefficient designs.

The antenna efficiency η_{ant} is generally expressed as the ratio of the (fictitious) radiation resistance R_{rad} and the sum of the ohmic losses R_{ohm} of the antenna and the radiation resistance R_{rad}, as depicted in Fig. 7.25. For a small circular loop antenna, often used in biotelemetry, the antenna efficiency, radiation and ohmic resistance can be approximated by [7]:

$$R_{\text{rad}} = 20\pi^2 N^2 \left(\frac{2\pi r}{\lambda}\right)^4 \tag{7.2}$$

$$R_{\text{ohm}} = \frac{Nr}{t} R_s \left(\frac{R_p}{R_O} + 1\right) \tag{7.3}$$

$$\eta_{\text{ant}} = \frac{R_{\text{rad}}}{R_{\text{rad}} + R_{\text{ohm}}} \tag{7.4}$$

N: number of turns
r: radius of a single coil winding
λ: wavelength of the carrier
t: radius of the coil wire
R_s: wire surface impedance $\sqrt{\frac{\omega\mu_0}{2\sigma}}$
R_p: resistance per unit length due to proximity effect
R_O: ohmic skin effect resistance per unit length $\frac{NR_s}{2\pi t}$

The ohmic losses are proportional to the skin and proximity effect. These depend on the wire thickness, the distance between the wires, the overall wire length and the frequency [7]. The increase of the ohmic resistance by the proximity effect can be retrieved from a graph [88], depending on the ratio of the spacing $2c$ between the windings and the wire diameter $2r$. The ohmic losses represent the amount of input power converted to heat, the radiation resistance the amount of input power effectively converted to EM waves.

To give the reader an idea about the (low) efficiency of a small antenna like in Fig. 7.25, see the following numerical example:

It is clear that for sub–GHz frequencies, the efficiency of small antennas is poor, even when there is an ideal power transfer from the source to the antenna. In order to find the best antenna size for the available space and frequency, the trade–off has to be made between the RF attenuation of the body (carrier frequency as low as possible) and the efficiency of the antenna (carrier frequency as high as possible). The plot in Fig. 7.26 shows the radiation resistance and maximal efficiency of the antenna vs. frequency, for the antenna example given in Table 7.6.

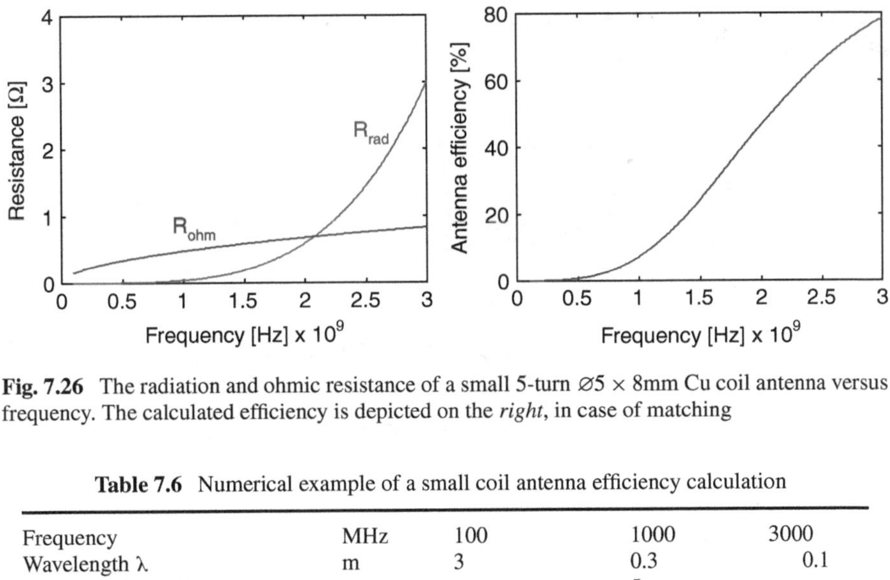

Fig. 7.26 The radiation and ohmic resistance of a small 5-turn ⌀5 × 8mm Cu coil antenna versus frequency. The calculated efficiency is depicted on the *right*, in case of matching

Table 7.6 Numerical example of a small coil antenna efficiency calculation

Frequency	MHz	100	1000	3000
Wavelength λ	m	3	0.3	0.1
Number of turns N	–		5	
Coil radius r	mm		2.5	
Coil spacing $2c$	mm		1.5	
Coil wire radius t	mm		0.25	
R_P/R_O [88]	–		0.15	
Free space permeability μ_0	H/m		$4\pi\,10^{-7}$	
Conductivity copper σ	$1/\Omega$m		5.9×10^7	
R_{rad}	Ω	3.7×10^{-6}	3.7×10^{-2}	3
R_{ohm}	Ω	0.153	0.479	0.83
η_{ant}	%	0.0025	7.2	78.3

7.6.2 Antenna Size vs. Bandwidth Requirements

Small antennas have an additional shortcoming—besides being inefficient radiators. A small antenna has difficulties obtaining a low quality (Q) factor, which translates to a small bandwidth. A general expression of Q is the ratio of the energy stored in the antenna and the power radiated per frequency cycle. A low Q means a large available bandwidth, allowing large bandwidths to be transmitted or received. A high Q means a small available bandwidth, which is a required feature for e.g. resonator tanks in oscillators: only a small fraction of the noise is allowed to be amplified, leading to oscillators with little phase noise. High Q antennas radiate little, and mainly store their energy.

An expression for the Q factor of small antennas was developed by Wheeler [109], relating the antenna volume to its radiation Q, see Eq. (7.5). The lower bound on the achievable Q for small antennas was described by Chu [21], Eq. (7.6). A recent review of the Chu and Wheeler formulas is described in [61] and confirmed in numerical simulations. To give the reader an idea about achieveable bandwidths for small antennas, the Wheeler and Chu limit Q factors for the small inductor antenna of Table 7.6 were calculated in Matlab, as well as the achievable Q of a lossy inductor, see Fig. 7.27. The bandwidth of this antenna was calculated using the approach of [114], Eq. (7.8).

$$Q_{\mathrm{Wheeler}} = 6\pi \frac{\left(\frac{\lambda}{2\pi}\right)^3}{\pi r^2 b \left(1 + 0.9\frac{r}{b}\right)} \tag{7.5}$$

$$Q_{\mathrm{Chu}} = \left(\frac{\lambda}{2\pi b}\right)^3 + \frac{\lambda}{2\pi b} \tag{7.6}$$

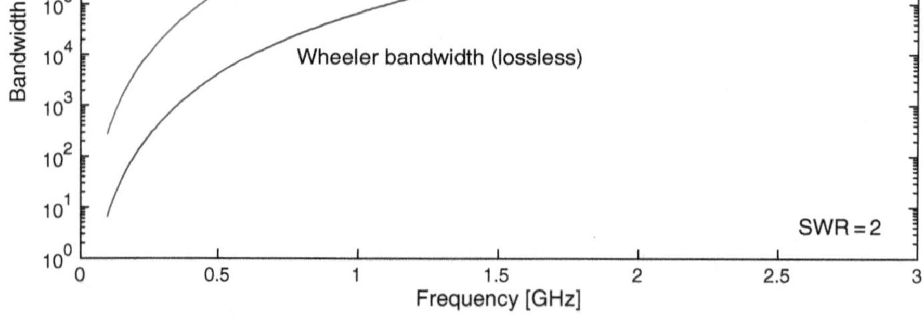

Fig. 7.27 The antenna bandwidth vs. frequency for the example of Table 7.6, a small 5-turn ⌀5 × 8mm, using the Chu, Wheeler and lossy inductor formulas, at a SWR = 2

$$Q_{\text{lossy}} = \frac{2\pi f_r L_{\text{ant}}}{R_{\text{Ohm}}} \tag{7.7}$$

$$\Delta f = \frac{f_r}{Q} \frac{\text{SWR} - 1}{\sqrt{\text{SWR}}} \tag{7.8}$$

r: radius of a single coil winding
b: axial coil length
λ: wavelength of the carrier
L_{ant}: the inductance of the coil antenna, calculated using the Wheeler formula [108]
f_r: resonance frequency of the coil
Δf: bandwidth of the coil antenna
SWR: standing wave ratio

The standing wave ratio (SWR) is the ratio between the maximal and the minimal voltage, caused by reflections on the transmission line to the antenna. A SWR $= 1$ agrees to a perfect impedance match: all the power from the source is transferred to the load without any reflection. A SWR of 2 or more is common for real life antenna applications.

Figure 7.27 shows the achievable bandwidths for a ∅5 × 8mm antenna, using the Wheeler, Chu and lossy inductor approach. Chu and Wheeler assume a lossless antenna (100% efficient), which is merely a theoretical case. At 100 MHz, the Wheeler bandwidth spans just 7 Hz, with a maximal theoretical bandwidth (Chu) of 270 Hz. At 1 GHz, the Wheeler bandwidth increases to about 7 kHz, with a maximal theoretical bandwidth (Chu) of 2.7 MHz. Basically, small antennas at low frequencies have a bandwidth issue.

Examining the bandwidth of a lossy antenna, taken into account the ohmic resistance as calculated in Eq. (7.3), 280 kHz is obtainable at 100 MHz, where a 900 kHz bandwidth is realistic at 1 GHz. Keep in mind that the efficiency is extremely poor at 100 MHz, hence there is little practical use for such a small antenna. More information on small antennas can be found in [69]. The antenna bandwidth can be increased by loading the antenna, at the cost of efficiency.

A limited antenna bandwidth means a limitation on the maximal information bandwidth that can be transmitted. This is not necessarily an issue, as e.g. in the case of the FSK modulator of Fig. 7.19, where the tank coil is used as antenna. In this case a large Q is advantageous, leading to a low phase noise oscillator [56]. The bandwidth of the antenna does not relate here to the transmittable information bandwidth, as the resonance frequency of the oscillator is adjusted when modulating the carrier. So for both modulation frequencies of a FSK modulator, the oscillator is working in a high Q regime, efficiently generating that frequency. Of course the radiation efficiency still largely depends on the antenna volume vs. the wavelength, but the bandwidth limitation is in this case not a limiting factor.

Depending on the required range—especially for low-range biotelemetry applications—this type of antenna will still work well, as not only the radiated power (which is in any case low) comes into play, but also the near-field. The near-field still enables acceptable link qualities at distances within one wavelength [97].

7.6.3 Regulations vs. Bandwidth Requirements

The useful bands for low power implantable RF biotelemetry are located in the sub–GHz band. Commercially accepted bands are the Industrial, Scientific and Medical (ISM) and Medical Implant Communication Services (MICS) bands. ISM and MICS usage basically do not require a license, although restrictions on bandwidth usage are often locally bound. MICS is not actually a true standard, it is a generic term for the US standard on Medical Device Radiocommunications Services [66] and the EU standard on Short Range Devices for Ultra Low Power Active Medical Implants [30]. The MICS band uses the same frequency band as meteorological satellites, for that reason the bandwidth and radiated power are limited. The ISM band is a generic term for different frequency bands all over the spectrum, including spectra available for inductive links. Table 7.7 gives a non-restrictive overview of the available bands and allowed bandwidths, as well as additional remarks. ISM and medical implant bands that have a spectrum access limitation have been omitted from the table, as they are not useful for continuous monitoring applications. Available unlicensed bands higher than 2.5 GHz have been omitted for their limited use in biotelemetry, related to their high attenuation. An overview of ISM-band frequencies in the US and EU is given in an application report from TI [98].

Table 7.7 Overview of available bands for RF biotelemetry

	Freq [MHz]	Max BW [MHz]	P_{out}/magnetic field limit
ISM a [46][1]	6.765–6.795	0.03	42 dBμA/m
ISM b [46][1]	13.553–13.567	0.014	42 dBμA/m
ISM c [46][1]	26.957–27.283	0.326	42 dBμA/m or 10 dBm
ISM d [46][1]	40.660–40.700	0.040	10 dBm
MICS [66, 30]	402–405	0.3	−16 dBm
ISM f1 EU [28][2]	433.05–434.79	1.74	0 dBm
ISM g4 [46]	869.7–870.000	0.3	7 dBm
ISM US [31]	902–928	26	30 dBm
ISM h [46][3]	2400–2483.5	83.5	10 dBm

[1]Useful for low data rate inductive links
[2]No audio or video
[3]WLAN, Bluetooth and Zigbee make use of this band

A conclusion to be drawn from Table 7.7: in the band below 1 GHz in the EU, little bandwidth is available, which automatically puts a limit to the maximal symbol rates that can transmitted. For low data rate applications this is no problem, but for high data rate application, like e.g. a swallowable endoscopic capsule, this poses a serious issue, pushing the designer to more complex modulation techniques. Achieving adequate bandwidth for a small antenna at low frequencies is one thing, but getting past governmental compliance is a seriously underestimated hurdle in the transfer from a working design to a commercially viable product.

7.7 Biomedical Telemetry Applications

Applications for short distance biomedical telemetry mainly use two methods for communication: near-field (inductive) links and far-field (radiofrequency) links. Other telemetry methods, like optical and ultrasound communication are applied mainly for research purposes.

Near-field applications are mostly used for combined power and data links, where an external coil generates an alternating magnetic field to power the implant. Inductive links make use of resonance-tuned closely-coupled coils, requiring a specific carrier frequency to work efficiently. This makes ASK the perfect candidate, as the carrier frequency remains constant, and the amplitude is modulated. At the implanted receiver, the data is demodulated with an envelope detector. Data from the implant to the powering module is transmitted using load modulation of the carrier, or sometimes BPSK subcarrier modulation. ASK load modulation will modulate the carrier at the transmitter, where it is demodulated using an envelope detector. Subcarrier BPSK modulation will modulate the data on a subcarrier derived from the powercarrier.

The use of far-field telemetry, relying on the propagation of EM waves, can be advantageous when higher ranges are required. Typical near-field applications have a range up to 1 m, while for a far-field transmitter the range depends on its output power and channel noise conditions, where ranges of 10–100 m are not an exception. As RF transmission for wearable electronics is usually generated at frequencies above 100 MHz, to enable the use of efficient antennas, the available bandwidth is higher than at lower frequencies, allowing higher bit rates for binary shift-keying modulation schemes. As treated in Sect. 7.6, the higher attenuation of EM waves at higher frequencies makes RF telemetry less favorable for communication with implanted devices.

7.7.1 Physiological Monitoring

7.7.1.1 Bladder Pressure Monitoring

An inductively powered bladder pressure monitoring system was conceived by Coosemans [25, 24]. A 132 kHz power carrier is used to induce power into the bladder monitoring system using three orthogonal external coils. The capacitive bladder pressure sensor is integrated in a ring oscillator configuration, translating the pressure change into a frequency change. The frequency is measured with a PIC microcontroller 12F629 and corrected for non-linearities, after which the pressure value is sent at 8.25 kbps to a BPSK modulator. This modulator uses a 66 kHz subcarrier frequency, derived from the 132 kHz power carrier, sending the pressure data via the inductive link to the base station. A picture of the setup, as well as a block diagram of the monitoring system is depicted in Fig. 7.28.

7.7.1.2 Wireless ECG Monitoring Integrated in Textile

A wireless ECG system was conceived by Coosemans and Hermans [26], intended for the continuous monitoring of the electrocardiogram (ECG) of children with

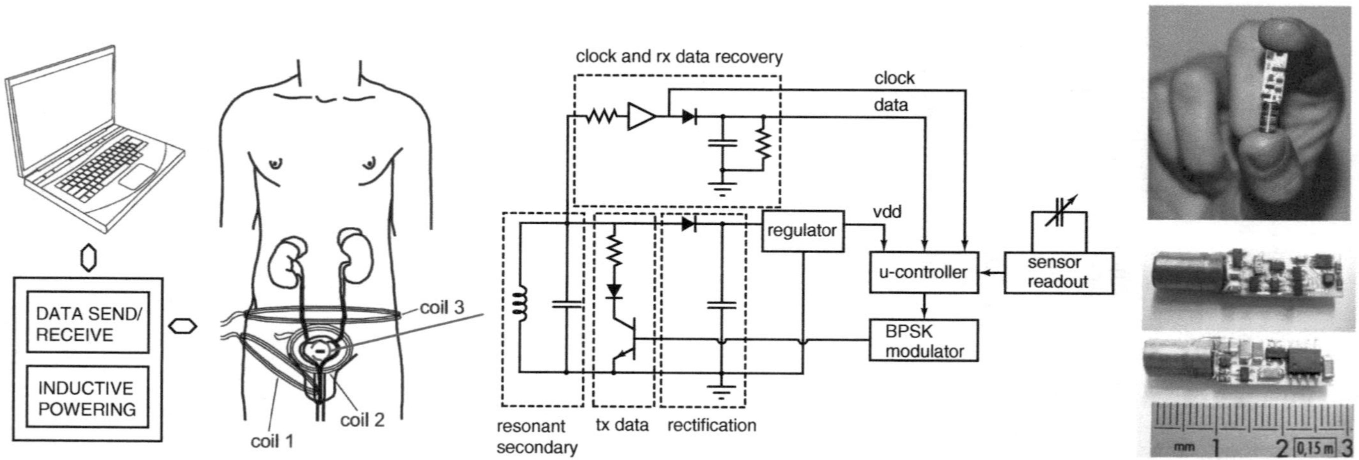

Fig. 7.28 System setup (*left*), block diagram of the monitoring system (middle) and pictures of the bladder pressure monitoring system (*right*) [25]

an increased risk of Sudden Infant Death Syndrome (SIDS). The sensors and the antenna are made out of textile materials, and all electronics are mounted on a flexible circuit to facilitate integration in the baby pajamas. The circuit diagram is similar as for the bladder pressure monitoring system in Fig. 7.28. The ECG data is transmitted with a 16.5 kbps BPSK modulated subcarrier. A picture of the textile coil, textile ECG electrodes and flexible circuit integrated in the baby pajama is depicted in Fig. 7.29.

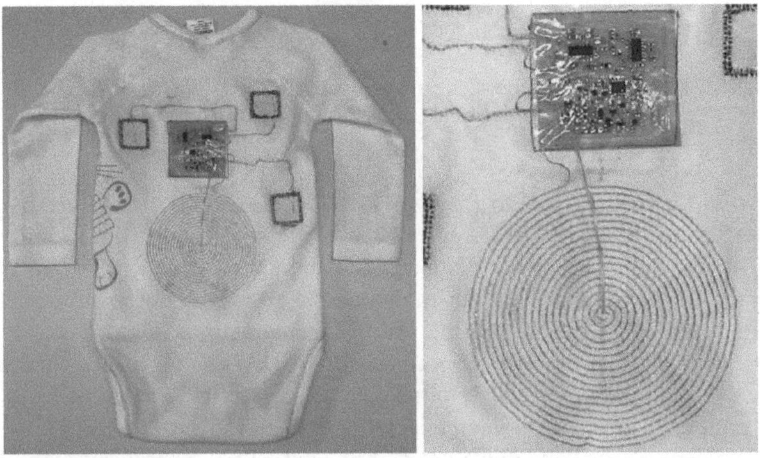

Fig. 7.29 Baby suit prototype for wireless ECG monitoring. Note the embroided coil antenna for power and data transmission, the three Textrodes and the flexible circuit board containing the readout electronics

7.7.1.3 Textile Integrated Breathing and ECG Monitoring System

A wireless system was designed by Jourand et al on flexible substrates to measure breathing rhythm and ECG [51]. A promising feature is the adoption of accelerometers to quantify the breathing cycle. The realised designs were placed on a T-shirt and tested on adults, keeping in mind the parameter variability when used with infants. The ECG and breathing rhythm data is sent to a computer wirelessly for further analysis, using the Nordic NR24L transceiver, at a carrier frequency of 2.4 GHz. A picture and block diagram of the system integration is depicted in Fig. 7.30.

7.7.1.4 Pacemaker Monitoring and Programming

Early pacemakers used internally placed reed switches, for transcutaneous programming by an externally modulated magnetic field [14, 94, 1]. This has been almost completely replaced by near field communication, enabling bidirectional data transfer from/to the pacemaker electronics. Sometimes a reed switch is still included for safeguarding, avoiding unintentional reprogramming.

A small coil inside the metal housing of a pacemaker is used to transmit programming information from the pacemaker to an external programming coil. A

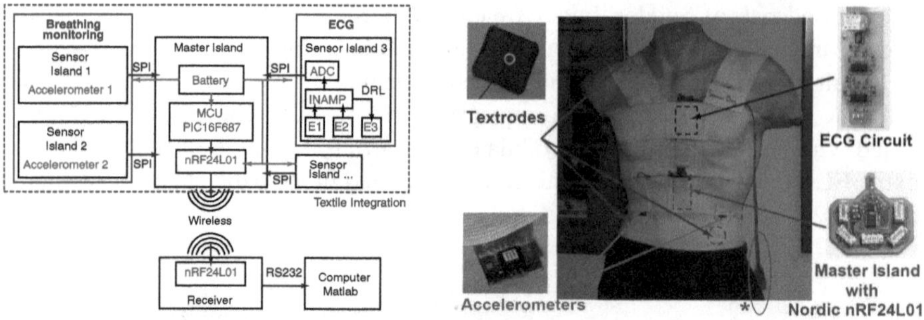

Fig. 7.30 Block diagram of a wireless breathing rhythm and ECG monitoring system, integrated into textile (*left*). A picture of the assembled textile integration on a dummy (*right*) [51]

carrier of 175 kHz is used, to transmit PPM encoded data to the pacemaker, as described in the patents of Medtronic [113, 35]. The carrier is generated by pulsing an LC circuit, generating a damped oscillation at the resonant frequency. The frequency of operation is small enough to couple through the metal case of the implant, with negligible attenuation through human tissue. The range of transmission is constrained to 5–10 cm and sensitive to orientation of the internal and external coils. Communication from the external coil to the implant is achieved by PWM modulated data, on-off keying a carrier at 175 kHz.

More recent pacemakers make use of FSK and BPSK for transmission of data, as recently published by Halperin et al. [37], examining the communication protocol used by a 2003 Medtronic implantable cardioverter defibrillator (ICD). For communication from the implant to the programmer, differential BPSK was used at a carrier of 175 kHz, at a bitrate of 87.5 kbps. The communication from the programmer to the implant was achieved by FSK, at a bitrate of 12.5 kbps, at a carrier frequency of 175 kHz, with a frequency swing of ±25 kHz. The principle of pacemaker communication is depicted in Fig. 7.31.

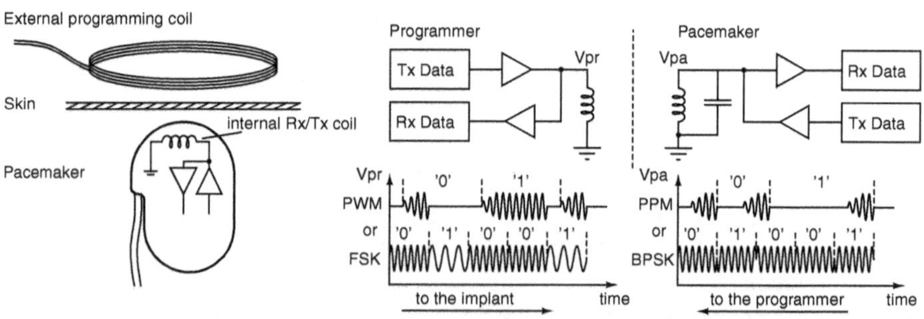

Fig. 7.31 Principle of near-field communication between a pacemaker and the programmer (*left*). The implant modulates a 175 kHz carrier with PPM or BPSK, the programmer uses PWM or FSK transmission to the implant

7.7.1.5 Inductive Power and Data Transmission for Wireless Endoscopy

Wireless capsule endoscopy is getting wider acceptance among specialists as a first examination of the gastro-intestinal (GI) tract, in case of anomalies [16, 29]. For understandable comfort reasons, the barrier for the patient to undergo a GI examination has dropped considerably, increasing the chances of early detection of GI cancers. However, capsule endoscopy does not offer yet a justified alternative for the traditional–wired–endoscopy, on ground of the capsule's limited frame rates, limited image resolution and the lack of active locomotion. The reason for these shortcomings is due to the limited available energy in a battery-powered capsule. Available energy in two button silver-oxide cells is typically 50 mAh, just enough to foresee passive locomotion, a frame rate of a few frames per second and low-resolution pictures [16]. Recently, Carta developed a new wireless powering technology based on 3-D inductive coupling. It is able to continuously supply up to 300 mW, independent of its orientation, in the same volume as two button cells [19]. This development enabled the use of a better and faster image sensor, requiring higher data rates to wirelessly transmit these images through the body.

It is possible to transmit commands at a speed of 9.6 kbps from the external coil to the 3D coil assembly, by ASK modulation of the power carrier. These low speeds command can be used to control the CMOS camera, LEDs, or in case of active locomotion, the actuators. An integration was made with a RF near-field transmitter developed by Thoné [97], to transmit the compressed image data a data rate of 2 Mbps at a carrier between 100–400 MHz, using FSK modulation. A picture of the integration by Thoné and Carta, along with the block diagram, is depicted in Fig. 7.32.

7.7.1.6 Wireless Capsule Endoscopy: Given Imaging Pillcam

The Zarlink ZL0081 is the transmitter used in the Given Imaging PillCam, the first commercial wireless endoscopy capsule. Two 256 × 256 pixel images are transmitted at 2 fps, and minimum-shift key (MSK) modulated with the Zarlink transmitter at a speed of 2.7 Mbps. A small coil antenna is used for transmission of the data to an array of coil antennas attached to the patient's body. The 0.35u CMOS IC is working at 2.6–3.2 V, in the ISM band of 433 MHz [116]. The MSK modulator is arranged as a PLL stabilized VCO, where the frequency is modulated by a separate DAC [10], as depicted in Fig. 7.33.

7.7.2 Orthopedic Implant Monitoring and Control

7.7.2.1 Distraction Nail Driver

For people suffering from limb-length discrepancy, an autonomous fully implanted distraction system, void from any percutaneous connections was developed by Van Ham et al. [102]. The implant houses a position sensing system on-board to feedback the accomplished extraction. Power and data transmission is achieved through

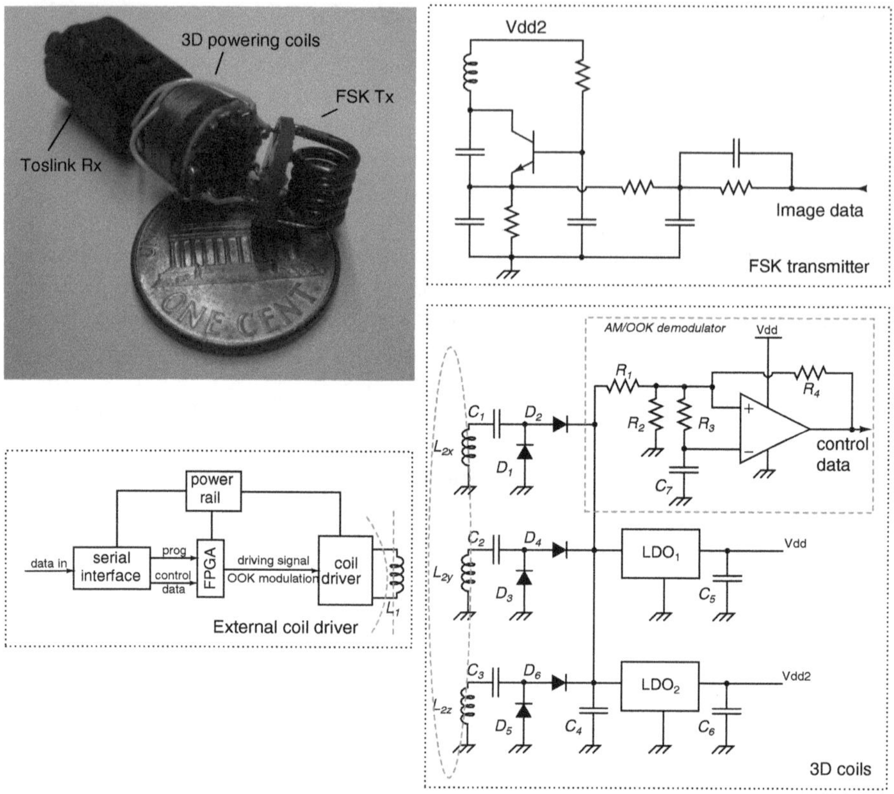

Fig. 7.32 3D inductive powering module of Carta [19] integrated with the FSK transmitter of Thoné [97], for low-speed uplink and high-speed downlink transmission and powering of a wireless endoscopic capsule (*top left*). *Top right*: circuit diagram of the FSK transmitter, *bottom right*: circuit diagram of the 3D power receiver and its low data rate receiver, *bottom left*: block diagram of the external coil driver

an inductive link, at a carrier frequency of 1 MHz, supplying up to 1 W in the implant. Transmission from the external power source to the distraction system is achieved by OOK. Data from the implant to the external world is achieved by load modulation of the power carrier. The system is depicted in Fig. 7.34.

7.7.2.2 Hip Prosthesis Fixation Analysis

Loosening is one of the major problems of hip replacement, causing pain and discomfort to the patient. Puers et al. [77] described a system for fixation detection of the femur prosthesis, by applying mechanical vibrations to the femur bone, and analysis of the the prosthesis response. A linear response shows a correct fixation, while a non-linear response points to loosening of the prosthesis. Monitoring

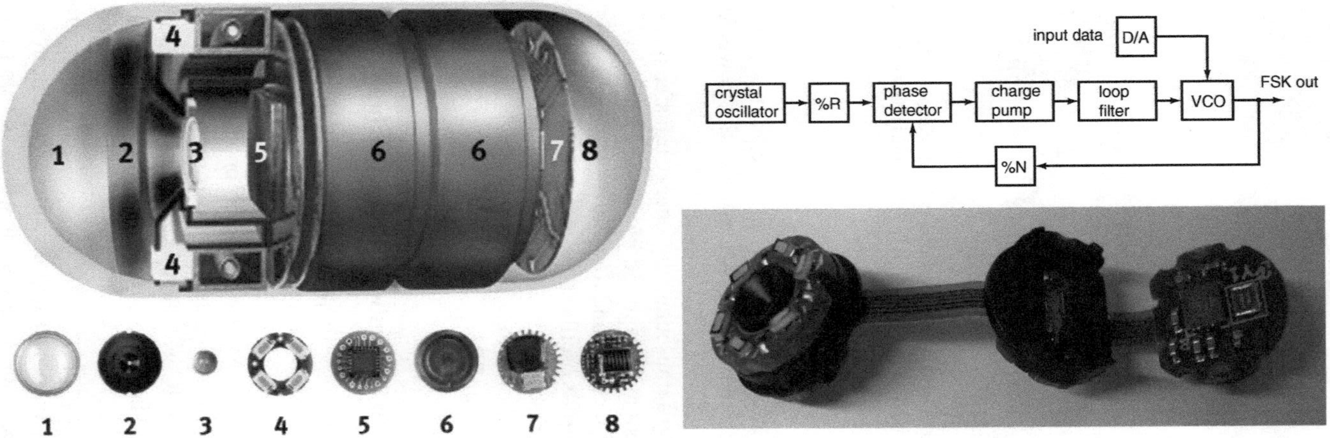

Fig. 7.33 Overview of internal structure of the Pillcam (www.givenimaging.com), dome (1), lens housing (2), lens (3), LED ring (4), CMOS camera (5), batteries (6), FSK transmitter (7), coil antenna (8) (*left*), FSK system block diagram of the Pillcam [10] (*top right*), and a picture of a recent Pillcam internal assembly

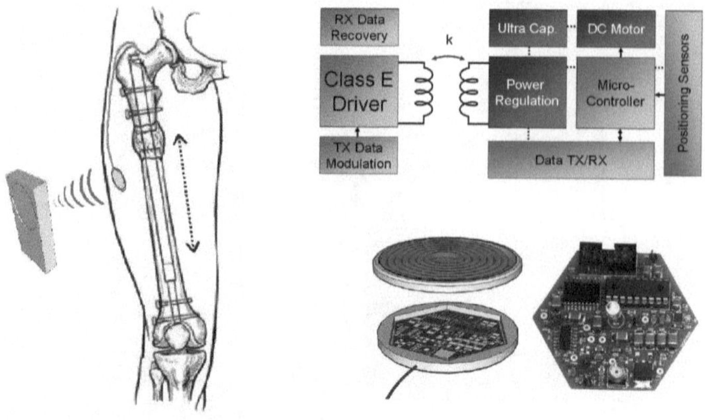

Fig. 7.34 Nail distraction system with combined power/bi-direction data communication (left). The system block diagram is depicted in top right, bottom right is showing the integrated system with coil assembly

is done by an accelerometer, inside the hip prosthesis head, enabling more precise measurements,although imposing restrictions to the power consumption and dimensions.

The system is powered by a transcutaneous energy transfer system, at a frequency of 750 kHz. The response signals, captured by the accelerometer, are converted by an implanted data conversion circuit and transmitted by a near-field transmitter. The transmitter consists of a Colpitts oscillator tuned at center frequency of 10 MHz, on-off keyed by PWM encoded sensor data. The transmitted signals are picked up by an external receiver analyzed on a PC. The system block diagram and photograph of the system is depicted in Fig. 7.35.

7.7.2.3 Telemetry IC Design for Orthopedic Monitoring

The design of an interface IC for orthopedic monitoring of implants was described by Van Ham [38]. It provides rectification and power supply regulation at the implant, and is able to recover ASK modulated on the power carrier. IC design for biomedical purposes differs greatly from the procedure commonly known to analog designers, both in terms of reliability as in terms of voltage requirements. This power and data front-end IC was therefore fabricated in a I3T80 HV extension on standard 0.35 m CMOS. The voltage regulation circuitry has a load regulation of 2.1% and a line regulation of 3.7 mV/V. The PSRR at the carrier frequency of 1 MHz is as high as 61 dB. A low transistor count clock extraction architecture delivers a digital clock for add-on digital processing circuitry. Two different data demodulation structures are presented, including one of which is able to detect a mere 4% external coil modulation depth ASK signal up to 9.6 kbps. The data rate can reach up to 17.2 kbps when deeper modulation is applied. The chip area is 6.76 mm^2 as

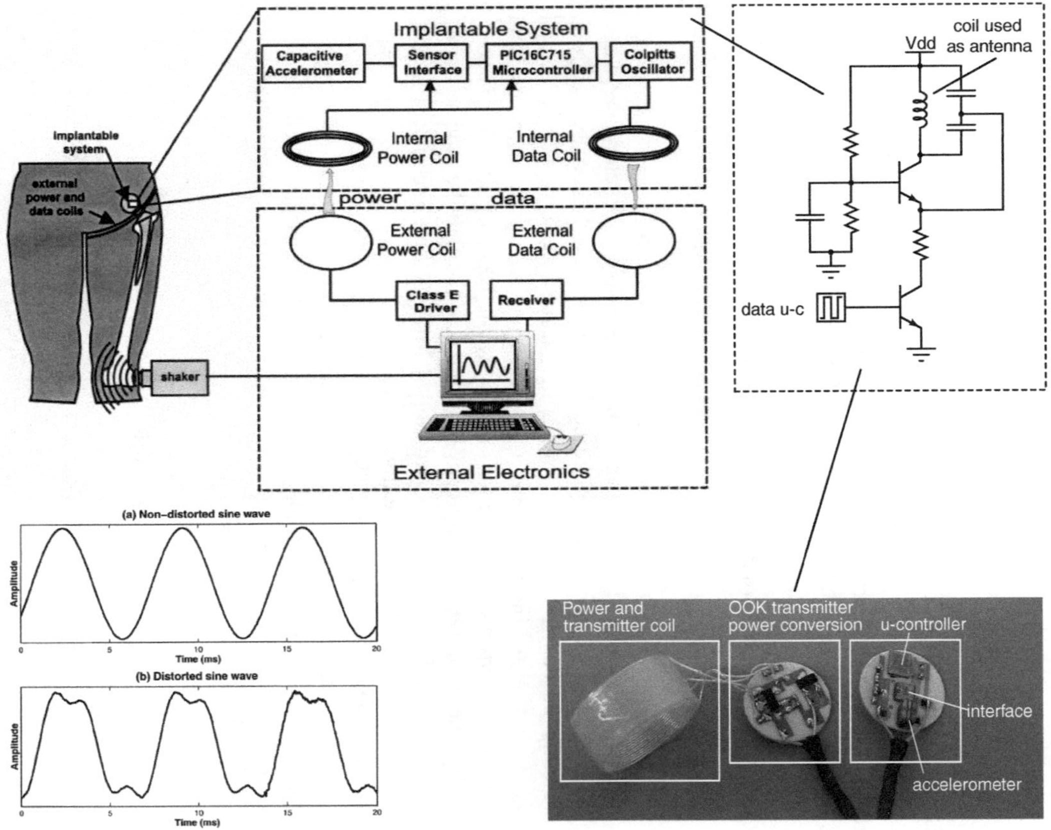

Fig. 7.35 The hip prosthesis fixation system by Puers [77] , with system block diagram (*top left*), linear and non-linear response (*bottom left*), photograph of the integrated sensor system with telemetry (*bottom right*) and circuit diagram of the OOK Collpits transmitter (*top right*)

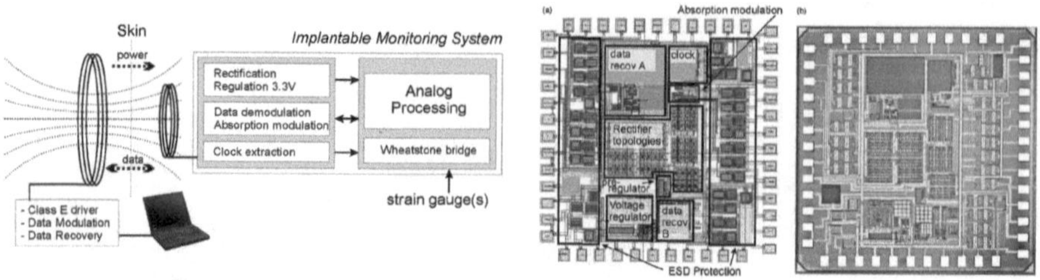

Fig. 7.36 Block diagram of the telemetry IC for orthopedic implant powering and monitoring (*left*), with the IC layout (*middle*) and IC photograph (*right*)

depicted on Fig. 7.36, of which ESD protection structures and bonding pads account for 4.32 mm^2. The total current consumption is 0.56 mA.

7.7.3 Nerve Implant Monitoring and Stimulation

7.7.3.1 Cochlear Implants

Cochlear implants have provided partial hearing to more than 120 000 persons worldwide; half of which being pediatric users who are able to develop nearly normal language [119]. The cochlear implant consists of a multiple-contact electrode that is surgically placed in the cochlea, stimulating the auditory nerve with processed audio data. A short range inductive link is used for transmission of power and data to the implanted electronics. The audio band between 100 and 8000 Hz is sampled, digitally filtered in several frequency bins and sent through the inductive link, ASK modulated at a data rate of approximately 1 Mbps. Often line encoding, e.g. PWM or Manchester is applied to the digital stream before modulation to increase the S/N ration or ease clock recovery. The carrier frequency ranges between 5 and 50 MHz. At the receiver, the data is demodulated, demultiplexed and converted to parallel current pulse-trains stimulating the different contacts of the electrodes. An implanted cochlear implant is depicted in Fig. 7.37, the system block diagram is depicted on the right.

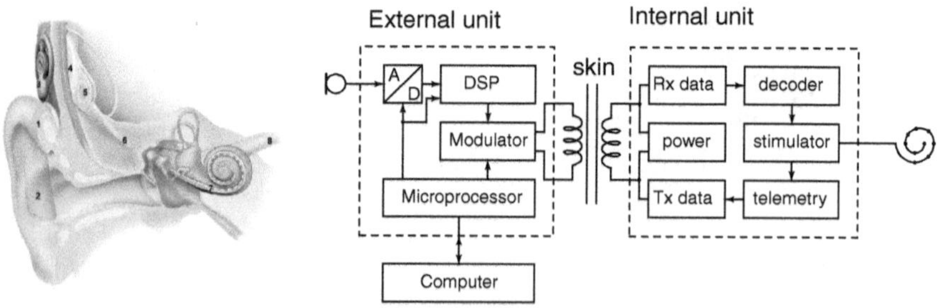

Fig. 7.37 Implanted cochlear system (www.cochlear.com), with external processor and microphone (1 and 2), external coil (3), internal coil (4), internal electronics (5), the cochlear electrode (7) and auditory nerves (8). A generic cochlear system block diagram is depicted on the right

7.7.3.2 Retinal Prosthesis

Retinal prosthesis, and other neural interfaces require a lot of data to be transmitted simultaneously, for interfacing with a large number of neurons. A minimum of approximately 600 pixels in the prosthesis is required to enable patients to read texts with large fonts. High data rate modulator and demodulator circuits have been developed by Ghovanloo [34] and Marin [64]. Ghovanloo uses 5/10 MHz FSK modulation through an inductive power/data link, reaching a maximal data rate of 2.5 Mbps. Wang et al. developed modulator and demodulator circuits for high speed inductive power/data link, using BPSK modulation, reaching maximally 4 Mbps [105] [106]. The modulator circuit of Wang is depicted in Fig. 7.38, as well as a system overview of the retinal implant.

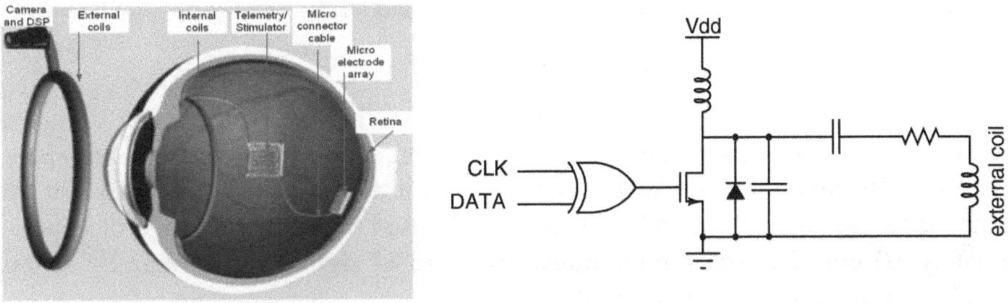

Fig. 7.38 Generalized retinal prosthesis system (*left*). Class E BPSK inductive power/data circuit of Wang [105] (*right*)

7.7.4 General Monitoring and Identification

7.7.4.1 RFID

Radio frequency identification (RFID) systems are a contactless, batteryless means to transfer identification data to a reader station. Typical applications include animal identification, credit cards and biometric passports. In most cases, the RFID tag is powered through an external magnetic field, returning its data content by load, phase or frequency modulation to the reader. Several RFID systems have been standardized, like the standard for identification of animals (ISO 11784, ISO 11785 and ISO 14233) and for contactless ID cards (ISO 10536, ISO 14443 and ISO 15693). These standards describe the required carrier frequency, modulation scheme and bitrates, as well as physical dimensions and implementation. An overview of several RFID standards can be found in Finkenzeller's book, Chapter 9 [32].

7.7.4.2 Portable Heart Rate Monitoring

More and more people work out in a controlled way, using a heart rate monitor. This provides feedback to the user for training at the right intensity for achieving a

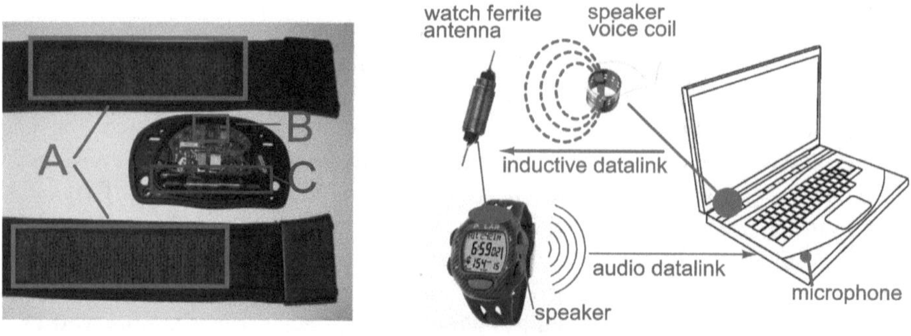

Fig. 7.39 On the *left*: Polar heart rate monitor chest belt with ECG electrodes (A), circuit board with ECG frontend and near-field transmitter (B) and ferrite antenna (C). The bidirectional data PC communication principle for Polar HRM is depicted on the *right*

certain performance goal, e.g. burning fat or improving their general performance. Such a system generally consists of a chest belt containing electrodes, a battery powered ECG frontend and a 5 kHz near-field transmitter, as depicted in Fig. 7.39. As antenna, a coil wound around a ferrite core is used, achieving a range of approximately 80 cm. The transmitter transmits its ECG signal to the wearer's watch, displaying and logging the heart rate.

Some heart rate monitors foresee the possibility to communicate bi-directionally with a PC, for training data analysis or configuration of training settings. The link from the watch to the PC uses the available PC audio microphone, to transmit data using (audible) audio modulation. The link from the PC to the watch uses the PC speakers to transmit information, but using the inductive field generated by the speaker, rather than the generated sound wave. The advantage is that this communication doesn't require any special hardware. The communication principle is depicted in Fig. 7.39.

7.7.5 Overview of Commercial Biomedical Transmitters

7.7.5.1 Zarlink ZL70101

The ZL70101 is a completely integrated transceiver, requiring only 3 external components. It operates in the 402–405 MHz MICS and 433–434 MHz ISM band. Data rates up to 800 kbps are claimed, although this is not the streaming data rate at the output of the IC. An integrated media access controller (MAC) takes care of the protocol handling, including Reed-Solomon forward error correction (FEC) and error detection. The device draws a current of 5 mA in active mode, and 250 nA in sleep mode. It makes sense to put the device in sleep mode when no continuous data transmission is required, to save power. There is a special wake-up feature, working at 2.4 GHz, as a lot more power can be transmitted in that band. Data communication with the ZL70101 is achieved with SPI. A block diagram of the internal structure

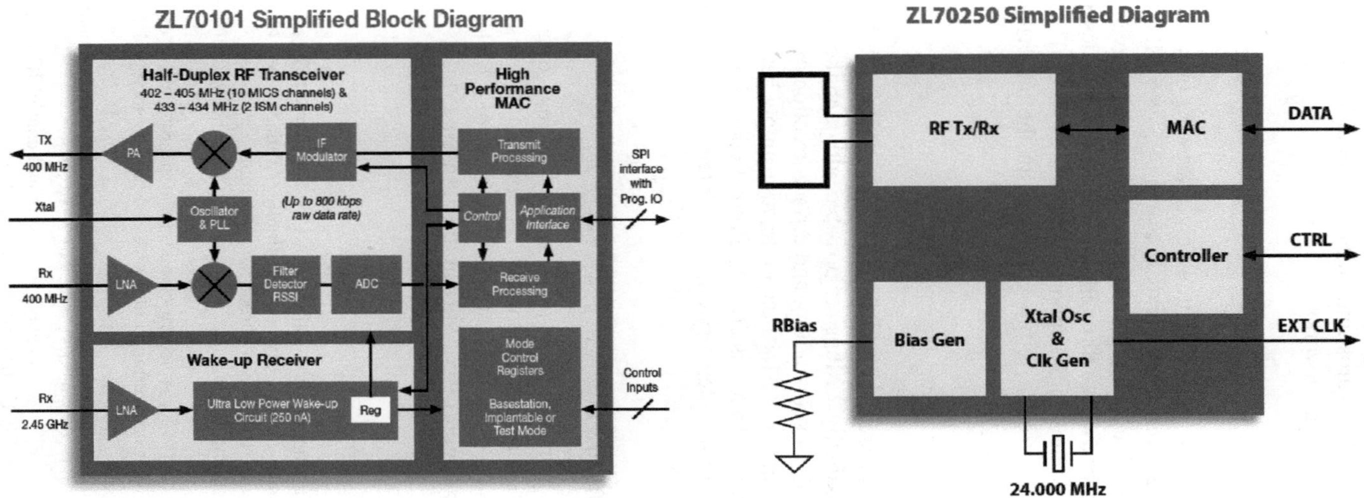

Fig. 7.40 Block diagram of the ZL70101 (*left*) and ZL70250 (*right*) medical transceivers of Zarlink. [117, 118]

of the transceiver is depicted in figure 40. Typical applications for this transceiver include, pacemakers, ICD's, neurostimulators, implantable insulin pumps, bladder control devices and body area networks.

7.7.5.2 Zarlink ZL70250

The ZL70250 is an ultra low-power RF transceiver, drawing only 2 mA at 1.2–1.8 V. It operates in unlicensed frequency bands between 795–965 MHz and offers data rates up to 186 kbps. Duty cycling can lower the consumption even more for applications that require lower average throughput. The device includes a RF transceiver and MAC that performs most link support functions. Using SPI allows easy interfacing with any microcontroller. A block diagram of the internal structure of the transceiver is depicted in Fig. 7.40.

7.7.5.3 Nordic NR24L01+

The Nordic NR24Lxx transceiver family works in the 2.4–2.4835 GHz ISM band. Burst data rates of maximally 2 Mbps are possible, although the the streaming data rate is lower because of packet handling and error correction overhead. The device consumes less than 15 mA at 2 Mbps data rate, at a supply between 1.9 and 3.6 V. The standby current consumption is 26 μA. Typical applications range from wireless PC peripherals to sensor telemetry and toys. Interfacing with the Nordic ICs is handled with the SPI bus.

7.7.5.4 Other Manufacturers

Plenty of other IC manufacturers provide similar low data rate ASK/FSK transceivers in their product range, mostly for low (<100 kbps) applications in the MICS or ISM band. A non-exhaustive list: Micrel [68], Melexis [67], Analog Devices [2], Texas Instruments [95], Maxim [65], Ansem [3], ...

References

1. Alferness C.A., Programmable body stimulator, U.S. Patent 4 066 086, Sep 1976
2. Analog devices short range devices. [Online]. Available: www.analog.com/en/rfif-components/short-range-transceivers/products/index.html
3. Ansem, RF cmos—pioneers from the early days. [Online]. Available: www.ansem.com/rf_cmos.html
4. Arabi K., Sawan M., (1995), A secure communication protocol for externally controlled implantable devices In: Engineering in medicine and biology society, 1995. IEEE 17th Ann Conf 2: 1661–1662 Sep 1995
5. Armstrong E., (1936), A method of reducing disturbances in radio signaling by a system of frequency modulation. Proc IRE 24(5):689–740, May 1936
6. Puers R., Sansen W., (1985) A dedicated processing and control IC for pressure telemetry. In: Proceedings of IEEE CICC Portland, 1985, pp 144–146
7. Balanis C., (2005) Antenna Theory, 3rd edn. Wiley, Hoboken, New Jersey, 2005

8. Bang S., Park J.Y., Jeong S., Kim Y.H., Shim H.B., Kim T.S., Lee D.H., Song S.Y. (2009) First clinical trial of the miro capsule endoscope by using a novel transmission technology: electric-field propagation. Gastrointest Endosc, 69(2): 253–259.
9. Barker P. (1966), Pressure sensor, U.S. Patent 3 240 207, Mar 1966
10. Bettesh I. (2006), Modulator and method for producing a modulated signal, U.S. Patent 2006/0 280 258, Dec 2006
11. Bluetooth compared to other technologies [Online]. Available: www.bluetooth.com/Bluetooth/Technology/Works/Compare/
12. Bluetooth SIG homepage. [Online]. Available: www.bluetooth.com
13. Bolton E.K., Sayler G.S., Nivens D.E., Rochelle J.M., Ripp S., Simpson M.L. (2002), Integrated cmos photodetectors and signal processing for very low-level chemical sensing with the bioluminescent bioreporter integrated circuit Sensors and Actuators B: Chemical, 85(1–2) 179–185
14. Bowers D.L. (1964), Controllable electric body tissue stimulators, U.S. Patent 3 311 111, Aug 1964
15. Bronzino J.D. (2000), The biomedical engineering handbook, 2nd edn. CRC Press, Florida
16. McCaffrey C., Chevalerias O., Mathuna C.O., Twomey K. (2008), Swallowable-capsule technology, Perv Comput, 7(1):23–29, Jan 2008
17. Callewaert L., Puers B., Sansen W., Jarvis J.C., Salmons S. (1991), Programmable implantable device for investigating the adaptive response of skeletal muscle to chronic electrical stimulation, Med Biol Eng Comput 29(5):548–553, Sep 1991
18. Carson J.R. (1922), Notes on the theory of modulation. Proc. IRE 10(1):57–64, 1922
19. Carta R., Tortora G., Thone J., Lenaerts B., Valdastri P., Menciassi A., Dario P., Puers R. (2009), Wireless powering for a self-propelled and steerable endoscopic capsule for stomach inspection, Biosensors Bioelectronics, 25(4):845–851
20. Casadei F.W., Gerold M., Baldinger E. (1972), Overview of biomedical telemetry techniques. IEEE Trans Biomed Eng 19(5):334–341
21. Chu L. (1948), Physical limitations in omnidirectional antennas. J Appl Phys 19(12): 1163–1175
22. Colpitts E. (1918), Oscillation generator, U.S. Patent 1 624 537, Feb 1918
23. Connel A., Rowlands E. (1960), Wireless telemetering from the digestive tract. Gut, 1: 266–272
24. Coosemans J., Puers R. (2005), An autonomous bladder pressure monitoring system. Sensors Actuators A: Physical 123–124:155–161
25. Coosemans J. (2008), Wireless and battery-less medical monitoring devices. Ph.D. dissertation, Katholieke Universiteit Leuven, Faculteit Ingenieurswetenschappen
26. Coosemans J., Hermans B., Puers R. (2006), Integrating wireless ECG monitoring in textiles. Sensors Actuators A: Physical, 130–131:48–53
27. Donaldson N.d.N., Perkins T. (1983), Analysis of resonant coupled coils in the radio frequency transcutaneous links. Med Biol Eng Comput 21:612–627
28. ECC(04)02 non-specific SRD in band 433.050-434.790 Mhz [Online]. Available: www.erodocdb.dk/doks/doccategoryECC.aspx?doccatid=1
29. Van Gossum A. et al., Capsule endoscopy vs. colonoscopy for the detection of polyps and cancer. New Engl J Med 361(3):264–270, July 2009
30. Etsi en 301-839-1; electromagnetic compatibility and radio spectrum matters (erm); short range devices (srd); ultra low power active medical implants (ulp-ami) and peripherals (ulp-ami-p) operating in the frequency range 402 mhz to 405 mhz; part 1: Technical characteristics and test methods." [Online]. Available: pda.etsi.org/pda/home.asp?wki id=hex-Tn6e2xhjnkjo08%27vB
31. FCC rules and regulations. [Online]. Available: http://www.fcc.gov/oet/info/rules/part15/PART15 07-10-08.pdf
32. Finkenzeller K. (2003), RFID Handbook, 2nd edn. Wiley, West Sussex, England

33. Forney G.D. (1973), The viterbi algorithm. Proc IEEE 61(3):268–278, March 1973
34. Ghovanloo M., Najafi K. (2004), A wideband frequency-shift keying wireless link for inductively powered biomedical implants. Circuits and Systems I: Regular Papers. IEEE Trans 51(12):2374–2383, Dec 2004
35. Grevious J.J. (1998), Efficient high data rate telemetry format for implanted medical device. U.S. Patent 5 752 977, May, 1998
36. Guillory K., Misener A., Pungor A. (2004), Hybrid RF/IR transcutaneous telemetry for power and high-bandwidth data. In: Engineering in Medicine and Biology Society, IEMBS'04. 26th Annual Int Conf IEEE 2:4338–4340, Sept 2004
37. Halperin D., Heydt-Benjamin T., Ransford B., Clark S., Defend B., Morgan W., Fu K., Kohno T., Maisel W., Pacemakers and implantable cardiac defibrillators: Software radio attacks and zero-power defenses. In: Security and Privacy, 2008. SP 2008. IEEE Sympos, 129–142, May 2008
38. Van Ham J., Puers R. (2008), A power and data front-end IC for biomedical monitoring systems. Sensors Actuators A: Physical 147(2):641–648. FCC
39. Hambrecht F.T. (1963), A multichannel electroencephalographic telemetering system. Massachusetts Institute of Technology, RLE Technical Report 413
40. Hartley R. (1915), Oscillation generator. U.S. Patent 1 356 763, June, 1915
41. Haykin S. (2001), Communication systems, 4th edn. Wiley, New York
42. Heetderks W. (1988), RF powering of millimeter- and submillimeter-sized neural prosthetic implants., IEEE Trans Biomed Eng, 35(5):323–327, May 1988
43. Huffman D.A. (1952), A method for the construction of minimum-redundancy codes. Proc. IRE. 40:1098–1101
44. ICNIRP. Guidelines for limiting exposure to time-varying electric, magnetic, and electromagneticfields (up to 300 Ghz). Health Phy 74(4):494–522, 1998
45. IEEE. IEEE std c95.1 - 2005 IEEE standard for safety levels with respect to human exposure to radio frequency electromagnetic fields, 3 khz to 300 ghz. IEEE Std C95.1-2005 (Revision of IEEE Std C95.1-1991), pp. 1–238, 2006
46. ERC recommendation 70-03 relating to the use of short range devices. www.erodocdb.dk/doks/doccategoryECC.aspx?doccatid=2#1622
47. Istepanian R., Woodward B. (1997), Microcontroller-based underwater acoustic ECG telemetry system. IEEE Trans Inform Technol Biomed 1(2):150–154, June 1997
48. Jaffe R., Rechtin E. (1955). Design and performance of phase-lock circuits capable of nearoptimum performance over a wide range of input signal and noise levels. IRE Trans Inform Theory 1(1):66–76, March 1955
49. Jeutter D.C. (1983), Overview of biomedical telemetry techniques, Engineering in Medicine and Biology Magazine. 2(1):17–24, 1983
50. Johnson C., Guy A. (1972), Nonionizing electromagnetic wave effects in biological materials and systems., Proc IEEE 60:692–720
51. Jourand P., De Clercq H., Corthout R., Puers R. (2009), Textile integrated breathing and ECG monitoring system Procedia Chemistry 1(1):722–725, proceedings of the Eurosensors XXIII conference
52. Judy J.W., Markovi D. (2009), Guest editorial special section on wireless neural interfaces. IEEE Trans Neural Sys Rehabil Eng 17(4):309–311, Aug 2009
53. Kim T.S. (2002), A method for communication between inside and outside of a transmission medium using the transmission medium as a communication line. U.S. Patent 7 307 544 B2, May 2002
54. Klein F., Davis D. (1976), A low power 4-channel physiological radio telemetry system for use in surgical patient monitoring. IEEE Trans Biomed Eng 23:478–481
55. Ko W., Liang S., Fung C. (1977), Design of radio-frequency powered coils for implant instruments. Med Biol Eng Comput 15:634–640
56. Lee T., Hajimiri A. (2000), Oscillator phase noise: a tutorial. IEEE J Solid-State Circuits 35(3):326–336, Mar 2000

57. Lenaerts B., Puers R. (2007), An inductive power link for a wireless endoscope. Biosens Bioelectron 22(7):1390–1395.
58. Liew H.C., Chan E. (1993), Biotelemetry of green turtles (chelonia mydas) in pulau redang, Malaysia, during the internesting period. Proc 12th Int Sympos Biotel. 157–163
59. Lin W.C., Pillay S.K. (1974), A micropower pulsewidth-modulation-pulse-position modulation two-channel telemetry system for biomedical applications. IEEE Trans Biomed Eng 21(4):273–280
60. Loeb G., Zamin C., Schulman J., Troyk P. (1991), Injectable microstimulator for functional electrical stimulation, Med Biol Eng Comput 29:13–19
61. Lopez A. (2006), Fundamental limitations of small antennas: validation of Wheeler's formulas. Antennas Propagat Mag. IEEE 48(4):28–36, Aug 2006
62. Mackay S.R. (1998), Bio-medical telemetry: sensing and transmitting biological information from animals and man, 2nd edn. Wiley, New york
63. Mackay R., Jackobson B. (1957), Endoradiosonde Nature 179:1239–1240, June 1957
64. Marin D., Troosters M., Martinez I., Valderrama E., Aguilo J. (1999), New developments for high performance implantable stimulators: first 3 Mbps up to 4.46 Mbps demodulator chip through a wireless transcutaneous link. In: Microelectronics for neural, fuzzy and bio-inspired systems. Proc Seventh Int Conf MicroNeuro '99, pp 120–126
65. Maxim wireless and RF product page. [Online]. Available: www.maxim-ic.com/products/wireless/
66. Medical devices radiocommunications service. [Online]. Available: wireless.fcc.gov/services/index.htm?job=service home&id=medical implant
67. Melexis product page. [Online]. Available: www.melexis.com/products.aspx
68. Micrel qwikradio range. [Online]. Available: www.micrel.com/page.do?page=product-info/qwikradio.shtml
69. Miron B.D., Small antenna design, 1st edn. Elsevier, Oxford, UK
70. Mueller J., Nagle H. (1996), Feasibility of inductive powering of miniature lowpower biotelemetry for use with microfabricated biomedical sensors. Proc 13th Int Sympos Biotelemetry 371–377
71. Nagumo J. (1960), Telemetering system for physiological measurements. U. S. Patent 3 229–684, Dec 1960
72. Nardin M., Ziaie B., Arx V.J., Coghlan A., Dokmeci M., Najafi K. (1996), An inductively powered microstimulator for functional neuromuscular stimulation. Proc 13th Int Sympos Biotelemet 99–104
73. Ohta J., Kagawa K., Tokuda T., Nunoshita M. (2005), Retinal prosthesis device based on pulse frequency-modulation vision chip. ISCAS 2005. IEEE Int Sympos Circuits Sys 3:2923–2926, May 2005
74. Oliver B., Pierce J., Shannon C. (1948), The philosophy of PCM Proc IRE 36(11):1324–1331
75. Olsson R.H.I., Wise K. (2005), A three-dimensional neural recording microsystem with implantable data compression circuitry. IEEE J Solid-State Circuits 40(12):2796–2804, Dec 2005
76. Pitsillides K., Symons J., Longhurst J. (1992), Biotelemetry of cardiovascular hemodynamic measurements in miniswine IEEE Trans Biomed Eng 39:982–986
77. Puers R., Catrysse M., Vandevoorde G., Collier J.R., Louridas E., Burny F., Donkerwolcke M., Moulart F. (2000), A telemetry system for the detection of hip prosthesis loosening by vibration analysis. Sensors Actuators A: Physic,85(1–3):42–47.
78. Rizk M., Obeid I., Callender H.S., Wolf D.P. (2007), A single-chip signal processing and telemetry engine for an implantable 96-channel neural data acquisition system. J Neural Eng 4(3):309.
79. Robrock R., Ko W. (1967), A six channel physiological telemetry system. IEEE Trans Biomed Eng 14(1):40–46

80. Schuder J., Stephenson H., Townsend J. (1961), High-level electromagnetic energy transfer through a closed chest wall. IRE Int Conv Record 9:119–126
81. Schuylenbergh K., Peeters E., Puers R., Sansen W., Neetens A. (1991), An implantable telemetric tonometer for direct intraocular pressure measurements. Abstracts 1st Eur Conf Biomed Eng 194–195
82. Schuylenbergh K.V., Puers R. (2009), Inductive Powering, 1st edn. Springer, Dordrecht
83. Sedra S.A., Smith C.K. (1998), Microelectronic Circuits, 4th edn. Oxford University Press
84. Shannon C.E. (1949), Communication in the presence of noise. Proc IRE 37(1):10–21
85. Simpson L.M., Sayler S.G., Patterson G., Nivens E.D., Bolton K.E., Rochelle M.J., Arnott C.J., Applegate M.B., Ripp S., Guillorn A.M. (2001), An integrated CMOS microluminometer for low-level luminescence sensing in the bioluminescent bioreporter integrated circuit. Sensors Actuators B: Chemic, 72(2):134–140
86. Skutt R., Fell B.R., Kertzer R. (1970), A multichannel telemetry system for use in exercise physiology. IEEE Trans Biomed Eng 17(4):40–46
87. Slater A., Bellet A., Kilpatrick G.D. (1969), Instrumentation for telemetering the electrocardiogram from scuba divers. IEEE Trans Biomed Eng BME-16(2):148–151, April 1969
88. Smith G. (1972), Radiation efficiency of electrically small multiturn loop antennas Antenna Propagat, IEEE Trans 20(5):656–657, Sep 1972
89. Sodagar A., Amiri P. (2009), Capacitive coupling for power and data telemetry to implantable biomedical microsystems. NER '09. 4th Int IEEE/EMBS Conf Neural Eng, 411–414, May 2009
90. Song S.J., Cho N., Yoo H.-J. (2007), A 0.2-mw 2-Mbps digital transceiver based on wideband signaling for human body communications. IEEE J Solid-State Circuits. 42(9):2021–2033, Sept 2007
91. Song Y.-K., Patterson W., Bull C., Borton D., Li Y., Nurmikko A., Simeral J. (2007), A brain implantable microsystem with hybrid RF/IR telemetry for advanced neuroengineering applications. Engineering in Medicine and Biology Society. EMBS 2007. 29th Ann Int Conf IEEE 445–448, Aug 2007
92. Sweeney P. (2002), Error control coding: from theory to practice, 1st edn. Wiley, West Sussex, England
93. Terman F.E. (1943), Radio Eng Handbook, 1st edn. McGraw-Hill, New York
94. Terry J. (1973), Implantable cardiac pacer having adjustable operating parameters. U.S. Patent 3 805 796, Jan 1973
95. Low power wireless RF Texas Instruments. [Online]. Available: focus.ti.com/analog/docs/rfifcomponentshome.tsp?familyId=367&contentType=4
96. Thoné J., Puers R. (2009), Implement a simple digital-serial NRZ data-recovery algorithm in an FPGA. Electron Des. News 42–43, June 2009
97. Thoné J., Radiom S., Turgis D., Carta R., Gielen G., Puers R. (2009), Design of a 2Mbps FSK near-field transmitter for wireless capsule endoscopy. Sensors Actuators A: Physical, 156(1):43–48
98. ISM-band and short range device regulatory compliance overview.[Online]. Available: www-s.ti.com/sc/techlit/swra048.pdf
99. Tjensvold M.J. (2007), Comparison of the IEEE 802.11, 802.15.1, 802.15.4 and 802.15.6 wireless standards. Sep 2007. [Online]. Available: janmagnet.files.wordpress.com/2008/07/comparison-ieee-802-standards.pdf
100. Turcza P., Zieliski T., Duplaga M. (2008), Hardware implementation aspects of new low complexity image coding algorithm for wireless capsule endoscopy. Proc Comput Sci ICCS 2008, 476–485
101. Turgis D., Puers R. (2005), Image compression in video radio transmission for capsule endoscopy. Sensors Actuators A: Physical 123–124:129–136
102. Van Ham J., Reynders Frederix P., Puers R. (2007), An autonomous implantable distraction nail controlled by an inductive power and data link. Actuators and Microsystems TRANSDUCERS 2007. Int Conf Solid-State Sensors 427–430, June 2007

103. Viterbi A. (1967), Error bounds for convolutional codes and an asymptotically optimum decoding algorithm. IEEE Trans Inform Theory 13(2):260–269, Apr 1967
104. Sansen W., Puers R., Govaerts R. (1984), Design and realisation of an ultra-miniature hybrid for a pressure telemetry capsule. In: Proceedings of International Conference on Microelectronics Tokyo pp. 406–409
105. Wang G., Liu W., Sivaprakasam W., Zhou M., Weiland J., Humayun M. (2005), A wireless phase shift keying transmitter with q-independent phase transition time. Engineering in Medicine and Biology Society. IEEE-EMBS 2005. 27th Ann Int Conf 5238–5241, Jan. 2005
106. —— A dual band wireless power and data telemetry for retinal prosthesis. Engineering in Medicine and Biology Society, 2006. EMBS '06. 28th Ann Int Conf IEEE, Sept 2006 pp 4392–4395
107. Watson B.W., Ross B., Kay A.W. (1962), Telemetering from within the body using a pressuresensitive radio pill. Gut 3:181–186
108. Wheeler H. Simple inductance formulas for radio coils. Proc IRE 16(10):1398–1400, Oct 1928
109. —— (1947) Fundamental limitations of small antennas. Proc IRE 35(12):1479–1484
110. Wimedia Alliance homepage. [Online]. Available: www.wimedia.com
111. Wolcott G.T., Hines H.A. (1989), Ultrasonic biotelemetry of muscle activity from free-ranging marine animals: a new method for studying foraging by blue crabs (callinectes sapidus) Biol Bull 176:50–56, Feb 1989
112. Wouters P., De Cooman M., Vergote S., Puers R. (1993), Simultaneous telemetric monitoring of body temperature and activity of herding mammals. Proc 12th Int Sympos Biotelemetry pp 128–136
113. Wyborny B.P. (1994), Telemetry system for an implantable medical device. U.S. Patent 5 354 319, Oct 1994
114. Yaghjian A., Best S. Impedance, bandwidth, and q of antennas. IEEE Trans Antennas Propag 53(4):1298–1324, April 2005
115. Yin M., Ghovanloo M. (2009), Using pulse width modulation for wireless transmission of neural signals in multichannel neural recording systems. IEEE Trans Neural Sys Rehabil Eng 17(4):354–363, Aug 2009
116. RF integrated circuits for medical applications (Zarlink). [Online]. Available: stf.ucsd.edu/presentations/2007-08%20STF%20-%20Zarlink%20ULP%20transceivers.pdf
117. Zl0250 product webpage. [Online]. Available: www.zarlink.com/zarlink/hs/82 ZL70250.htm
118. Zl7010 product webpage. [Online]. Available: www.zarlink.com/zarlink/hs/82 ZL70101.htm
119. Zeng F-G., Rebscher S., Harrison W., Sun X., Feng H. (2008), Cochlear implants: system design, integration, and evaluation. IEEE Rev Biomed Eng 1:115–142
120. Zigbee Alliance homepage [Online]. Available: www.zigbec.org

Chapter 8
Bio-Medical Application of WBAN: Trends and Examples

Julien Penders, Chris van Hoof, and Bert Gyselinckx

8.1 The New Wave of Healthcare Systems

Many national health services struggle in the face of financial resource constraints and shortages of skilled labor. The cost of healthcare delivery is steadily on an upward trend. US health care spending is estimated at approximately 16% of the GDP [1]. This upward trend is expected to continue, with projections that the healthcare share of the GDP reaches 19.5% by 2017. Health care spending in other OECD countries is projected to consume up to 16% of GDP. As a result, the pressure on healthcare systems to step up efforts in cost containment and efficiency improvement keeps growing. Consensus about the main determinants of expenditure is not complete but revolves generally around cost drivers such as rising income and patient expectations; demographic change, in particular the aging of population; and new technologies.

Established healthcare systems are often inappropriate to deal with new healthcare requirements in an efficient way. The Continua Health Alliance identifies three major trends driving the change in healthcare systems [2]. The first is a change in lifestyle, shifting towards busier time schedules with little motivation and time left for fitness and health management. As a result, the number of overweight people worldwide is growing, and is now estimated to 1 billion, out of which 300 million of those are clinically obese. Without action, more than 1.5 billion people are expected to be overweight by 2015. Second is the epidemiologic transition from episodic to chronic healthcare needs. Over 600 million people worldwide have chronic diseases. In the US alone, spending is expected to increase from $500 billion a year to $685 billion by 2020. Future healthcare systems should thus focus on prevention, on effective provision of continuous treatment, on integrating lifestyle parameters, and should be customized to individual needs of each patient. The last, demographic,

J. Penders (✉)
Holst Centre, IMEC, Eindhoven, The Netherlands
e-mail: julien.penders@imec-nl.nl

H.-J. Yoo, C. van Hoof (eds.), *Bio-Medical CMOS ICs*,
Integrated Circuits and Systems, DOI 10.1007/978-1-4419-6597-4_8,

trend is an aging society. Globally, the number of persons 60 and older was 600 million in 2000, and is expected to double to 1.2 billion by 2025. This calls for radical changes in how care will be provided for the elderly and how technology may assist.

Alternative ways of providing health services are being sought, targeting increased efficiency, productivity and usability while controlling cost. One strategy that is gaining major attention consists of offloading healthcare institutions by shifting the health management outside the expensive formal medical institutions. Other strategies seek to improve the appropriateness of treatment or emphasize preventive care rather than treatment. For example, the field of chronic diseases is a vast domain in which the provision of real-time data from and to the patient anywhere and at any point in time may hold significant potential for cost reduction.

The supporting role of an adequate technology platform is critical here. E-health technology, enabling wireless and mobile based healthcare services, is increasingly coined as the revolutionizing enabler for the next decades to come. As defined by the World Health Organization, e-health refers to the use in the health sector, of digital data transmitted, stored and retrieved electronically for clinical, educational and administrative purposes, both at the local site and at a distance. E-health is claimed to offer the potential to reduce costs and better addresses the requirements of this new wave of healthcare. It enables personalized healthcare, delivers remote health services and increases the delivery efficiency in real-time. However, at this early stage today it is only through the many pilot projects on e-health ongoing in different countries around the world that evidence will be gathered to determine the economic viability, and answer the question how e-health can enhance the healthcare system efficiency and resolve the associated cost burden.

8.2 An Enabling Technology: Body Area Networks

One key component of e-health is the Wireless Personal Area Network (WPAN), consisting of a network of wireless devices centered around an individual person's life and work space. A WPAN can serve to interconnect all the ordinary computing and communicating devices that a person uses on an every-day basis, or it could serve more specialized purposes such as the monitoring of health and bodily functions. In this case, the WPAN is often called a Wireless Body Area Network (WBAN) [3, 4]. A personal WBAN comprises a series of miniature sensor/actuator nodes each of which has its own energy supply, consisting of storage and energy scavenging devices. Each node has enough intelligence to carry out its task. Each node is able to communicate with other sensor nodes or with a central node worn on the body. The central node communicates with the outside world using a standard telecommunication infrastructure such as a wireless local area network or cellular phone network. Experts might then provide services to the individual wearing the BAN, such as management of chronic disease, medical diagnostic, home monitoring, biometrics, and sport and fitness tracking. Next generation of BAN will include feedback

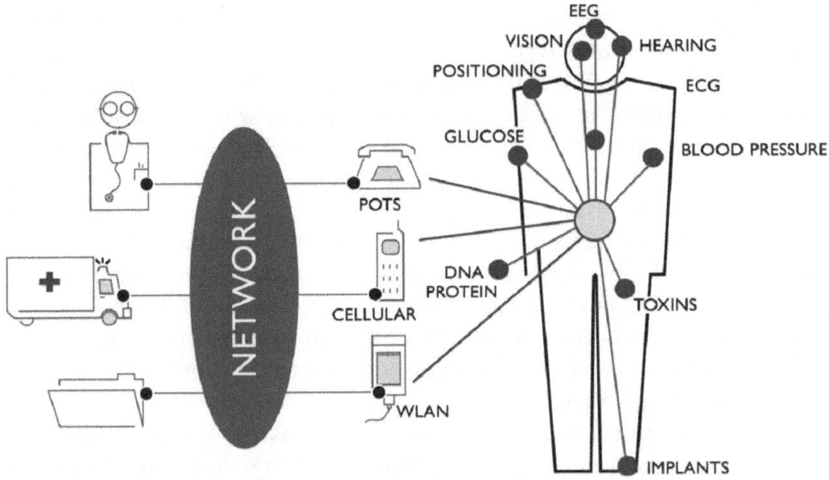

Fig. 8.1 IMEC vision for the year 2015: people will be carrying their personal body area network and be connected with service providers regarding medical, lifestyle, assisted living, sports and entertainment functions

loops for disease management or drug and treatment delivery within so-called closed-loop systems, and will provide feedback to the individual about her lifestyle and health status, eventually leading to human-in-the-loop systems. The vision for Body Area Networks is illustrated on Fig. 8.1.

Recent years have seen the multiplication of body sensor network platforms and one can today find a panel of wireless sensor nodes for the monitoring of various biological and physiological signals [5]. Among others, *Crossbow* offers a generic wireless sensor platform to facilitate research and development in multiple application domains [6]. The IMOTE, TELOS, MICA or IRIS sensor nodes differ by their radio technology and embedded computing power, and address different application needs. *Shimmer Research* has developed a wearable, miniaturized sensor platform for real-time kinematic motion and physiological sensing. It relies on standard wireless communication technologies and offers a large storage capacity which facilitates wearable wireless sensing in both connected and disconnected modes [7]. *Quasar* has developed a wireless sensor platform for monitoring physiological and cognitive state. The advantage of their platform comes from a proprietary non-invasive biosensor technology, enabling dry measurement of bio-potential signals [8]. Recently, *Toumaz* introduced Sensium^{TM}, an ultra low power sensor interface and transceiver platform opening new applications in healthcare and lifestyle management [9].

The last few years have also seen early market adopters of body sensor network technology, and the introduction of first wearable systems for wireless health and lifestyle monitoring. Although monitoring systems featuring WLAN connectivity have been around for a couple of years already, they are usually very bulky systems, therefore impossible to carry along. The adoption of BAN technology allows to significantly reduce the size and multiply the number of wireless sensing units, then worn on the body. Table 8.1 provides a list of some of these early market adopters. This list is non-exhaustive, and only includes products which are available on the market today. A couple of products have recently been announced, mainly in the field of ambulatory cardiac monitoring (e.g. *Toumaz* and *Corventis*), and are currently in clinical trials. These will likely grow this list in the near future. Although few examples of truly networked systems begin to be seen, most of the products available today still rely on a single wireless sensor node communicating to a gateway: cell phone, PDA or dedicated data aggregator. The use of wireless body area networks is still in its infancy. Nevertheless, this growing list emphasizes the trend in health and lifestyle monitoring, towards low-cost systems providing meaningful information at the right time.

Overall, these various body area network platforms and sensor nodes differ by their form-factor, their autonomy, the inherent building-blocks (micro-controller, radio, sensors) and their portability. However, they are all facing the same technological challenges, including: autonomy, functionality, intelligence, miniaturization and manufacturing cost. Addressing these challenges will contribute to enabling novel eHealth solutions. Three examples of applications relying on Body Area Networks are addressed in this chapter. Prototypes of wireless health monitoring systems will be presented for these applications. Evaluation of these early prototypes in real-life situations enables to further characterize application requirements, and identify emerging technology challenges that shall be addressed to eventually enable widespread deployment of body area networks.

8.3 Ambulatory Cardiac Monitoring

8.3.1 Trends

Cardiovascular disease is the number one of death and disability in United States and most European countries. Heart diseases, among which cardiac arrhythmias, are estimated to account for 30% of all death in the United States. Nowadays, diagnosis of cardiac arrhythmias is performed by point-of-care ECG monitoring, or using Holter devices. Very recently, a new wave of portable patient monitors has been introduced, targeting out-patient monitoring with embedded detection of arrhythmia. It is recognized that treatment and prevention of many cardiovascular diseases would benefit from a wireless ECG system for long term continuous monitoring, but technology barriers have so far prevented the wide-spread use and acceptance of such a continuous ECG monitor on a daily basis. Such a device shall indeed be very small, shall not require battery replacement, shall be able to analyze the data on-line and take appropriate actions in case of emergency, shall be wirelessly connected to a network, and shall not affect the wearer in his daily life activity while constantly looking over his health.

Table 8.1 Early market adopters of Wireless Body Sensor Networks

Application domain	Product	Company	URL	# wireless nodes	Signals recorded
Health	VitalSense	Respironics (Philips)	www.vitalsense.respironics.com	2	Core body temperature and skin temperature; additional wired sensors available
	Bioharness^{TM}	Zephyr	www.zephyr-technology.com	1	ECG, respiration, skin temperature, 3D-acceleration
Fitness	Nike+	Nike & Apple	www.nikeplus.nike.com/nikeplus	1	Activity
	Polar	Polar	www.polar.fi	2	Heart rate, Activity
	Bodymedia FIT	Bodymedia	www.bodymedia.com	1	Motion, Heat flux and skin temperature
	ActiPed	FitSense Technology	www.fitlinxx.com	1	Activity
Wellness	DirectLife	Philips	www.directlife.philips.com	1	Activity
	FitBit	FitBit	www.fitbit.com	1	Activity
Sleep	AXbo	AXbo	www.axbo.com	1	Activity
	SleepTracker	SleepTracker	www.sleeptracker.com	1	Activity
	Personal sleep coach	Zeo	www.myzeo.com	1	EEG/EOG
Gaming/BCI	Wii console	Nintendo	www.nintendo.com/wii	1	3D-accelerometers & 3D-gyroscopes
	Epoc	Emotive	www.emotive.com	1	Multi-channel EEG
	MindSet	Neurosky	www.neurosky.com	1	1ch-EEG

8.3.2 Snapshot on the State-of-the-Art

A few players on the market already offer mobile ECG holter systems, with the ambition to provide long-term cardiac activity monitoring. *Cardionet* has recently introduced its Mobile Cardiac Outpatient Telemetry^{TM} (MCOT^{TM}), which enables heartbeat-by-heartbeat, ECG monitoring, analysis and response, at home or away, 24/7/365. *Cardionet*'s MCOT^{TM} consists of a wireless sensor device worn around the neck and measuring 3-lead ECG. The ECG signals are wirelessly transmitted to a portable cardiac monitor where they are analyzed. In case of detection of a possible cardiac issue, the data is sent to expert centers where it is inspected by clinicians. *LifeWatch* offers a similar wireless cardiac telemetry system for remote arrhythmia monitoring in any location. A small transmitter worn on the patient measures 1–3-channel ECG and sends the ECG data to a portable handheld device where it is analyzed. If an arrhythmia is identified, the data is automatically transmitted to a Monitoring Center for immediate review.

Allowing 24/7 monitoring of cardiac activity, these mobile ECG holters provide great promises for people diagnosed with cardiac problems. They however remain too bulky for wide-spread use as prevention devices in the larger population of people at risk of cardiovascular problems. In an attempt to decrease size, *Curvus* is developing a disposable sensor, branded as an ECG "electronic electrode". The smaller size of the sensor shall improve ease of use, and enable the patient to carry out daily life activities in his own environment. It can operate for 72 hours without any user interaction. It incorporates a short range wireless radio connection to a handheld device, where the data is stored and analyzed for irregular events. Similarly, *Corventis* is developing the Piix^{TM} patch-like monitoring system, an unobtrusive, water-resistant, patient-worn device that adheres to the skin and automatically collects physiological information. The Piix^{TM} system processes the data to detect exceptions in the cardiac activity. It then sends the information to a small portable device that relays the information to *Corventis*' web services.

Also, *Toumaz* is pursuing the vision of a disposal, digital plaster for ECG monitoring. It has recently developed Sensium^{TM}, a System-on-chip solution for digital plaster featuring custom sensor interfaces, 10-bit ADC, 8051 micro-processor core, RF transceiver and custom hardware MAC [9, 10]. The Sensium^{TM} chip has been integrated into the Sensium Life Peble, a small size vital signs monitor optimized for ambulatory conditions which continuously monitors heart rate, physical activity and skin temperature. Furthemore, *Toumaz* is developing a digital plaster relying on the same technology, to be used as a disposable system in hospital environment.

Finally, *KAIST* has been developing a disposable, wirelessly powered ECG patch sensors with self-configured wearable health monitoring system that continuously monitor ECG at the selected locations on body Fig. 8.2 shows a continuous health monitoring system with the wearable Body Sensor Network (BSN) [11, 4, 12–13]. It is composed of two parts: the adhesive bandage type Planar-Fashionable Circuit

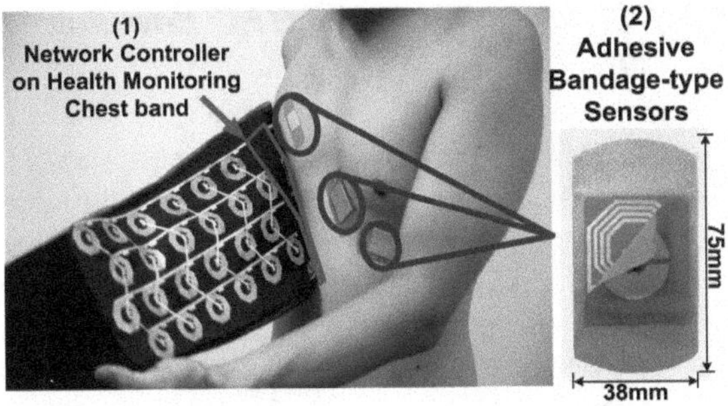

Fig. 8.2 Implemented wearable body sensor network for continuous health monitoring with wirelessly powered adhesive bandage-type sensor and health monitoring chest band containing network controller

Board (P-FCB)^{TM} sensor and the health monitoring chest band. The health monitoring chest band, with a 12×4 inductor array and the network controller, is worn over the chest, where the sensors are attached at arbitrary locations. The network controller on the chest band automatically finds the locations and types of sensors (self-configuration) and provides power only to selected sensors; it can continuously operate and collect data from sensors for up to eight days without replacing the battery (1/2 AA type, 1000 mAh capacity). For convenience and safety, the battery is not integrated with the sensor; instead, power is wirelessly provided by the surrounding health monitoring chest band. The basic concept is to take the power overhead from the sensors, moving it to the relatively power-sufficient health monitoring chest band. To ensure continuous monitoring without irritating the skin, dry P-FCB fabric electrodes are used at the sensors. Motion artifacts are minimized by an adhesive bandage patch that tightly sticks to the skin. The sensor patch is disposable, so it is convenient to use. The components of an adhesive bandage patch sensor. A 4-turn, octagonal P-FCB inductor ($Q = 10.2$ and $L = 0.98\ \mu H$) is screen printed on a cloth with electrodes. Then, the sensor IC is wire bonded on to a cloth, and the molding is applied. When the molding hardens, the cloth is folded in half and attached to an adhesive bandage base. The sensor chip adopts a CMOS-only Adaptive-Threshold Rectifier (ATR) with the dual-mode power transmission scheme to harvest power from electromagnetic waves at 13.56 MHz and 400 MHz. It has a sensor readout circuit capable of handing various types of vital signals. The specific version of the chip in Fig. 8.2 is optimized for ECG monitoring application. The controller chip activates only the selected inductors and transacts data with the selected sensors automatically for health monitoring, so careful alignment between the reader and sensor is not necessary. The health monitoring chest band is composed of an array of 12 4 P-FCB inductors and a network controller chip. P-FCB allows the chest band to closely adhere to the body, so a subject will feel comfortable during everyday monitoring. The pitch between the inductors is determined to maximize the coverage. With the zigzag inductor array configuration shown in Fig. 8.2, all of the chest area around the body is covered. Power is transmitted to the patch sensors through these inductors, and the monitored health data is captured by the inductors at the same time.

8.3.3 Detailed View on IMEC Low-Power Ambulatory ECG Prototypes

In 2008, *IMEC* reported the development of a first generation wireless ECG patch [14], illustrated on Fig. 8.3. Low-power and high performance ECG monitoring is achieved through the use of a proprietary single channel ASIC for biopotential read-out [15]. The ASIC consists of AC coupled chopped instrumentation amplifier, a spike filter, and amplification stage with constant gain, and a variable gain amplifier stage. Power consumption is 60 μW. The ECG patch also integrates a low-power micro-controller, low-power radio, antenna, battery and optimized power management circuitry. Depending on the application, the ECG patch streams the 1-channel ECG data to a receiver within 10m range, or performs local analysis on the data to extract R-peak and other fiducial points. Power consumption of the system has been reported to be 1.17 mW in data streaming mode, and 1.74 mW if the data is processed locally and sent at every beat [14]. The latter value is due to a heavy usage of the micro-controller resources, pointing to an important limitation of today's off-the-shelf micro-controllers. Targeted at very low processing duty-cycles, they become quickly limiting for applications requiring quite advanced embedded digital signal processing capabilities.

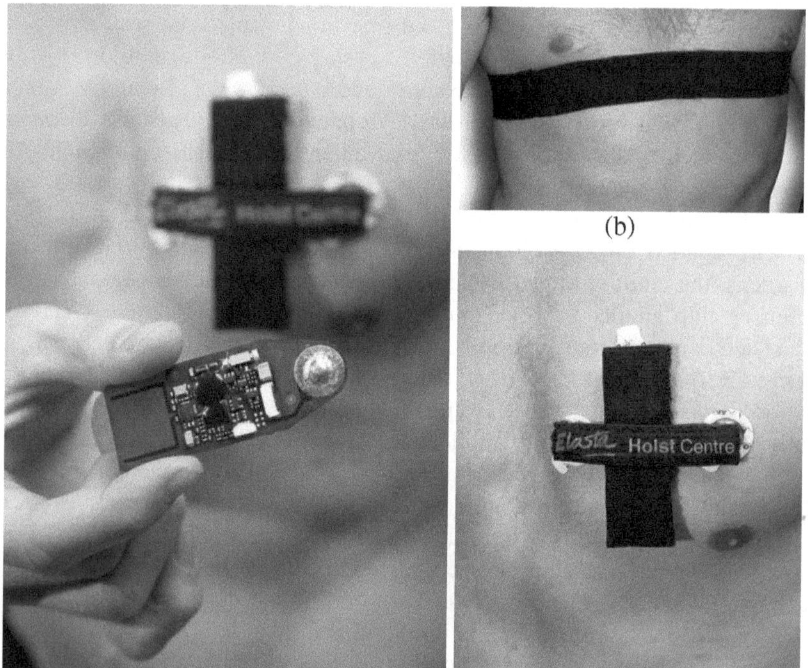

(a) (c)

Fig. 8.3 **a** Wireless ECG patch integrated on flexible substrate; **b** chest belt package; **c** textile pocket package

The wireless ECG patch has been tested in ambulatory settings, to evaluate the influence of physical activity on the quality of ECG recordings. 10 healthy volunteers have been involved in the study. Three levels of physical activity are considered: at rest (low level of activity), biking on a static bike at 100 W and 70 RPM (moderate level of activity), and running on a treadmill at 7.5 km/h (high level of activity). Two packaging means are investigated, as illustrated in Fig. 8.3: a textile pocket containing the ECG patch and attached to the skin via standard Ag/AgCl electrodes (non-standard ECG lead configuration), and a chest belt integrating the ECG patch (lead I configuration). Recordings of 10 minutes are obtained for each level of activity, for each packaging and for each patient. Beats are manually annotated whenever possible. Not all recordings can be annotated due to degraded signal quality at higher levels of physical activity. Overall, these experiments have lead to an annotated database of 45 ambulatory ECG recordings, as summarized in Table 8.2.

Table 8.2 Average performance of beat detection algorithm when applied to database of ambulatory ECG recordings; results highlighted in gray are considered non-significant

Activity level / package	# recordings	# annotated recordings	(Se ± P)
Low / pocket	10	10	100–100%
Moderate / pocket	10	4	99.7–99.3%
High / pocket	10	3	99.6–100%
Low / belt	10	10	100–100%
Moderate / belt	10	10	99.9–100%
High / belt	10	8	99.8–100%

Two criteria are used for evaluation of the system: visual inspection of the data, and the performance of a beat detection algorithm on the monitored data. The latter is used as a quantitative measurement for the possibility of using the data as an input to automated ECG analysis algorithms. An enhanced version of Romero's algorithm [15] is used for this study. Extracts from the signals obtained at different levels of activity and using the two different system packaging means are illustrated on Fig. 8.4, for one of the subject. Average performances of Romero's algorithm on the database are given in Table 8.2. These results suggest that the ECG patch provides signal of excellent quality in resting conditions, where the two packages lead to 100% sensitivity and 100% positive predictivity. As physical activity increases, the ECG signal is considered to be acceptable for automated analysis at both moderate and high levels of physical activity in case the chest belt is used. The algorithm indeed maintains a high performance, characterized by 99.9% (resp. 99.8%) sensitivity and 100% (resp. 100%) positive predictivity while biking (resp. running). In case the textile pocket is used to attach the electronics to the chest, the signal quality quickly deteriorates as physical activity increases, as it can be observed on Fig. 8.3. Consequently, respectively six and seven ECG recordings were unusable for analysis (see Table 8.2). For the ones that could be analyzed, the algorithm gives slightly lower performances: 99.7% sensitivity and 99.3% positive predictivity at moderate physical activity, 99.6% and 100% at high activity. These performance numbers are,

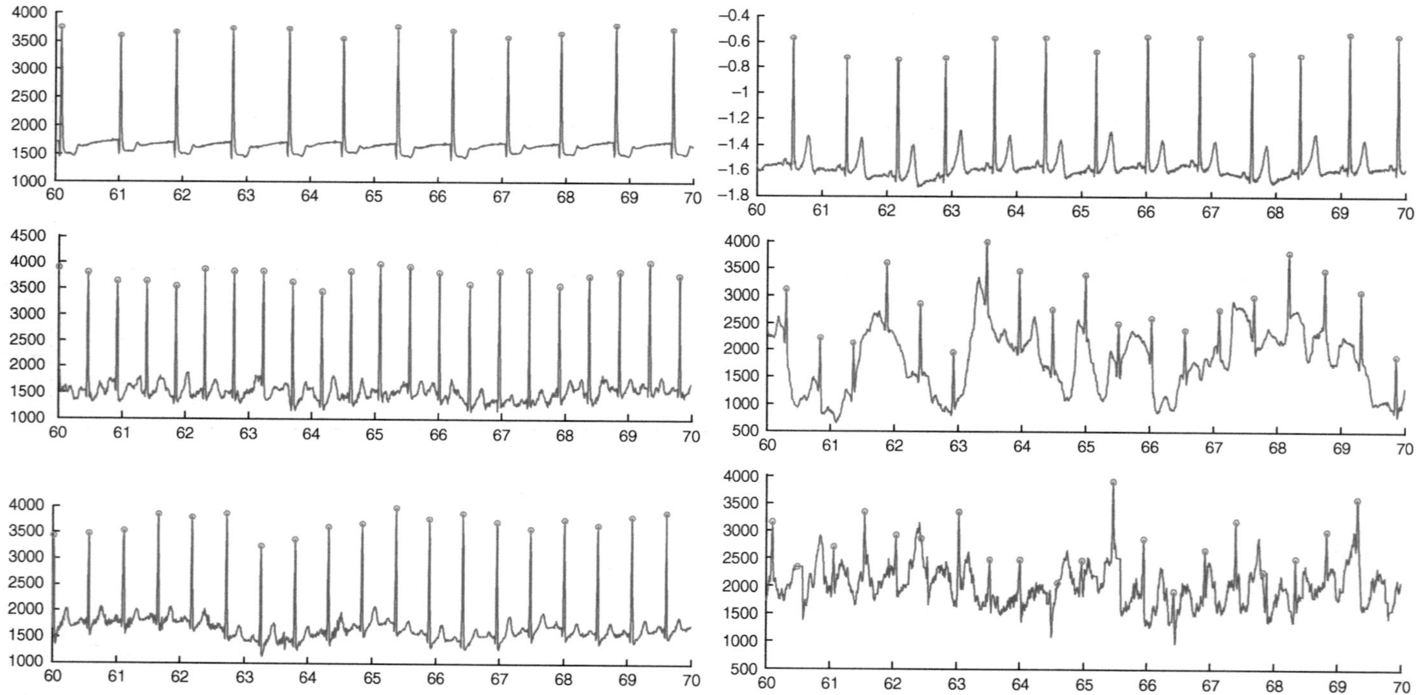

Fig. 8.4 Recording samples extracted from the database of ambulatory ECG recordings. The three left graphs correspond to the ECG patch integrated in a chest belt in case of low, moderate and high level of physical activity (from *top to bottom*). The three right graphs correspond to the ECG patch integrated in a textile pocket

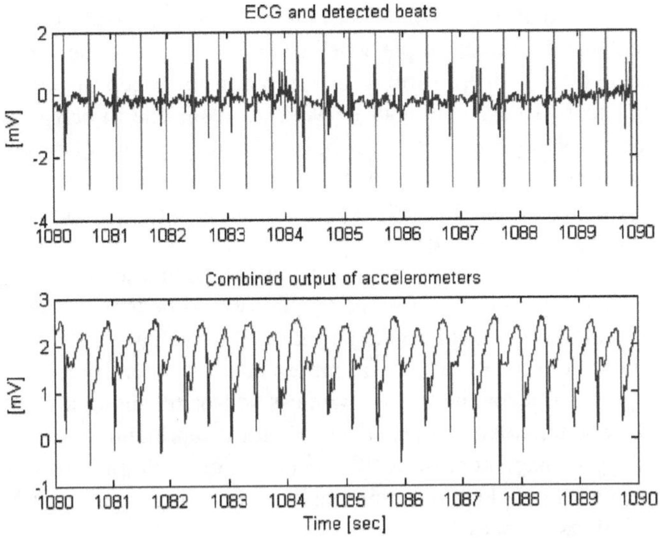

Fig. 8.5 Combined output of multi-axis accelerometers and ECG allows classifying motion-related artifacts and removing those from beat detection

however, not significant due to the small numbers of recordings in these cases, and should, thus, be interpreted with caution.

The stronger deterioration of the signal in case of the textile pocket versus the chest belt shows the importance of system packaging and integration in achieving high quality recordings in ambulatory environments. Furthermore, it suggests that relative motion between patch and body is causing larger artifacts than absolute motion. The reduction in signal-to-noise ratio has to be prevented as it will hamper feature extractions such as QRS detection and rhythm analysis. Motion detection is one approach to determine motion-related artifacts in ECG recordings. Figure 8.5 shows the simultaneous recording of accelerometers and ECG signal, suggesting that some misdetections of the heart beat correlate with irregularities in the acceleration signal. This could be exploited for efficient removal of motion-related artifacts. One issue with accelerometer-based motion detection, however, is that accelerometers mainly detect absolute motion and not relative motion. For this reason, alternative methods are being pursued, which uses other sensors, or a combination of several ECG leads.

8.4 Wireless Sleep Monitoring

8.4.1 Trends

The prevalence of sleep disorders has increased significantly over the last few years. Severe sleep disorders affect between 10 and 13% of the American population, and between 2% (female) and 4% (male) population in Europe. Moreover, it has been estimated that 1 billion people worldwide experience some kind of chronic nasal congestion during sleep, snoring or sleep apneas. Sleep disorders have dramatic consequences as they may lead to work accident due to daily sleepiness, to cardiovascular problems and may even have neurologic consequences leading to psychological breakdowns. Yearly, 38000 deaths are associated with complications from sleep apneas, and it has been shown that those who suffer from apnea are 3 to 6 times more likely to suffer a stroke.

The main drivers for wireless sleep monitoring are: increase monitoring capacity, reduce cost of investigation, improve sleep quality, and better acceptance from the patients [16]. On the other hand, the expectations for such a system are: high diagnostic sensitivity, high reliability of equipment and low interference with patient's habits. The opportunity for the use of wireless sensor technologies in the area of sleep diagnosis and treatment spreads over the entire continuum of care, from smart systems for sleep management in healthy people (smart alarms), to sleep disorder detection systems for people at risk, and finally as enabling technologies to improve therapeutic solutions for patients.

8.4.2 Snapshot on the State-of-the-Art

Typical diagnosis of sleep disorders is performed using polysomnography tests at the point-of-care. During this test, the patient is equipped with a wide variety of sensors wired to the PSG system to monitor all relevant parameters such as EEG, ECG, EMG, EOG, airflow, oxygen saturation, and respiratory effort. There have been several attempts from sleep monitoring equipment manufacturers to design portable sleep monitoring systems intended for home use, which review falls beyond the scope of this discussion. Typical issues encountered by these portable systems include: size and weight of the system, need for wires going from the electrodes to the data acquisition box, and issues with reproducibility of the tests due to misplacement of the electrodes by the patients themselves or electrode falling off during the night.

Body area networks have the potential of improving sleep diagnostics by reducing the number of wires. This is achieved by spreading miniaturized and wireless sensor units all in different locations on the body. *Cardinal Health* has been a precursor in exploiting this advantage with the NOX-T3 portable sleep diagnostics system [17]. The NOX/T3 integrates sensor for respiratory efforts (abdominal and thorax), 2 bi-polar channels, microphone and accelerometer. In addition, a wireless pulse oximeter module connects to it via Bluetooth, providing all the signals required for sleep diagnostics. Going further in this approach, *Fraunhofer* has developed the SleepBee® system for sleep diagnostics [18]. The system consists of 4 wireless units, or "concentrators", located at the head, chest, arm and two legs. Local sensors are connected with wires to these concentrators instead of the main gateway, which reduces the total amount of wires in the system.

Smart alarms are now becoming available in the consumer market. The concept of the smart alarm is simple: the alarm monitors an individual's sleep and only switches on during the good part of the sleep cycle—that is, during light sleep. The SleepTracker® and Axbo® systems use actimetry to monitor sleep stages. *Zeo*'s personal sleep coach monitors sleep stages based on a single-channel EEG/EOG measurement.

8.4.3 Detailed View on IMEC Wireless Sleep Staging Prototype

The development of a prototype body area network for wireless sleep staging was reported in 2007 [10]. The system relies on a proprietary single channel ASIC for biopotential read-out [15]. The ASIC low-power consumption (60 μW) allows to dramatically reducing the size of the battery, hence of the entire system, while maintaining autonomy suitable for sleep analysis (>12 hours). The system, illustrated on Fig. 8.6, consists in a body sensor network composed of three wireless sensor nodes, collecting data from 2-channel EEG, 2-channel EOG and 1-channel EMG, and sending it wirelessly to a receiver located in the patient's room. These particular signals were selected as per the Rechtschaffen and Kales standards for sleep staging [19]. The data can then be analyzed, on- or off-line, by the clinical staff. Each node achieves a power consumption of 15 mW, for a sampling rate of 200 Hz. Thanks to their small size and light weight, the three sensor nodes can easily be integrated in a headband, hence increasing patient comfort and acceptance.

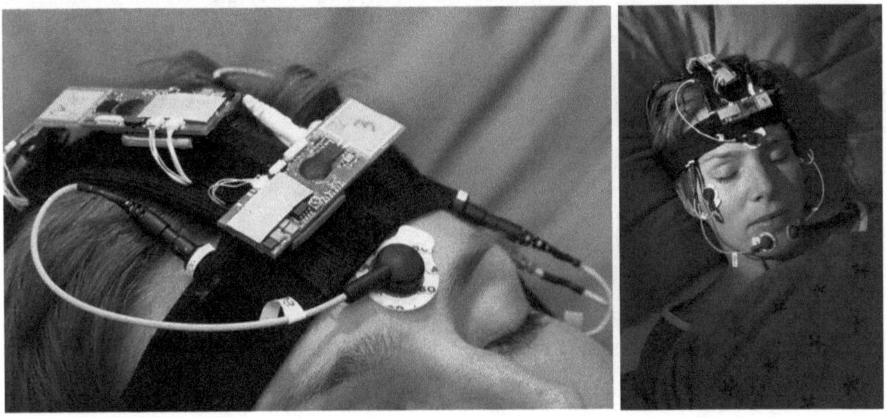

Fig. 8.6 Prototype body area network for wireless sleep staging

The system has been benchmarked against a commercial off-the-shelf portable PSG system, in collaboration with the Sleep Laboratory of the Vesale Hospital in Charleroi, Belgium. 12 healthy volunteers participated in the study, from which 9 are used for the benchmarking analysis. Each volunteer is monitored for a complete night, during which the 5 physiological signals (2ch-EEG, 2ch-EOG and 1ch-EMG) are recorded using the wireless BAN prototype and a reference system. The two systems are set-up in parallel, and all electrodes are duplicated. Once all subjects have been recorded, the recordings are manually and blindly analyzed by a sleep expert to obtain the hypnograms for both systems. The two systems are compared using three criteria: qualitative feedback from the clinician, hypnogram similarity analysis and hypnogram feature correlation analysis. Hypnogram similarity is quantified as the percentage of time for which the two hypnograms—inferred from the wireless and reference systems—are similar, and is computed individually for each stage. A set of features can be extracted from each hypnogram, namely: time spent in each stage, number of stage changes, number of micro-awakenings (or arousals) and sleep efficiency. Hypnogram feature correlation is quantified using correlation coefficient, slope and confidence interval.

Qualitatively, the two sets of signals looked very similar. Visual inspection performed by clinical experts lead to the qualitative observation that the signals obtained using the wireless system are of equivalent quality when compared to the signals acquired using standard portable PSG equipment. In addition, the wireless system is found much easier to set-up, provides enhanced comfort for the patient, and reduces artifacts due to extensive wiring. The hypnogram similarity analysis leads to an average similarity across all stages and patients of 80% (standard deviation of 7.85%). This is comparable to inter-rater similarity—that is, the similarity between hypnograms scored by two different medical experts, typically included between 80 and 90% for healthy subjects. Figure 8.7 illustrates the percentage of similarity per stage, in which Stage 1 and awake are grouped. REM sleep and stage 3/4, very important in sleep disorder diagnosis, both have a similarity percentage over 80%. Finally, the feature correlation analysis show that the features extracted from the hypnograms inferred from the WBAN system correlate with those inferred

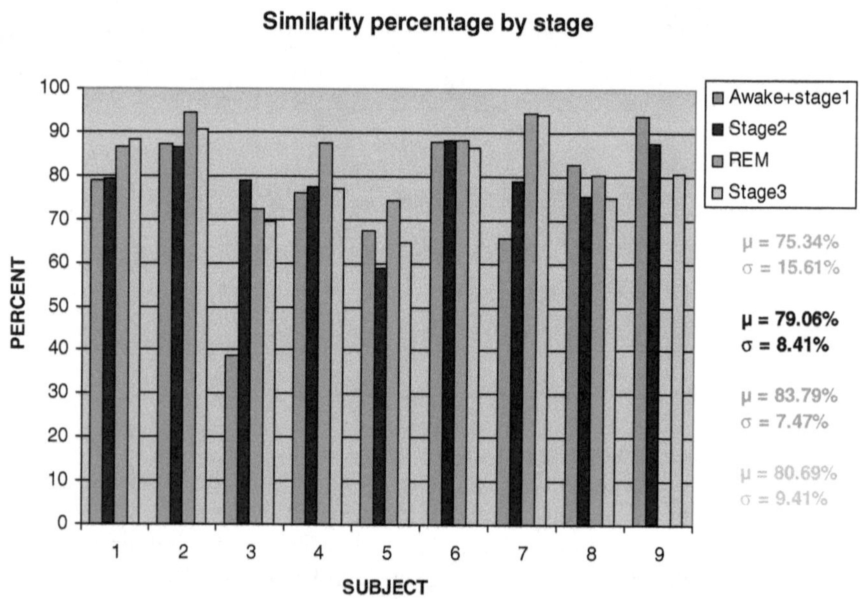

Fig. 8.7 Benchmark of wireless sleep staging system: hypnogram similarity per sleep stage, for the 9 volunteers included in the study

Table 8.3 Hypnogram feature correlation

Feature	Correlation coefficient	Confidence interval	Slope
Time in stage 1	0.41	[−0.36;0.85]	0.48
Time in stage 2	0.89	[0.53;0.98]	0.75
Time in stage ¾	0.95	[0.76;0.99]	1
REM	0.98	[0.90;0.99]	0.77
# state changes	0.92	[0.66;0.98]	1
Sleep efficiency	1.00	[0.99;1.00]	1.19
Arousals	0.72	[0.10;0.94]	0.55

from the reference system. Table 8.3 summarizes the correlation coefficients, slopes and confidence intervals for each feature.

8.5 Mental Health and Emotion Monitoring

8.5.1 Trends

Monitoring of emotions or mental health has received a special interest in the last few years. In this context, emotion is usually defined as a mental and physiological state associated with a wide variety of feelings, thoughts, and behaviors. Following this definition, monitoring physiological and mental states should enable to understand and "read" the emotional state of an individual in a particular situation.

The vision for emotion monitoring is a world adapting to individual's emotions and feelings, thus enabling a better living, enhanced mental performances and an improved society for better life-style experience. Individual's emotional responses are monitored using sensors worn on the body, implanted or disseminated in the environment. These smart sensors are able to interpret emotions and feelings, and wirelessly transmit this information to objects in the surrounding environment, enabled and authorized to interact with the emotion monitoring system. The environment then adapts based on individual's emotional and mental status.

A number of groups have reported a wide range of studies to the objective evaluation of emotions, investigating varying modalities such as facial expressions [20], vocal patterns [21, 22], physiological responses [23, 24] or combinations of the above [25]. In the following discussion, the focus will be on monitoring physiological responses from the Autonomic Nervous System.

8.5.2 Snapshot on the State-of-the-Art

Although there is not yet a well defined market for emotion monitoring, a few companies have introduced products relying on wireless sensor technologies. Early adopters of these technologies are seen in the entertainment and infotainment industry, which can be explained by lower entry barriers than for health related applications. An example is *Nemesysco*, which offers technology for voice-based emotion detection systems. This technology has been integrated in a few products such as employees monitor or love detector. Another example is *Exmocare*, which is developing a bio-sensor watch for emotion and stress monitoring. To the best of our knowledge, performance and validation of these systems have not yet been reported.

The *MIT Media Lab* has been very active in emotion monitoring and affective computing [24], and has developed several wireless systems for emotion monitoring. Their latest system, iCalm^{TM}, consists in a wearable wireless bio-sensor platform integrated in a wrist-band. The main functionality is to monitor and analyze electro-dermal activity. Validation of this prototype has shown agreement with reference systems [26].

8.5.3 Detailed View on IMEC Wireless ANS Monitoring Prototype

The realization of a low-power body area network for monitoring ECG, respiration, skin conductance and skin temperature has been reported in 2009 [27]. Each of these modalities is known to be regulated by the Autonomic Nervous System, and thus represent interesting candidates to capture ANS responses to external stimuli. The system, illustrated on Fig. 8.8, consists of two low-power miniaturized body sensor nodes which communicate with a receiver connected to a PC or to a data logger.

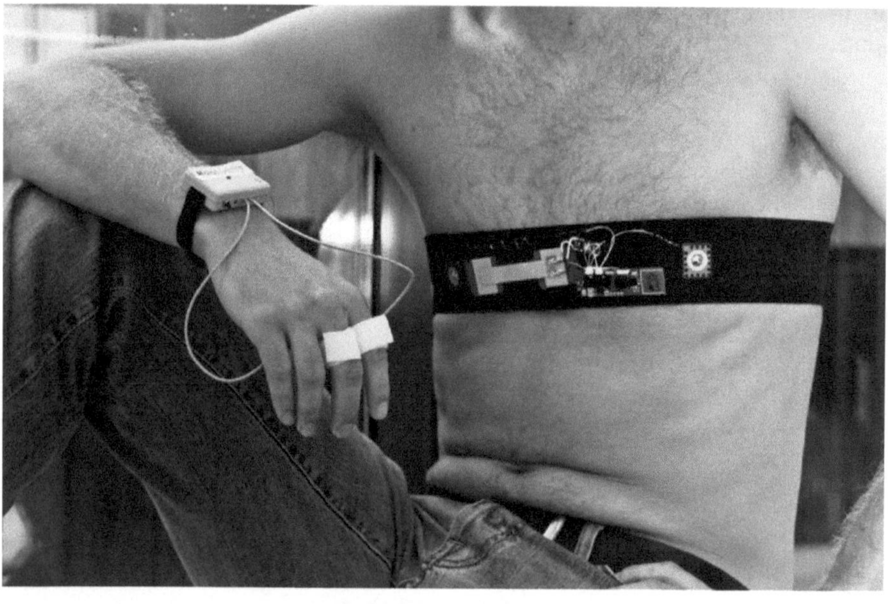

Fig. 8.8 Integrated body area network for ambulatory monitoring of physiological responses from the Autonomic Nervous System

The first node is integrated in a wireless chest belt and monitors ECG (lead-I) and respiration. The second node is integrated in a wireless wrist sensor and monitors skin conductance and skin temperature. The total size of each individual node is approximately $40 \times 25 \times 8$ mm^3, including battery, sensors and read-outs. Power consumption of the ECG/respiration node is 2.5 mA, whereas the wrist-based sensor consumes 4 mA, mainly due to the use of an infra-red temperature sensor.

This body area network for ANS responses monitoring has been tested in controlled environment to evaluate its potential usage for monitoring emotional states. 10 subjects (mean age 29.3, 3 females, 7 males) are involved in the study. They are asked to watch 5 emotionally arousing film clips to elicit sadness, happiness, fear, disgust and neutrality, while wearing the wireless monitoring equipment. At the end of each clip, the subjects fill a self-report questionnaire. The 5 film clips are grouped into 3 categories in function of their expected arousal level: fear and disgust, happiness and neutral, and sadness. The four physiological signals are analyzed offline, and a set of 13 features are extracted based on general physiology considerations and previous studies on emotion recognition. The 13 features are then mapped to 2 axis using Fisher Mapping. Linear Discriminant Classification is finally used to classify the data. This process eventually leads to error rates of 0.36 computed using leave-one-out cross-validation on the data-set. This compares to previous studies on emotion classification [24, 28].

A second study was designed to evaluate the possibility to use the proposed BAN system for performing real-time measurement of an individual's arousal level [29]. 20 healthy volunteers are involved in the experiment. A movie extract is chosen as the arousal stimulus, characterized by a calm beginning followed by a building-up phase culminating to a frightening event. A reference or target arousal function is defined as being zero during most of the movie, except in a region surrounding the frightening event (see [29] for all details on the choice of the target function). Volunteers are asked to watch the movie while their physiological signals are monitored using the wireless system. All tests are performed in a controlled laboratory environment, in order to minimize the sources of distraction that may eventually lead to unexpected and uncontrolled increases in arousal. A set of features is extracted from the ECG and skin conductance signals, found to be the most responsive parameters to the tests. In a second step, these features are combined in an optimal arousal estimator using linear regression against the target arousal level.

The resulting estimator can then used to monitor the arousal level of individuals wearing the system. Several tests have been performed in various environments, varying from laboratory to small public audience. As much as possible, the test subject is isolated from the outside world, for instance, using headphones. The test protocol used for these tests consists in four parts: a short movie to get acclimatized, a modified Stroop test, an audio extract and a movie fragment. The Stroop test is modified to induce confusion (and hence mental stress) in the second part of the test. The audio extract is a 3-minute very relaxing piece of classical music abruptly disturbed by noises of several kinds after 120 and 150 seconds, expected to trigger startling responses. The movie clip is identical to the one used to develop the arousal monitor. An example of the estimated arousal level over the test sequence is given

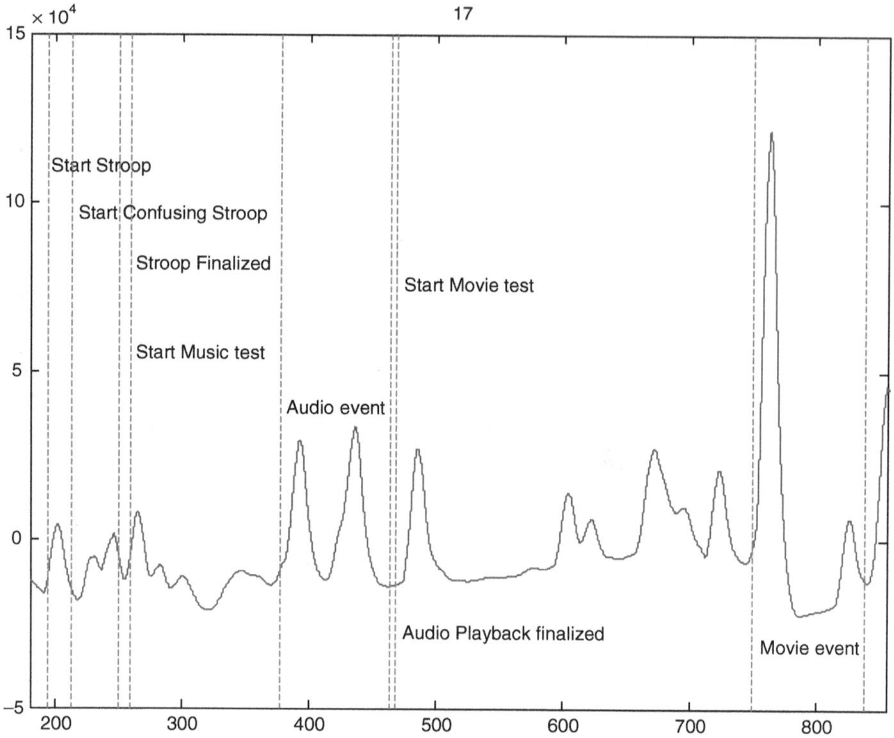

Fig. 8.9 Estimated arousal level over the part of the test protocol: modified Stroop test, audio extract and movie clip. *Dashed vertical lines* give the timing of the events

in Fig. 8.9 for one of the test subject. In this figure, the solid line gives the estimated arousal over time. The dashed, vertical lines represent the events, as specified by the name shown directly to the right of these lines. It can be seen from this picture that the subject did not show a significant increase in estimated arousal during the modified Stroop test. There was, however, a sharp and large increase in estimated arousal level just after the audio events and the movie event. Apart from the expected responses, there were also some responses that clearly do not origin in any of the events. These false positives can be due to anything that triggers the subject's mind, such as an arousal triggering thought, or something surprising in the surrounding environment.

A last study explored the feasibility of using ANS responses to monitor stress in healthy individuals and in patients with psychiatric disorders. The population here consists of 15 healthy volunteers and 15 patients. The healthy volunteers are all female, with a mean age of 32 (±10). The patients are all female, with a mean age of 35 (±13), diagnosed with traumatic stress disorder, generalized anxiety disorder, panic disorder or major depression. The pilot study was run under the supervision of the psychiatric department of the University Hospital from Leuven. The experimental protocol consists of a Trier Social Stress Test (TSST). During this test, participants are standing in front of a jury, and are asked to perform two assignments: a presentation for an imaginary job application, and an arithmetic task. The TSST lasts for 15 minutes, after which the participants are asked to relax for 45 minutes. Before the tests, the participants are equipped with the wireless BAN prototype for ANS responses monitoring. Physiological signals are recorded before (15 minutes), during and after (45 minutes) the test. Features are extracted from the ECG, galvanic skin conductance and respiration signals, and averaged over the stressor period and the subsequent relaxation period, leading to two data points per participant.

The possibility to detect stress using ANS responses is treated as a classification problem, in which the accuracy in classifying the data points corresponding to the stressor event and to the following relaxation period is evaluated. Patients and healthy controls are all included in the analysis. Results show that heart rate solely leads to an estimated classification error of 10%, for a guessing rate equals to 50%. The error rate is here computed by averaging estimated error rates with several classifiers (Linear discriminant, quadratic discriminant and support vector machine), after application of Fisher mapping to the feature set. Although heart rate may also be influenced by the increase of physical effort associated to the TSST, this result has been shown to extend to other stressors in which physical activity was removed (non-published results).

The possibility to distinguish patients from controls using ANS responses is a second classification problem, in which the accuracy in classifying patients and controls during the stressor event is evaluated. The number of features is reduced to two using Fisher mapping. The top scoring features are the standard deviations (computed over the TSST duration) of the heart rate and skin conductance kurtosis—that is, the fourth moment of the skin conductance. The corresponding scatter diagram is illustrated on Fig. 8.10. The estimated classification error—computed by averaging estimated error rates with several classifiers—is here equal to 25%, for a 50% guessing rate.

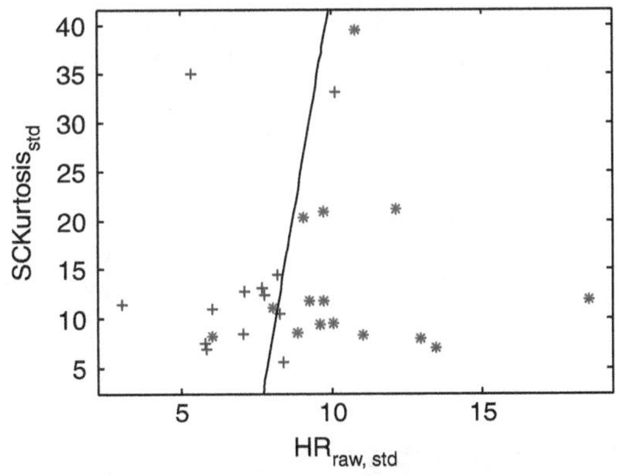

Fig. 8.10 ANS responses as a measurement for stress: scatter diagram for patients (*stars*) and healthy controls (*crosses*). X-axis: standard deviation of heart rate, computed over the TSST period. Y-axis: standard deviation of skin conductance kurtosis, computed over the TSST period. Data points corresponding to patients and healthy controls cluster in two different groups, leading to an estimated classification error of 25%

In summary, the first study confirmed that physiological responses from the Autonomic Nervous System may be used to obtain information about the emotional state of an individual, as reported in other studies. The second study showed that it is possible to estimate an individual's arousal level quite reliabile in controlled environment—that is, in a lab setting where distraction opportunities are minimized. Experimentation in other environments indicated that these conclusions can be generalized from movie to other arousing stimuli. Finally, the third study suggested that monitoring ANS responses using a WBAN prototype may provide a more quantitative and objective way to monitor stress in healthy individuals and patients with psychiatric disorders. These conclusions should however be confirmed on larger populations before generalizing. Furthermore, ANS responses may need to be complemented with additional sensors measuring the cognitive state of an individual, providing information about the appraisal direction of the emotion (valence). Finally, further experiments are needed before conclusions can be drawn about the extension of the results to non-controlled environments. The availability of Wireless Body Area Networks will facilitate the transition from lab to real-life environments.

8.6 Remaining Challenges

This chapter has highlighted a few application examples in which WBAN opens new perspectives for better a more efficient health services. The evaluation of these early prototypes in real-life situations enables to further characterize application requirements, and identify important technology challenges that shall be addressed to eventually enable widespread deployment of body area networks.

8.6.1 Ultra-Low-Power Technologies

When integrated in prototypes of wireless health systems, the best low-power commercial off-the-shelf electronic components lead to typical power consumption ranging from 1 to 10 mA, depending on the application. Most of the power is usually drawn in the wireless transmission of the data, or in local signal processing. In some cases, sensors are also found to consume a significant part of the power. Further research is needed on ultra-low-power analog interfaces, sensors, DSP and radios to reduce power dissipation. The target shall be to reach 100 μW per body sensor node.

8.6.2 Increasing Functionality

Most of today's body-worn sensors act as simple gateways, passing on the information to a central hub where the data is converted into actionable information. In the case of bio-potential monitoring, the evaluation studies reported in this paper have shown the high impact of motion artifact on the quality of the acquired signals. Advanced techniques for embedded, real-time motion artifact need to be further developed. Many methods have already been introduced on this topic with variable performances [30], emphasizing the need for a systematic way to tackle the problem of motion artifacts. Furthermore, by adding intelligence to the sensors they can take decisions locally and the signaling overhead in terms of data and latencies can be reduced. Additionally, intelligence may be required for the system to make decisions depending on the status of the environment, thus enabling context-awareness. Compromises between local processing of the data versus data streaming or data storage exist. A rational approach to distributed processing shall allow achieving optimal performances while minimizing power consumption.

8.6.3 Autonomous Systems

The prototypes presented here can run from a few days to a week at full functionality. Breakthroughs in ultra-low-power technologies will eventually enable months or years of autonomy. To come to a truly autonomous system however, it should be able to operate over its full lifetime without maintenance. Early demonstrators of autonomous wireless health monitors have shown that harvesting energy from the environment during the operation of the system will allow the system to run eternally with a battery or a super-capacitor acting only as a temporary energy buffer [31]. Further research in miniaturization of energy harvesters system using micromachining techniques is needed to decrease the form factor and achieve micro-scale harvester devices that can be integrated in patch-like systems.

8.6.4 Multi-Parameter Sensors

Much effort has been put on monitoring bio-potential signals. Although these are certainly very important for assessing cognitive and physiological health of individuals, extending the functionality range to include new sensing modalities will be crucial in fostering research for body area networks. Novel sensing technologies are needed to reliably measure more complex parameters such as chemical compounds, hormones and proteins in body fluids, whilst pursuing ultra-low-power consumption. Continuous measurement of cortisol in saliva would for instance open new perspectives in stress monitoring, whereas monitoring melatonin concentration may turn useful for insomnia diagnosis and treatment.

8.6.5 Dry Electrodes

Most of current systems for bio-potential monitoring require wet or gel electrodes to be attached to the skin. Although they have the major advantage of providing good quality signals, gel electrodes exhibit significant drawbacks with regards to long-term use and ease of set-up. Dry electrode technology is necessary to solve these issues, and eventually enable set-up of the system by the patient itself. Several research groups have explored the area of dry electrode for ECG monitoring applications, for instance [13]. Further research is required to systemically tackle signal quality, robustness to motion artifact and bio-compatibility.

8.6.6 Integration and Packaging Technology

Prototypes such as the wireless ECG patch point to current limitations in state-of-the-art micro-electronic integration technologies. As body sensor nodes shrink in size and power consumption, end-user acceptance and compliance will eventually be bound to how comfortable technology can become. Advanced integration technology is needed to achieve electronic integration in bi-dimensional flexible and stretchable foils. Pioneering research in this area has led to first functional prototypes of ultra thin chip packages [32, 33] and stretchable interconnects [34]. Technology for printing electronics on organic foils has materialized in early prototypes of smart bandage systems [35]. Furthermore, packaging technology is required to encapsulate the whole system in a bio-compatible package, which can be attached to the body for months. When interaction with the body is required, the properties of the body-electronics interface also need to be considered. As integration and packaging technology matures and becomes commercially available, the cost of sensor nodes will also be reduced.

8.7 Conclusions

The pressure of raising health care costs, new demographic, health and societal trends, are shaping a new wave of healthcare systems. This chapter shows how wireless body area networks contribute in enabling this future. Three applications are discussed in details: ambulatory cardiac monitoring, wireless sleep monitoring and emotion monitoring.

The concept of a wireless ECG patch offers a path towards a novel, patient-centered, delocalized, monitoring paradigm for ambulatory cardiac monitoring and cardiovascular prevention. The evaluation of the wireless sleep staging system demonstrates that wireless body area networks have the potential to provide medical information of equivalent quality when compared to standard recording equipment, while offering unique advantages. In the emerging field of emotion monitoring, the use of body area networks to monitor ANS responses paves the way towards a more objective measurement of emotions and stress.

Early technology deployment in these application environments leads to the identification of key technology challenges that need to be addressed in terms of ultra-low-power radios, DSPs and analog interfaces, ultra-low-power sensors extending the set of monitored parameters, dry electrode research and 2D flex/stretch electronic integration. The race to overcoming these new technology barriers is open, targeting smaller and smarter systems, running on ever decreasing power, integrated in stretchable patches and available for a few dollars.

References

1. The U.S. Healthcare System: Best in the world, or just the most expensive, Available: http://dll.umaine.edu/ble/U.S.%20HCweb.pdf
2. Continua Health Alliance: www.continuaalliance.org
3. Schmidt R et al. (2002) Body area network BAN, a key infrastructure element for patient-centered medical applications Biomed Tech (Berl) 2002;47 suppl 1 pt 1:365–368
4. Yoo J, Yan L, Lee S, Kim Y, Yoo H-J A 5.2 mW Self-configured wearable body sensor network controller and a 12 μw wirelessly powered sensor for a continuous health monitoring system. IEEE J Solid-State Circuits 45, January 2010
5. Yang, Guang-Zhong (Ed.) 2006, XXVIII, 500p. 275 illus., 32 in color., Hardcover ISBN: 078-1-84628-272-0
6. http://www.xbow.com/
7. http://shimmer-research.com/
8. Matthews R, et al. (2007) A Wearable physiological sensor suite for unobtrusive monitoring of physiological and cognitive state. Proceedings of the IEEE EMBS conference
9. Wong A.C.-W, McDonagh D, Kathiresan G, Omeni OC, El-Jamaly O, Chan TC-K, Paddan P, Burdett AJ (2008) A 1 V, Micropower system-on-chip for vital-sign monitoring in wireless body sensor networks. Proceedings of the IEEE ISSCC conference, 2008
10. Wong AC-W, McDonagh D, Omeni OC, Nunn C, Silveira M, Burdett AJ (2007) Sensium: an ultra-low-power wireless body sensor network platform: design & application challenges. Proceedings of the IEEE EMBS conference, 2007
11. YooJ , LeeS, Yoo H-J (2010) A 1.12pJ/b Inductive transceiver with a fault-tolerant network switch for multi-layer wearable body area network applications. IEEE J Solid-State Circuits. 45(1):178–188, Jan 2010
12. Yoo J, Yan L, Lee S, Kim H, Yoo H-J (2009) A Wearable ECG Acquisition system with compact planar-fashionable circuit board-based shirt. IEEE Trans Inf Tec Biomed vol 13, Nov 2009
13. Yan L, Yoo J, Kim B, Yoo H-J (2009) A 0.5 μV_{rms} 12 μW patch type fabric sensor for wearable body sensor network. IEEE Asian Solid-State Circuits Conference, Taiwan, pp.105–108, Nov 2009
14. Penders J, Gyselinckx B, et al. (2008) Human++: from technology to emerging health monitoring concepts. In: Proceedings of the 5th International workshop on wearable and implantable body sensor networks, The Chinese University of Hong Kong, China, Jun 1–3, p 94–98
15. Yazicioglu RF, Merken P, Puers R et al. (2007) A 60 μW 60 nV/Hz Readout front-end for portable biopotential acquisition systems. IEEE J. Solid-State Circuits 1100–1110, May 2007
16. Escourrou P, Luriau S, Rehel M, Nédelcoux H, Lanoë JL (200) Needs and costs of sleep monitoring. Stud Health Technol Inform 78:69–85.
17. NOX/T3 Portable sleep monitor, http://www.carefusion.com/
18. Mobile System SleepBee® for sleep diagnosis, http://www.ipms.fraunhofer.de/
19. Rechtschaffen A, Kales A (eds) (1968) A manual of standardized terminology, techniques and scoring system for sleep stages of human subject. US Government Printing Office, National Institute of Health Publication, Washington DC
20. De Silva CL, Miyasato T, Nakatsu R (1997) Facial emotion recognition using multimodal information. Proceedings of IEEE international conference on information, communications and signal processing (ICICS'97), Singapore, pp. 397–401 September 1997

21. Cowie R, E-Cowie D, Tsapatsoulis N, Votsis G, Kollilas S, Fellenz W, Taylor J (2001) Emotion recognition in human-computer interaction. IEEE Signal Process Mag 18(1):32–80
22. Fernandez R, Picard WR (25) Classical and novel discriminant features for affect recognition from speech. Proceedings of Interspeech 2005, September, 4–8 2005.
23. Kreibig DS, Wilhelm HF, Roth TW, Gross JJ (2007) Cardiovascular, electrodermal, and respiratory response patterns to fear- and sadness-inducing films. Psychophysiology 44(5): 787–806, September 2007
24. Picard R, Vyzas E, Healey J (2001) Toward machine emotional intelligence: analysis of affective physiological state. IEEE Trans Pattern Anal Mach Intell 23(10):1175–1191
25. Bailenson NJ, Pontikakis DE, Mauss BI, Gross JJ, Jabon EM, Hutcherson AC, Nass C, John O (2008) Real-time classification of evoked emotions using facial feature tracking and physiological responses. Int J Human-Computer Studies 66:303–317
26. Poh MZ, Swenson NC, Picard RW (2009) Comfortable sensor wristband for ambulatory assessment of electrodermal activity. 1st Biennial Conference of the Society for Ambulatory Assessment, Greifswald, Germany. June 25–28, 2009
27. Brown L, Grundlehner B, van J de Molengraft, Penders J, Gyselinckx B (2009) Body area networks for monitoring autonomous nervous system responses. In: Proceedings from the international workshop on wireless pervasive health, London, United Kingdom, April 2009
28. Kim HK, Bang WS, Kim RS (2004) Emotion recognition system using short-term monitoring of physiological signals. Med Biol Eng Comput 42(3):419–427, May 2004
29. Grundlehner B, Brown L, Penders J, Gyselinckx B (2009) The design and analysis of real-time, continuous arousal monitor. In: Proceedings from the IEEE international workshop on body sensor network conference, San Francisco (CA), June 2009
30. Suchs O (2007) Motion tolerance in wearable sensors: the challenge of motion artifact. Proceedings of the IEEE EMBS Conference, 2007 1542–1545
31. Leonov V, Gyselinckx B, Van Hoof C, Torfs T, Yazicioglu F, Vullers RMJ, Fiorini P (2008) Wearable self-powered wireless devices with thermoelectric energy scavengers. In: Conference on Smart Systems Integration, Barcelona, 9–10 April 2008, pp 217–224
32. Christiaens W, Loeher T, Pahl B, Feil M, Vandevelde B, Vanfleteren J (2008) Embedding and assembly of ultrathin chips in multilayer flex boards. Circuit J World 34(3):3–8
33. Torfs T, Christiaens W, Vanfleteren J, Huwel W, Perdu W, Firat Yazicioglu RF, Brebels S, Van Hoof C (2009) Flexible wireless biopotential system with embedded ultra-thin chip. Proc Smart Sys Integration conf, 11–13 March 2009
34. Brosteaux D, Axisa F, Gonzalez M, Vanfleteren J (2007) Design and fabrication of elastic interconnections for stretchable electronic circuits. IEEE Electron. Device Lett 28(7): 552–554.
35. Koetse M, et al. (2008) In plane optical sensor based on organic electronic devices. Proceedings of SPIE, vol 7054, 70541I

Chapter 9
Body Channel Communication for Energy-Efficient BAN

Seong-Jun Song and Hoi-Jun Yoo

9.1 Introduction

9.1.1 Motivation

Recent advances in semiconductor technologies and computing systems have led to the proliferation of mobile and portable electronic devices opening the ubiquitous mobile computing environment. A wearable computing technology is an example for the user to facilely place such devices around the human body. The wearable electronic devices (e.g., wrist-type computers, earphones, video eyeglasses, and head-mounted displays) and sensors offer potentials for wide range of applications from the health management to ambient intelligence [1–2]. Since such devices are distributed on the human body, a body area network (BAN) can provide the connectivity between each wearable device with the communication range of the human body, corresponding to 1–2 m. Moreover, it should get powered by a very small battery in order to minimize its physical size and get connected through simple interfaces for the convenience of the use. Since the wearer utilizes wearable electronic devices and sensors continuously anytime and anywhere, the devices require a low power data transceiver employing energy-efficient communication schemes. Moreover, the high data rate operation is needed for exchanging multimedia data such as audio or video over BANs.

There are two approaches to implementing the BAN: one exploiting the human body itself as a transmission medium and the other using external medium such as wire and air. Traditional wireline technologies can provide high data rates but needs long copper wires which are generally cumbersome for human body applications. The radio-frequency (RF) short-range personal area connection using Bluetooth can provide more practical usability. However, it has potential problems such as low data

S.-J. Song (✉)
Samsung Electronics Co., LTD, Suwon 443-742, Republic of Korea
e-mail: sj33.song@samsung.com

H.-J. Yoo, C. van Hoof (eds.), *Bio-Medical CMOS ICs*,
Integrated Circuits and Systems, DOI 10.1007/978-1-4419-6597-4_9,

rate, high power consumption, and vulnerability to interference. Even the Zero-IF receiver [3] to achieve significant power savings over the Bluetooth radios consumes still too high power. Alternatively, the UWB receivers have been widely developed to achieve higher data rate operation, but must operate at vast bandwidth inherently leading to increase of power dissipation. For example, according to [4], a UWB receiver based on impulse radio was reported with the power consumption of about 30 mW at 3-to-5 GHz bandwidth for wearable and wireless body area networks. In addition, it faces increasing cost for additional RF process, severe interference problem through the air at very short range around human body, and FCC regulation.

9.1.2 Human Body Communications

Recently, several human body communication (HBC) schemes have been suggested for BAN and the data is transferred through the skin of the human body. The near-field electrostatic coupling scheme using a narrowband low frequency signal was firstly introduced by Zimmerman [5] and expected to significantly reduce power consumption. His coupling scheme is dependent on the conditions of the surrounding environment such as the earth ground for the return path and has limited data rate of 2.4 kb/s due to the narrow bandwidth of 400 kHz. Another scheme employing an electromagnetic wave of 10 MHz also suffers from the bandwidth limitation of conventional FM and FSK. Recently, another group reported a transceiver adopting electroopic conversion method to achieve higher data rate of 10 Mb/s by using a special off-chip sensor. However, it leads to high cost, high power consumption, and large physical size for human body applications. Moreover, it must have both signal and ground electrodes which make it inconvenient to use.

This chapter presents a novel HBC scheme exploiting wideband signaling (WBS) technique with a direct-coupled interface (DCI) over the optimized HBC channel. The HBC channel is optimized for high speed operation on the human body, which is identified as the *Bodywire* channel. The DCI is an interface method connecting the silicon chip with the human body directly. It uses only a single electrode for data transmission without ground electrode, in contrast to other methods which require the off-chip sensor to detect the feeble electric field and the earth ground path. In addition, the 2 Mb/s WBS transceiver chip is implemented by a 0.25 μm standard CMOS technology. The transceiver including the receiver analog front-end (AFE) consumes only 5 mW from a 1 V supply. Therefore, the proposed HBC scheme can achieve lower power consumption with high data rate operation than other HBC schemes in [5–7], which make it suitable for the application to energy-efficient point-to-point transmission around the human body using the BAN. The proposed HBC using the DCI around the human body is illustrated in Fig. 9.1. The HBC transceiver with the DCI does not need an off-chip component or a sensor to detect electromagnetic (EM) field. Thus, it can fully integrate all functional blocks on a silicon chip excluding a signal electrode, thereby achieving low cost and small physical size. Also, the WBS with the optimized HBC channel can provide energy-efficient transmission to extend the life time of the battery.

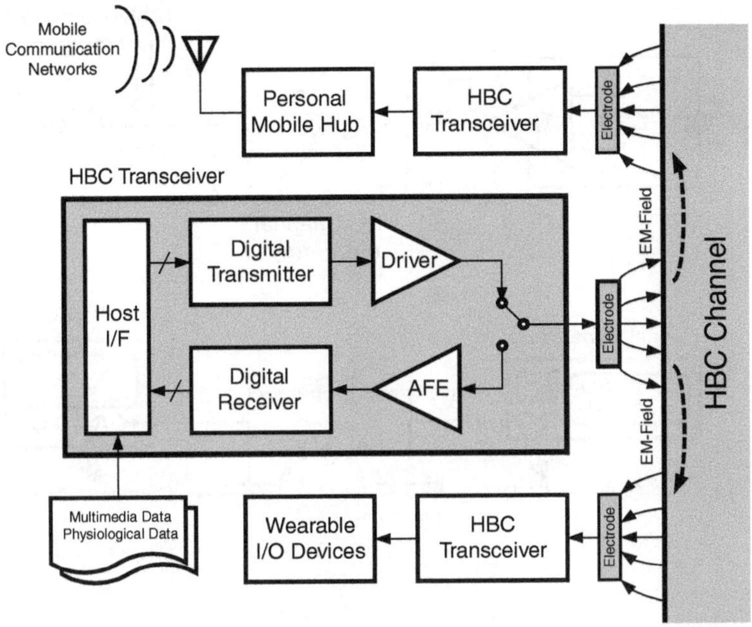

Fig. 9.1 Conceptual block diagram of human body communication with a direct-coupled interface through a human body

9.2 Channel Characteristics

The transmission characteristics of HBC channel have been investigated by using RF signals from 1 to 40 MHz [6] and from 1 MHz to 3 GHz [8]. However, the frequency characteristics presented in [6] was obtained by using a pair of signal and ground electrodes and only the frequency domain response was reported in [8]. In this paper, for the potential optimization of the WBS transceiver with the DCI, the characteristics of the HBC channel are investigated by using only a single signal electrode without any ground electrode or path in the time and frequency domains. Figure 9.2 illustrates the measurement setup for the time and frequency domains.

The distance between a transmitter and a receiver is fixed to 15 cm. As shown in Fig. 9.2(a), a battery-powered crystal-based transmitter is connected to the forearm with a single Ag/AgCl electrode. An electrode as the receiver is connected to a digital oscilloscope and its ground is floated to isolate it from the signal ground of the transmitter in consideration of its usage in the real situation. In this setup, the transmitter transfers the electromagnetic pulse only through the forearm to the oscilloscope without the earth ground path.

Figure 9.3(a) shows the measured output waveform for the time-domain characteristics. For a square wave of 3 V at 2 MHz, the channel outputs is measured to be the positive and negative pulse signals with no DC offset. Each pulse signal exhibits a narrow small pulse signal with a width of about 8 ns corresponding to a

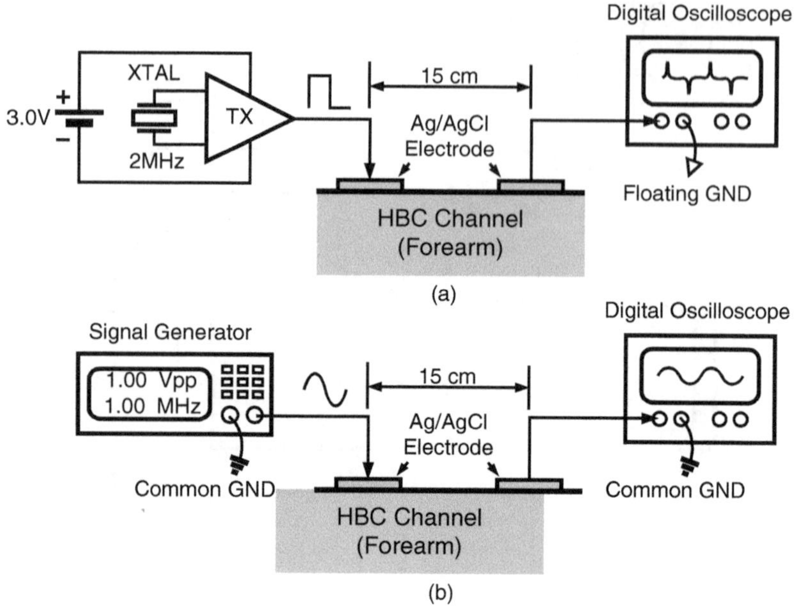

Fig. 9.2 Measurement setup for the investigation of the HBC channel (**a**) in the time domain (**b**) in the frequency domain

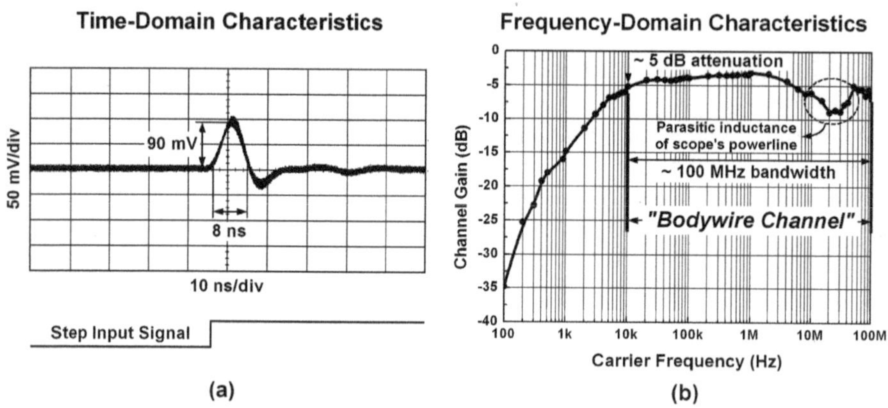

Fig. 9.3 Measured characteristics of the HBC channel (**a**) in the time domain (**b**) in the frequency domain

bandwidth of 125 MHz and the amplitude of 90 mV. In order to obtain the characteristics of the human body over the frequency sweep without the effect of the ground, a signal generator is exploited as a transmitter and its ground is connected to the common ground of the oscilloscope as shown in Fig. 9.2(b). The signal generator generates the sinusoidal waves with 1 V peak-to-peak, corresponding to the transmitting power of about −17 dBm, in the frequency range from 100 Hz to 100 MHz.

Fig. 9.3(b) indicates the measured frequency-domain characteristics. The human body behaves as a band-pass filter with a bandwidth of about 100 MHz and shows approximately 5 dB attenuation. This behavior is attributed to the open loop configuration exploiting the DCI without the ground path for a return signal. Dip of about 10 dB in the channel gain at near 20 MHz may be attributed to the parasitic inductance of the powerline cable of the oscilloscope. According to this investigation, the suitable frequency for the WBS over the HBC channel exists in the range of 10 kHz to 100 MHz, which is named as the *"Bodywire channel."* Therefore, the *bodywire* channel enables the WBS transceiver to operate at the high data rate.

Along with the channel investigation, a HBC channel model can lead to the optimized performance of the WBS transceiver over the HBC channel. A lumped electrical model for the WBS with the DCI is obtained by modification of a previous simplified electrical model (R_{ext}, C_{int}, and R_{int}) for the biological tissues [9] as shown in Fig. 9.4. The capacitance of C_{air}, a parasitic capacitor model associated with the electrostatic coupling via air as the return path, is very small due to the feeble air electrostatic coupling. R_{loss} and C_{loss} are added as the transmission lossy models associated with the electrostatic coupling to the receiver's ground path above about 10 MHz.

According to the safety study of the human exposure to the RF signals, the minimum electric field intensity is 28 V/m over the frequency range up to 300 GHz for general public exposure to time-varying electric fields [10]. The maximum

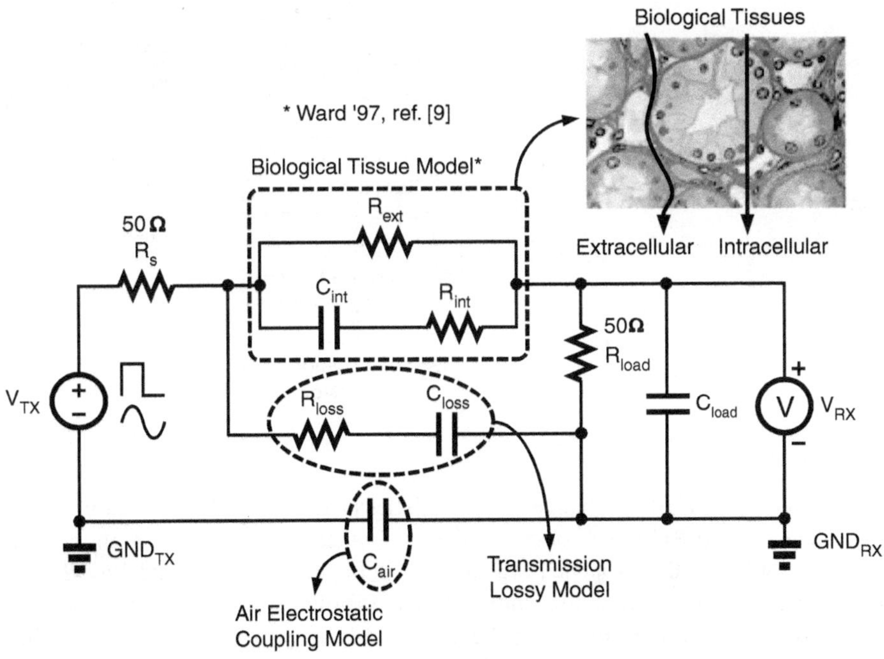

Fig. 9.4 Proposed lumped electrical model for the Bodywire channel

electric field intensity is roughly estimated at 20 V/m and the induced displacement current is much less than 45 mA, because the impedance of the human body exhibits the range of 300–500 Ω over the frequency range of 0.1–100 MHz [11]. All measurement conditions meet the ICNIRP guidelines [10] as well as the IEEE recommendations to provide an electrical safety for the human body [11].

9.3 Design of Wideband Signaling Communication Link

The time and frequency behaviors of the HBC channel have been investigated in Chapter 9.2. In order to improve the communication performance, the analytical model for a communication link, WBS link in this work, is necessary to the design and implementation of a transceiver. Figure 9.5 illustrates the design flow diagram for the WBS communication link.

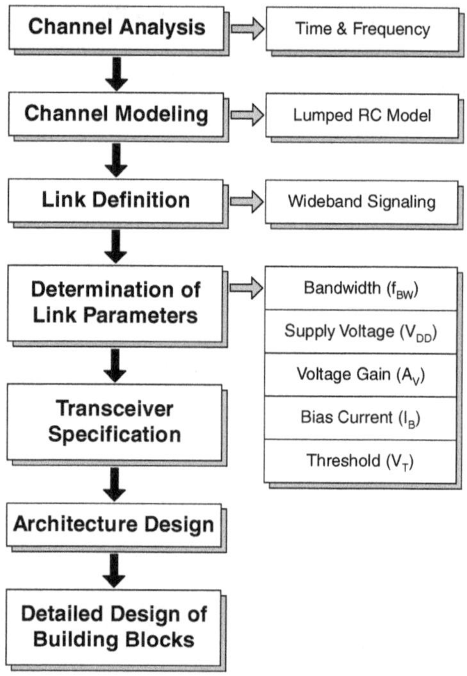

Fig. 9.5 Design flow diagram of the WBS communication link

First, the analysis of the channel characteristics is investigated and the physically-equivalent lumped RC model is established as in Chapter 9.2, then simplified link model suitable for the proposed signaling is defined. After that, the link design parameters such as bandwidth, supply voltage, voltage gain, bias current, and threshold are determined by using the theoretical analysis of the simplified link model. According to the determined link parameters, the design specification and the architecture of the transceiver can be defined in regard to energy-efficient data

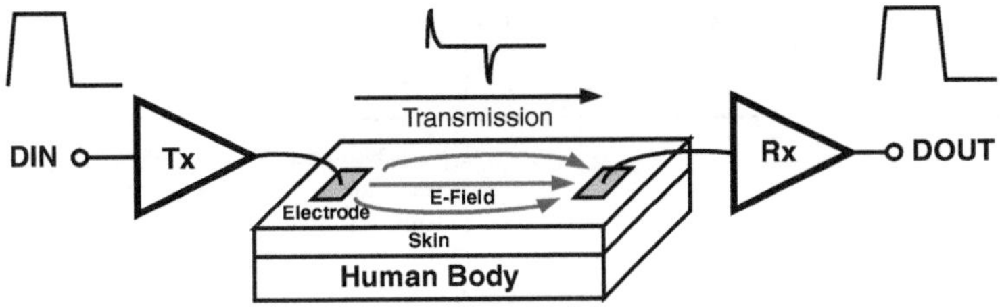

Fig. 9.6 Conceptual diagram of the WBS communication link

transmission. Finally, the detailed design of each building block is carried out by exploiting circuit techniques to minimize its power consumption.

Figure 9.6 illustrates the conceptual diagram of the WBS communication link. The feeble electric field that is induced by the output voltage of the transmitter creeps toward the receiver's electrode on the surface of the skin. The transmitted signal on the skin is measured to be the wideband pulse signal consisting of positive and negative pulse with no DC offset. The receiver recovers the wideband pulse to the digital signal. Since the link behavior is dominant to the near-field effect around the human body where the wavelength of the transferred signal is longer than the transmission distance, the link model can be simplified as shown in Fig. 9.7. The WBS transmitter model consists of a voltage source with step function V_{TX} and an output resistance R_{TX}. The HBC channel model is simplified to a high-pass filter with a coupling capacitance C_{HB} and a load resistance R_{HB}. The WBS receiver model is considered to be a preamplifier with a pole of $1/R_{RX}C_{RX}$ and a trigger with trigger point of V_T.

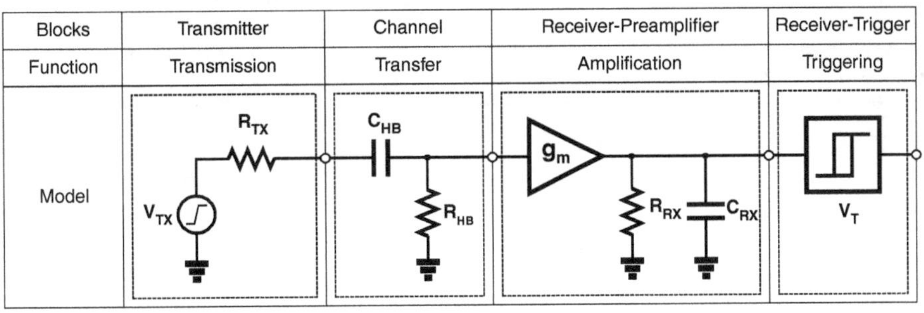

Fig. 9.7 WBS communication link model

The design parameters of the link are summarized in Table 9.1, which will be used for the theoretical analysis with the link model of Fig. 9.7.

The lumped electrical model has been driven from physical point of view in Chapter 9.2. However, it is difficult to determine the link parameters through the theoretical analysis using such the lumped model due to its complexity. Thus, the simplified electrical model is suggested as shown in Fig. 9.8. Figure 9.9 compares simulated channel characteristics with the models in Figs. 9.4 and 9.8 for V_{TX} and

Table 9.1 Link design parameters

Parameter	Notation	Description
Distance	d	Distance between the transmitter and the receiver
TX Output Voltage	V_{TX}	Output voltage of the transmitter
TX Output Resistance	R_{TX}	Output resistance of the transmitter
Channel Resistance	R_{HB}	Equivalent resistance of the HBC channel
Channel Capacitance	C_{HB}	Equivalent capacitance of the HBC channel
Slope of V_{TX}	K	Slope of the output voltage of the transmitter
RX Input Voltage	V_{RX}	Input voltage of the receiver
Unit Time Constant of Channel	τ_0	Time constant for d=15 cm
Time Constant of Channel	τ_d	Time constant of the channel as a function of d
Maximum of RX Input Voltage	V_{RXMAX}	Maximum input voltage of the receiver
Pulse Width Time	T_P	Pulse width time at the half of V_{RXMAX}
Rising Time	t_r	Rising time for step response at a one-pole system
Closed-Loop Gain	A_V	Voltage gain of the preamplifier
Closed-Loop Bandwidth	f_{BW}	Operational bandwidth of the preamplifier
Gain-Bandwidth Product	GBW	Gain-bandwidth product of the preamplifier
Open-Loop Gain	A_{VO}	Voltage gain of the feedforward amplifier
Open-Loop Bandwidth	f_{BWO}	Operational bandwidth of the feedforward amplifier
Transconductance	g_m	Transconductance of the operational amplifier
Output Resistance	R_{RX}	Output resistance of the preamplifier
Output Capacitance	C_{RX}	Output capacitance of the preamplifier
Supply Voltage	V_{DD}	Supply voltage of the preamplifier
Overdrive Voltage	V_{ov}	Overdrive voltage of the feedforward amplifier
Threshold Voltage	V_{th}	Threshold voltage of the MOS transistor
Bias Current	I_B	Bias current of the feedforward amplifier
Triggering Threshold	V_T	Threshold level of the trigger
Bit Energy	E_b	Energy consumption per a bit for reception
Power Consumption	P_b	Power consumption for reception
Bit Period	T_b	Data bit period

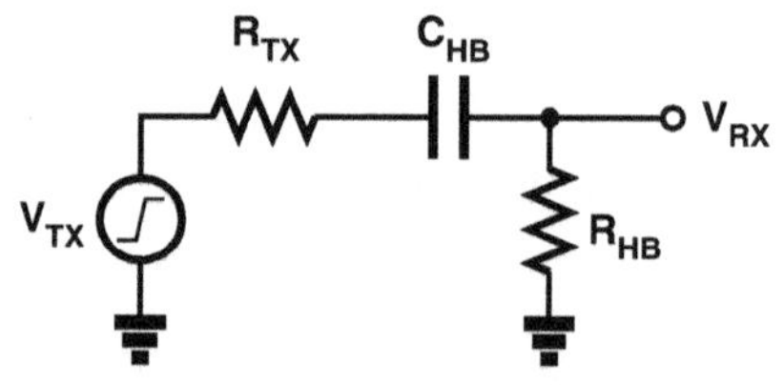

Fig. 9.8 Simplified electrical model for the channel

V_{RX}. It can be seen that the shape of the wideband pulse simulated by using the simplified model is very similar with that of the lumped model.

The simplified model is verified for the unit distance of 15 cm. In order to establish the model as a function of the distance, the distributed RC model is exploited with the simplicity of the HBC channel model. The proposed distributed model as

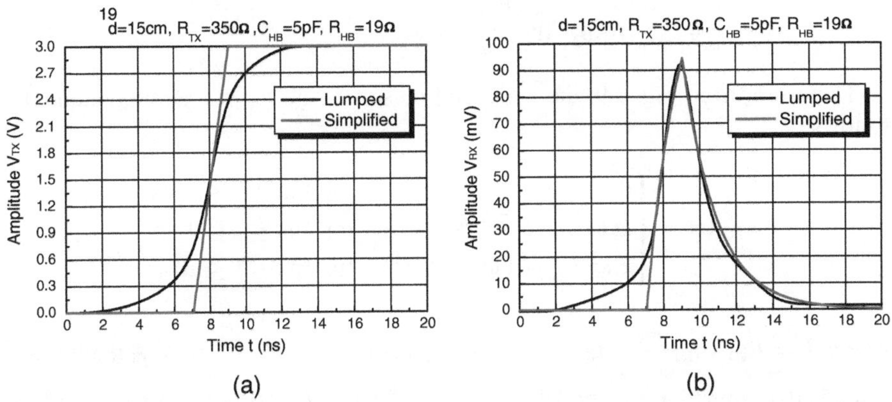

Fig. 9.9 Comparison of channel characteristics with the lumped and simplified models for (**a**) the transmitted signal and (**b**) the received wideband pulse signal

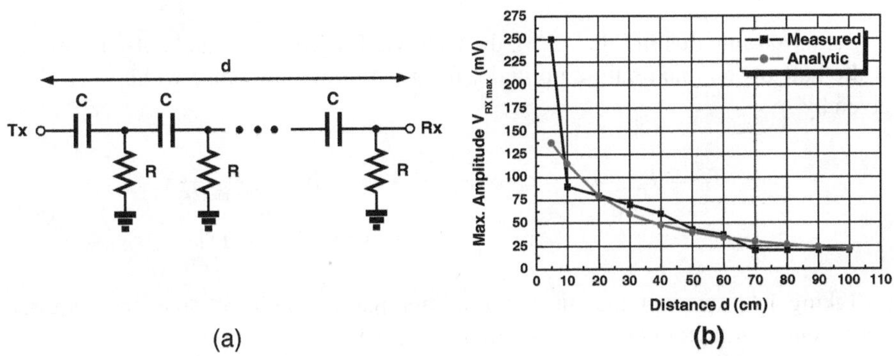

Fig. 9.10 (**a**) Distributed model and (**b**) the comparison characteristics for the HBC channel

a function of *d* depicts in Fig. 9.10(a). It is made up of the series of the unit model. The comparison of the measured and analytic results as the height of the received wideband pulse is shown in Fig. 9.10(b).

In this section, we have determined the link parameters through the theoretical analysis. Let us consider the simplified electrical model depicted in Fig. 9.8. The relation equation between the input voltage of the receiver and the output voltage of the transmitter is given by

$$
\begin{aligned}
V_{RX} &= R_{HB} C_{HB} \frac{dV_{CHB}}{dt} \\
&= R_{HB} C_{HB} \frac{d(V_{TX} - R_{TX} I - V_{RX})}{dt} \\
&= R_{HB} C_{HB} \left[\frac{dV_{TX}}{dt} - \left(1 + \frac{R_{S}}{R_{HB}}\right) \frac{dV_{RX}}{dt} \right]
\end{aligned}, \qquad (9.1)
$$

and hence

$$(R_{\mathrm{TX}} + R_{\mathrm{HB}})C_{\mathrm{HB}}\frac{dV_{\mathrm{RX}}}{dt} + V_{\mathrm{RX}} = R_{\mathrm{HB}}C_{\mathrm{HB}}\frac{dV_{\mathrm{TX}}}{dt}. \tag{9.2}$$

Therefore, from Fig. 9.9, the wideband pulse voltage $V_{RX}(t)$ can be obtained as follows:

$$V_{\mathrm{RX}}(t) = \begin{cases} 0 & \text{for } 0 \text{ ns} \le t \le 7 \text{ ns} \\ k(1 - e^{-(t-7n)/\tau_0}) & \text{for } 7 \text{ ns} \le t \le 9 \text{ ns} \\ k(e^{2n/\tau_0} - 1)e^{-(t-7n)/\tau_0} & \text{for } 9 \text{ ns} \le t \le 20 \text{ ns,} \end{cases} \tag{9.3}$$

where, $k = R_{\mathrm{HB}}C_{\mathrm{HB}}\frac{dV_{\mathrm{TX}}}{dt}$, $\tau_0 = (R_{\mathrm{TX}} + R_{\mathrm{HB}})C_{\mathrm{HB}}$, and $\tau_d = \sum_{i=0}^{N} \tau_i = \frac{d}{15}\tau_0$, the each value is determined at the model in Fig. 9.8. That is, $k = 0.143$ and $\tau_0 = 1.85$ ns. Also, maximum voltage of the V_{RX} as a function of d is expressed as

$$V_{\mathrm{RX\,max}} = k(1 - e^{-2n/\tau_d}). \tag{9.4}$$

Let us assume that in Fig. 9.9(b), the bandwidth of the wideband pulse is defined at the half of the maximum amplitude, from (3) to (4), the pulse width time T_P is given by

$$\begin{aligned} T_P = T_2 - T_1 &= \tau_d \ln \frac{k(e^{2n/\tau_d - 1})}{V_{\mathrm{RX\,max}}/2} - \tau_d \ln \frac{k}{k - V_{\mathrm{RX\,max}}/2} \\ &= \tau_d \ln \left[\left(\frac{2k}{V_{\mathrm{RX\,max}}} - 1\right)(e^{2n/\tau_d} - 1) \right] \end{aligned}. \tag{9.5}$$

Taking into account that the preamplifier has a dominant pole, the required operational bandwidth f_{BW} as a function of d is calculated as follows:

$$f_{\mathrm{BW}} = \frac{0.35}{t_r} = \frac{0.7}{T_P} = \frac{0.7}{\tau_d \ln \left[\left(\frac{2k}{V_{\mathrm{RX\,max}}} - 1\right)(e^{2n/\tau_d} - 1) \right]}. \tag{9.6}$$

Meanwhile, gain-bandwidth product of the preamplifier is equal to

$$GBW = A_V \cdot f_{\mathrm{BW}} = A_{V0} \cdot f_{\mathrm{BW0}} = g_m R_{\mathrm{RX}} \cdot \frac{1}{2\pi R_{\mathrm{RX}} C_{\mathrm{RX}}} = \frac{g_m}{2\pi C_{\mathrm{RX}}} \tag{9.7}$$

The preamplifier in the link model is configured with an operational amplifier (op amp). In the typical configuration of the op amp as shown in Fig. 9.11, the minimum supply voltage to have proper operation is given by

$$V_{\mathrm{DD}} = \max\{3V_{\mathrm{ov}}, V_{\mathrm{th}} + 2V_{\mathrm{ov}}\} \tag{9.8}$$

where, V_{ov} is the overdrive voltage and V_{th} is the threshold voltage of the MOS transistor.

For $V_{\mathrm{th}} > V_{\mathrm{ov}}$, $V_{\mathrm{ov}} = \frac{V_{\mathrm{DD}} - V_{\mathrm{th}}}{2}$ and op amp's transconductance g_m is given by

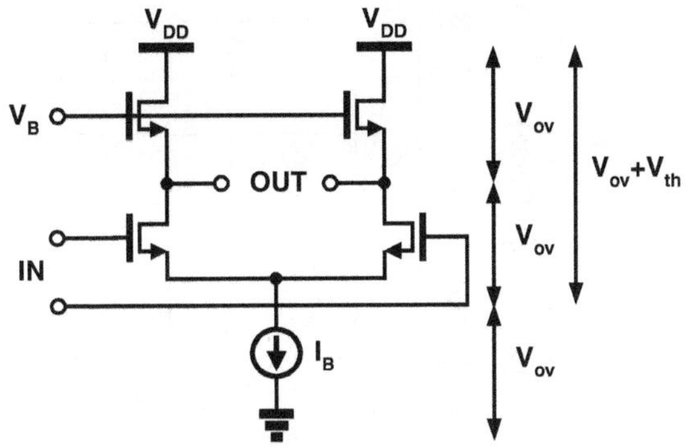

Fig. 9.11 Typical configuration of a feed forward amplifier

$$g_m = \frac{I_B}{V_{\text{ov}}} = \frac{2I_B}{V_{\text{DD}} - V_{\text{th}}}. \tag{9.9}$$

Hence, the minimum supply voltage follows from (7) to (9) that

$$V_{\text{DD}} = \frac{I_B}{\pi C_{\text{RX}} A_V f_{\text{BW}}} + V_{\text{th}}. \tag{9.10}$$

Substituting (9.6) into (9.10) and for given bias current I_B, we obtain the minimum supply voltage V_{DD} as a function of d. Also, for given supply voltage V_{DD}, the bias current I_B as a function of d is can be obtained from (9.10) as follows:

$$I_B = \pi C_{\text{RX}} A_V f_{\text{BW}} (V_{\text{DD}} - V_{\text{th}}). \tag{9.11}$$

Triggering threshold voltage V_T as a function of d is defined as the half of the maximum amplitude of the amplified wideband pulse. Therefore,

$$V_T = \frac{V_{\text{RX max}}}{2} A_V = \frac{k(1 - e^{-2n/\tau_d})}{2} A_V. \tag{9.12}$$

The trigger is configured as the same op amp of the preamplifier. Its power consumption is equal to the preamplifier. Accordingly, the energy consumption required for recovering a bit to the digital signal is expressed as

$$E_b = P_b T_b = 2V_{\text{DD}} I_B T_b. \tag{9.13}$$

From (9.10) to (9.11), the bit energy E_b as a function of d is considered for the given supply voltage V_{DD} and the given bias current I_B.

Figures 9.12, 9.13, 9.14, 9.15 and 9.16 show the characteristics between each link parameter as a function of d, illustrated by using (9.6), (9.10), and (9.13), respectively.

Table 9.2 summarizes the design specification of link parameters extracted by the analysis.

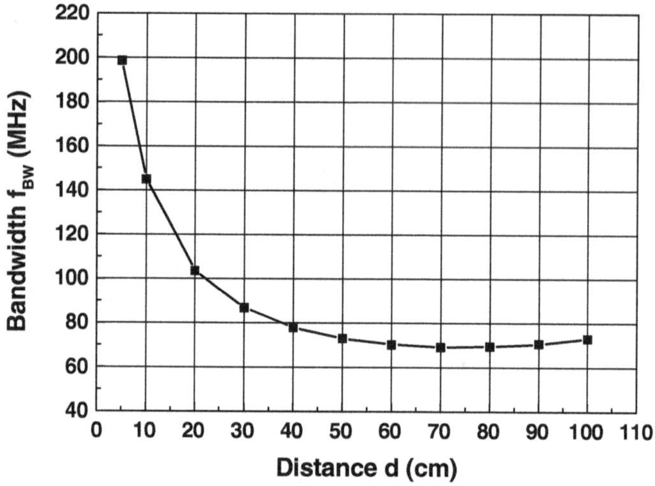

Fig. 9.12 Required operational bandwidth of the preamplifier

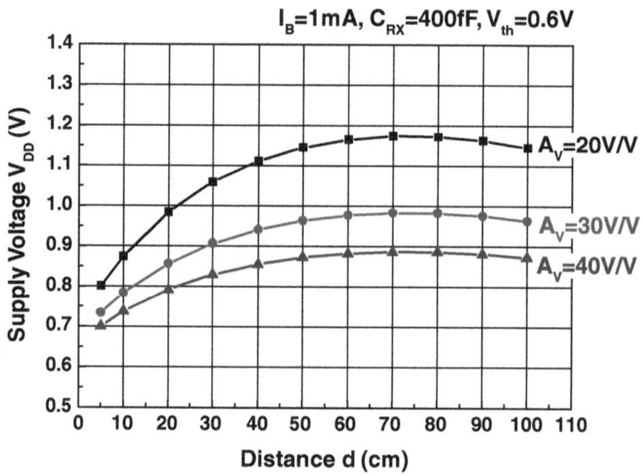

Fig. 9.13 Minimum supply voltage of the preamplifier for three voltage gains

9.4 Wideband Signaling Transceiver

A HBC transceiver is based on the WBS that directly transmits binary digital signal through a transmitter into the human body, transfers wideband pulse signals over the HBC channel, and then recovers the binary data at a receiver. The WBS has two features: simple interface and high data rate capability. The WBS enables the use of only a single electrode for the data transmission. A transceiver exploiting the WBS is connected to the human body with a single Ag/AgCl or metal electrode. The grounds of the transmitter and the receiver are completely isolated from each other. The earth ground path for a return signal is not necessary as well. In addition, since

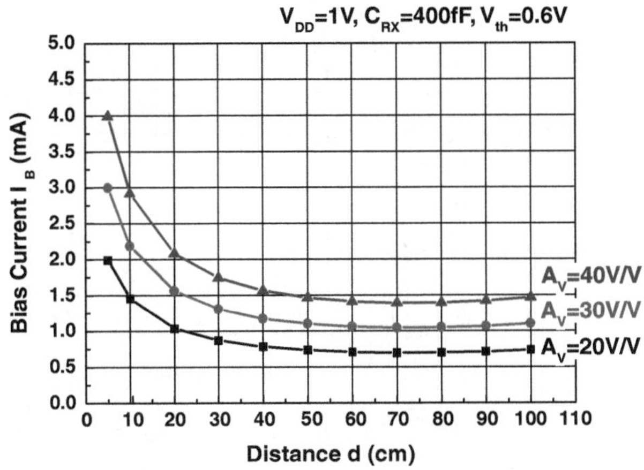

Fig. 9.14 Bias current consumption of the preamplifier for three voltage gains

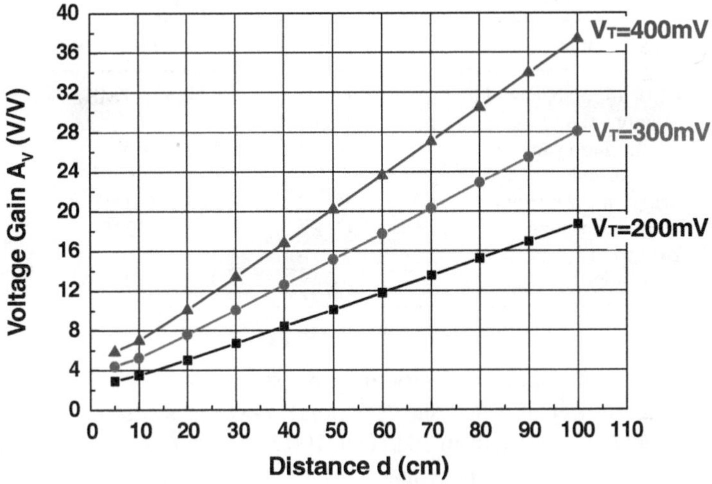

Fig. 9.15 Voltage gain of the preamplifier for three thresholds of the trigger

the WBS is independent to the conditions of surrounding environments such as the earth ground, the data transmission using the WBS enables stable operation for the situation even if the human body is touched on the ground. The WBS provides the maximum data rate of 125 Mb/s. According to the time-domain characteristics as shown in Fig. 9.3(a), the data rate can go up to about 125 Mb/s because the pulse bit signal has a width of about 8 ns. Theoretically, the HBC channel capacity is limited by the Hartley-Shannon law given as follows.

$$C = B\log\left(1 + \frac{S}{N}\right), \tag{9.14}$$

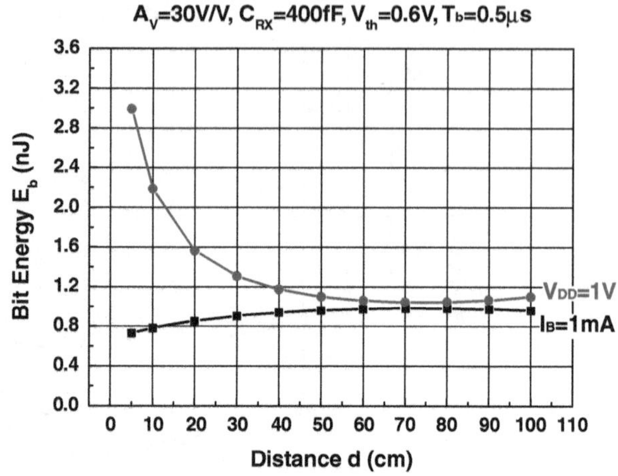

Fig. 9.16 Reception bit energy of the receiver

Table 9.2 Design specification of link parameters

Link parameter	Notation	Specification
Closed-loop bandwidth	f_{BW}	200 MHz
Supply voltage	V_{DD}	1 V
Closed-loop gain	A_V	30 V/V
Bias current	I_B	3 mA
Triggering threshold	V_T	300 mV
Output capacitance	C_{RX}	400 fF
Bit period	T_b	0.5 μs
Bit energy	E_b	3 nJ

where C is maximum channel capacity in bit per second, B is channel bandwidth in Hertz, *S* is signal power in watts, and N is noise power in watts. The bandwidth of the HBC channel is 100 MHz from 10 kHz to 100 MHz. Thus, for the signal-to-noise ratio of 10, the maximum channel capacity is approximately 104 Mb/s.

According to the channel investigation, several design requirements can be defined in realizing a HBC transceiver with WBS technique. A WBS transceiver operating at the high data rate of 2 Mb/s can transfer the real-time multimedia data streams such as audio signals without any encoding and decoding. It should consume a power less than 10 mW with a 1 V supply for longer lifetime of a very small battery. The supply voltage of 1 V is below the sum of the threshold voltage of NMOS and PMOS. This eliminates short circuit currents for the digital circuits, thereby minimizing unnecessary power consumption [12]. For transmission of binary data, the nonreturn-to-zero (NRZ) is chosen for embedded clocking to enable clock recovery from data transitions with no additional timing reference at a receiver. Accordingly, the transmitter is designed to drive the binary data over the human body channel. The receiver requires input impedance of 50 Ω to maximize power transfer and to provide impedance matching with respect to signal integrity for high-speed received signals. The 3 dB operational bandwidth of about 200 MHz at the receiver is chosen to sufficiently sustain the channel bandwidth without intersymbol interference (ISI) effects. To achieve the transmission between the fingertip and the ear, corresponding to the distance of 1m, the voltage gain should be larger than 30 dB, and the minimum input sensitivity is to be −27 dBm. Since the channel output signal is comprised of positive and negative pulses without DC offset, the symmetric operation is demanded. A clock recovery function for embedded clocking in NRZ data service requires a bit error rate (BER) less than 10^{-5}.

In order to meet the requirements of a HBC transceiver, a WBS transceiver is proposed to use the *Bodywire* channel. Figure 9.17 shows the block diagram of the WBS transceiver that comprises a direct digital transmitter and a CDR-based WBS receiver.

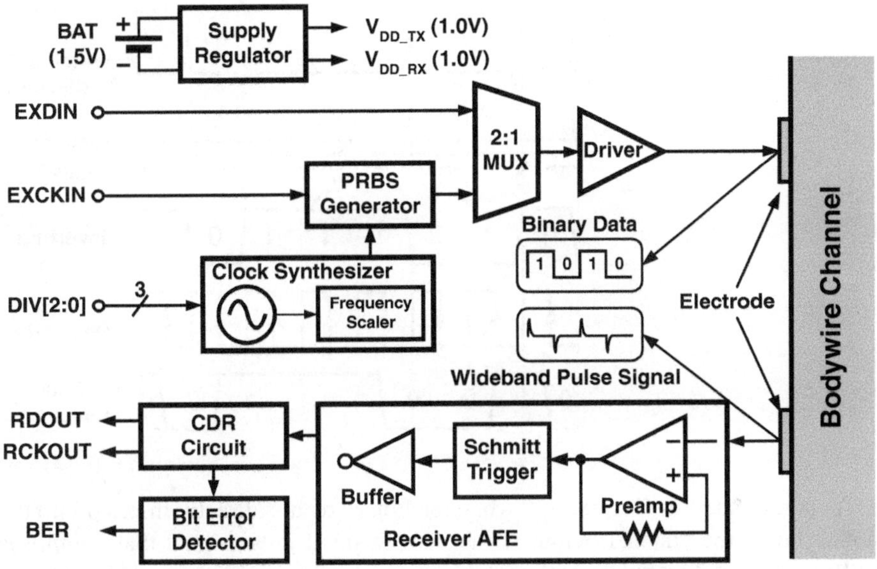

Fig. 9.17 Proposed architecture of the wideband signaling transceiver

The direct digital transmitter consists of a clock synthesizer, a pseudo random binary sequence (PRBS) generator, a 2-to-1 multiplexer (MUX), and a driver. The clock synthesizer has a ring oscillator structure and oscillates the clock signal through frequency scaling to activate the PRBS generator. The PRBS generator generates $2^7 - 1$ PRBS data and transmits them through the driver to the human body for on-chip link testing. Also, external binary data such as digitally converted audio data or baseband data can be directly transmitted to the human body by the 2-to-1 MUX. The driver is connected to a single electrode and induces the electric field on the skin of the human body. The CDR-based WBS receiver consists of a receiver AFE, a CDR circuit, and a bit error detector. The receiver AFE amplifies, triggers, and inverts the received wideband pulse signal in order to recover the binary data. Next, the schmitt trigger generates positive and negative states, and then the signal is inverted for the next CDR. The CDR circuit extracts a clean clock signal from the recovered binary data for embedded clocking and latches the data. The bit error detector is integrated for on-chip bit error rate (BER) testing and detects error bits of the recovered data from the extracted clock for $2^7 - 1$ PRBS. The transceiver is powered by a supply regulator, which generates the supply voltage of 1 V from a 1.5 V battery.

Figure 9.18 illustrates the timing diagrams for the operation of the WBS transceiver

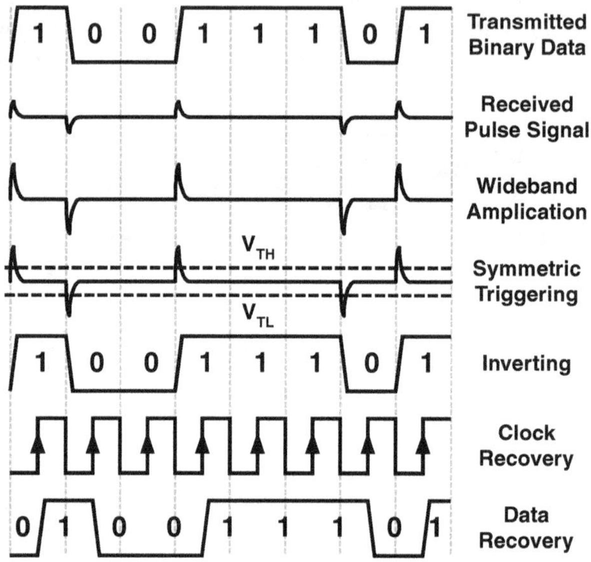

Fig. 9.18 Timing diagrams for the operation of the WBS transceiver

As investigated in Chapter 9.2, when the binary data is directly inserted into the human body, the channel output is the narrow small pulse signal that comprises positive and negative pulses with no DC offset. The received pulse signal corrupted by the channel is sufficiently amplified for wide bandwidth, and subsequently, the signal is triggered to positive and negative states by using two symmetric thresholds, V_{TH} and V_{TL}, where the symmetric operation provides the duty cycle of 50%. Consequently, the binary data can be recovered by inverting the triggered signal. For the CDR at the receiver, the full-rate clock signal is locked at the center point of the bit interval window. The binary data is recovered by latching the inverted signal at the rising edge of the clock signal.

The WBS transceiver should operate at wide bandwidth of 200 MHz in order to recover binary data from the wideband pulse signals. However, it leads to the increase of power dissipation inherently. Particularly, since the receiver AFE and the CDR circuit are the crucial elements and power-hungry building blocks, the low power techniques are necessary for their designs. For these reasons, the WBS transceiver incorporates four low power techniques: low power op amp, all-digital CDR architecture, low-voltage digitally-controlled oscillator (DCO), and quadratic sampling technique. These techniques allow the WBS transceiver to significantly reduce power consumption along with high data rate operation, which results in more energy-efficient transmission than other transceivers [5–7] over the HBC channel.

9.4.1 WBS Receiver AFE

According to the channel investigation, when the binary data is directly applied to the human body by a single electrode, the output of the channel exhibits narrow small pulse signals with no DC offset. These characteristics have some analogy with the AC-coupled chip-to-chip interconnect using 50 Ω transmission lines on a FR-4 PC board shown in [13]. The pulse receiver used in the AC-coupled interconnect achieves high data rate of several Gb/s. However, it should be based on differential structure with poor receiver sensitivity more than 100 mV from a 1.8 V supply. In the HBC, the supply voltage is limited to 1 V for longer battery life. From the transceiver's requirements, the input sensitivity of a receiver AFE needs to be less than 10 mV, which corresponds to −27 dBm for the 50 Ω input impedance. In order to recover binary data from the narrow small pulse signals over the HBC channel, a WBS receiver AFE exploits wideband symmetric triggering that separately accomplishes wideband amplification and symmetric triggering. Thus, the proposed AFE achieves not only high data rate operation and also lower input sensitivity with a 1 V supply. Figure 9.19 shows the block diagram of the proposed WBS receiver AFE consisting of four blocks: an on-chip symmetric bias circuit, a wideband preamplifier, a schmitt trigger, and an inverting buffer.

The AC coupling capacitor is connected between the electrode and the input of the AFE. This capacitor eliminates the conductive current path to the body and

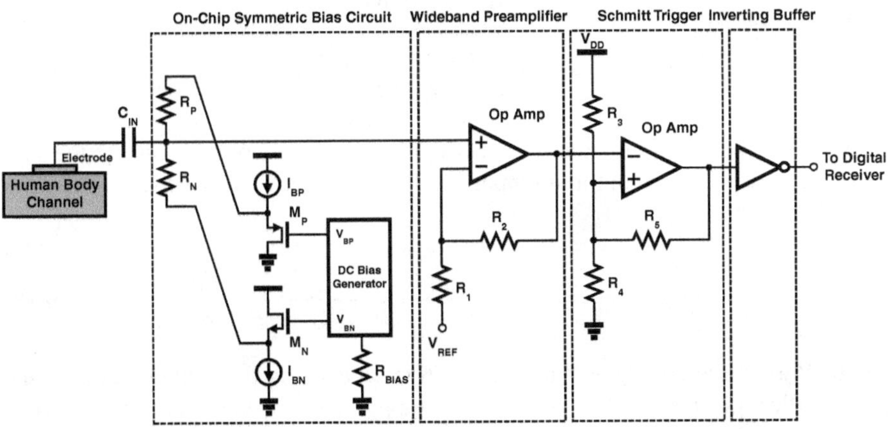

Fig. 9.19 Block diagram of the WBS receiver AFE

provides DC biasing for the input of the preamplifier regardless of the body's potential. The capacitance of C_{IN} with the input impedance of the on-chip bias circuit determines low 3 dB frequency of the AFE. The on-chip bias circuit is designed as a complementary configuration to acquire both symmetric impedance of 50 Ω and DC biasing with no external bias circuit. Hence, the on-chip symmetric bias circuit comprises pull-up resistor R_P, pull-down resistor R_N, a pair of complementary source followers($I_{\mathrm{BP}}-M_P$, $I_{\mathrm{BN}}-M_N$), and a DC bias generator controlled by a resistor R_{BIAS}. The input impedance of the AFE R_{IN} can be expressed as:

$$R_{\mathrm{IN}} = \left(\frac{1}{g_{\mathrm{mP}}} + R_P\right) // \left(\frac{1}{g_{\mathrm{mN}}} + R_N\right) \approx \frac{1/g_m + R}{2}, \tag{9.15}$$

where g_{mP} and g_{mN} are the transconductance of M_P and M_N, respectively, $R_P = R_N = R$ and $g_{\mathrm{mP}} = g_{\mathrm{mN}} = g_m$ for symmetric design. From (9.15), the effect of $1/g_m$ and R variations can be alleviated by half. Addition of the resistor R_P and R_N achieves not only high rejection to power supply noise that may corrupt the received signal, but also reduction of the variation of $1/g_{\mathrm{mP}}$ and $1/g_{\mathrm{mN}}$ due to large swing of the received signal. According to the simulations, the on-chip symmetric bias circuit dissipates 1.8 mA, which is 36% of the power consumed by using only the pull-up and pull-down resistors without M_p and M_N. The wideband preamplifier is designed as a non-inverting amplifier where the op amp incorporates the low power wideband configuration. The high 3 dB frequency of the AFE is constrained by the 3 dB bandwidth of the preamplifier, 200 MHz, because the power spectrum of the received signal remains below 200 MHz. The schmitt trigger consists of three resistors (R_3–R_5) and the same op amp as that of the preamplifier and produces positive and negative triggering thresholds (V_{TH} and V_{TL}), which can be controlled by varying the resistance of R_5. V_{TH} and V_{TL} are expressed as

$$V_{\mathrm{TH}} = \frac{R_4}{R_3//R_5 + R_4} V_{\mathrm{DD}}, \tag{9.16}$$

$$V_{TL} = \frac{R_4//R_5}{R_3 + R_4//R_5} V_{\mathrm{DD}}, \tag{9.17}$$

respectively. With $R_4 = R_3$, the sum of two triggering thresholds is equal to the supply voltage. Hence, the minimum input sensitivity of the receiver AFE $(V_{\mathrm{RXSEN}})_{\mathrm{min}}$ is given by

$$(V_{\mathrm{RXSEN}})_{\mathrm{min}} = \frac{V_{\mathrm{TH}} - V_{\mathrm{DD}}/2}{A_V} = \frac{R_3}{2(R_3 + 2R_5)} \frac{V_{\mathrm{DD}}}{A_V}, \tag{9.18}$$

where V_{DD} is the supply voltage of the AFE and A_V is the voltage gain of the preamplifier, given as $1 + R_2/R_1$. As the supply voltage decreases and the voltage gain of the preamplifier increases, the input sensitivity of the AFE can be minimized.

The wide bandwidth operation in the AFE leads to the increase of the power consumption. Since the op amp is a crucial element in the receiver AFE as shown in Fig. 9.19, a low power op amp is mandatory to reduce the power consumption. To reduce the power consumption drastically, the use of the low supply voltage is very attractive [1]. But, the supply voltage, 1 V in this design, is lower than the sum of the threshold voltages of the NMOS and PMOS. This makes it difficult to design a low-voltage and wideband op amp. To alleviate this difficulty, the low-voltage fully complementary folded cascode topology is proposed in this work. Figure 9.20 shows the proposed topology, consisting of an input stage with the source follower pairs, an intermediate gain stage employing fully complementary input pairs with low-voltage folded cascode loads, and a cross-coupled class-AB output stage.

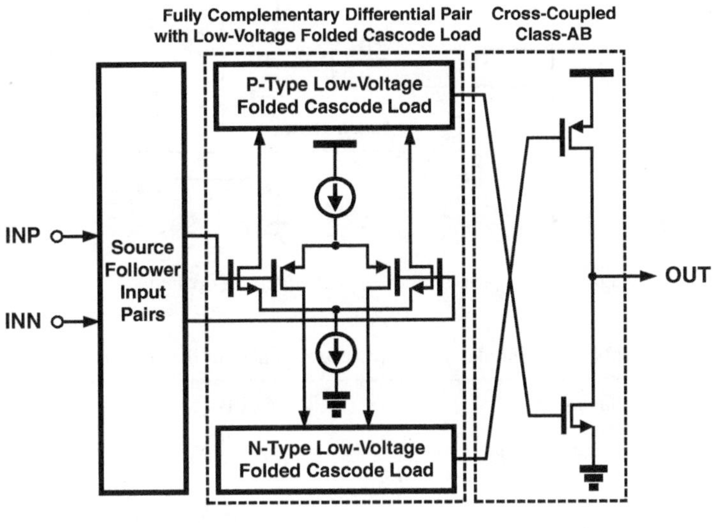

Fig. 9.20 Low-voltage fully complementary folded cascode op amp topology

All configurations of the op amp are based on complementary structure for symmetric operation. The source follower input pairs need to obtain the sufficient overdrive voltage and to minimize the area overhead at the gain stage. Particularly, the complementary low-voltage loads with each output can further reduce the number of the stacked transistors than the configuration with single output load. A class-AB output stage has been widely chosen to minimize the quiescent power consumption leading to a fast operation with high slew rate. The cross-coupled type of the class-AB is adopted in accordance with the fully complementary gain stage. Figure 9.21 shows the transistor-level implementation of the proposed low power wideband op amp. According to the HSPICE simulations, it exhibits 60-dB DC gain and 684 MHz gain bandwidth product while dissipating only 1.5 mW with a 1 V supply. With the help of the cross-coupled class-AB, its slew rate is 880 V/μs. Its performance is summarized in Table 9.3.

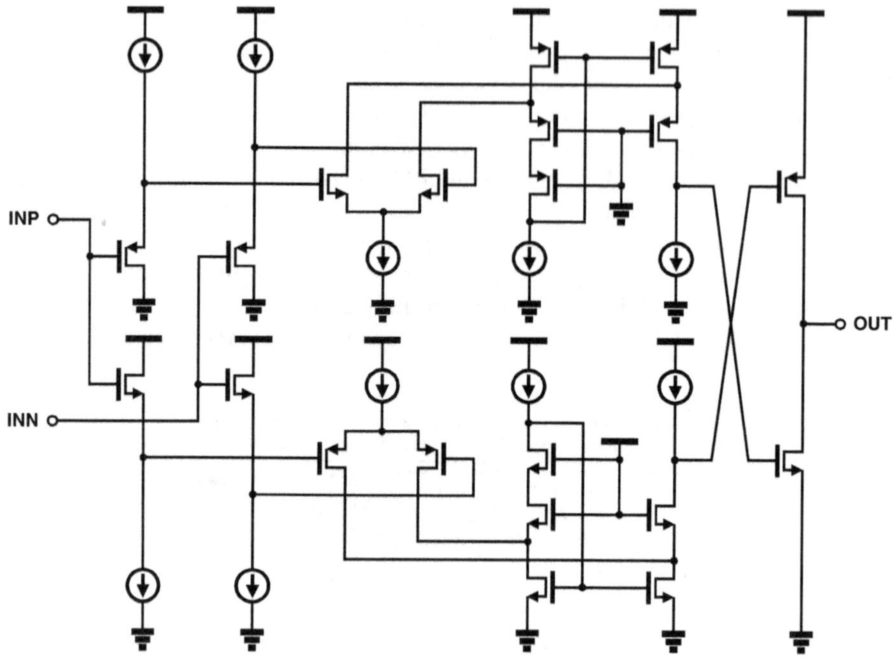

Fig. 9.21 Circuit diagram of the proposed low power wideband op amp

Table 9.3 Simulated performance summary of the op amp

DC voltage gain	60 dB
3-dB Bandwidth	12 MHz
Unity-gain bandwidth	937 MHz
Phase margin	59 °
Slew rate	± 880 V/μs
Power consumption	1.5 mW @ 1.0 V
Core area	0.0064 mm^2
Technology	0.18-μm standard CMOS

9.4.2 All-Digital Quadratic Sampling CDR Circuit

In human body communications, a stream of binary data transfers over a single human body with no accompanying clock, but a receiver must process the data synchronously. Thus, to extract a clean clock from the stream of binary data and synchronize the data by the extracted clock, the WBS transceiver employs a CDR circuit that is exploited in areas such as optical communications and high-speed serial links or interconnects. Since the transceiver operates at relatively low data rates for the recovered binary data, differently from gigabit serial links [14], bandwidth-limited effects such as ISI or amplitude noise and clock frequency offset between a transmitter and a receiver are tolerable in the WBS transceiver. Hence, it is more desirable for the CDR circuit to be focused on low power consumption rather than high speed operation. However, under the condition of $V_{thn} + |V_{thp}| > 1$ V, an analog CDR circuit faces performance degradation. But, the power dissipation of digital circuitry is proportional to the square of the supply voltage. Therefore, the transceiver incorporates all-digital sampling CDR architecture based on a low-voltage DCO. To further reduce power consumption and complexity, a quadratic sampling technique is adopted. The proposed CDR circuit takes all-digital PLL configuration as shown in Fig. 9.22.

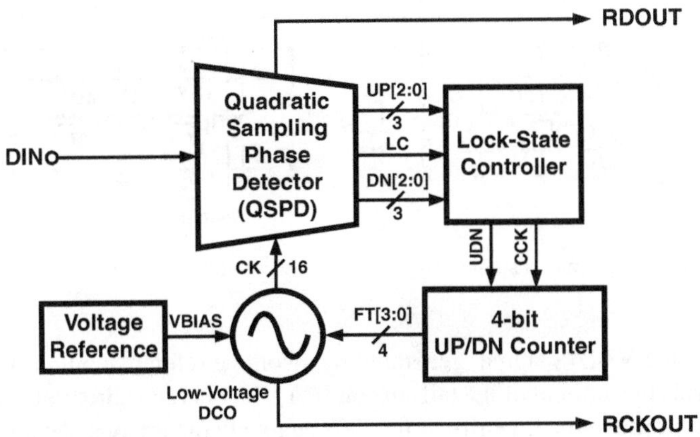

Fig. 9.22 Quadratic sampling CDR architecture based on a low-voltage DCO

It consists of a quadratic sampling phase detector (QSPD), a lock-state controller, a 4-bit UP/DN counter, and a low-voltage DCO with voltage reference. The QSPD samples the incoming data at every rising edge of the clock and detects the transition of the data. The 4-bit FT signals can be generated by averaging and low-pass filtering the sampled values through the lock-stage controller and the 4-bit UP/DN counter. When the LC signal goes high, the loop will be locked. The DCO controlled by FT[3:0] is designed to generate 16 multi-phase clock signals for quadratic sampling. In order to further reduce power consumption, the low-voltage DCO and the quadratic sampling technique are exploited.

Figure 9.23 shows the circuit diagram of a single delay stage of the low-voltage DCO with 8 delay stages.

The PMOS differential input pair (M_1–M_2) is used to achieve lower flicker noise and large input capacitance. The delay stage takes a fully differential structure based on the switched NMOS capacitor arrays (M_5–M_8 and M_9–M_{12}) that are digitally controlled by FT[3:0] for frequency tuning [15]. To secure stable oscillation against the variation of the resistor loads (R_1–R_2), the cross-coupled pair (M_3–M_4) is adopted in parallel with M_1–M_2. The incremental timing delay T_d of the unit capacitance (1C), 25 fF in this design, should be smaller than a sampling time interval T_s for the stability of the capture behavior. For fine tuning of the oscillation frequency

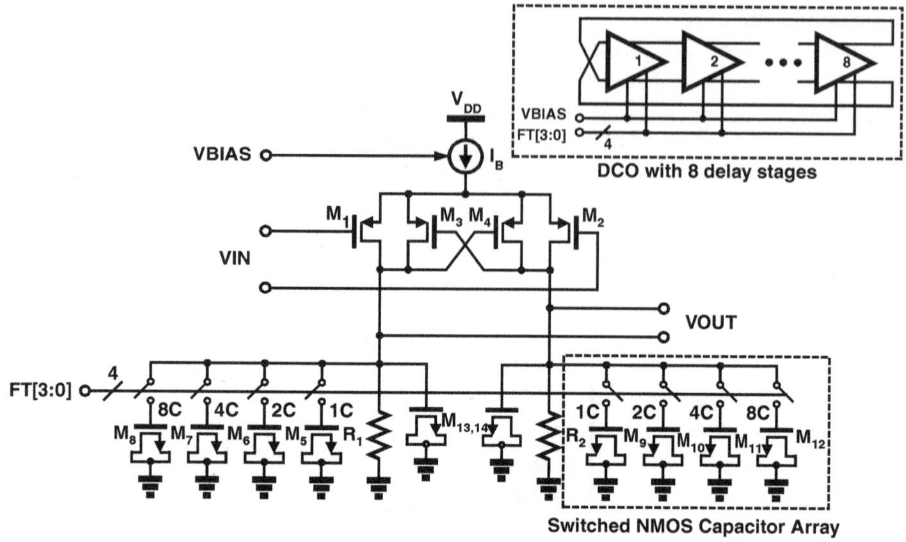

Fig. 9.23 Circuit diagram of a single delay stage of the low-voltage DCO

of the DCO, the VBIAS signal, generated by a voltage reference with off-chip trimming, controls the amount of the tail current (I_B). The bias capacitor pair ($M_{13}-M_{14}$) is needed to improve the linearity of the DCO gain characteristics. According to the simulations, the tuning range of the DCO shows 800 kHz with the tuning gain of 50 kHz/bit. The power consumption of the single delay stage is only 14 μW.

Figure 9.24 illustrates the proposed quadratic sampling algorithm and the characteristic curve of the QSPD.

The QSPD generates the 3-bit UP, LC, and 3-bit DN signals that are produced by XORing two consecutive bits of the DIN signal with the quadratic interval window. All of them are converted to 4-bit 2's complement values before weighting and averaging functions. To avoid potential loop instability, the averaged values is low-pass filtered by a simple digital filter. Then, the low-pass filtered values are used to generate the UDN and CCK signals. Based on this algorithm, the QSPD provides the quadratic gain over a bit interval window as shown in Fig. 9.24(b). With the help of the quadratic gain characteristics, the number of sampling clock signals can be reduced by a factor of two. Therefore, it can further reduce power consumption and the area overhead.

9.4.3 Direct Digital Transmitter

Figure 9.25 shows the detailed block diagram of the direct digital transmitter. The clock signal is generated from the ring oscillator and scaled by the frequency scaler. The ratio of the frequency scaling is chosen by the 7-bit thermometer decoder. The

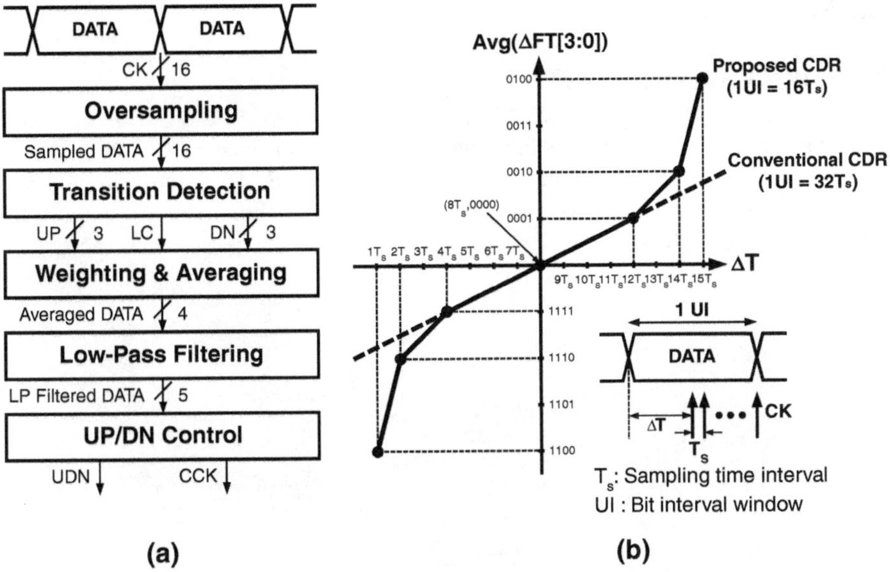

Fig. 9.24 Quadratic sampling technique (**a**) quadratic sampling algorithm (**b**) characteristic curve of the QSPD

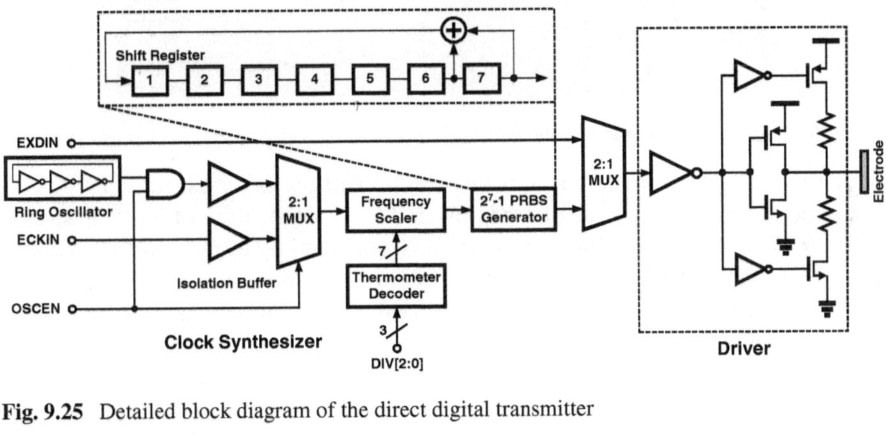

Fig. 9.25 Detailed block diagram of the direct digital transmitter

PRBS generator is based on the generation polynomial of $1 + x^6 + x^7$. The voltage-mode driver uses large-size transistors to drive the large human body capacitance and controls the output impedance with the digital signal. The transmitted power is 13 dBm for the output impedance of 50 Ω.

9.5 Measurement Results

9.5.1 WBS Receiver AFE

The proposed receiver AFE was fabricated with 0.18 μm standard CMOS technology, where the threshold voltage for NMOS and PMOS at saturation region are 0.58 V and −0.58 V, respectively. Figure 9.26 shows the chip microphotograph. Its active area is about 0.04 mm^2.

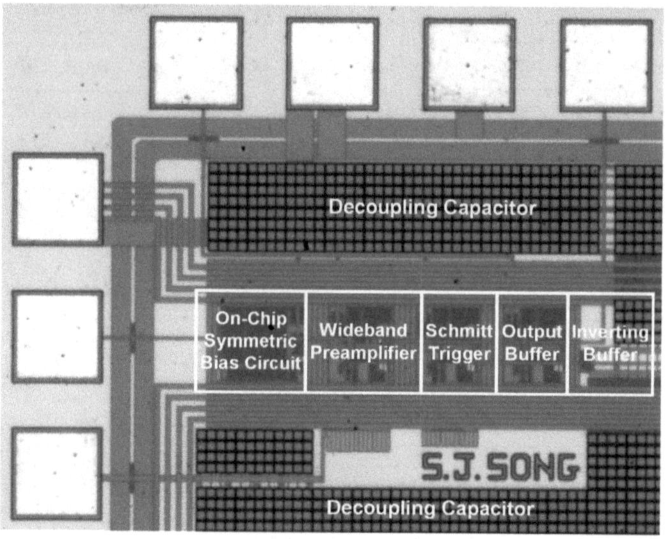

Fig. 9.26 Chip microphotograph of the AFE

The test chip was mounted on a FR-4 PC board by using chip-on-board. The prototype of the wrist-type test board powered by a coin battery of 3.0 V is shown in Fig. 9.27. The supply voltage of the AFE is generated by using an off-chip linear regulator on the board. The board size is about 3.5 cm by 4.5 cm. All measurements were conducted between the wrist and the fingertip that corresponds to the distance of about 25 cm. The electrode on the backside of the board is touched to the wrist. When the fingertip touches the electrode of a transmitter board, the binary data is transmitted through the body to the wrist-type test board. Figure 9.28 shows the measured timing diagram of the input and output signals of the AFE exploiting wideband symmetric triggering technique for 10 Mb/s $2^7 - 1$ PRBS transmitted data. The received input signal is measured to be about 50 mV amplitude. The preamplifier amplifies the received input signal with the voltage gain of about 30 dB before triggering at the following schmitt trigger. The recovered output exhibits almost 50% duty cycle due to the symmetric operation. The measured 3-dB frequency spectrum of the AFE is in the range of 1–200 MHz.

Table 9.4 summarizes the performance of the AFE. The AFE achieves the reception bit energy of 0.48 nJ/bit, which is 20 times more efficient than the UWB

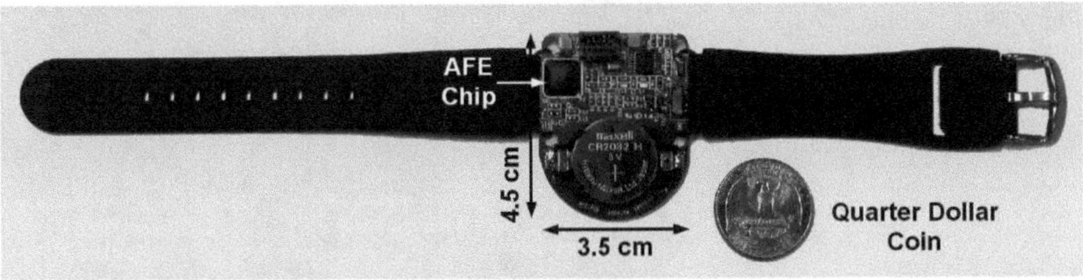

Fig. 9.27 Prototype of wrist-type test board

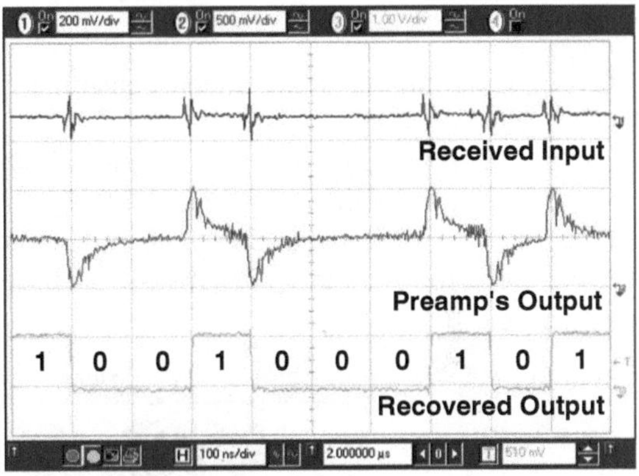

Fig. 9.28 Measured timing diagram of the AFE for 10 Mb/s $2^7 - 1$ PRBS

Table 9.4 Performance summary of the AFE

Data rate	10 Mb/s
Input impedance	50 Ω
3-dB bandwidth	$1 \sim 200$ MHz
Voltage gain	30 dB
Input sensitivity	−27 dBm
Power consumption	4.8 mW
Supply voltage	1.0 V
Core area	0.04 mm^2
Technology	0.18 μm standard CMOS

receiver in [4]. The achieved performance of the AFE is used for the HBC and show 10 Mb/s data transmission successfully. Therefore, the proposed AFE provides energy-efficient communications around the human body.

9.5.2 WBS Transceiver

The proposed WBS transceiver is fabricated with a 0.25 μm standard 1P4M CMOS technology, where the threshold voltage for NMOS and PMOS at saturation region are 0.6 V and −0.65 V, respectively. Figure 9.29 shows the microphotograph of the

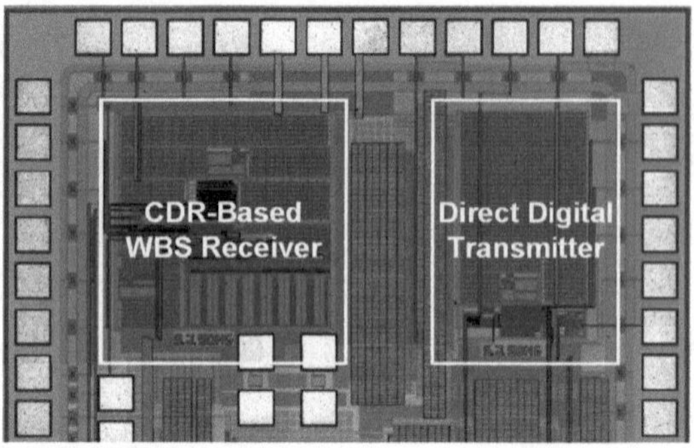

Fig. 9.29 Microphotograph of the test chip

test chip, including the direct digital transmitter and the CDR-based WBS receiver. Its core area is 0.85 mm^2.

A test board powered by a 1.5 V battery is shown in Fig. 9.30. The metal electrode on the test board is composed of a gold plate with the size of 5 mm × 7 mm. The Ag/AgCl electrode, 2 cm in diameter, is also connected to the board. The electrode for the signal transmission is chosen between the metal and Ag/AgCl

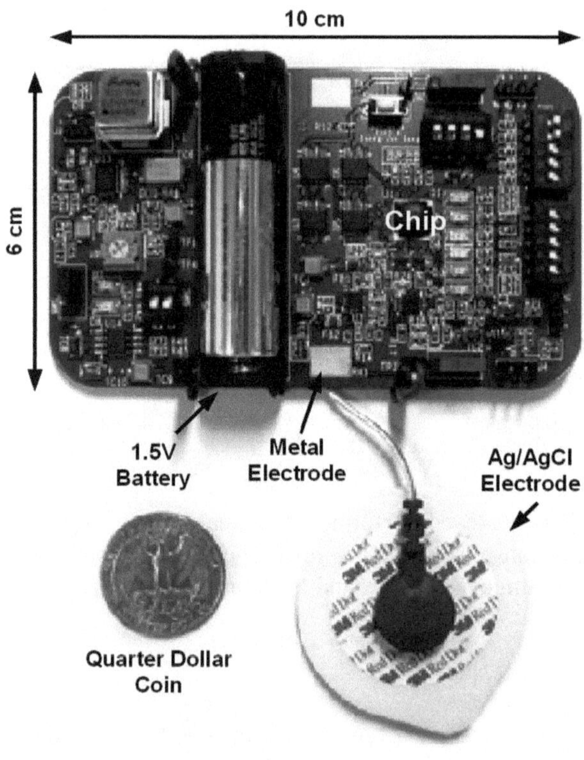

Fig. 9.30 Photograph of the test board

electrodes by using a slide switch on the board. The transceiver chip was mounted on the FR-4 PC board by using chip-on-board assembly. The board size is about 6 cm by 10 cm. All measurements are conducted between the wrist and the fingertip, which corresponds to the distance of about 25 cm as shown in Fig. 9.31. A WBS transmitter board is attached to the wrist using a single Ag/AgCl electrode on the backside. When the fingertip touches the metal electrode on a WBS receiver board, the stream of binary data is transferred through the body to the WBS receiver board.

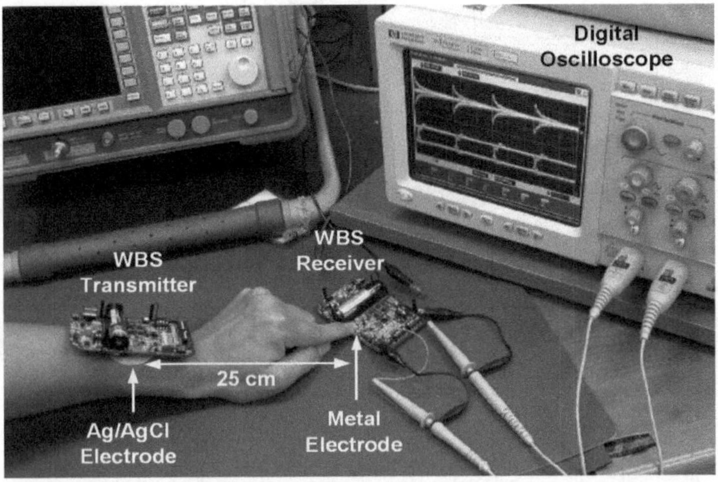

Fig. 9.31 Photograph of the test measurement setup

Figure 9.32(a) shows the measured eye diagrams of each output of the direct digital transmitter, the HBC channel, the receiver AFE, and the recovered clock and data of the CDR circuit for 2 Mb/s $2^7 - 1$ PRBS data, respectively. The channel output of the binary data exhibits the narrow small pulse signals with the amplitude of about 50 mV and no DC offset. The recovered clock jitter is measured to be 1.4 ns rms. Figure 9.33(b) shows the measured eye diagrams of each output of the wideband preamplifier, the schmitt trigger, and the inverting buffer at the WBS receiver AFE for 2-Mb/s $2^7 - 1$ PRBS data, respectively.

To prove the feasibility of the audio data transmission over a human body, the measurement for the receiver input power is conducted between a fingertip and an ear. The measured receiver input power as a function of the communication distance from the fingertip to the ear for 2 Mb/s $2^7 - 1$ PRBS data is shown in Fig. 9.33, indicating that the receiver input power rapidly decreases to the minimum input sensitivity of −27 dBm.

The measured 3 dB operational spectrum is in the range of 1–200 MHz. The BER is estimated to be less than 10^{-7} by the on-chip bit error detector by counting error bits for $2^7 - 1$ PRBS data. The transceiver chip including the receiver AFE consumes the total power of 5 mW from a 1 V supply.

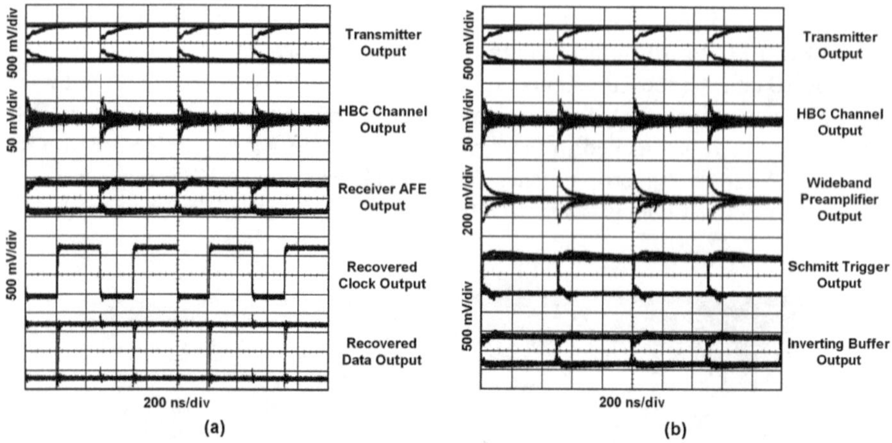

Fig. 9.32 Measured eye diagrams of (**a**) the WBS transceiver (**b**) the WBS receiver AFE for 2 Mb/s $2^7 - 1$ PRBS

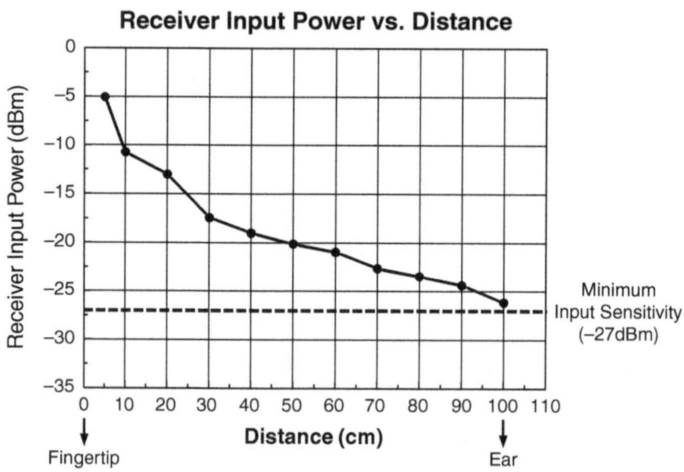

Fig. 9.33 Measured receiver input power as a function of the communication distance

Table 9.5 summarizes the performance of the proposed WBS transceiver. The performance comparison with previous works is summarized in Table 9.6. The WBS transceiver achieves the transmission bit energy of 2.5 nJ/bit, which is 26 times more efficient than other HBC transceivers introduced in [5–7]. Hence, the proposed WBS transceiver achieves lower power along with high data rate operation and is suitable for energy-efficient communications for BANs.

The proposed WBS transceiver chip with the DCI optimized for the *Bodywire* channel is used for MP3 music playing. The fabricated prototype of the HBC MP3 player comprises a HBC audio transmitter board connected to a MP3 player and a HBC audio receiver board into an earset. The receiver electrode is mounted on the

Table 9.5 Performance summary of the proposed WBS transceiver

Data rate		2 Mb/s
3-dB operational bandwidth		1–200 MHz
Receiver input impedance		50 Ω
Receiver input sensitivity		−27 dBm
Receiver voltage gain		30 dB
Recovered clock		2 MHz
DCO gain		50 kHz/bit
Clock Jitter for $2^7 - 1$ PRBS		1.4 ns rms
BER for $2^7 - 1$ PRBS		$<10^{-7}$
Power consumption	Transmitter	0.05 mW
	Receiver AFE	4.80 mW
	CDR circuit	0.15 mW
	Overall	5.0 mW
Supply voltage		1.0 V
Core area		0.85 mm^2
Technology		0.25 μm 1P4M CMOS

Table 9.6 Performance comparison with other HBC transceivers

	Zimmerman 1995 [5]	Hachisuka 2003 [6]	Shinagawa 2004 [7]	This Work
Communication method	Narrowband modulation	Narrowband modulation	Electrooptic conversion	Wideband signaling
Electrode (#)	Signal, Ground (2 EA)	Signal, Ground (2 EA)	Signal, Ground (2 EA)	Signal (1 EA)
Modulation	OOK/DSSS	FM/FSK	No	No
Carrier Frequency	330 kHz	10.7 MHz	0–10 MHz	1–200 MHz
Data rate	2.4 kb/s	9.6 kb/s	10 Mb/s	2 Mb/s
Supply voltage	9 V	3 V	5 V	1 V
Power consumption	400 mW	Not reported	650 mW	5 mW
Bit energy	170 mJ/bit	N/A	65 nJ/bit	2.5 nJ/bit

hanger of the earset. Real-time MP3 music playing through the human body without use of any earphone wire is successfully demonstrated by the prototype transceiver chip. The detailed explanation of its demonstration will be described in Chapter 9.6.

9.6 System Operation Demonstration

9.6.1 Introduction

Wearable computing is emerging as a key technology that enables users to exchange multimedia information such as high quality audio and to enjoy them anywhere and anytime in ubiquitous computing environments [16]. This Chapter describes a prototype for the digital audio streaming through the wearer's body by exploiting the transceiver chip of [17]. Figure 9.34 shows an example for audio music listening by only touching the prototype system without any wire. When the finger touches the electrode on the backside of the audio player, the audio signal goes through the skin of the body to the earset on the user's ear. It can remove earphone wire and enjoy audio music comfortably and continuously. Therefore, this work shows possibility for touch-based digital audio streaming services in ubiquitous computing environments.

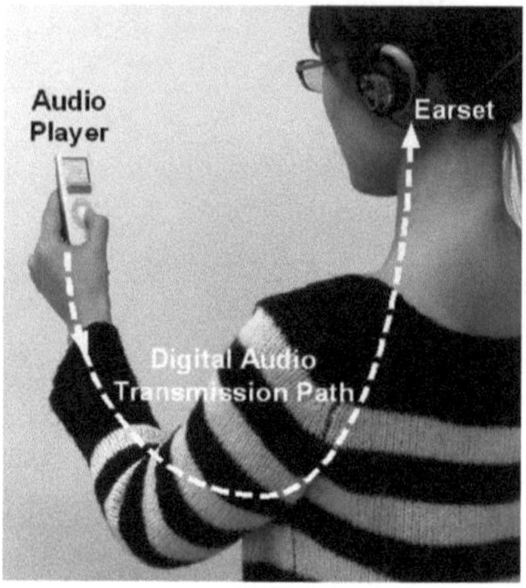

Fig. 9.34 Photo of real-time audio playing

9.6.2 Related Works

There are three approaches for wearable digital audio streaming:Bluetooth, smart textiles, human body communication (HBC). The Bluetooth transceiver [18] operates at 2.4 GHz ISM band that is very congested and prone to interference, and also consumes a huge power. Recently, a wearable digital audio player system integrated into fashionable clothing was presented and all components are connected via the conductive textiles, but it needs additional manufacturing processes. Meanwhile, touch-based HBC schemes introduced in [6] and [7] face the difficulty to apply for wearable audio player applications due to limited low data rate or bulky size with high power consumption.

9.6.3 Design Architecture

The design architecture for our prototype system is illustrated in Fig. 9.35, which exploits wideband signaling (WBS) for the digital audio streaming over the human body.

Its architecture consists of five parts: physical medium, interface layer, signaling link layer, transceiver module, and application. The human body as the physical

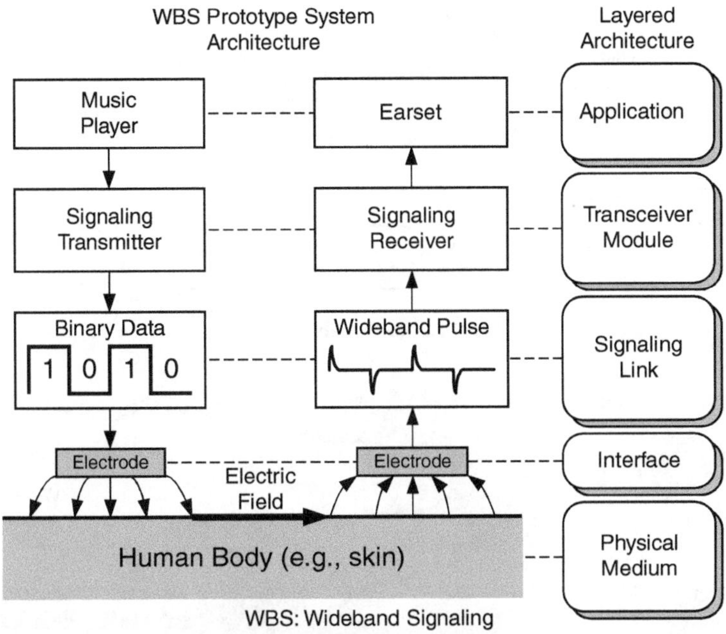

Fig. 9.35 Design architecture for the digital audio streaming based on WBS

medium shows a wide band-pass operational spectrum and sustains the channel capacity sufficient to transfer real-time multimedia data streams as investigated in [17]. The electrical signal can be flowed by the feeble electric field induced on the skin of the body. The interface layer provides the connection with the physical medium. After the connection is established, the digital audio data transfers to the wearable I/O devices over a WBS link. Transceiver module performs the packet generation of the streaming data and supports the WBS link that enables a point-to-point transfer or broadcasting between each application layer.

9.6.4 Realization

The prototype system is realized by using the WBS transceiver chip developed of [17]. The system hardware excluding the transceiver chip consists of off-chip components: 24-bit sigma-delta analog-to-digital converter (ADC) and digital-to-analog converter (DAC), digital audio interface transmitter and receiver as shown in Fig. 9.36(a). The S/PDIF standard is chosen for the digital audio interface. The analog audio signals for two stereo channels are sampled by the 24-bit sigma-delta ADC. The 24-bit sampled audio data is converted to a 32-bit packet consisting of a 4-bit preamble, the 24-bit data, and a 4-bit channel status and error detection bits. Subsequently, the data is transmitted with biphase-mark encoding at twice of the bit

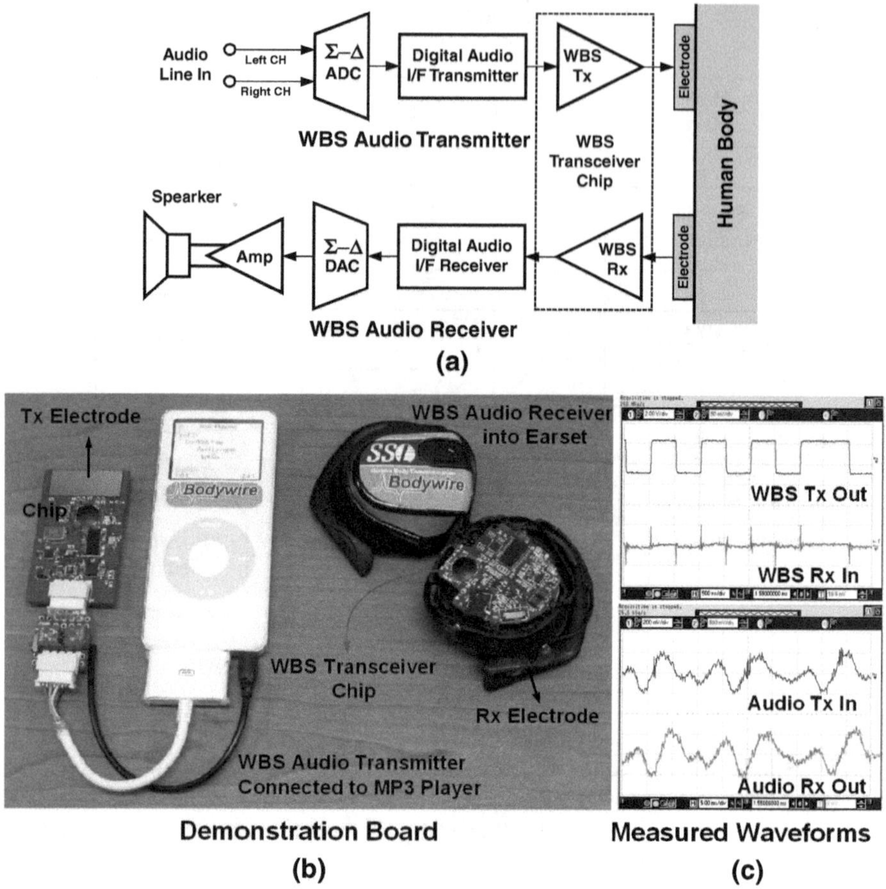

Fig. 9.36 (**a**) Block diagram, (**b**) prototype, and (**c**) waveforms of the WBS audio player system

rate, which allows clock recovery from the data at the receiver. The effective data rate is 2.048-Mb/s at the sampling rate of 16-kHz over the channel. The audio sound is played back by the 24-bit sigma-delta DAC. No audio decoder is required at the earset. Thus, it leads to the reduction of the cost and the power consumption. The transceiver chip controls the WBS link for the transmission of digital audio data. The 0.25-μm CMOS WBS transceiver chip including a receiver AFE consumes only 5-mW from a 1-V supply. Compared with a Bluetooth transceiver [18], it exhibits lower power consumption at high data rate. Figure 9.36(b) shows the photo of the prototype system consisting of a WBS audio transmitter connected to a MP3 player and a WBS audio receiver assembled into an earset. The transmitter attaches to the back of the audio player. The Rx electrode is mounted on the hanger of the earset. Measured waveforms are shown in Fig. 9.36(c).

9.6.4.1 Summary

A wearable real-time audio player system is implemented and demonstrated to show the feasibility of the transmission of the multimedia data through the human body. The system supports the WBS link controlled by the developed 0.25-μm CMOS WBS transceiver chip. It achieves low cost, and low power dissipation along with high data rate operation, compared with fabric or short-range RF solutions.

9.7 Conclusion

A novel HBC scheme is proposed and implemented for energy-efficient communications using the human body as a data transmission medium. From the investigation of the HBC channel using the DCI, the human body behaves as a band-pass filter with a bandwidth of 100-MHz and the channel output exhibits the narrow small pulse signals with a width of 8-ns, which is identified as the *Bodywire* channel. The proposed WBS communication link model provides the simplicity for the extraction of the design parameters and the theoretical link performance. The proposed HBC scheme exploits WBS technique with the DCI over the *Bodywire* channel optimized for high speed operation. For the convenience of usage, the DCI uses only a single electrode without ground return path for data transmission. With the help of four low power techniques such as low power op amp, all-digital CDR architecture, low-voltage DCO, and quadratic sampling technique, the WBS transceiver achieves the 2-Mb/s operation with the power consumption of 5-mW from a 1-V supply, resulting in the transmission bit energy of 2.5-nJ/bit. Real-time MP3 audio data transmission through the human body is successfully demonstrated by using the fabricated 0.25-μm CMOS WBS transceiver chip on the prototype MP3 player.

References

1. Kim S, Lee J-Y, Song S-J, Cho N, Yoo H-J (2006) An Energy-efficient analog front-end circuit for a sub-1-V digital hearing chip. IEEE J Solid State Circuits 41:876–882, Apr 2006
2. Ashok RL, Agrawal DP (2003) Next-generation wearable networks. IEEE Computer Mag 31–39, Nov 2003
3. Verma S, Xu J, Hamada M, Lee TH (2005) A 17-mW 0.66-mm^2 Direct-conversion receiver for 1-Mb/s cable replacement. IEEE J Solid State Circuits 40:2547–2554, Dec 2005
4. Ryckaert J, Badaroglu M, Heyn VD, Plas DGV, Nuzzo P, Baschirotto A, S. D'Amico, Desset C, Suys H, Libois M, Poucke BV, Wambacq P, Gyselinckx B (2006) A 16 mA UWB 3-to-5 GHz 20 Mpulses/s Quadrature analog correlation receiver in 0.18 μm CMOS.ISSCC Dig Tech Papers, 114–115, Feb 2006
5. Zimmerman TG (1995) Personal area networks (PAN): near-field intra-body communication. M.S. Thesis, MIT, Sept 1995
6. Hachisuka K, Nakata A, Takeda T, Terauchi Y, Shiba K, Sasaki K, Hosaka H, Itao K (2003) Development and performance analysis of an intra-body communication device. transducers '03, 1722–1725, June 2003
7. Shinagawa M, Fukumoto M, Ochiai K, Kyuragi H (2004) A near-field-sensing transceiver for intrabody communication based on the electrooptic effect. IEEE Trans Instrum Meas 53:1533–1538, Dec 2004

8. Ruiz JA, Xu J, Shimamoto S (2006) Propagation characteristics of intra-body communications for body area networks. IEEE Consumer Commun Network Conf Jan. 2006, pp 509–513
9. Ward LC, Byrne NM, Rutter K, Hennoste L, Hills AP, Cornish BH, Thomas BJ (1997) Reliability of multiple frequency bioelectrical impedance analysis: an intermachine comparison. Amer J Hum Biol 9:63–72
10. International Commission on Non-Ionizing Radiation Protection (ICNIRP) Guidelines for limiting exposure to time-varying electric, magnetic, and electromagnetic fields (up to 300 GHz). Health Phys 74:494–522
11. IEEE Standard for safety levels with respect to human exposure to radio frequency electromagnetic fields, 3 Hz to 300 GHz, IEEE C95.1-1991, (1999 ed.).
12. Neuteboom H, Kup JBM, Janssens M (1997) A DSP-based hearing instrument IC. IEEE J Solid State Circuits 32:1790–1806, Nov 1997
13. Luo L, Wilson JM, Mick SE, Xu J, Zhang L, Franzon PD (2006) 3 Gb/s AC coupled chip-to-chip communication using a low swing pulse receiver. J Solid State Circuits 41:287–296, Jan 2006
14. Song S-J, Park SM, Yoo H-J (2003) A 4-Gb/s CMOS clock and data recovery circuit using 1/8-rate clock technique. IEEE J Solid State Circuits 38:1213–1219, July 2003
15. Anand SB, Razavi B A 2.75 Gb/s CMOS clock recovery circuit with broad capture range. IEEE Int. Solid-State Circuits Conf. Dig. Tech. Papers, Feb 2001 pp 214–215
16. Billinghurst M, Starner T (1999) Wearable devices: new ways to manage information. IEEE Computer 32:57–64, Jan 1999
17. Song S-J, Cho N, Kim S, Yoo J, Yoo H-J (2006) A 2 Mb/s wideband pulse transceiver with direct-coupled interface for human body communications. IEEE ISSCC Dig Tech Papers, 558–559, Feb. 2006
18. Cojocaru C (2005) Low power bluetooth for headset applications.IDEAS'05, 223–226, July 2005

Part III
Examples of Bio-Medical ICs

Chapter 10
Wearable Healthcare System

Jerald Yoo and Hoi-Jun Yoo

10.1 Introduction

With the world aging, chronic diseases are becoming the major causes of death around the world. For example, according to US National Center for Health Statistics, major chronic disease such as heart disease, cerebrovascular disease, and diabetes mellitus account for 35.6% of death in US in the year 2005 [1] (Fig. 10.1). An important perspective of the chronic diseases is that the sooner they are detected, the better prognoses the patients will have. However, asymptomatic or intermittent properties of many chronic problems lead to difficulties: therefore, long-term continuous health monitoring is essential in detecting and treating with the diseases [2, 3].

Unfortunately, current chronic disease monitoring method suffers from a fundamental issue: monitoring vital signal at specific time may not accurately reflect the patients' actual status.

Figure 10.2 is the graph showing the blood glucose level of a patient with diabetes mellitus during a day. The dotted line is the normal band of the blood glucose level. If the patient samples the glucose level at point A or B, he/she will be considered as normal, which is not true. The problem is that the most of the chronic diseases have exactly the same situation. A good case in example is arrhythmia: abnormal electrical activity in the heart that may result in life-threatening medical emergencies. Many times, the irregular heartbeat comes without any notice. Therefore, it is essential to *continuously monitor* health to detect and treat with the chronic diseases. Also, continuous health monitoring during normal life will satisfy the National Institute of Health (NIH)'s 4Ps' of future medicine: Predictive, Personalized, Preemptive and Participatory [4, 5].

Body Sensor Network (BSN), or wearable network is a good candidate to form a continuous health monitoring system [6]. This chapter will cover pros and cons

J. Yoo (✉)
Microsystems Engineering, Masdar Institute of Science and Technology (MIST), Abu Dhabi, United Arab Emirates
e-mail: jyoo@masdar.ac.ae

H.-J. Yoo, C. van Hoof (eds.), *Bio-Medical CMOS ICs*,
Integrated Circuits and Systems, DOI 10.1007/978-1-4419-6597-4_10,

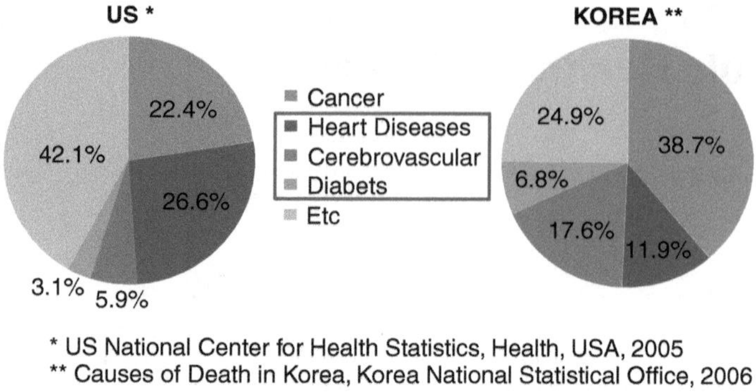

Fig. 10.1 Causes of death in US (2005) and in Korea (2006)

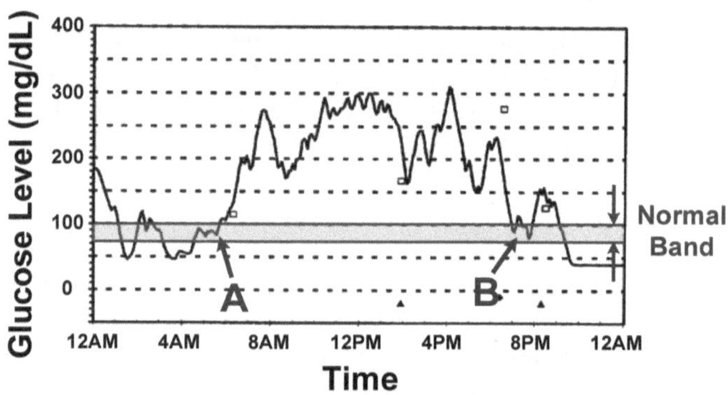

Fig. 10.2 Blood glucose level of a diabetes patient during a day

of BSN for health monitoring system and provide a prototype continuous health monitoring system using the BSN: *a wearable healthcare system.*

10.1.1 Issues on Continuous Wearable Healthcare Using BSNs

Continuous health monitoring requires several system issues: reliability, convenience, and safety. First of all, since it is related to health, reliability is a very important factor. Data loss and distortion is not allowed. It must be tolerant to physical and logical faults. Secondly, convenience is another issue; continuous monitoring is conducted over 24 hours, and the system should be easy to use and as pervasive as possible. Lastly, it must be safe to use.

There are two approaches to form a BSN; one uses wireless communication, and the other uses wireline. For wireless technologies, ZigBee, Bluetooth, and Wireless LAN (WLAN) are used. However, these wireless-based technologies suffer from interference, fading, and low data rates [7]. In addition, they are not suitable for healthcare applications where inelastic data transmission must be guaranteed for "always reliable" communication. Especially, ZigBee and Bluetooth are vulnerable to interferences at 2.4 GHz [8], because they share the same frequency range with WLAN, so that they have difficulty in communication under WLAN environments [9, 10]. In addition, Bluetooth is influenced significantly by requirement to establish the line-of-sight (LoS) [11]; when reflection is absent such as in open area, the LoS requirement becomes more stringent, because human body plays as an obstacle for 2.4 GHz signals. Therefore, considering the "always reliable" requirements of healthcare applications, ZigBee and Bluetooth are not sufficient.

As an alternative to the wireless approaches, wireline-based technologies are introduced to meet the BSN requirements. When it comes to sensor interface, low noise operation guarantees stable data to be captured. Also, low power operation is necessary to withstand long-term monitoring.

10.1.2 Snapshots of Previous Works in Health Monitoring

There are several methods to continuously monitor health in everyday life (Fig. 10.3). Holter monitor system records electrocardiogram (ECG) data on 24-hour basis, and currently this is the most powerful method to detect irregular heart activities during everyday life. However, typical Holter monitor has 6 to 10 wires that connect the electrodes to the recording system; the wires will be worn around the body all day, and it is uncomfortable and inconvenient for patients. Moreover, for chronic use, conventional Ag/AgCl wet electrodes may stimulate skin, and signal quality degrades as the electrolyte gel dehydrates [14]. Therefore, it is unsuitable for long term, continuous health monitoring. Meanwhile, [15] proposed a novel wearable healthcare system based on knitted integrated sensors. With the help of conductive and piezoresistive yarns, it successfully presented the feasibility of fabric sensors to capture biomedical signals such as ECG, respiration and activity. Nevertheless, sophisticated knitting and interconnection of conductive yarn result in relatively high production and maintenance cost.

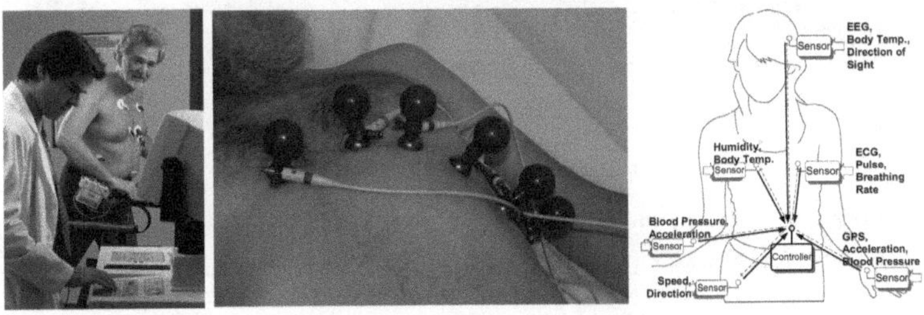

Fig. 10.3 Conventional Body Sensor Network for health monitoring

Another good approach is the Digital plaster for Body Area Network (BAN) by *Toumaz Technology* [16]. The small patch includes an ECG monitor, a battery, a processor and an ISM band wireless transceiver that will transmit captured data to central base-station. This smart approach meets the convenience requirement of the continuous health monitoring: the patient can monitor the ECG data, and throw the sensor away afterwards. *Toumaz* developed the Sensium^{TM}, a System-on-chip solution for digital plaster, which integrates custom sensor interfaces, a 10-bit ADC, a 8051 micro-processor core, a RF transceiver and a full-custom MAC. However, security and interference-resilient requirements are stringent for health monitoring, and therefore, using open-air wireless ISM band is not suitable for health monitoring body sensor network (BSN). Also, and most importantly, some people are reluctant to patch batteries directly on their skin all day.

IMEC has been actively working in the field of ambulatory health monitoring. In 2008, the first generation wireless ECG patch was reported [17] (Fig. 10.4). It exploits a proprietary single channel ASIC for biopotential read-out [18]. The ASIC consists of a AC coupled chopped instrumentation amplifier, a spike filter, and amplification stage (constant gain), and a variable gain amplifier (VGA). The ECG patch deals with an 1-channel ECG data to a receiver within 10m range at 1.17 mW (data streaming mode) / 1.74 mW (local data processing mode).

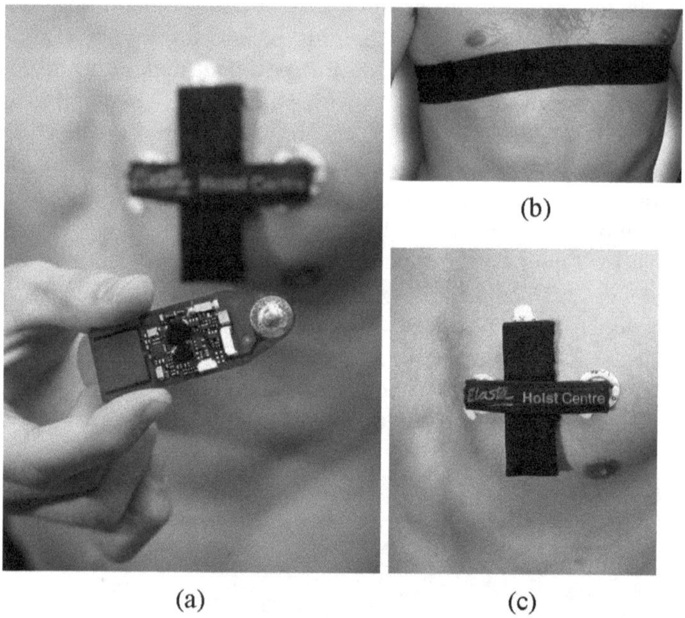

Fig. 10.4 IMEC's wireless ECG patch: (**a**) wireless ECG patch integrated on flexible substrate; (**b**) chest belt package; (**c**) textile pocket package

When it comes to BSN, several works have shown the concept and implementations of wearable BSN. A novel concept of Fabric Area Network (FAN) was introduced in [19], but the size of antenna was bulky and the data rate (<125 kbps) was low. A breakthrough fault-tolerant wearable network was shown in [12, 13], but the work was limited to intra-layer (within a layer) network, and there was no chip implementation to integrate on. Moreover, [12, 13] used torus topology, which requires continuous monitoring of link faults even if there is no data transaction, resulting in waste of energy.

10.1.3 An Example Wearable Healthcare System

This chapter focuses on exploring and designing a wearable healthcare system using a self-configured wearable BSN controller SoC and efficient wirelessly powered adhesive bandage sensors.

The self-configured BSN controller SoC automatically configures and provides power to sensors attached at the arbitrary locations on chest with low power consumption [22, 23]. Wirelessly powered, adhesive bandage-like sensor that continuously monitors ECG and other vital signals is also designed. Fabric circuit board technology enables making adhesive bandage-like sensors. Dry electrode is implemented using fabric circuit board, and its characteristics are analyzed, and compared with conventional wet type electrodes. A novel rectifier is designed to achieve high efficiency power recovery at low cost, and it is compared with conventional rectifiers. Fabric dry electrode and batteryless configuration enables long-term vital signal monitoring with the small patch sensor. These design methods lead to an energy-efficient solution to realize continuous health monitoring system using wearable BSN.

10.2 Reliable and Low Cost BSN for Wearable Healthcare

10.2.1 Self-Configured Wearable BSN

An approach to implement a wearable healthcare system with BSN is the dynamic reconfigurable method [22, 23]. Sensor can be deployed as attachable bandage type, and the array based dynamic reconfigurable BSN is proposed (Fig. 10.5).

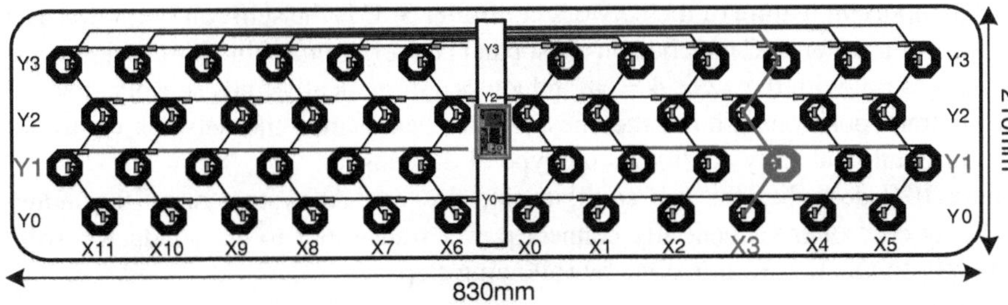

Fig. 10.5 Arrayed health monitoring band structure

The proposed health monitoring chest band is composed of a network controller SoC and an inductor array. Figure 10.6 shows the architecture of the proposed network controller for array based dynamic reconfigurable BSN. The main role of

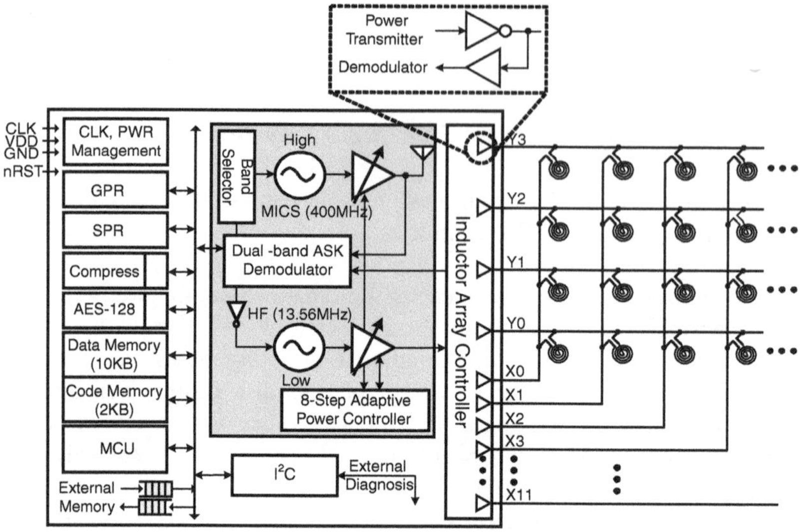

Fig. 10.6 Architecture of the proposed network controller

the network controller is to automatically configure the sensor location around the body, provide power to the configured sensors, and finally, collect the sensor data to storage. It is composed of an inductor array controller, a dual-band power transmitter with an 8-step adaptive power controller, an ASK demodulator, a MCU, a compression block, an AES–128 accelerator, SRAM, and an I^2C interface. It can communicate with up to 48 sensors. In every programmed period (0.2 to 24 h), the array controller scans the inductor array for self-configuration. According to the signals from the sensors, the network controller automatically selects the HF (for patch-type sensors such as ECG, temperature, skin conductivity, etc) or MICS (for implantable sensors) band. Data rate can be varied from 10 kb/s to 120 kb/s, depending on the sensor types.

An important feature of the network controller SoC is the self-configuration. The network controller SoC attached on chest band is worn over a subject's chest, and the inductor array with the $12 \times 4 = 48$ inductors automatically finds sensors attached on arbitrary positions around the chest. At the same time, the network controller SoC also automatically configures the type of the sensors.

Figure 10.7 describes the self-configuration process. At system reset, the inductor array controller sequentially connect power transmitter to each inductor, from (X0, Y0) to (X11, Y3). It is done by following steps

Step 1) Scan begins, start from (X0, Y0). Check the current Y position. If this is the first iteration, or the current Y value is smaller than Y3, go to Step 2. If the current value is Y3, then move to Step 5.

Step 2) Provide power to the current position. It is done by connecting the power transmitter to the inductor in current position. An inductor is "selected"

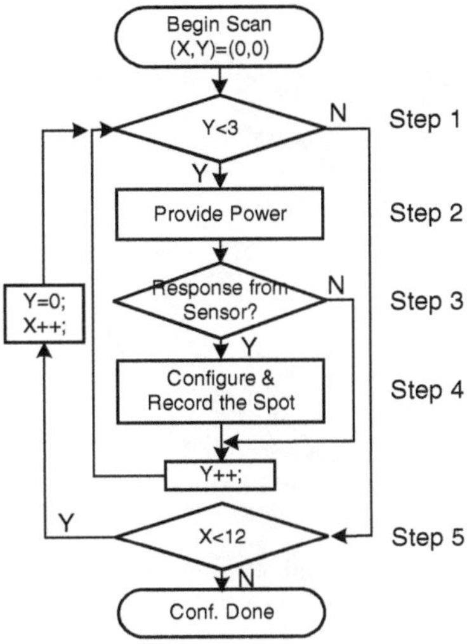

Fig. 10.7 Self-configuration process

when a X address and a Y address are connected to the power transmitter. Then the power transmitter drives the connected inductor; for example, if (X2, Y3) links are activated (as shown in bold line in Fig. 10.20), carrier signal is injected through the inductor in (X2, Y3) position.

Step 3) Wait for a response from sensor. If a sensor exists beneath the current position, it will generate power and respond with an acknowledgment packet (ACK) including the sensor information; in this case, the network controller immediately moves on to Step 4. If there is no response, the network controller waits for 168 ms for time expiry before moving on to next Step 4. In case more than an inductor detects a sensor, first-come, first-served fashion is applied: The first inductor that detects the sensor takes the control of the sensor.

Step 4) The network controller configures the current position that there is a sensor existing beneath the current position. Increase the Y position by 1 and return to Step 1.

Step 5) Check if the current X position. If it is smaller than X11, then go to Step 1. If the current position is equal to X11, then the configuration is done.

Figure 10.8 shows an example of the self-configuration process at (X3, Y1) position of the health monitoring band. At Step 1, the inductor at (X3, Y1) position in the health monitoring band is selected, and Step 2, the carrier signal is provided through the inductor. For Step 3, the adhesive bandage sensor placed underneath the inductor is activated, and responds with ACK packet via ASK load modulation. In Step 4, the backscattered information is demodulated and the network controller configures the position. Lastly, in Step 5, the network controller moves on to the next position (X3, Y2).

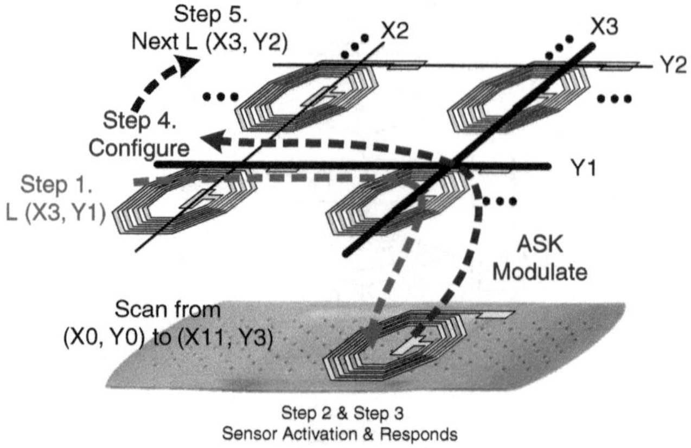

Fig. 10.8 An example of the self-configuration: scanning (X3, Y1) position

The network controller is powered by a single 3.6 V, 1000 mAh, 1/2–AA size battery; it is then regulated down to 1.8 V as a supply. With 5.2 mW measured minimum power consumption of the network controller, this translates into 8 days of continuous ECG monitoring without replacing the battery. This is sufficient for 7-day ambulatory monitoring that is essential for tracking heart disease [30].

Sensor signal and the configuration information (time, position, type of sensors) are compressed by quadratic level compression algorithm with average compression ratio of 8.4:1 to reduce the data memory size [31]. To protect privacy of the personal health information, the sensor data are encrypted by AES-128 accelerator [32] before stored into data memory. Through high-speed mode I^2C diagnosis interface, a healthcare expert can access the collected data for healthcare assistance.

10.2.2 Adaptive Power Transmission

Received power level at sensors may vary over time due to misalignment between sensors and power transmitter, or convexity of skin that may distort the shape of printed inductors. The adaptive power controller can control the transmitting power signal strength in real time to compensate the fluctuation in received power at sensors, based on signal strength information already programmed into data packet previously by the sensor nodes. Figure 10.9 shows (a) the adaptive power transmission scheme and (b) the 8-level power transmitter. At the beginning, highest power is provided to the sensor. As the sensor detects generated voltage level, it marks the level into the packet header; the network controller then determines an appropriate power level. That is, the adaptive power transmission prevents excessive power transmission or power starvation at sensor. This is done in real-time.

Figure 10.10 shows the adaptive power transmission simulation result. Misalignment between TX and RX inductors degrades bit error rate (BER) considerably. As adaptive power control is applied, BER performance is improved.

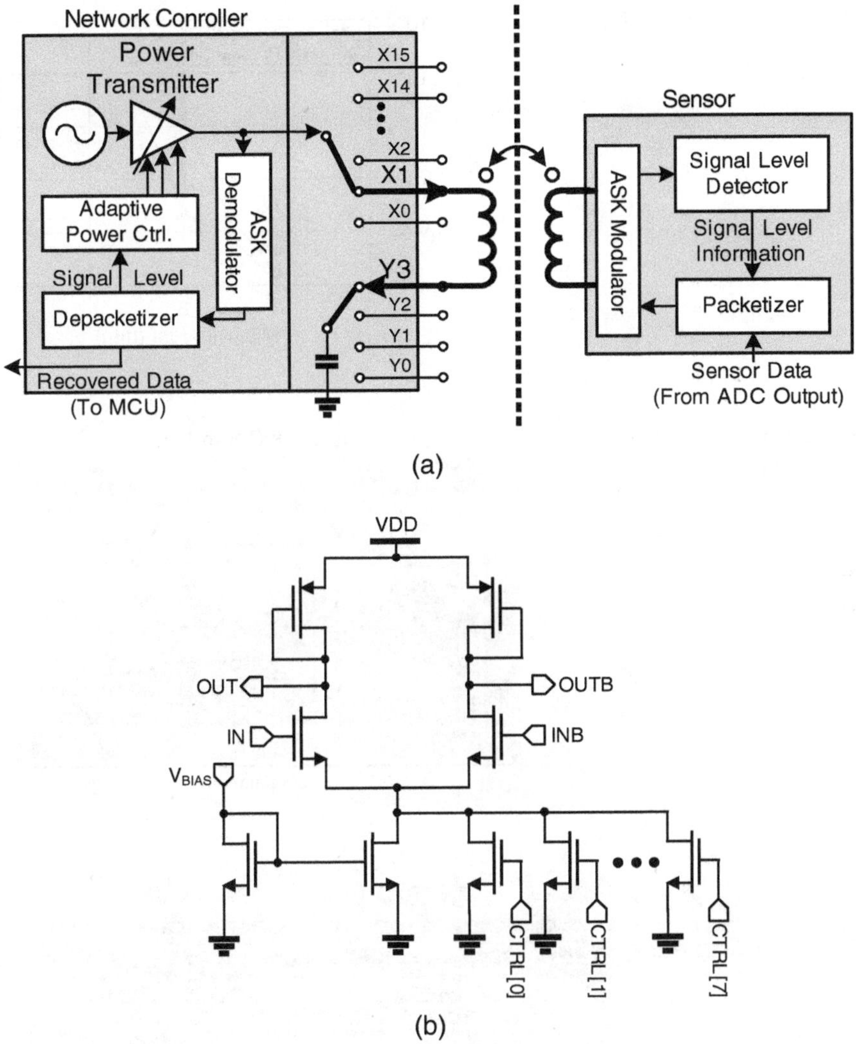

Fig. 10.9 Adaptive power transmission

10.2.3 Network Controller SoC

The self-configured wearable BSN is demonstrated with a wearable network controller BSN SoC implemented in 0.18 μm 1P6 M standard CMOS technology. It occupies 15.0 mm^2 including pads. Figure 10.11 shows the chip micrograph of the self-configured wearable BSN controller SoC. Inductor array controller, VCO, CDR, and power transmitter are full-custom designed, and MCU and SRAM are synthesized.

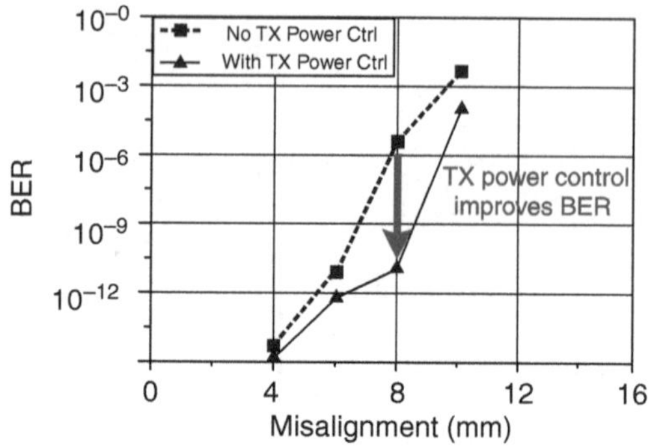

Fig. 10.10 Adaptive power transmission

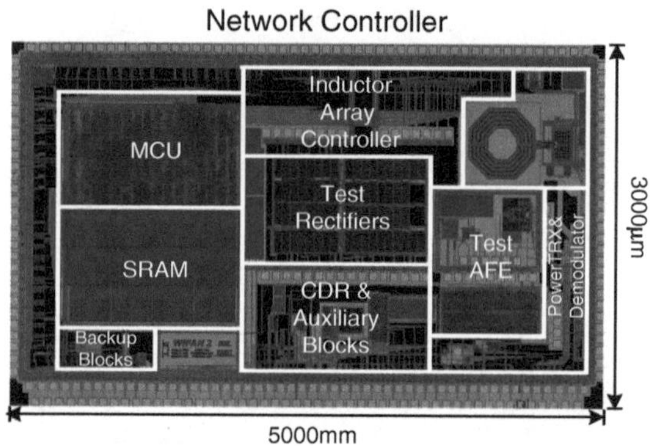

Fig. 10.11 Chip micrograph of the self-configured wearable BSN controller SoC

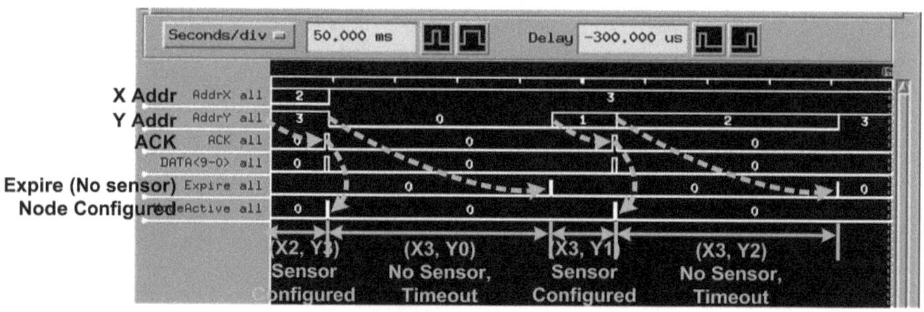

Fig. 10.12 Measured self-configuration operation

Measurement results of Fig. 10.12 show that the network controller scans (X2, Y3), (X3, Y0), (X3, Y1) and (X3, Y2) positions sequentially. In this specific time window during the self-configuration phase, (X3, Y0) and (X3, Y2) do not reply until timeout (168 ms), whereas (X2, Y3) and (X3, Y1) respond, and are successfully configured.

10.2.4 Summary

In this section, we have investigated a self-configured wearable BSN. The self-configured wearable BSN controller SoC is made up of a network controller and wirelessly powered adhesive bandage-like sensors. The sensors are attached to arbitrary positions around a patient's body; the network controller automatically locates, provides power to, and collects vital signal from the sensors. The network controller SoC controls a 12×4 inductor array; 8-level power transmission scheme adapts to sensor power level in real time to minimize power consumption and to prevent power starvation at sensor. The controller SoC also includes digital backend (an MCU, data and code memory, encryption engine, compression engine) to process recorded vital signal on chip.

10.3 Fabric Circuit Board

10.3.1 Introduction

In this section, a fabric circuit board technology, namely the Planar-Fashionable Circuit Board (P-FCB), is introduced. Also, dry P-FCB electrode characteristic is described in detail.

P-FCB is the planar printed circuit technology on fabric. Silver paste is directly screen printed on fabric to form patterned electrodes, passive elements or circuit boards [32, 33]. Figure 10.13(a) shows the SEM photograph of the P-FCB cross section (1500× magnified), and Fig. 10.13(b) shows the surface of the P-FCB (100× magnified). Silver paste is applied at around 10–30 μm thick, with down to 200 μm resolution. Mechanical strength of the P-FCB is tested through washing tests. After 50+ times, change in its electrical characteristics is negligible. Chip can be directly wire bonded and molded on fabric as well (Fig. 10.13(c)). Therefore, patch sensors can be easily made at low cost, and wearability and flexibility are improved at once.

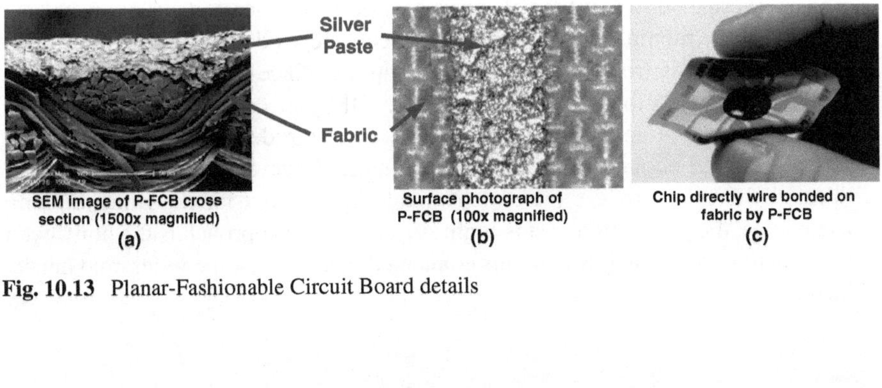

Fig. 10.13 Planar-Fashionable Circuit Board details

10.3.2 Dry Electrodes by P-FCB

Ag/AgCl electrodes with electrolyte gel (Fig. 10.14(a)) are commonly used as biopotential electrodes in ambulatory vital signal monitoring, such as electrocardiogram (ECG), electroencephalogram (EEG), electrooculogram (EOG), and electromyogram (EMG) [34].

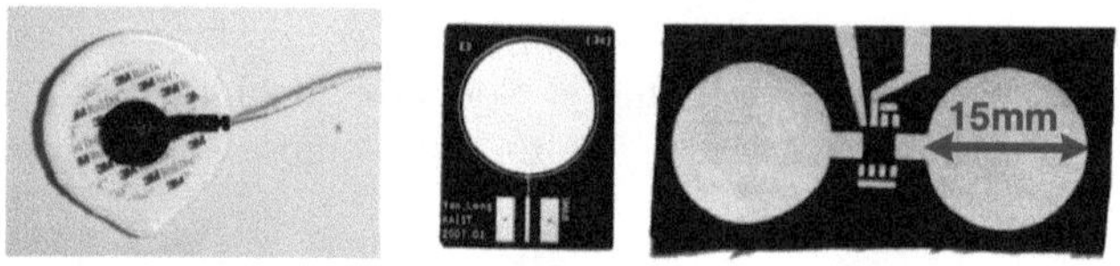

(a) (b) (c)

Fig. 10.14 The biopotential electrodes: (**a**) Ag/AgCL (wet), (**b**) PCB (dry), and (**c**) FCB (dry)

The electrolyte gel forms a conductive path between skin and electrode, and stable reception of signal is obtained. However, for chronic use, the wet type Ag/AgCl electrode suffers from signal quality degradation as the electrolyte gel dehydrates. Moreover, there are toxicological concerns regarding electrolyte gel: for example, skin irritation [14, 34]. Therefore, in the proposed system, a dry electrode made out of fabric is used to ensure continuous monitoring without any harm. There are three reasons why the wet type is inappropriate for ambulatory health monitoring over several days

1) *Safety.* Toxicological concerns exist regarding the electrolyte gel within the Ag/AgCl electrodes. [35] indicates the gel ingredients may cause dermatitis.
2) *Signal degradation.* Electrolyte gel of a wet electrode dries out over time. Consequently, the skin-electrode impedance tends to increase, and signal quality degrades. Re-applying the gel may mitigate the problem, but it is time and effort consuming.
3) *Inconvenience.* Applying and removing the gel type electrode is unpleasant for both the patient and the clinician. This activity results in material cost and time implications.

As an alternative to wet type electrodes, various dry electrodes were proposed for long term monitoring [36–38]. The metal electrode formed on PCB [36] (Fig. 10.14 (b)) is free from skin stimulation, but since PCB is stiff, its wearability is low. The Biopotential Fiber Sensor (BFS) successfully solves stiffness by adopting metal-coated fabric strands as an electrode [37]; the performance of BFS is good enough to capture vital signals. However, because of the complicated thin-film process (electrochemical deposition with the rapid dipping technique), the production cost is high. Another smart approach is the nonwoven fabric active electrode [38], but it is connected by cumbersome wires, and hence, inconvenient to use.

P-FCB electrode solves the above mentioned problems at once: since the electrode is a dry type, it is safe and free from skin irritation. Figure 10.14(c) shows the proposed P-FCB electrodes. Due to flexibility of the fabric in nature, it is attached on well with respect to convexity of skin, and effective contact area is increased. As a result, the signal quality is better than that of [36]. In addition, the thick-film printing does not concern any complicated process, so the production cost can be lowered when compared with [37].

10.3.2.1 Electrode Impedance

Of course, there is also a drawback in using dry electrodes. The skin-electrode contact impedance is an important factor in electrodes, and if the impedance is too large, loading effect degrades the signal strength that will be fed into the amplifier. Also, more noise will be induced. Unfortunately, dry electrodes have higher contact impedance than wet type has. Figure 10.17 shows the measured skin-electrode contact impedance of P-FCB and wet electrodes. Most of the vital signal lies between 1–200 Hz, and the wet type impedance is around 10–30 kΩ within the band. On the other hand, dry P-FCB electrode impedance ranges between 60–800 kΩ, where the smaller the electrode size, the bigger the impedance. Therefore, to minimize the loading effect, the proposed system is designed to have higher input impedance at the amplifier input than the conventional system, and this will be discussed in the following Chapter 10.4.

10.3.2.2 Impedance Versus Frequency

Figure 10.15 shows the measured skin-electrode contact impedance of P-FCB and wet (Ag/AgCl) electrodes, without pressure on it. Most of the vital signals lie between 1–200 Hz, where the impedance of wet type electrodes is around 10–30 kΩ. On the other hand, dry P-FCB electrode impedance ranges between 60 and 800 kΩ, where the smaller the electrode size, the bigger the impedance

Figure 10.16 shows the measured skin-electrode contact impedance of P-FCB and wet (Ag/AgCl) electrodes, with mild pressure (40 mmHg) on it. Pressure cuff is worn around arm to apply pressure over the electrodes.

Again, we can observe P-FCB electrode has about 10 times higher skin-electrode impedance than the conventional Ag/AgCl electrode; frequency response of the electrode shows that the dry P-FCB electrodes are applicable for vital signal monitoring.

10.3.2.3 Impedance Over Time

When it comes to time variation, dry electrode shows significantly different performance; Ag/AgCl electrodes have relatively stable skin-electrode impedance over time. On the contrary, dry P-FCB electrodes skin-electrode impedance, regardless of size and shape, drops as time passes by. At least 900 seconds is required for dry P-FCB electrode's skin-electrode impedance to settle down to a stable value

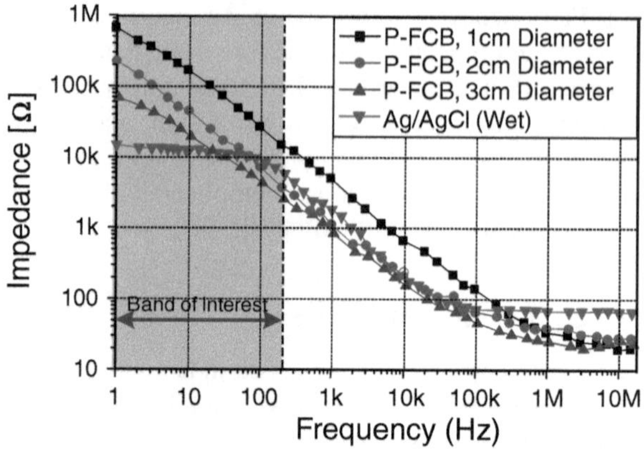

Fig. 10.15 Electrode impedance versus frequency (no pressure)

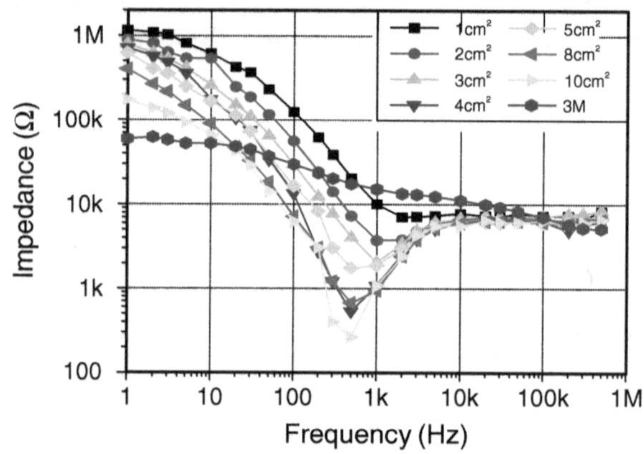

Fig. 10.16 Electrode impedance versus frequency (40 mmHg)

(Fig. 10.17). This measurement implies the dry P-FCB electrode is not suitable for short term monitoring that is done within several minutes; in the short term case, captured vital signal may fluctuate too much over time.

However, if the goal is to measure the vital signal over a long period, for example, over a week, then the first several minutes may be ignored; the long term monitoring is exactly what continuous health monitoring is focusing on. Also, wet type electrode may dry out when attached over a day, which results in signal degradation. Therefore, dry P-FCB electrode is a good alternative to Ag/AgCl (wet) electrodes for long-term monitoring.

10.3.3 Inductors by P-FCB

Figure 10.18 shows two types of P-FCB inductors that are used in the proposed adhesive bandage sensor and the health monitoring chest band. Type A measures to

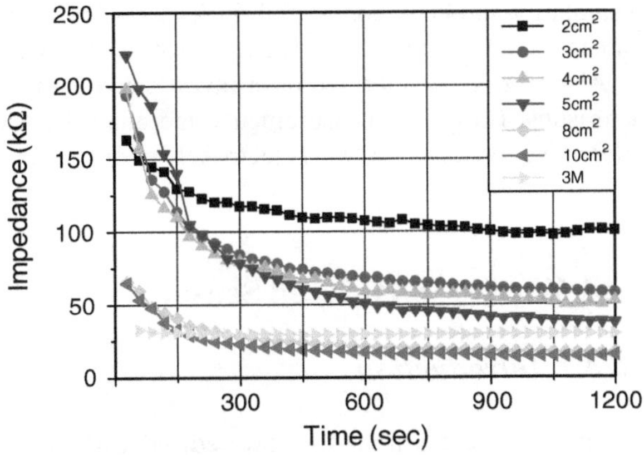

Fig. 10.17 Impedances characteristics of P-FCB electrodes

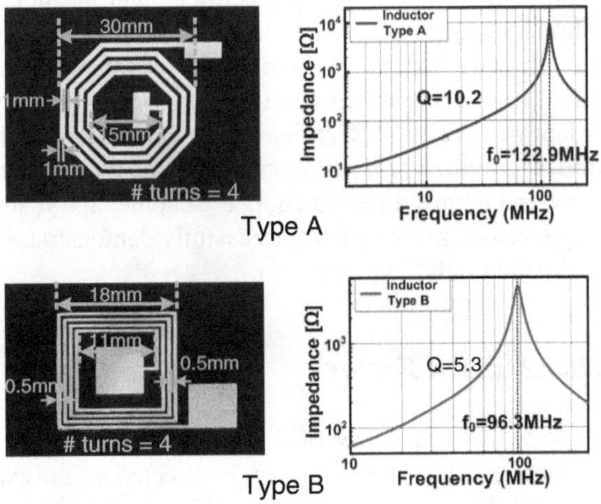

Fig. 10.18 Inductors by P-FCB

have Q = 10.2 and L = 0.98 μH, whereas the smaller type B has Q = 5.1 and L = 1.02 μH.

The inductors are used in the adhesive bandage sensors, or the health monitoring chest band. With the flexible fabric inductors, wearability is increased, while the production cost is lowered.

10.3.4 Summary

Fabric circuit board, or Planar Fashionable Circuit Board (P-FCB) technology, is a direct chip integration method on fabric. Screen printing process is exactly the same as conventional textile producing process, and no additional process or change is necessary. Therefore, low cost production is possible. P-FCB enhances wearability and body compatibility. Adhesive bandage type sensors, dry electrodes, and an inductor array can be made with ease at low cost.

Also in this Chapter, we have investigated P-FCB electrode and inductor characteristics. Based on measurements and analysis, dry P-FCB electrodes are good alternative to Ag/AgCl (wet) electrodes in long term vital signal monitoring.

10.4 Wirelessly Powered Sensor

10.4.1 Introduction

In this section, a wirelessly powered, adhesive bandage vital sensor for BSN is described in detail. The sensor patch improves convenience at low cost; it is composed of a pair of dry electrodes and an inductor directly screen printed on fabric by using Planar-Fashionable Circuit Board (P-FCB). The sensor chip is also directly wire bonded on fabric. The sensor patch generates power from incoming HF (13.56 MHz) or MICS (400 MHz) band signal to remove battery for safety; the Adaptive Threshold Rectifier (ATR) enables high efficiency rectification. Dry electrodes minimize skin irritation and enable long term monitoring. The sensor readout front-end adopts a nested chopper scheme, and it suppresses noise effectively. The implemented sensor patch successfully demonstrates capturing of ECG signal while dissipating only 12 μW power.

10.4.2 Form Factor

To show the feasibility of the proposed wirelessly powered vital sensor, electrocardiogram (ECG) monitoring is selected as an example. The ECG sensor form factor is determined by topology (2-electrode/3-electrode), electrode size and distance between electrodes. Since we want to keep the size of the sensor patch is small to maximize convenience, we do not have room for conventional 3-electrode topology. Instead, reference-free electrodes are used; to suppress noise and increase CMRR, analog readout front-end adopts the nested chopper topology. Details on this will be described in Chapter 10.4.5. Distance between electrodes is determined by measurement.

Figure 10.19 shows the measurement setup, and Fig. 10.20 shows the result for 20 mm diameter circle type electrode and 20 $\times$ 20 mm square type P-FCB dry electrodes: when it comes to ECG around chest, the maximum signal strength is obtained at distance between electrodes of 10 cm, and if the distance is close enough, the distance effect is negligible [39]. The maximum signal magnitude (after amplification) is around 400 mV with electrodes apart by 10 cm, which is about twice the magnitude of minimum value (electrodes apart by 2 cm). Considering the chest size, the distance between the electrodes set as 25–35 mm. Form factor of the sensor patch is decided to sufficiently cover the electrodes.

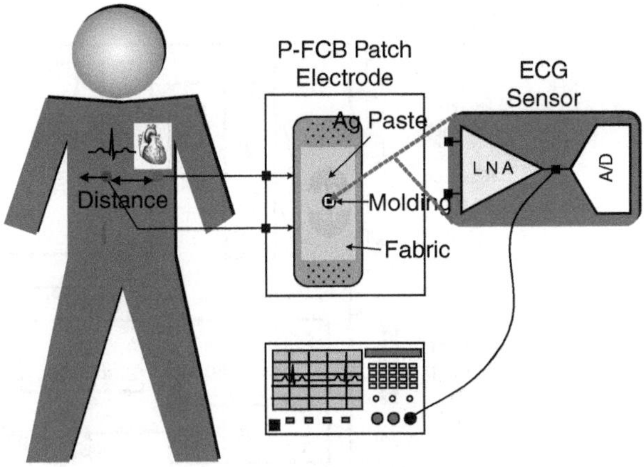

Fig. 10.19 Determining sensor form factor: distance between electrodes

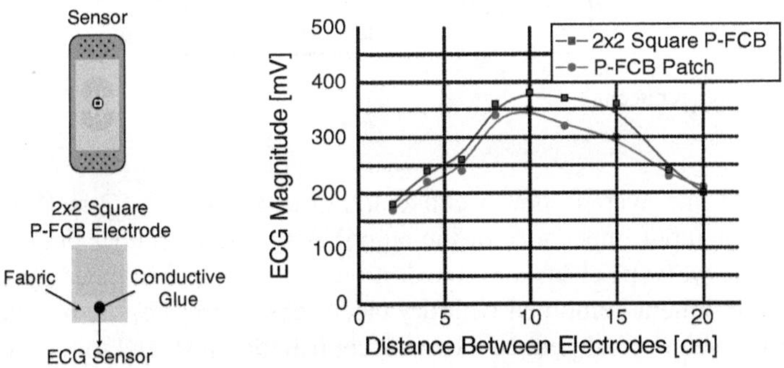

Fig. 10.20 ECG signal magnitude versus distance between electrodes

10.4.3 Sensor Design

The adhesive bandage sensor patch integrates a sensor IC. Figure 10.21 the architecture of the sensor IC. It is composed of an Adaptive Threshold Rectifier (ATR), a nested chopper ECG Instrumentation Amplifier (IA), a 10 b SAR ADC, an ASK modulator, an oscillator, and a signal level detector. The signal level detector stores the information, such as received signal strength and VDD level, in its data packet transmitted to network controller by the ASK modulator.

A pair of P-FCB electrodes is capacitively coupled to the IA input. The ATR generates power from incoming carrier signal, either in HF (13.56 MHz) or MICS (400 MHz) band; HF band is for patch sensors, whereas MICS band is supported for implantable sensors. Generated voltage is regulated and dispatched to clock generator, the IA and the SAR ADC. Due to low efficiency of P-FCB inductors, the amount of generated power is very limited, so the overall power budget of the sensor is designed to stay below 20 μW. As a result, the 10b 2-phase SAR-ADC consumes 2 μW, clock generator 2 μW, the regulator 3 μW, and the AFE consumes only

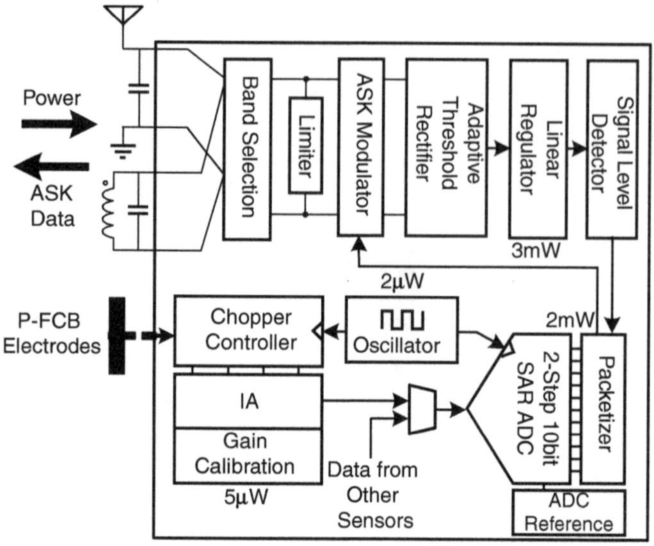

Fig. 10.21 Sensor architecture

5 μW. Overall, the average power consumption of the sensor IC is 12 μW when active. Also, as input impedance of the amplifier should be higher than that of the amplifier using wet type electrodes. In addition, more noise is induced by wirelessly powered environment combined with dry electrodes; therefore, the nested chopper scheme is adopted in IA to reduce the noise contribution [40] and to minimize loading effect. Although the sensor itself can process various vital signals, the IA in this version of the sensor is optimized for ECG to verify its operation. Captured ECG signal is amplified at IA and fed into 10b SAR ADC; the converted data is packetized and transmitted to network controller by backscattering using ASK.

10.4.4 Wireless Power Transmission

For sensor power source, the system uses wireless power transmission using inductive coupling and the ATR. There are several ways to provide power to a sensor patch

1) *Battery*: Small batteries such as a flexible, paper-like battery [41] can be used at sensor. However, the "chronic" health monitoring over long-term forces to employ large capacity battery, making it inconvenient and expensive. Moreover, and perhaps most importantly, when it comes to sensors that will be attached directly to skin, some people are reluctant to use batteries.
2) *Thermal energy harvester*: Body heat is a good candidate for power source, since human body temperature is stable at 36–38°C. Several studies have

demonstrated the feasibility of scavenging thermal energy from the body [42, 43]. However, generated power level is around a few μW/cm^2 with a load attached [42], which is too small to operate a sensor IC that will consume at least tens of μWs. [43] shows sufficient power levels of up to 30 μW/cm^2, but thermopiles are too expensive to be used in disposable sensors. Therefore, thermal energy harvester alone would not be sufficient for the sensors.

3) *Vibration energy harvester*: Vibration is another candidate for sensor power source. The electrostatic transducers or piezoelectric harvesters [44] can generate as much as 40 μW at 1.8 kHz. Although the power level is large enough to operate sensor IC, continuous vibration is required to generate power, and therefore, vibration energy harvester is not appropriate for the sensors.
4) *Solar cell*: Solar energy cannot be used in this case, since the patch sensor will be surrounded by clothes almost all the time.
5) *Wireless power transmission*: Both inductive coupling or RF can provide power to the sensor. RF link can transmit power over longer distance than the inductive coupling does, but it is more sensitive to interferences in the back-scattering-based uplink [21]. For example, RF in ISM band such as ZigBee and Bluetooth may interfere with Wireless LAN (WLAN) because they share the same frequency range [21]. On the other hand, inductive coupling link is less sensitive to interference because it requires proximity between transmitter and receiver. Since the healthcare system requires high security and interference-resilience, we choose to adopt inductive coupling to provide power to patch sensors. Even if unintended signal activates the patch sensor, the sensor response is automatically blocked by the network controller since it is physically not connected to the network controller.

10.4.4.1 Conventional Rectifier

Maximizing the rectification efficiency is of primary importance in implementing wirelessly powered rectifiers, and in this system, sensitivity is not a major problem due to the short distance (<cm) between sensor and reader coil. Typically, CMOS rectifiers are preferred because Schottky diodes or ferroelectric capacitors are expensive and difficult to integrate with standard CMOS process.

Efficiency drop of a rectifier mainly comes from the threshold (V_{th}) drop of the diodes within a rectifier (Fig. 10.22(a)). As a result, dead zone occurs, and ideal maximum output voltage of the conventional CMOS rectifier is limited to $2(V_{in}-V_{th})$, where V_{in} is the maximum input signal amplitude (Fig. 10.22(a)). To increase rectification efficiency, various methods were proposed. A battery-assisted rectification scheme [45] (Fig. 10.22(b)) compensates V_{th} drop at diode-connected M3 and M4 by adding a small battery with voltage (V_{BAT}) enough to override the V_{gs} drop. This simple-but-powerful method increases the maximum output voltage is now 2 V_{in}, and as a result, the efficiency is increased as well. However, it has fundamental limitation: it requires batteries at sensor, which is inappropriate in our case.

In [46], a rectifier with ferroelectric capacitors was introduced (Fig. 10.23). It cancels out the threshold (V_{th}) drop of diode-connection MP and MN by adding

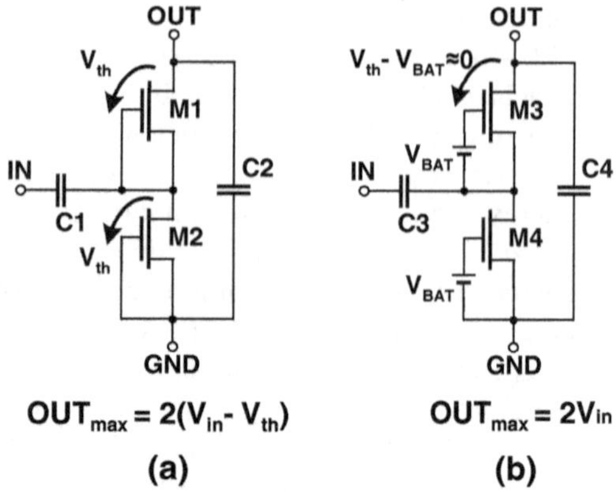

Fig. 10.22 (**a**) Conventional and (**b**) battery-assisted rectifier

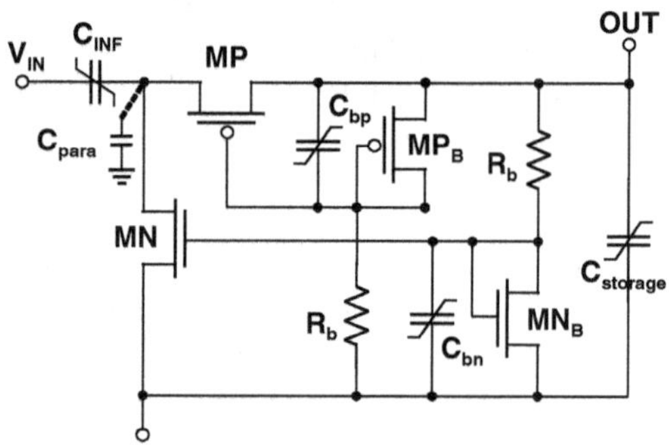

Fig. 10.23 Ferroelectric capacitor-based rectifier

pre-applied voltage between gate and source of the diode-connected transistors, using ferroelectric capacitors and biased transistors MP_B and MN_B. The ferroelectric capacitor has 10-fold better permittivity than the oxide cap, consequently, the area reduction is achieved and parasitic capacitance is reduced. However, as in the previous case, this smart idea requires additional ferroelectric capacitors, and cannot be integrated into standard CMOS process; thus too expensive to adopt in patch sensors.

10.4.4.2 Adaptive Threshold Rectifier (ATR)

To improve rectification efficiency while not using any expensive additional process, nor using battery, the Adaptive Threshold Rectifier (ATR) is proposed.

Figure 10.24 shows the schematic diagram of the proposed ATR. In ATR, the V_{th} drop of diode-connected transistor is minimized to improve the rectification efficiency. M1–M4 transistors form a CMOS bridge rectifier. At startup, SW1–SW4 are connected to the drain of each transistor (ATR off). As the incoming signal is applied to ANT+ and ANT– nodes, the VP node voltage gradually increases. When the generated VOUT goes above V_{th}, the power-on-reset (POR) signal triggers SW1–SW4.

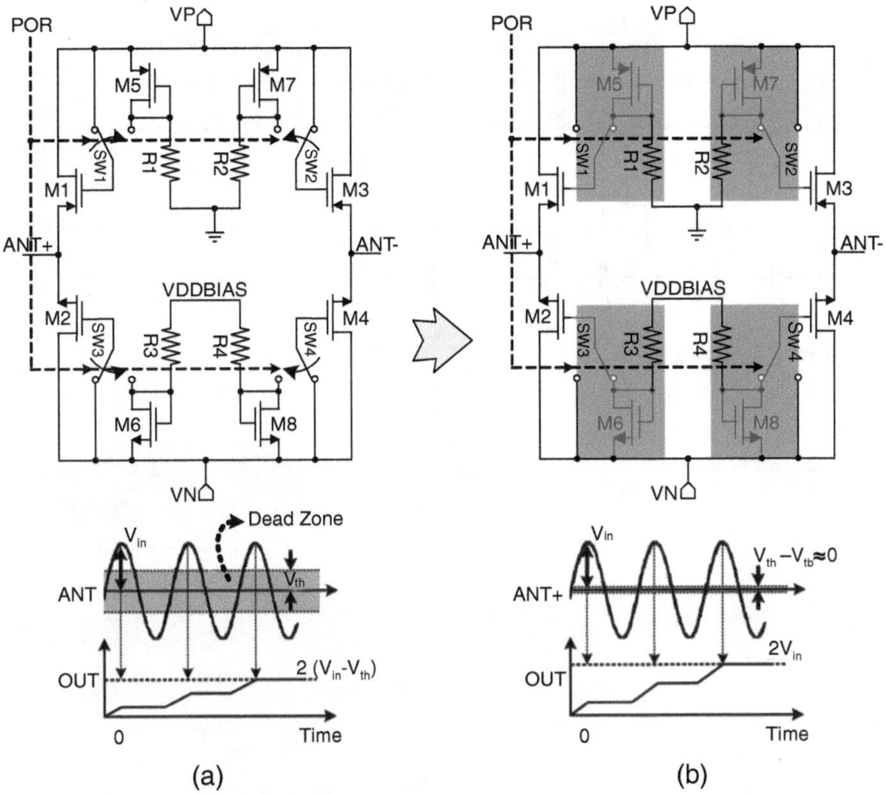

Fig. 10.24 The Adaptive-Threshold Rectifier (ATR): (**a**) ATR OFF, and (**b**) ATR ON

Then the diode-connected devices M5–M8, which have voltage difference of V_{th} between drain and source, are attached to gate of M1–M4 (ATR on). Since M5–M8 are matched to M1–M4, V_{th} drop of M1–M4 are cancelled out, making dead zone virtually zero. As shown in Fig. 10.25, 4 ATR blocks are cascaded to generate 1.8 V at high frequency (HF, 13.56 MHz) band or 1.6 V at MICS (402 to 405 MHz) band.

Figure 10.26 shows the simulation results for 1 stage ATR. When ATR is off, maximum generated voltage is 200 mV, whereas the generated voltage is boosted up when ATR is on.

10.4.5 Sensor Readout Front-End

Figure 10.27 shows the architecture of the sensor readout front-end with nested chopper instrumentation amplifier (IA). It is composed of an input chopper stage, a preamplifier, a demodulator, and a Programmable Gain Amplifier (PGA). Due to the large dry P-FCB electrode impedance, the sensor is vulnerable to noise; large bias resistors (10 MΩ each) are used at the input to minimize the loading effect.

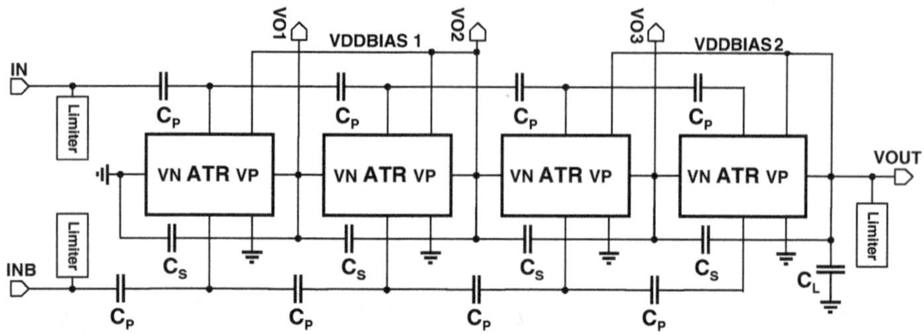

Fig. 10.25 Cascaded ATRs

Fig. 10.26 Simulation results of ATR

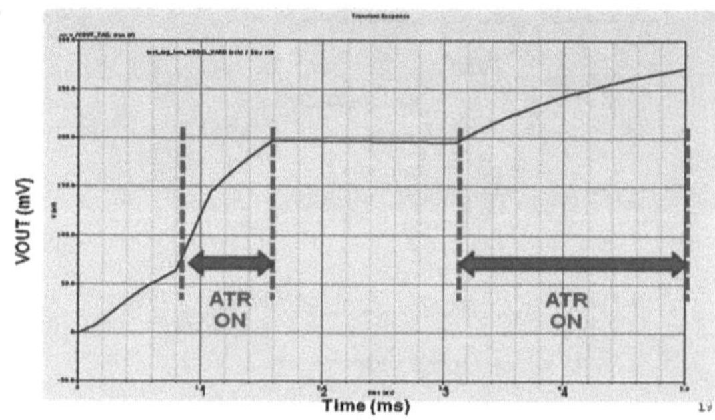

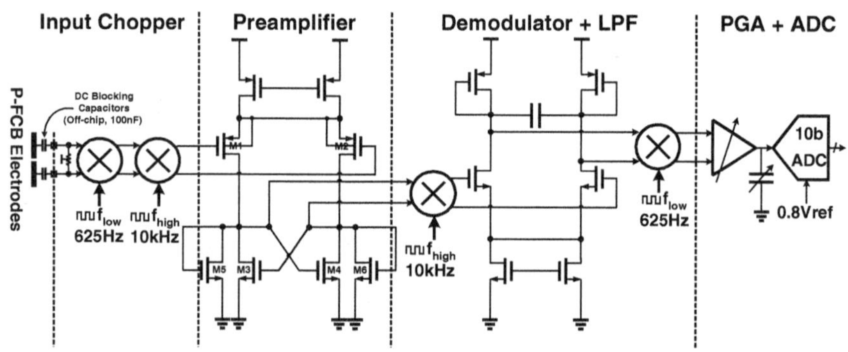

Fig. 10.27 Sensor readout front-end

Thermal noise induced by the large bias resistors is low pass filtered by two off-chip DC blocking capacitors (100 nF each). In the IA, the nested chopper scheme [40] is adopted to achieve low noise and high SNR operation. At the preamplifier, the M3, M4 pair configures a positive feedback loop to enhance the gain of the preamplifier. The transconductance of this pair is designed to be 0.9 of the M4, M5 pair and

0.16 of the M1, M2 pair so that the M3, M4 pair contributes only 5% of the total noise, which is equal to a 16% noise contribution of the input pair. Consequently, the positive feedback adds a negligible amount of noise, and, in addition, it does not degrade the power noise efficiency. The measured input impedance of the open-loop topology IA is 9.8 MΩ, which is sufficiently high to avoid loading effects for the dry electrodes.

As the vital signal level at the P-FCB electrode is 10–500 μV, the readout front-end is designed to have the preamplifier gain of ~60 dB. Since the reference voltage of the 10 b ADC is 0.8 V, the resolution is equal to 800 μV, and this translates into a maximum tolerable 0.8 μV input referred noise level. To suppress the *1/f* noise, chopping frequency should be sufficiently higher than the *1/f* corner frequency, so a 10 kHz clock is used at the inner chopper. However, chopping spikes caused by the charge injection at the inner chopper (operating at 10 kHz) leave the residual offset after the preamplifier and the inner chopper (Fig. 10.28). Therefore, another chopper operating at lower frequency (625 Hz) is added to form a nested chopper topology, and as a result, the residual offset is removed. The preamplifier with diode-connected load pairs M5 and M6, followed by demodulator with LPF have gain of 40 dB, and the PGA have 6 to 20 dB gain.

Figure 10.29 shows the noise power reduction with the nested chopper IA. After the nested chopper IA, noise is suppressed down to target range.

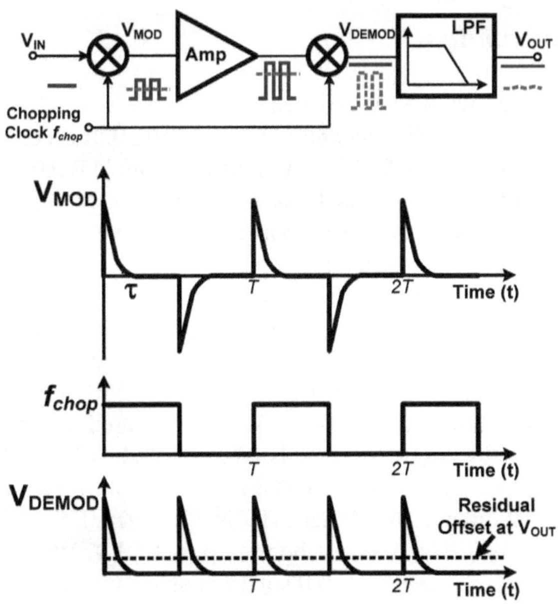

Fig. 10.28 Chopper-stabilized noise reduction and its residual offset

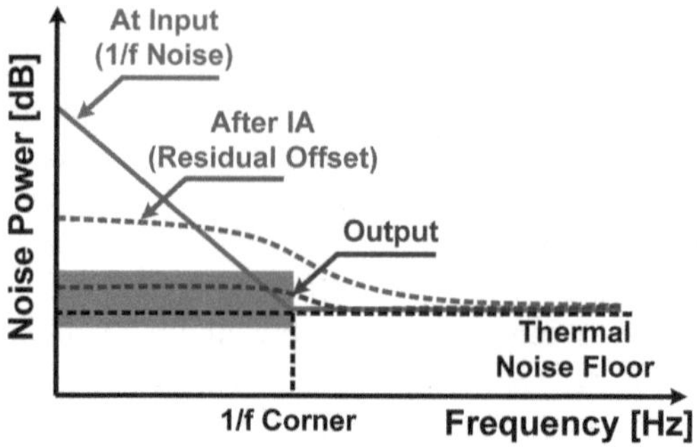

Fig. 10.29 Noise power of nested chopper IA

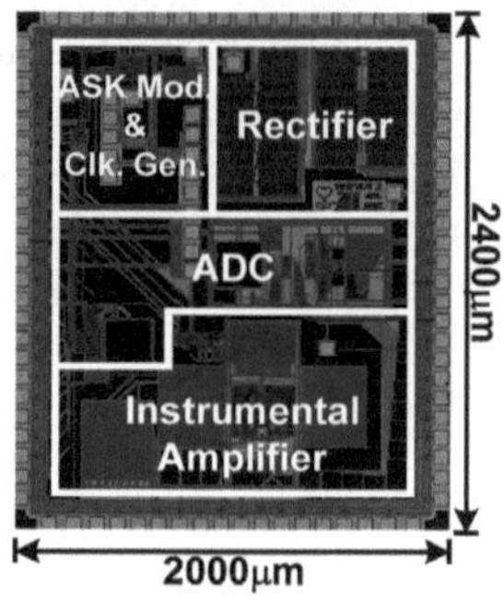

Fig. 10.30 Chip micrograph of the sensor IC

10.4.6 Implementation

To verify the proposed wirelessly powered sensor, the sensor IC is implemented in 0.18 μm 1 P6M standard CMOS technology of Fig. 10.30. It is composed of a ASK modulator, clock generator, an ADC, an Adaptive Threshold Rectifier (ATR), and an instrumentation amplifier; the sensor IC takes 4.8 mm^2 including pads. Simulation shows that the sensor IC consumes less than 20 μW when the sensor interfaces, the ECG IA and the ADC are in operation.

The transceiver uses HF band (13.56 MHz) for patch sensors, and MICS band (400 MHz) for implantable sensors. Figure 10.31 shows the ATR rectification efficiency in (a) HF band, and in (b) MICS band. It shows the peak efficiency for HF band is 54.9% at 800 mV input signal magnitude. For MICS band, the peak efficiency is 45.2% at −2 dBm input power. When compared to [46], at same input power (−6 dBm), rectification efficiency is 43.2%, which is 18.1% higher.

Figure 10.32 shows the generated voltage (VDD) with ATR versus frequency band, at 0 dBm fixed input power. When ATR is off, peak VDD is 1.15 V at around 50 MHz, whereas when ATR is on, it boosts up to 1.80 V. When we observe the frequency response of the generated VDD, it drops sharply at around 400 MHz. This is due to parasitic capacitance at the ATR input; however, as our band of interest is below MICS band (around 400 MHz), this is not a problem in this system. If more

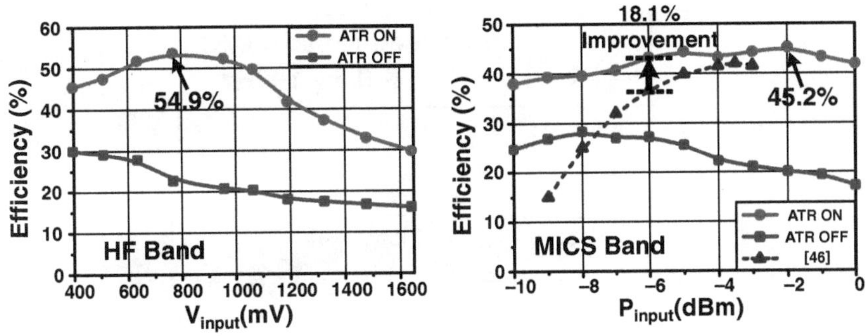

Fig. 10.31 ATR efficiency: (**a**) HF band, (**b**) MICS band

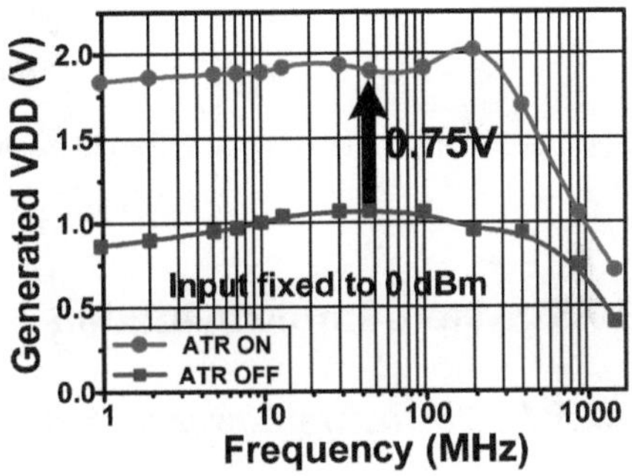

Fig. 10.32 Frequency response of the ATR at MICS band (0 dBm input)

high frequency performance is needed, then using ferroelectric capacitor as in [46] may be a good solution at increased cost.

Figure 10.33 shows the input referred noise power spectral density (PSD) of the implemented IA. The noise floor is 47 nV/$\sqrt{}$Hz, and this is equivalent to 0.51 μV_{rms} over vital signal bandwidth (0.5–100 Hz).

10.4.7 Summary

A wirelessly powered, adhesive bandage vital sensor for BSN is designed. The sensor patch improves convenience at low cost. The patch is composed of a pair of dry electrodes and an inductor directly screen printed on fabric by using Planar-Fashionable Circuit Board (P-FCB).

The Adaptive Threshold Rectifier (ATR) boosts up power recovery efficiency while not using any special process. The patch generates power from incoming HF (13.56 MHz) or MICS (400 MHz) band signal to remove battery for safety. Dry electrodes minimize skin irritation and enable long term monitoring.

Nested chopping topology is adopted in the sensor readout front-end, and it suppresses noise effectively. The implemented sensor patch successfully demonstrates capturing of ECG signal while dissipating only 12 μW power.

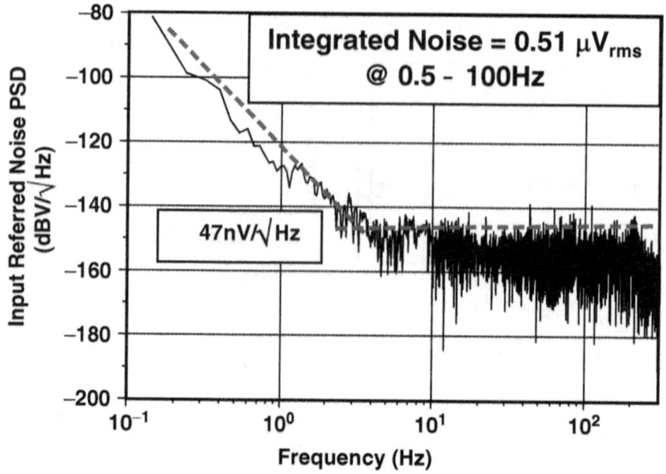

Fig. 10.33 Input referred noise PSD of the sensor readout front-end

10.5 System Implementation

10.5.1 Wirelessly Powered Adhesive Bandage Sensor

The adhesive bandage type sensor patch (Fig. 10.34) is implemented using P-FCB technology. There are two types of sensors: one is 75 × 38 mm, and the other is 68 × 25 mm large. On an adhesive bandage, a P-FCB inductor, electrodes and the sensor IC are integrated together by P-FCB technology. The sensor IC is wire bonded directly on P-FCB electrode layer, and molding is applied for protection.

10.5.2 Health Monitoring Chest Band

The health monitoring chest band is implemented by P-FCB process. The health monitoring band is 210 × 830 mm large, and it has an array of 12 × 4 inductors on it (Fig. 10.35). In this implementation, the network controller SoC is not directly wire bonded on the chest band; instead, a separate PCB test board is implemented and connected to the 12 × 4 inductor array. Implementing a complete wearable system with the network controller SoC also attached to the chest band or innerwear, as shown in Fig. 10.36, will be the further work.

The measured ECG waveforms by the P-FCB sensor are also shown in Fig. 10.37. The VDD ripple after regulation is suppressed down to 30 mV (1.6% of generated VDD) while the sensor chip is in operation.

Table 10.1 summarizes the performance and key features of the proposed self-configured wearable BSN system with the wirelessly powered sensor patch. Average power consumption is 5.2 mW for the network controller chip with 12 × 4 inductor array.

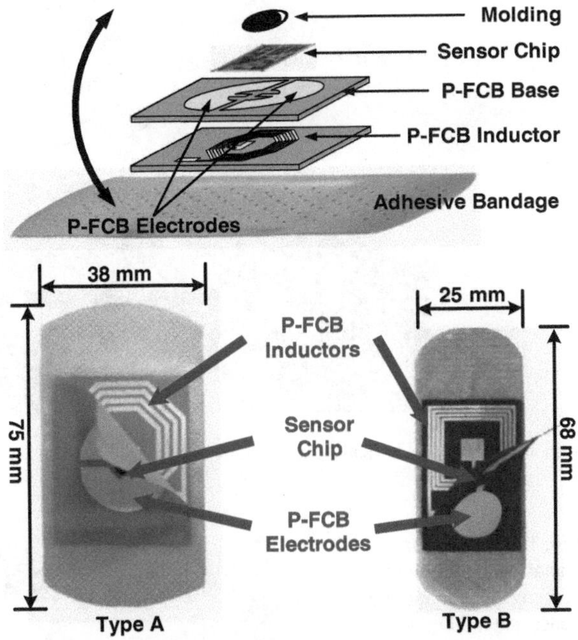

Fig. 10.34 Adhesive bandage sensor integration

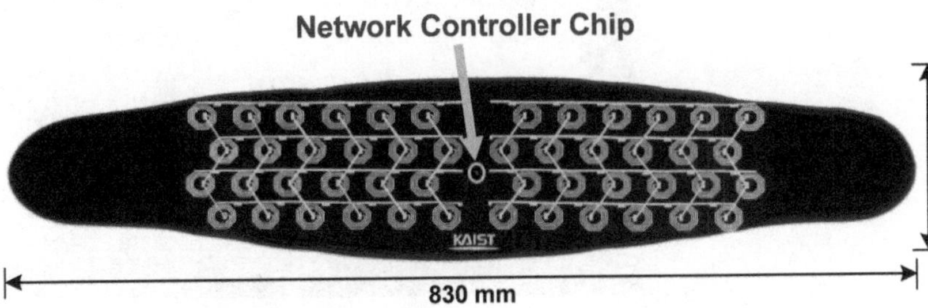

Fig. 10.35 Health monitoring chest band

Table 10.2 compares previous works and this system. The proposed system is highly immune to interference, and the rectifier efficiency is the highest, with over 50% (based on simulation) at HF band. Since the network controller takes the sensor power overhead, the sensor only consumes 12 μW under operation. Also, the proposed system is implemented using standard CMOS process only, without any expensive additional process.

10.6 Conclusion

In this chapter, an example wearable healthcare system using Body Sensor Network (BSN) for continuous health monitoring is examined in detail. The BSN SoC is a self-configured wearable BSN SoC for continuous health monitoring over long

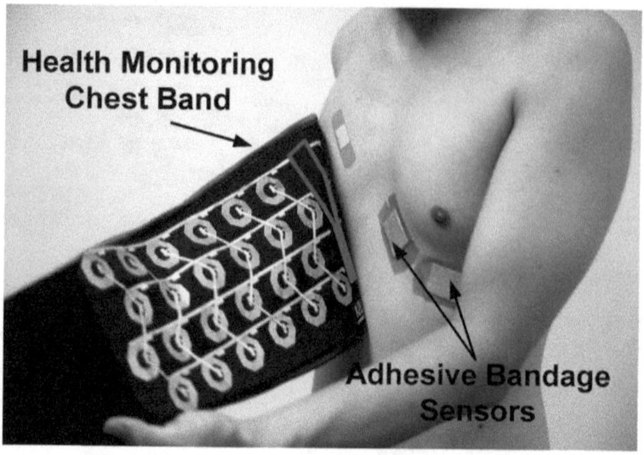

Fig. 10.36 Wearable healthcare system

Fig. 10.37 Measured ECG signal by the wirelessly powered sensor and the generated VDD

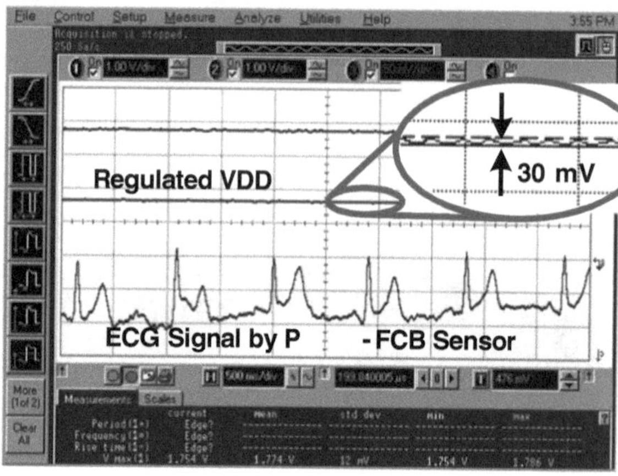

Table 10.1 Performance summary of the self-configured wearable BSN controller

Process	TSMC 0.18 μm 1P6M CMOS	
Die size [mm × mm]	5.0×3.0 (Network Controller) 2.4×2.0 (Sensor)	
Supply voltage	1.8 V	
Average power consumption	Network controller	5.2mW
	Sensor	12μW
Clock frequency [MHz]	Transceiver	13.56 (HF) 402–405 (MICS)
	Sensor node	N/A
Rectifier type	4-Stage ATR CMOS	
Rectifier efficiency	54.9% (HF)/45.2% (MICS)	
Modulation	15% or 100% ASK	
Array dimension	12×4 Inductor Array	

Table 10.2 Performance comparison of the self-configured wearable BSN controller with the previous works

Subject	A. Wong, ISSCC2008 [18]	H. Nakamoto, JSSC2007 [46]	T. Umeda, JSSC2006 [45]	J. Yoo, This Work
	Healthcare	Rectifier		Both
Immunity to interference	Low (Wireless)	–	–	Very High
Communication channel	Wireless (RF)	Wireless (RF)	Wireless (RF)	Wireless+Wireline
Modulation type	FSK	ASK	ASK	ASK
Carrier frequency [MHz]	862–870 902–928	953	950	13.56 (Patch) 402–405 (Implantable)
Rectifier efficiency	–	36.6% (UHF)	16.6% (UHF)	>50% (HF Band)
Sensor power consumption	2928μW (All active)	–	–	<20μW (All active)
Cost of production	(Battery+MCU at Sensor node)	Hight (Additional process)	Moderate (Battery at Sensor node)	Low
Technology	0.13μm CMOS	0.3μm CMOS with FeRAM	0.30μm Triplewell CMOS	018μm CMOS

period. The adhesive bandage type sensor IC was implemented using standard 1P6M 0.18 μm CMOS process, as well. The wirelessly powered ECG sensor is implemented with Planar-Fashionable Circuit Board (P-FCB) technology. The proposed system exploits adhesive bandage type patch sensor that is wirelessly powered by health monitoring chest band. Dry P-FCB electrode is used at sensor, and unlike conventional wet electrode, it is free from skin stimulation problem, thus suitable for long term monitoring. The sensor IC that is attached to the adhesive bandage patch exploits the Adaptive Threshold Rectifier (ATR) to improve the rectification efficiency by over 50% at HF band (simulation), and with the help of the nested chopper instrumentation amplifier, the integrated noise PSD (simulation) of the sensor is below $1\,\mu V_{rms}$ for vital signal bandwidth. The network controller integrated within the health monitoring band adopts self-configuration scheme to automatically locate the position of sensors worn on arbitrary location around the body, and it provides power to the selected sensors only. The network controller and the sensor consume only 5.2 mW and 12 μW, respectively. With the implemented self-configured wearable BSN controller and the sensor IC, a continuous ECG monitoring system was implemented.

References

1. Kung H-C, Hoyert DL, Xu J, Murphy SL (2008) Deaths: final data for 2005. National vital statistics reports 56(10):5–9, Apr. 2008
2. Korhonen I, Parkka J, Van Gils M (2003) Health Monitoring in the Home of the Future. IEEE Eng Med Biol Mag 22: 66–73, May–Jun 2003
3. Jabaudon D, Sztajzel J, Sievert K, Landis T, Sztajzel R (2004) Usefulness of ambulatory 7-Day ECG monitoring for the detection of atrial fibrillation and flutter after acute stroke and transient ischemic attack. Stroke J Am Heart Assoc 35: 1647–1651, May 2004
4. Zerhouni EA (2006) NIH in the post-doubling era: realities and strategies. Science 314: 1088–1090, Nov 2006
5. Zerhouni EA (2008) NIH aims to transform findings into clinical changes—future medicine: predictive, personalized, preemptive, and participatory. A Column in U.S. Medicine, Jan 2008
6. Ashok RL, Agrawal DP (2003) Next generation wearable networks. IEEE Comput Mag 36(11):31–39, Nov 2003
7. Gutiérrez JA, Callaway EH Jr, Barrett RL Jr (2003) Low-rate wireless personal area networks: enabling wireless sensors with IEEE 802.15.4, IEEE Press, pp 4–5
8. Komurasaki H, Sano T, Heima T, Yamamoto K, Wakada H, Yasui I, Ono M, Miwa T, Sato H, Miki T, Kato N (2003) A 1.8-V operation RF CMOS transceiver for 2.4-GHz-band GFSK applications. IEEE J Solid-State Circuits 38(5):817–825, May 2003
9. Alliance Z (2007) ZigBee and wireless radio frequency coexistance. ZigBee white paper Jun 2007
10. Golmie N, Van Dyck RE, Soltanian A, Tonnerre A, Rébala O (2003) Interference evaluation of bluetooth and IEEE 802.11b Systems. Wireless Networks 9(3):201–211, May 2003
11. Bilstrup U, Wiberg P-A (2000) Bluetooth in industrial environment. Dig IEEE Int Workshop Factory Commun Sys 239–246, Sep 2000
12. Nakad Z, Jones M, Martin T (2003) Communications in electronic textile systems. Proc Int Conf Commun (CIC) 37–43, Jun 2003
13. Nakad Z, Jones M, Martin T (2004) Fault-Tolerant networks for electronic textiles. Proc Int Conf Commun (CIC), Jun 2004

14. Searle A, Kirkup L (2000) A direct comparison of wet, dry and insulating bioelectric recording electrodes. Physiol Meas 21(2):271–283, May 2000
15. Paradiso R, Loriga G, Taccini N (2005) A wearable health care system based on knitted integrated sensors. IEEE Trans Inf Technol Biomed 9(3):337–344, Sep 2005
16. Wong AC-W, McDonagh D, Kathiresan G, Omeni OC, El-Jamaly O, Chan TC-K, Paddan P, Burdett AJ (2008) A 1 V, Micropower system-on-chip for vital-sign monitoring in wireless body sensor networks. IEEE Int Solid-State Circuits Conf (ISSCC) Digest of technical papers 138–139, Feb 2008
17. Penders J, Gyselinckx B, et al. (2008) Human++: from technology to emerging health monitoring concepts. In: Proceedings of the 5th International workshop on wearable and implantable body sensor networks, The Chinese University of Hong Kong, China, Jun 1–3, pp 94–98
18. Yazicioglu RF, Merken P, Puers R, et al. (2007) A 60 μW 60 nV/$\sqrt{\text{Hz}}$ Readout front-end for portable biopotential acquisition systems. IEEE J Solid-State Circuits 1100–1110, May 2007
19. Hum A (2001) Fabric Area Network—A new wireless communications infrastructure to enable ubiquitous networking and sensing on intelligent clothing. Comput Net 35:391–399, Mar 2001
20. Yoo J, Lee S, Hoi-Jun Yoo (2008) A 1.12pJ/b Resonance compensated inductive transceiver with a fault-tolerant network controller for wearable body sensor networks. Proc IEEE Asian Solid-State Circuits Conf (A-SSCC) 313–316, Nov 2008
21. Yoo J, Lee S, Yoo H-J (2009) A 1.12pJ/b Inductive transceiver with a fault-tolerant network switch for multi-layer wearable body area network applications. IEEE J Solid-State Circuits 44(11): 2999–3010, Nov 2009
22. Yoo J, Yan L, Lee S, Kim Y, Kim H, Kim B, Yoo H-J (2009) A 5.2 mW Self-configured wearable body sensor network controller and a 12 μw 54.9% efficiency wirelessly powered sensor for continuous health monitoring system. IEEE Int Solid-State Circuits Conf (ISSCC) Digest of technical papers 290–291, Feb 2009
23. Yoo J, Yan L, Lee S, Kim Y, Kim H, Kim B, Yoo H-J (2010) A 5.2 mW Self-configured wearable body sensor network controller and a 12 μw 54.9% efficiency wirelessly powered sensor for continuous health monitoring system. IEEE J Solid-State Circuits, 45(1): 178–188, Jan 2010
24. Arnon S, Bhastekar D, Kedar D, Tauber A (2003) A comparative study of wireless communication network configuration for medical applications. IEEE Wireless Commun, 56–61, Feb 2003
25. Yoo J (2007) Design and analysis of an energy-efficient human body channel network for healthcare applications. M. S. thesis, Korea Advanced Institute of Science and Technology, Feb 2007
26. Miura N, Mizoguchi D, Inoue M, Niitsu K, Nakagawa Y, Tago M, Fukaishi M, Sakurai T, Kuroda T (2006) A 1 Tb/s 3 W Inductive-coupling transceiver for inter-chip clock and data link. IEEE Int Solid-State Circuits Conf (ISSCC) Digest of technical papers 424–425, Feb 2006
27. Miura N, Ishikuro H, Sakurai T, Kuroda T (2007) A 0.14pJ/b Inductive-coupling inter-chip data transceiver with digitally-controlled precise pulse shaping. IEEE Int Solid-State Circuits Conf (ISSCC) Digest of technical papers 358–359, Feb 2007
28. Guermandi D, Gambini S, Rabaey J (2007) A 1 V 250 Kpps 90 nm CMOS Pulse based receiver for cm-range wireless communications Proc IEEE Eu Solid State Circuits Conf (ESSCIRC) 135–138, Sep 2007
29. Song S-J, Cho N, Kim S, Yoo J, Choi S, Yoo H-J (2007) A 0.9 V 2.6 mW Body-coupled scalable phy transceiver for body sensor applications. IEEE Int Solid-State Circuits Conf (ISSCC) Digest of technical papers 366–367, Feb 2007
30. Jabaudon D, Sztajzel J, Sievert K, Landis T, Sztajzel R (2004) Usefulness of ambulatory 7-day ecg monitoring for the detection of atrial fibrillation and flutter after acute stroke and transient ischemic attack stroke. J Am Heart Assoc 35:1647–1651, May 2004

31. Kim H, Kim Y, Yoo H (2008) A Low Cost Quadratic Level ECG Compression algorithm and its hardware optimization for body sensor network system. Proc IEEE Eng Med Biol Conf (EMBC) 5490–5493, Aug 2008
32. Kim H, Kim Y, Yoo H-J (2008) A 1.12 mW Continuous healthcare monitor chip integrated on a planar-fashionable circuit board. IEEE Int Solid-State Circuits Conf (ISSCC) Digest of technical papers 150–151, Feb 2008
33. Kim Y, Kim H, Yoo H-J (2010) Electrical characterization of printed circuits on the fabric. IEEE Trans Adv Package, 33(1): 196–205, Feb 2010
34. Webster JG (1998) Medical instrumentation: application and design, 3rd edn. Wiley, 1998, ch. 5
35. Connolly M, Buckley DA (2004) Contact dermatitis from propylene glycol in ECG electrodes, complicated by medicament allergy. Contact Dermatitis 50(1):42, Apr 2004
36. Yan L, Cho N, Yoo J, Kim B, Yoo H-J (2008) A Two-Electrode 2.88 nJ/Conversion Biopotential Acquisition System for Portable Healthcare Device. IEEE Asian Solid-State Circuits Conference (A-SSCC) Proc. Tech Papers 329–332, Nov 2008
37. Lobodzinski SS, Laks MM (2008) Comfortable textile-based electrocardiogram systems for very long-term monitoring. Cardiology J 15(5):477–480, Feb 2008
38. Kang T-H, Merritt CR, Grant E, Pourdeyhimi B, Nagle HT (2008) Nonwoven fabric active electrodes for biopotential measurement during normal daily activity. IEEE Trans Biomed Eng 55(1), Jan 2008
39. Yoo J, Yan L, Lee S, Kim H, Yoo H-J (2009) A Wearable ECG Acquisition system with compact planar-fashionable circuit board based shirt. IEEE Trans Inf Tech Biomed 13(6):897–902, Nov 2009
40. Bakker A, Thiele K, Huijsing JH (2000) A CMOS Nested-chopper instrumentation amplifier with 100-nV Offset. IEEE J Solid-State Circuits 35(12):1877–1883, Dec 2000
41. PowerPaper, http://www.powerpaper.com/, accessed on Apr. 20, 2009
42. Lauterbach C, Strasser M, Jung S, Weber W (2002) Smart clothes self-powered by body heat.avantex symposium, 2002
43. Leonov V, Fiorini P, Sedky S, Torfs T, van Hoof C (2005) Thermoelectric MEMS genrerators as a power supply for a body area network. Proc 13th Int Conf Solid-State Sensors, Actuators and Microsys (Transducers 2005) 291–294, Jun 2005
44. Renaud M, Karakaya K, Sterken T, Fiorini P, van Hoof C, Puers R (2008) Fabrication, modelling and characterization of MEMS piezoelectric vibration harvesters. Sensors Actuators A: phys. 145–146:380–386, Jul–Aug 2008
45. Umeda T, Yoshida H, Sekine S, Fujita Y, Suzuki T, Otaka S (2006) A 950-MHz Rectifier circuit for sensor network tags with 10-m distance IEEE J Solid-State Circuits 41(1):35–41, Jan 2006
46. Nakamoto H, Yamazaki D, Yamamoto T, Kurata H, Yamada S, Mukaida K, Ninomiya T, Ohkawa T, Masui S, Gotoh K (2007) A Passive UHF RF identification CMOS tag IC using ferroelectric RAM in 0.35-μm Technology. IEEE J Solid-State Circuits 42(1):101–110, Jan 2007

Chapter 11
Digital Hearing Aid and Cochlear Implant

Sunyoung Kim and Hoi-Jun Yoo

11.1 Introduction of the Digital Hearing Aid

11.1.1 Population Trends of the Hearing Aids

Approximately 70 million individuals worldwide suffer from hearing loss, which makes it the most common sensory disorder in the world [1–3]. There are estimated 28 million individuals with hearing loss in the United States. Hearing loss affects 17 in 1,000 children under the age of 18, with the incidence increasing with age. Approximately 314 in 1,000 people over the age of 65 have hearing loss, and 40–50% of people 75 and older have hearing loss.

There are numerous causes of hearing loss, acoustic trauma, acoustic tumors, ototoxic medication, disease, and other environmental factors. Of the most prominent causes of hearing loss, noise-induced hearing loss has produced irreversible damage to 10 million Americans, and another 30 million Americans are exposed to dangerous noise levels every day [4, 5].

According to the World Health Organization International Classification of Functioning [6], the objective of an auditory rehabilitation program is to reduce the individual's activity limitations and participation restrictions resulting from the hearing impairment. The World Health Organization International Classification of Functioning acknowledges that exogenous factors found in the individual's communication environment play a role in either enhancing or hindering the rehabilitation program. With the passage of the Americans with Disabilities Act in 1990, society became more involved in creating a more facilitative environment for those with hearing loss. The passage of this act along with the rapid change in technology has encouraged researchers to develop ways to reduce functional impairment experienced by those with hearing loss through the use of sensory aids such as amplification devices and assistive listening devices as well as environmental modifications.

S. Kim (✉)
Interuniversity Microelectronics Center (IMEC), Leuven, Belgium
e-mail: sunyoung@imec.be

H.-J. Yoo, C. van Hoof (eds.), *Bio-Medical CMOS ICs*,
Integrated Circuits and Systems, DOI 10.1007/978-1-4419-6597-4_11,

In 2002, the American National Standards Institute [7] published "Guidelines for Acoustics in Educational Environments," which gave specific details regarding maximum limits of background noise, reverberation times, and signal to noise ratios (SNRs). These recommendations make sense for all environments in which communication will take place. Engineers and architects comply with these guidelines through the use of acoustic tiles to absorb reverberation, the size and structure of the room, and certain large-area listening systems that provide better SNRs. Although environmental engineering has a direct impact on communication for individuals with hearing loss, the focus of this article will be on the current benefits and limitations of personal amplification systems for individuals with hearing loss rather than universal design, which is also an important aspect of inclusion [8] (Fig. 11.1).

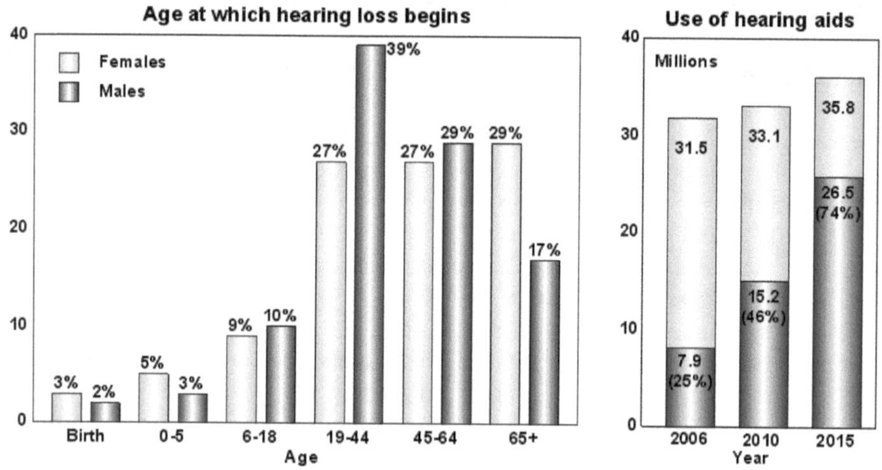

Fig. 11.1 Population trends

11.1.2 Future of the Hearing Aids

Hearing aids are not just for deaf people in the future. The much-maligned ear implants also hold the key to a new era in personal audio technology, if only they can make them as fashionable as spectacles. By adopting these future hearing aids, people can control their hearing ability.

First of all, the future hearing aids should perform as personal hearing devices, such as aids that enhance conversational speech or filter out ambient noise in crowded circumstances. One of the conceptual devices uses four microphones built into a pair of glasses to amplify sound depending on which direction the wearer is facing. In addition, there are a lot of attempts to meet the demand for universal hearing-assistance products with the development of new communication technology. Another conceptual appliance is a set of earphones that would repeat the previous 10 seconds of conversation in case the wearer missed a snippet. In addition, an earphone-linked remote control that can mute sounds coming from whatever it is pointing at could also be just a couple of years away [9].

11.2 Conventional Digital Hearing Aids

11.2.1 Types of the Digital Hearing Aids

There are several types of digital hearing aids [10]. Figure 11.2 shows the different types of digital hearing aids.

Behind the ear type (BTE) hearing aid is composed of a case behind the ear, an ear mold which locates inside the ear and an ear tube which connects between them. The hearing aid rests behind the ear and a plastic tube connects it to the ear mold. BTE is the most common type of hearing aid and can be used for mild to profound losses. BTE has several advantages over other types of hearing aids. One advantage is that they tend to be more durable. This is because the electrical components are located outside the ear. This reduces the amount of earwax and moisture that the electrical components are subjected to. Another advantage is that BTE can be connected to assistive listening devices, such as classroom FM systems. Lastly, if the earmold no longer fits the user, the earmold can be replaced for a fraction of the price of a new hearing aid.

In the ear type (ITE) and in the canal type (ITC) hearing aid fits completely in the outer ear bowl (call the concha) and is used for mild to severe hearing loss. The case, which holds the components, is made of hard plastic or silicon and is custom made to fit each individual's ear. ITE aids can accommodate added technical mechanisms such as a telecoil, a small magnetic coil contained in the hearing aid that improves sound transmission during telephone calls. The main disadvantage of the

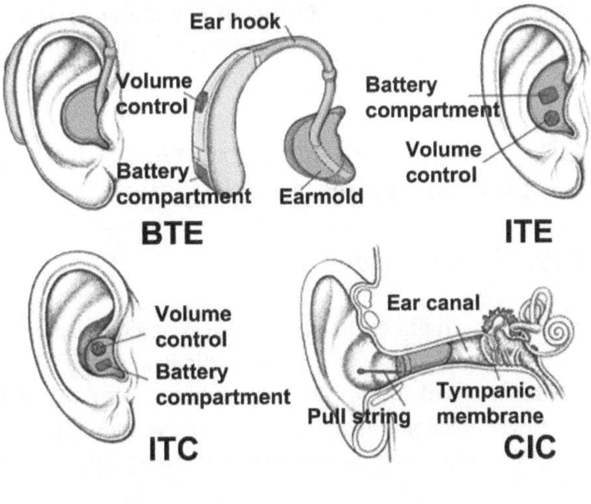

Fig. 11.2 Types of digital hearing aids

ITE aid is an acoustic feedback, a squealing/whistling caused by sound (particularly high frequency sound) leaking and being amplified again, especially for severe hearing losses. Therefore, some modern circuit designs are able to provide feedback regulation or cancellation.

A Completely-in-Canal (CIC) hearing aid is largely concealed in the ear canal and is used for mild to moderately severe hearing loss. Because of their small size, CIC hearing aids may be difficult for the user to adjust and remove, and may not be able to hold additional devices, such as a telecoil. In the Canal hearing aids can also be easily damaged by earwax and ear drainage. These aids are intended for mild to moderately-severe losses. Although the advantage of the CIC aid style is the small size, the disadvantages of higher cost and extreme difficulty in adjusting are often not worth the effort. In addition, CICs are usually not recommended for people with good low frequency hearing, as the occlusion effect is much more perceivable.

11.2.2 Design Issues of the Digital Hearing Aids

Hearing aids have advanced significantly over the past decade, primarily due to the maturing of digital technology. The next decade should see an even greater number of innovations to hearing aid technology [11]. The main innovation trend in digital hearing aids is a human-centered intelligent system design, such as high connectivity between user and environment and design to achieve user-friendly and autonomous functionality.

New digital wireless technology between digital hearing aids and various audio applications will allow enhancing the functionality of digital hearing aids. Moreover wireless ear-to-ear communication gives high flexibility to the user which describes the situation where the left and right hearing aids of a bilaterally-fit wearer communicate wirelessly with each other. In addition, a brand new hearing aid technology known as ADRO (adaptive dynamic range optimization) is starting to become available from some manufacturers. This is one of the most significant changes in the recent history of hearing aids, as it is a major update from traditional compression circuits that were most often used with digital hearing aids [12].

11.3 An Adaptive Digital Hearing Aid Chip with On Chip Human Factors Consideration

11.3.1 Introduction

Recently, the rapid expansion of the biomedical-electronic market has necessitated low-power and low-voltage biomedical systems [13, 14]. Since battery power is used for most of the portable biomedical devices, expanding battery lifetime with low-power dissipation systems is very crucial. In the digital hearing aid applications, the battery is typically made of zinc-air and should offer a life span of at least

2 weeks at 10 hours use per day [15]. Moreover, the digital hearing aid requires wide dynamic range, high performance, more programmability, and small form factor. Hence, it is necessary to achieve low-power dissipation, high-performance, and programmability to expand battery lifetime and to offer convenient hearing to the users.

Adopting extremely low supply voltage is an attractive solution to reduce power dissipation because the power dissipation of a system is strongly dependent upon its supply voltage. However, low supply voltage generally causes significant degradation of the system performance and complicates the analog circuit design. For example, the accuracy and the dynamic range of the analog front-end circuit may suffer from the low supply voltage due to reduced voltage headroom. By adopting a smart power management unit, the power consumption can be reduced. However, extra power dissipation due to the power management unit can occupy a great part of total system power because a digital hearing aid consumes extremely low total power. In addition, it is very hard to design a low-power high-performance power management unit.

Moreover, user convenience and system flexibility have been more emphasized in the design of biomedical and healthcare devices in addition to traditional issues of low power and high performance [16, 17]. Especially, the digital hearing aid chip takes special consideration on human factors so that it can give convenience to individual users without requiring additional effort from users.

When a hearing aid is worn, the remaining volume of the ear canal after the earmold is stuffed generates a resonance gain, which is ignored in the previous design of hearing aid. To estimate the difference of the resonance gain due to the volume variation, the remaining volume of the external ear canal should be calculated precisely after a hearing aid is inserted into it. In addition, there are many signal paths that introduce unpredictable additional gains after a hearing aid is worn.

Figure 11.3(a) shows the architecture of a conventional hearing aid. It uses an open loop gain fitting and a verification algorithm which requires long fitting time without optimization for individual users. However, conventional gain verification methods have many limitations. An insertion gain verification method with a probe-tube microphone generates measurement errors due to the different position of the probe-tube microphone in the ear canal. In addition, this method is unsuitable for a completely-in-the-canal (CIC)-type hearing aid because of the accuracy problem [18]. Furthermore, this method is hard to use with babies or children because it limits the body movement when the gain measurement is processed. The functional gain verification method which measures the sound field threshold includes the coupler method and the ear simulator method. This method measure the performance of a hearing aid with regard to the residual ear canal volume by using a standard coupler. The use of a standard coupler and an ear simulator enables high test-retest reliability and convenient measurement. However, the volume of a standard coupler, 2 cc, is not a good approximation of the average residual adult ear canal volume when a hearing aid is worn. The ear simulator enables various residual volumes to be tested instead of a single 2 cc volume. Unfortunately, the simulator fails to consider accurate characteristics of the residual ear canal volume because the simulator has a

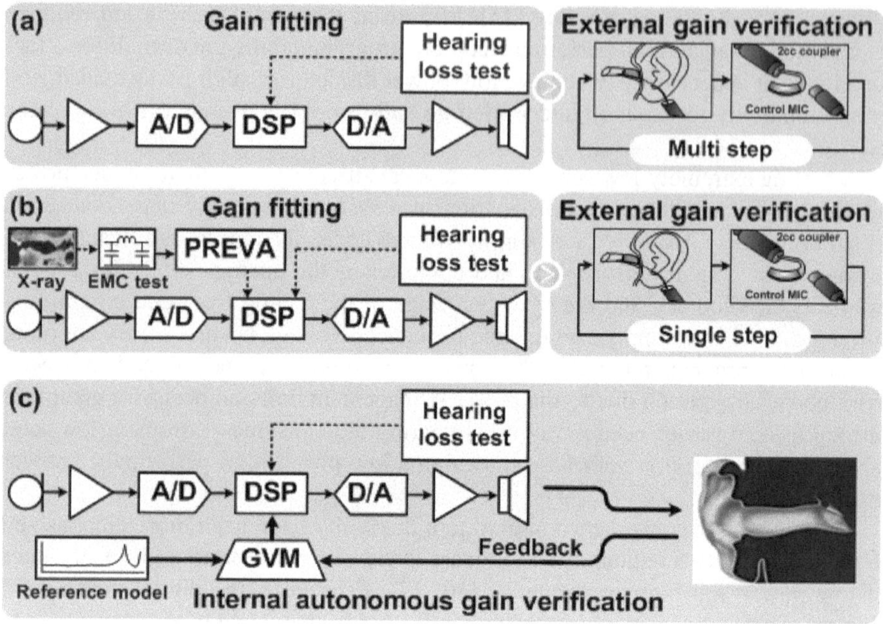

Fig. 11.3 Architecture of the various hearing aids

limited number of cavities. In addition, these methods fail to reflect any effects due to a vent or leakage path.

One way of reducing the number of iterations for fast gain verification and enhancing the performance optimization is to use a pre fitting verification algorithm and an ear modeling circuit as shown in Fig. 11.3(b): Therefore, human factors can be incorporated into a hearing aid before the chip is fabricated [19, 20]. In spite of its fast gain fitting and verification, the hearing aid in Fig. 11.3(b) still requires multiple external gain verification processes because of its open loop gain verification structure.

In this chapter, a real-time autonomous gain verification algorithm using feedback control is presented. Four different internal signal paths that cause an unpredictable gain variation are modeled and analyzed. This method minimizes the resonance gain error due to the residual volume differences to achieve a user-optimized performance with a single hearing loss test result.

11.3.2 An Internal Gain Verification Algorithm

11.3.2.1 Conventional Gain Verification Method

To achieve proper performance of the hearing aid according to the patient's condition, a personal hearing loss test should be performed for the first time. Then second hearing test is carried out to verify the operation of the hearing aid when the hearing aid is worn with the inserted gain. However, users usually complained that there exist significant differences between the gain from a first hearing loss test results and the gain from a second hearing loss test results which are patient actually demanded. These gain differences are originated from a different residual ear canal volume after the hearing aid is worn and a various sound path effects such as vent effect and leakage path effect. Therefore gain verification methods have been proposed to reduce the gain differences before and after the hearing aid are worn.

A functional gain verification method compares sound field thresholds of a psychoacoustic measurement before and after the hearing aid is worn. This method has lots of restrictions such as poor test-retest reliability, limited frequency resolution, inefficiency, and inability to provide any information about real-ear maximum output levels [21]. An insertion gain verification method uses a probe-tube microphone to measure the internal gain after the hearing aid is worn. However severe errors of the measured gain still exist due to the position of the probe-tube microphone which reduces the gain accuracy. Moreover, it is hard to apply this method to the babies or the kids because it limits the body movement during the gain measurement period. Due to these significant problem, a completely in the canal (CIC) type hearing aid tents to use the functional gain verification method instead of the insertion gain verification method [21].

Since these gain verification methods adopt open loop architecture, they require external control to apply the revised gain to the hearing aid which interrupts fast and accurate gain verification. In addition, the residual ear canal volume and the sound path have significantly different values according to the user's condition which makes conventional gain verification hard. That means the conventional gain verification methods with an open loop gain verifying algorithm cannot optimize the performance of the hearing aid quickly and accurately according to the different ear characteristics.

An alternative method, a real time internal gain verification algorithm which adopts closed loop mechanism is proposed to solve these problems. The proposed method enables an accurate measurement of the resonance gain due to the residual volume of the external ear canal by using an internal sound feedback through various sound traveling paths. Moreover, this method also extracts additional gains due to the vent effect and the leakage path effects. To prove the co-operation between the proposed gain verification algorithm and the hearing aid, a new system, hearing aid processor, is designed and implemented.

11.3.2.2 Autonomous Gain Verification Algorithm

Various Sound Paths in the Ear Canal

When a patient wears a hearing aid, the remaining ear canal is bound by the tympanic membrane, the end of the earmold, and the walls of the ear canal. The residual volume of the ear canal generates an additional internal sound pressure whenever

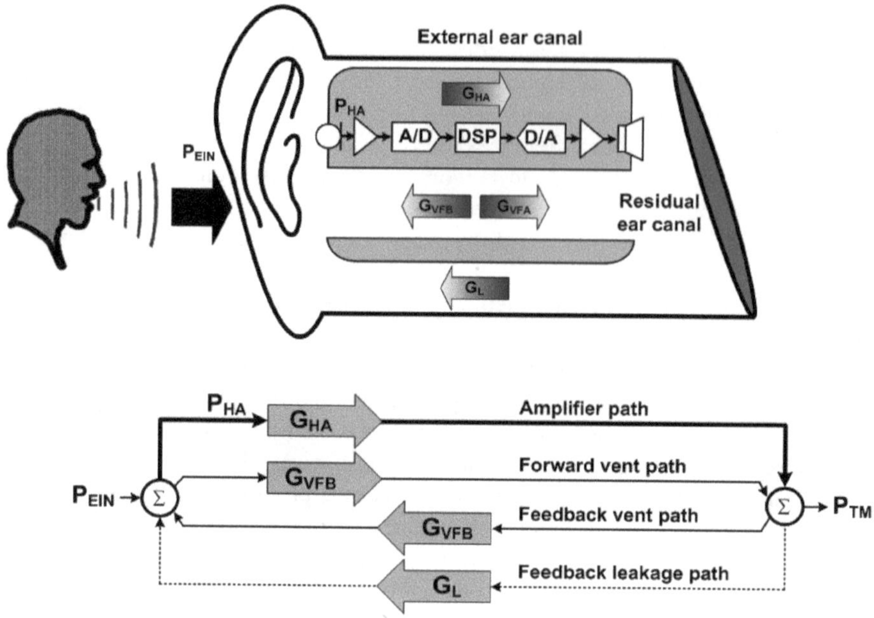

Fig. 11.4 Modeling of the sound paths and internal gains

these boundaries vibrate for various reasons. One way of reducing this type of occlusion effect, is to use keep the residual ear canal volume open by means of a vent. However, the vent affects the low frequency gain of a hearing aid by forcing sounds out of the ear canal and by enabling sounds to enter the residual ear canal volume [22]. Figure 11.4 shows simplified sound paths that cause feedback and feed forward additional gains. The paths and their gains can be listed as follows:

Sound travels from a source to the eardrum via a forward amplifier path: $G_{HA.}$ The hearing aid gain of the amplifier path is presented as a product of preamplifier gain and a DSP gain given by

$$G_{\mathrm{HA}} = G_{\mathrm{PREAMP}} \cdot G_{\mathrm{DSP}} = \frac{W_1 L_2 \left(1 + \dfrac{V_x}{V_{\mathrm{dd}} - V_{\mathrm{VC}}}\right)}{W_2 L_1 \left(1 - \dfrac{V_x}{V_{\mathrm{dd}} - V_{\mathrm{VC}}}\right)}, \quad \begin{array}{l} V_1 = V_{\mathrm{VC}} - V_x \\ V_2 = V_{\mathrm{VC}} + V_x \end{array} \tag{11.1}$$

For the preamplifier gain, G_{PREAMP}, V_{VC} is a volume control voltage, V_1 and V_2 are resistance control voltages, and W_i and L_i are width and length of the MOS resistive circuits, respectively [23]. The DSP gain, G_{DSP}, is acquired from the initial hearing loss test results and various fitting formulas.

Sound travels from a source to the eardrum via a forward vent path: $G_{\mathrm{VFA.}}$ The forward vent gain due to the commercially used vent with the hearing aid is derived by

$$G_{\mathrm{VFA}} = \frac{1 + e^{-\pi f_1/F_n}}{2} \cdot \frac{1 - z^{-1}}{1 - e^{-\pi f_1/F_n}} \qquad (11.2)$$

where f_l is a filter corner frequency, and F_n is a Nyquist rate.

Sound travels from a receiver to the microphone via the feedback vent path: G_{VFB} The backward vent gain due to the commercially used vent with the hearing aid is derived by

$$G_{\mathrm{VFB}} = \frac{1 - e^{-\pi f_1/F_n}}{1.12} \cdot \frac{1 + 0.12z^{-1}}{1 - e^{-\pi f_1/F_n}} \qquad (11.3)$$

where f_l is a filter corner frequency, and F_n is a Nyquist rate.

Sound travels from a receiver to the microphone via the feedback leakage path: $G_{L.}$ The leakage path is composed of a number of sub-sections which are presented as distributed RLC network. These sub-sections are regarded as cylindrical tube with rigid walls to apply the transmission line matrix method [24]. The transmission matrix T_n of the nth section and the reflection matrix $R_{n,n+1}$ between neighboring sections "n" and "n + 1" is given by [25]

$$T_n = \begin{pmatrix} e^{-\gamma_n x} & 0 \\ 0 & e^{-\gamma_n x} \end{pmatrix}, \quad R_{n,n+1} = \frac{1}{2\sqrt{Z_n \cdot Z_{n+1}}} \begin{pmatrix} Z_n + Z_{n+1} & Z_n - Z_{n+1} \\ Z_n - Z_{n+1} & Z_n + Z_{n+1} \end{pmatrix} \qquad (11.4)$$

where γ_n is a propagation constant, χ is the section length, and Z_n is a characteristic impedance. Therefore, the leakage path gain G_L is given by

$$G_L = \left(\prod_{i=0}^{n-1} (Z_{i,\,i+1} \cdot T) \right) \cdot R_{n,\,n+1} \qquad (11.5)$$

Conventionally, an internal feedback sound signal due to the vent or leakage path generates feedback oscillation or howling, which is a major problem for hearing aids. As a result, many techniques have been employed to reduce the vent effect or leakage path effect by controlling the sound phase or by shifting the sound frequency. In addition, the vent and the leakage paths generate additional internal gain in relation to parameters such as the locations, the lengths, and the diameters of the paths.

However, the vent and leakage path also facilitate the internal gain estimation. In the proposed gain verification method, various sounds that travel through the vent and the leakage path, as shown in Fig. 11.4, are used to estimate the verification gain from the residual ear canal volume. To establish the gain verification method, we first need to model an internal sound path, and a transfer function of acoustic feedback and a feed forward path.

Figure 11.4 also shows the simplified gain modeling of a hearing aid system. The hearing aid includes a microphone, a preamplifier, a ΣΔ ADC, a DSP, a ΣΔ DAC, and a receiver. To model the acoustic responses of the hearing aid system, we first need to define the pressure level. In this system, P_{TM} represents the pressure at the tympanic membrane, P_{HA} is the pressure in front of the hearing aid microphone, and P_{EIN} denotes the external sound input pressure of the overall system. Two different kinds of insertion gain are generally used to verify whether the hearing aid gain corresponds to each individual user.

A *real-ear aided gain* can be extracted by using a probe-tube when the hearing aid is inserted into the external ear canal and operated normally. In this case, the gain can be modeled as follows.

$$\frac{P_{\mathrm{TM}}}{P_{\mathrm{EIN}}} = \frac{G_{\mathrm{HA}} + G_{\mathrm{VFA}}}{1 - (G_L + G_{\mathrm{VFB}}) \cdot G_{\mathrm{HA}}} \tag{11.6}$$

A conventional *real-ear aided gain* can be calculated from Eq. 11.6 with only feed forward vent effect by setting $G_L = G_{\mathrm{VFB}} = 0$:

$$\frac{P_{\mathrm{TMCON}}}{P_{\mathrm{EINCON}}} = \left.\frac{P_{\mathrm{TM}}}{P_{\mathrm{EIN}}}\right|_{G_L = G_{\mathrm{VFB}} = 0} = G_{\mathrm{HA}} + G_{\mathrm{VFA}} \tag{11.7}$$

A *real-ear occluded gain* can be measured when the hearing aid is inserted into the external ear canal without any operation. From Fig. 11.4, the gain can be expressed as follows when $G_{HA} = 0$.

$$\left.\frac{P_{\mathrm{TM}}}{P_{\mathrm{EIN}}}\right|_{G_{\mathrm{HA}}=0} = \frac{G_{\mathrm{HA}} + G_{\mathrm{VFA}}}{1 + (G_L + G_{\mathrm{VFB}}) \cdot G_{\mathrm{VFA}}} = \frac{G_{\mathrm{VFA}}}{1 + (G_L + G_{\mathrm{VFB}}) \cdot G_{\mathrm{VFA}}} \tag{11.8}$$

A conventional *real-ear occluded gain* can be calculated from Eq. 11.8 with regarding only feed forward vent effect can be calculated by setting $G_L = G_{\mathrm{VFB}} = 0$:

$$\frac{P_{\mathrm{TMCON}}}{P_{\mathrm{EINCON}}} = \left.\frac{P_{\mathrm{TM}}}{P_{\mathrm{EIN}}}\right|_{G_L = G_{\mathrm{VFB}} = G_{\mathrm{HA}} = 0} = G_{\mathrm{VFA}} \tag{11.9}$$

Autonomous Internal Gain Verification

Figure 11.5 shows a block diagram of the proposed internal feedback verification method. The system has two operational modes: a gain verification mode, for achieving an internal gain due to various leakage effects and a residual ear canal volume and a hearing aid mode, for when the system operates as a low power hearing aid.

During the gain verification mode, a test signal generator produces a pure sinusoidal test signal ranging from 0.5 to 8 kHz in steps of 500 Hz, and each sinusoidal signal is maintained for 0.5 s. The sound oscillation due to the internal positive feedback only occurs when the amount of amplification through the hearing aid is greater than the amount of attenuation from the ear canal back to the microphone. Thus, the

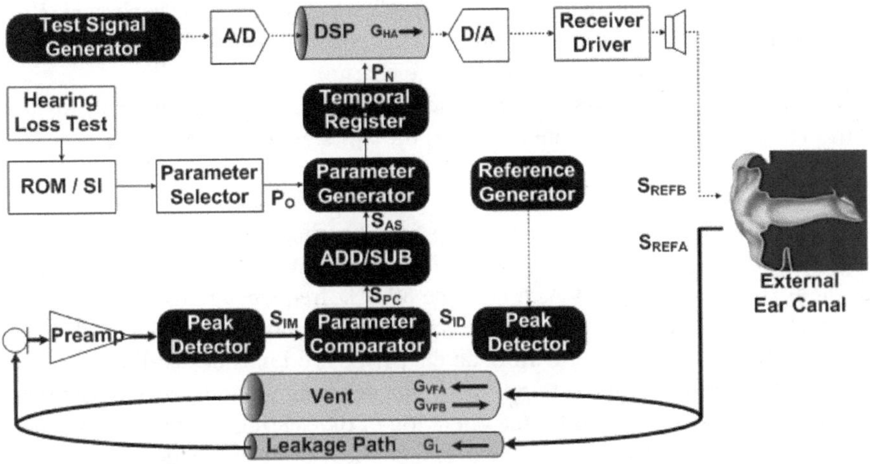

Fig. 11.5 Block diagram of the internal feedback verification methodology

amplitude of the test signal is small enough to prevent any internal oscillation due to the large input signal [26].

After the single test signal is injected and processed via the hearing aid part, the test signal has a specific DSP gain, P_O, through the DSP function. The parameter P_O is the initial gain from the hearing loss test results for each individual user. The hearing aid output signal, S_{REFB}, ahead of the external ear canal contains only hearing aid gain, G_{HA}, whereas the input signal of the vent, S_{REFA}, contains the internal gains G_{VFA}, G_{VFB} and G_L as well as G_{HA}. Due to the internal transfer function from Eqs. 11.6 and 11.9, the S_{REFB} signal and the S_{REFA} signal have a different value which vary for each individual user and each frequency range. The S_{REFA} signal includes the gain from the hearing loss test results and the gain from the vent effects, the leakage effects, and the residual ear canal volume effect for each individual user. The main purpose of the internal gain verification algorithm is to obtain a verification gain that generates the S_{REFA} signal for each frequency range without any external gain fitting verification methods, such as functional gain method or an insertion gain method.

The S_{REFA} signal is fed back through a vent and another leakage path to the microphone and injected into the preamplifier. In this case, the preamplifier has a unit gain which is used to extract the pure internal gain of G_{VFA}, G_{VFB} and G_L. The S_{IM} signal from the preamplifier output includes the overall gain of G_{HA}, G_{VFA}, G_{VFB} and G_L, whereas the S_{ID} signal from the $\Sigma\Delta$ DAC output includes only the hearing aid gain, G_{HA}. A parameter comparator subsequently compares the S_{IM} signal with the S_{ID} signal to extract the internal gain of G_{VFA}, G_{VFB} and G_L. When the S_{IM} signal has a higher level than the S_{ID} signal, the parameter comparator generates a value of 1. Otherwise, the parameter comparator generates a value of 0. The S_{PC} signal controls the ADD/SUB block to generate the S_{AS} signal, which increases or decreases the value of the parameter P_O. The first parameter to be generated, P_{N_1}, which denotes the DSP parameter for the first frequency range of 500 Hz frequency range, is used as new DSP parameters of the hearing aid part.

With the iteration process, DSP parameter set value can be continuously updated until the S_{IM} signal has the same value as the S_{ID} signal. Hence, finally generated parameter, P_{N_1F}, includes the gain of G_{HA}, G_{VFA}, G_{VFB} and G_L through numerous internal iterations. The P_{N_1F} value is stored in a temporal register as a DSP parameter and, before the power supply of the hearing aid is turned off, it can be used as a hearing aid gain for a 500 Hz frequency range. After the iteration cycle for the first frequency step, which is in the 500 Hz frequency range, the next iteration cycle for a 1 kHz frequency range starts and continues for 0.5 s with the same process as mentioned before. During an 8 s period, 16 kinds of iteration cycles are completed according to the different 16 frequency ranges and the newly generated DSP parameter, P_N, is stored for the operation of the hearing aid. For a stable mode change between the gain verification mode and the hearing aid mode, a 2 s period is supplemented as extra time.

After the gain verification mode, the hearing aid mode starts. A 10 s counter controls the switch block, and the switch block connects the preamplifier to the $\Sigma\Delta$ ADC to perform a hearing aid function. In the hearing aid mode, all the parts of the gain verification mode are turned off to reduce the extra power consumption. An external audio sound is injected into the preamplifier, which has a specific gain value. This gain value corresponds to the condition of the user and the DSP gain parameters, P_N, which is generated in the gain verification mode. A feedback reduction algorithm uses the external internal gain parameters fro the gain verification mode to prevent any internal feedback oscillation via the vent path, the leakage path, and the residual ear canal volume. Because the leakage path and the residual ear canal volume of the user are changed in time whenever the user wears the hearing aid, the internal gain of G_{VFA}, G_{VFB} and G_L should be recalculated every time the hearing aid is worn and the battery should be turned on. Thus, the hearing aid mode with the parameter P_N is maintained until the battery power is turned off and the new parameter is achieved the next time the hearing aid is worn.

11.3.2.3 Simulation Results

The simulation results of the real-time gain verification algorithm are shown in Fig. 11.6. In this simulation, the input audio signal has 500 Hz and 1 kHz frequency component, respectively. Because the reference signal has a smaller value than the feedback signals through the first iteration, the next iterations are performed to reduce the value of the feedback signals by decreasing the DSP parameters. Hence, the additional gain parameters, which are the first DSP parameter, are continuously updated through repeated iteration until the feedback signals have the same value as the reference signal. During the assigned time for each frequency range, 0.2 s, we simulate the responses of the gain verification algorithm for a 500 Hz and a 1 kHz frequency range and the achieved additional gain parameter, respectively. The numbers of iterations vary between a 500 Hz and a 1 kHz input signal since the hearing

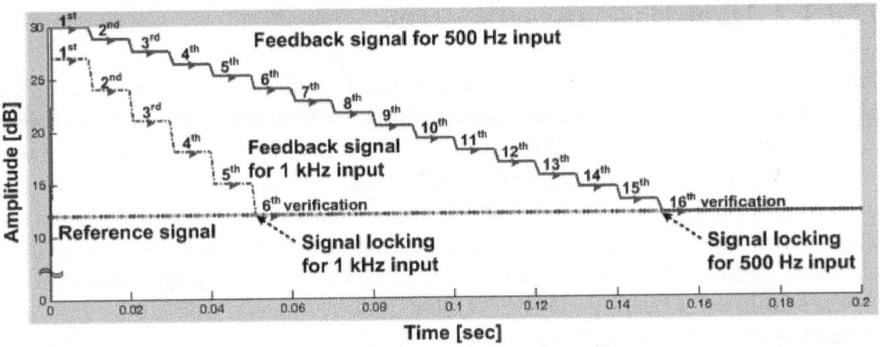

Fig. 11.6 Simulation results of the real-time gain verification algorithm

loss values are changed at different frequency ranges. In addition, the number of iteration steps depends on the amplitude difference between the reference signals and the feedback signals. Therefore, six and sixteen internal iterations are required for a 1 kHz and 500 Hz input signals, respectively in this simulation. By analyzing and considering the gain differences between the reference signals and the feedback signals at a various frequency ranges, a 0.2 s time step is determined for each frequency range.

Figure 11.7 shows a simulation result of the hearing aid chip with the proposed gain verification algorithm. During the internal gain verification period, the test signal generator creates 16 pure sinusoidal test signals from 0.5 to 8 kHz frequency range in steps of 500 Hz. In each frequency steps, additional gain parameter is updated to regard an additional gain. In this simulation, the vent specifications, such as a diameter and a length, are achieved from the conventional completely-in-the-canal (CIC)-type hearing aid.

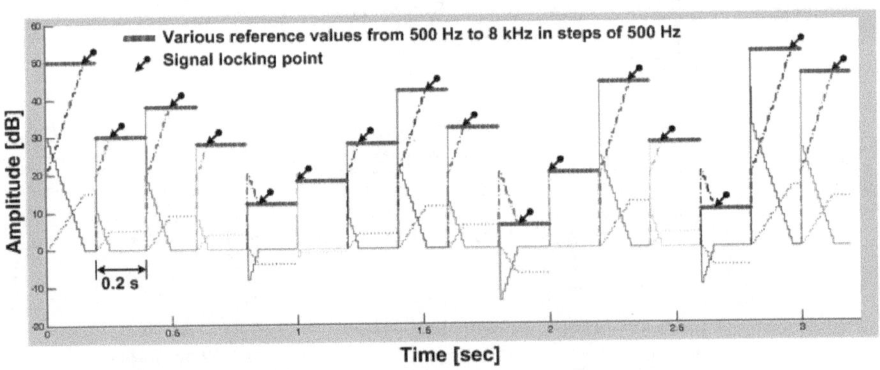

Fig. 11.7 Simulation results of the hearing aid with the proposed algorithm

11.3.3 A Multi Mode Audio Processor

Traditional hearing aid devices only concentrated on an exact amplification of the external sound signal and lossless transmission to the middle ear or inner ear according to patient's requirement. All functions of the hearing aid were designed and implemented to relieve pathological symptoms. Since the hearing aid was usually considered as a medical device which reveals user's disease, the patient who did not want to show their private disease avoids wearing of the hearing aid in spite of their serious symptoms. Therefore, there have been lots of attempts to fine revolutionary new possibilities in hearing aid for everyone, not just people who are deaf or hard of hearing as a personal audio assistant device [27].

To satisfy these requirements multi mode audio processor is proposed and implemented in Fig. 11.8. The fabricated processor is composed of three different modes; a hearing aid mode, a smart earphone mode and a direction perception mode. The hearing aid mode offers a conventional hearing aid operation with accurate external sound amplification and transmission according to the user's demands. The hearing aid processor also operates as a smart earphone by receiving an electrical sound signal from an external audio source such as MP3 player. Finally the direction perception mode enables a discrimination of a sound source location which is essential to enhance an understanding of speech. The fabricated audio processor introduces the possibilities of hearing aid as a personal audio assistant device to allow everyone to control and enhance the sounds around us.

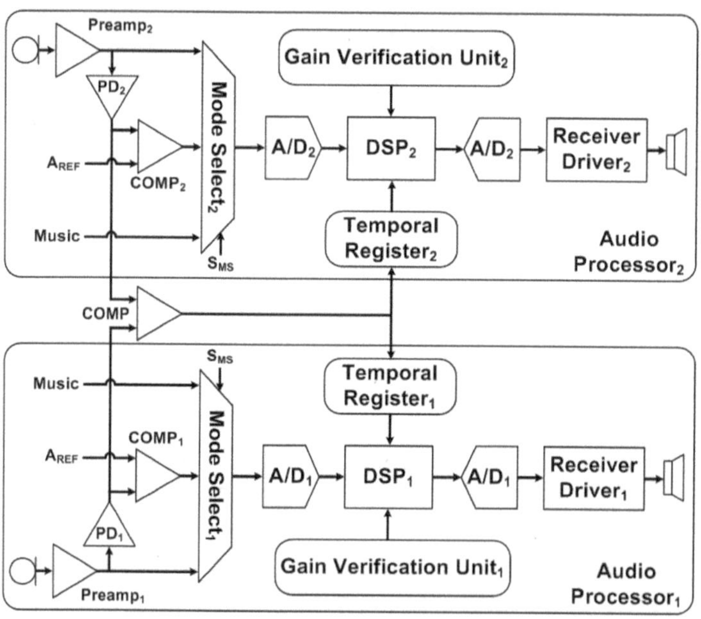

Fig. 11.8 Multi mode audio processor

11.3.3.1 Hearing Aid Mode Operation

Figure 11.9 shows a hearing aid mode operation of the proposed audio processor. It is composed of a microphone, a multi-threshold preamplifier, an adaptive-SNR $\Sigma\Delta$ ADC, a multi-channel DSP, a heterogeneous $\Sigma\Delta$ DAC, and a receiver. The mode selection signal S_{MS} determines the operation between the smart earphone and the hearing aid. When the signal S_{MS} is set to 1, the preamplifier amplifies the speech sound from an external surrounding to allow a hearing aid operation. In this case, a peak detector (PD) and a comparator (COMP) are turned off to reduce extra power dissipation.

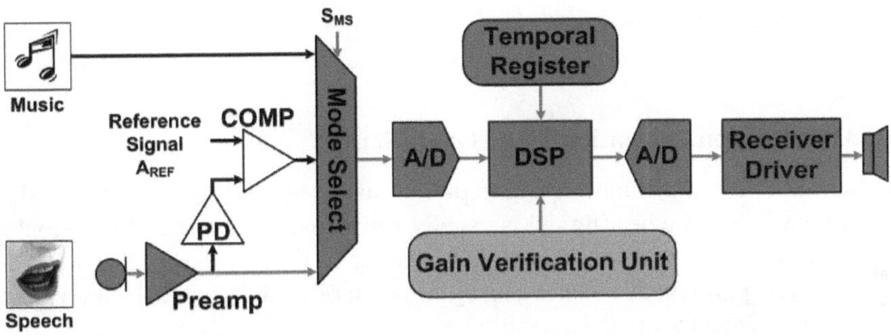

Fig 11.9 Hearing aid mode operation

11.3.3.2 Smart Earphone Mode Operation

The requirement of the hearing aid as a personal audio assistant device has been increased to control an auditory ability and offer a convenience for each individual user [27]. The proposed multi mode hearing aid processor can satisfy these necessities by operating as a personalized smart earphone and a convenient hearing aid applications at time same time.

Figure 11.10 represents a smart earphone mode operation of the hearing aid processor. When the signal S_{MS} is set to 0, the hearing aid processor can perform a smart earphone operation by receiving an external audio signal from electrical devices such as MP3 player. Since an internal path from a preamplifier to the ADC is disconnected when the smart earphone operation is activated, a user cannot hear an external speech sound or an urgent sound. To protect a patient from an urgent situation and offer a proper conversation environment when the earphone mode is activated, the preamplifier always senses an external speech and environment sound and compares its value with an A_{REF} value, a variable threshold voltage to avoid an urgent situation. When the external sound level is higher than the A_{REF} value, the hearing aid mode operation is started regardless of the value of S_{MS} to protect a patient from environmental accidents and provide conversation circumstances. In this work, the A_{REF} value is determined to recognize somewhat higher speech conversation level, an 80 sound pressure level (SPL).

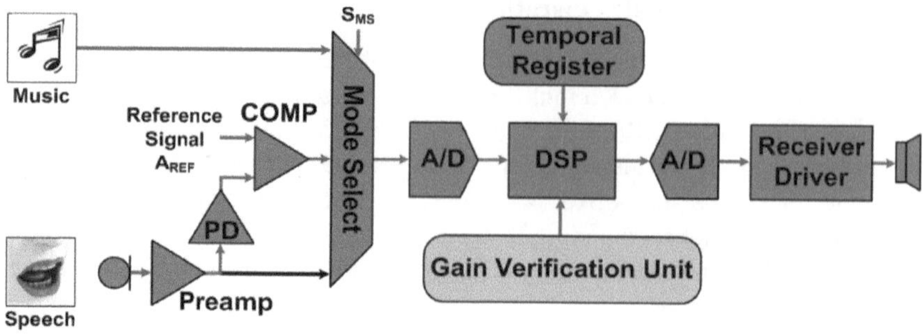

Fig. 11.10 Smart earphone mode operation

11.3.3.3 Direction Perception Mode Operation

The normal one can distinguish a direction of sound by sensing an external sound with two ears. Since a hearing aid is usually worn solely, there are lots of troubles to sense a sound localization with single wearing of hearing aid. In this work, the direction perception mode enables a discrimination of a sound source location which is essential to enhance an understanding of speech.

Figure 11.11 shows a direction perception mode operation of the audio processor. When an external sound source is injected into the hearing aid, two peak detectors, PD_1 and PD_2, generate an envelope of the external sound from each preamplifier of the hearing aid. The generated output amplitude of the peak detector is compared by using a common comparator, COMP to determine accurate sound localization. The output of the COMP controls the gain parameters of the DSP to inform the localization of the sound by increasing the volume of the hearing aid between the right side and the left side of the digital hearing aid. By adopting this operation, a user can distinguish the location of the sound through the different sound level.

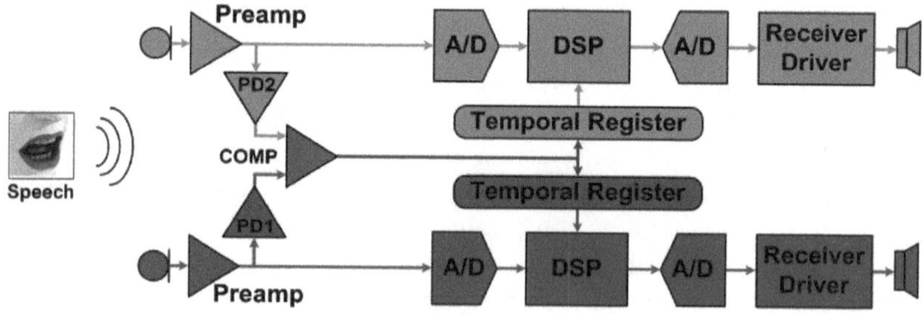

Fig. 11.11 Direction perception mode operation

11.3.4 Low Power Analog Front-End

11.3.4.1 System Design Considerations

There have been continuous attempts to design a hearing aid system which satisfies both low-power and high-performance characteristics. However, a high-performance hearing aid system inevitably allows excessive power dissipation. As such, this trade-off has been an obstacle to the design of a low-power, high-performance hearing aid system.

In order to achieve high-performance hearing aid system, novel circuit techniques have been suggested such as adaptive noise reduction [28]. This technique is attractive to achieve wide dynamic range with personal calibration by attenuating noise level independently in each frequency band. However, this algorithm is only focused on performance increase and needs distinct power management unit which operation is usually based on a control of digital part. In the hearing aid system, power consumption of the digital part is insignificant compared with analog part. Therefore, in order to design low–power, high–performance hearing aid, the power reduction of the analog front–end, which is the most power–consuming part in a hearing aid system, is inevitable.

To alleviate these design difficulties, the new design method which contains dynamically varying structure is proposed. Figure 11.12 shows a proposed hearing aid system wherein the SNR changes dynamically.

By controlling both the structure and clock frequency adaptively, proposed hearing aid system acquires both the low-power dissipation and high-performance using environmental monitoring. Further, modifying the parameters adaptively, whereby the gain and threshold voltage of the preamplifier are determined according to the input amplitude, offers more power reduction. By extending these low-power techniques to the overall hearing aid system proposed here, system power dissipation is reduced drastically with little degradation of system performance.

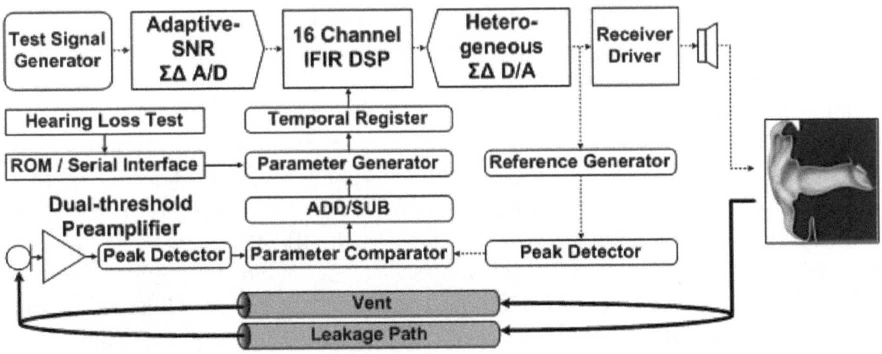

Fig. 11.12 Block diagram of proposed hearing aid system

11.3.4.2 Overall Architecture of the Analog Front-End

The architecture of the proposed analog front–end is shown in Fig. 11.13. It consists of a multi-threshold preamplifier and an ΣΔ ADC.

The proposed analog front-end architecture achieves not only high–performance but also the power optimization according to the control parameters obtained from the external environmental monitoring. To classify and extract these parameters, the proposed analog front-end includes an off-chip DSP.

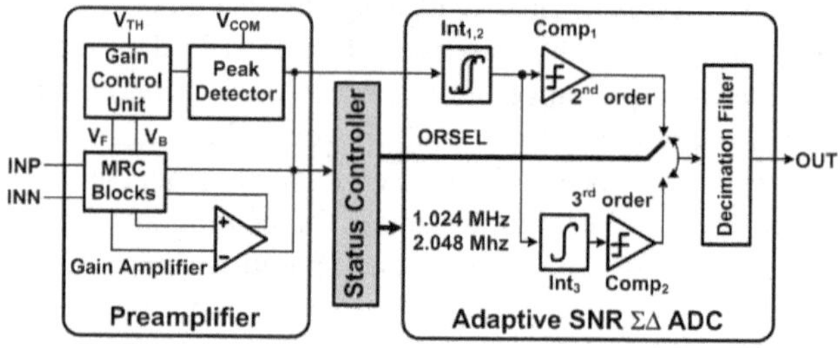

Fig. 11.13 Proposed analog front-end architecture

The multi-threshold technique is employed for accurate preamplification and high flexibility and the Adaptive-SNR technique for wide dynamic range with low power consumption, respectively. In the conventional preamplifier, an automatic gain control (AGC) and an exponential gain control (EGC) are designed separately because of design difficulties in standard CMOS technology [29, 30]. In this work, however, the multi-threshold technique integrates AGC and EGC techniques into a single block to reduce power consumption and to expand its dynamic range with accurate preamplification.

In the proposed ΣΔ ADC, the Adaptive-SNR technique which varies an order and a clock frequency adaptively by adopting control parameters is introduced. This technique offers various SNR values and results the energy-efficient analog front-end according to the aware of environmental conditions.

To generate the control parameters from the external environment, the proposed analog front-end uses the input amplitude of a microphone. Figure 11.14 shows the relationship between the input of the microphone and the input of a preamplifier. The preceding study revealed that a normal sound level common in human daily life ranges from 30 to 90 dB SPL, which corresponds to Range 1 in Figure 11.14 [31]. In this range, the sound amplitude is so small that a high performance analog front-end must be used. On the other hand, above 90 dB SPL, Ranges 2 ~ 4, a high performance analog front-end is not necessary since the sound amplitude is sufficiently large. If we use a high performance analog front-end for all sound regions, the analog front-end produces excessively high performance and dissipates power needlessly. In the design of the proposed analog front-end, the input sound level is divided into four parts as described in Fig. 11.14 to control the SNR separately at each range so as to optimize power and performance.

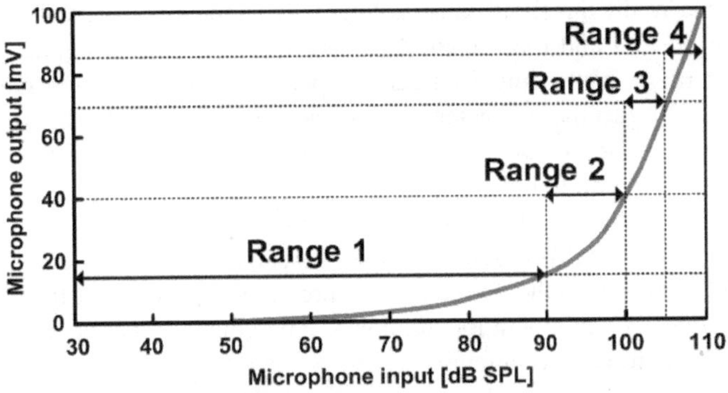

Fig. 11.14 Characteristics of microphone for digital hearing aid

11.3.4.3 Adaptive Analog Front-End Design

Architecture of the Multi-Threshold Preamplifier

Non-linear characteristics of the human auditory system allow the native ability of the automatic gain control to enable the wide dynamic range of the human ear. In order to imitate the capability of the normal human ear which covers the wide dynamic range, the conventional preamplifier usually adopts the single threshold knee voltage with the high compression ratio of the dynamic range. However the high compression of the input dynamic range causes a serious edge distortion at the conversation region. It requires an arduous and laborious training of a patient when he or she wears the hearing aid for the first time. Moreover, the high compression of the conversation region degrades the articulation index of the hearing.

To eliminate these problems and allow a wide linearity in the normal conversation range, the programmable multi-threshold preamplifier is implemented. Figure 11.15 shows the principle of the low power multi-threshold preamplifier with

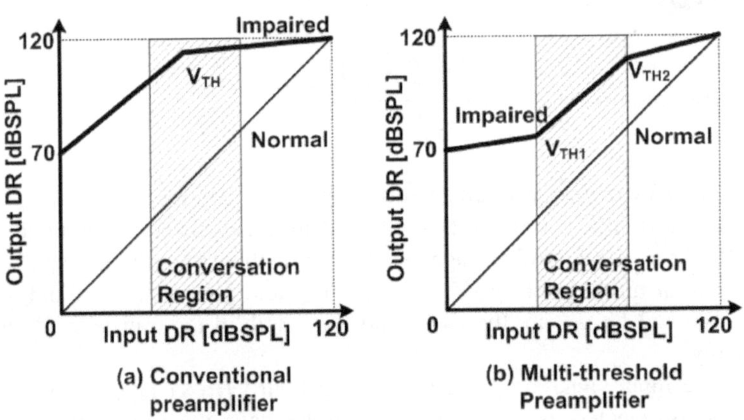

Fig. 11.15 Principle of the multi-threshold preamplifier

programmable multi-threshold knee voltages. It enables the same dynamic range between the normal and the impaired ear in the conversation region. In addition, the dynamic range of the impaired ear mapped to the conversation region can be changed according to the characteristics of the each individual user by controlling the two threshold knee voltages, V_{TH1} and V_{TH2}. It enhances the articulation index and flexibility of the digital hearing aid.

Figure 11.16 shows the block diagram of the low power programmable multi-threshold preamplifier. To reduce the power dissipation, only two MOS–resistive circuit (MRC) units with the gain amplifier are adopted instead of the conventional three MRC units to implement the automatic gain control function and exponential gain control function at the same time.

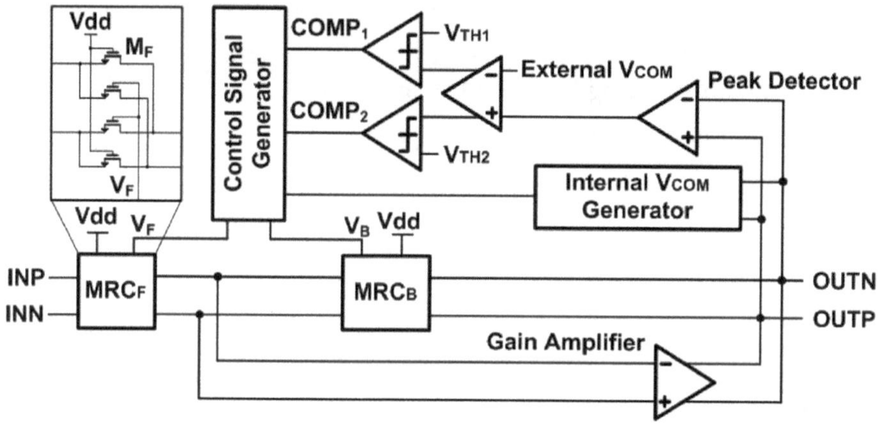

Fig. 11.16 Architecture of the fabricated multi-threshold preamplifier

The voltage gain of the implemented multi-threshold preamplifier is expressed in the form of (10) as

$$A_{PREAMP} = \frac{W_1L_2\left(1+\frac{V_{DIF}}{V_{dd}-V_{COM}}\right)}{W_2L_1\left(1-\frac{V_{DIF}}{V_{dd}-V_{COM}}\right)},\quad \begin{array}{l} V_F = V_{COM} - V_{DIF} \\ V_B = V_{COM} + V_{DIF} \end{array} \tag{11.10}$$

where V_F and V_B are negative and positive resistance control voltages respectively and V_{COM} is a volume control voltage which determines the common mode level of V_F and V_B. V_{DIF}, the control voltage determined by the preamplifier output, decides the differential mode level of V_F and V_B. V_{TH1} and V_{TH2} are threshold voltages and W_F, W_B and L_F, L_B are the width and the length of the transistor M_F and M_B respectively.

When the input signal amplitude is smaller than V_{TH1} or larger than V_{TH2}, the V_{DIF} value is selected as 0 and the V_F and V_B have the same value of V_{COM}. Therefore the voltage gain of the preamplifier is determined from the dimension of the MRC units and the value of the V_{COM}. If the input signal has an amplitude value between the V_{TH1} and V_{TH2}, the V_{DIF} has different values to provide a various gain according to the each individual user. Moreover the threshold gain is changed with the variation of the V_{COM} to prevent the howling problem due to the positive feedback of the input sound.

Adaptive-SNR Analog Front-End with an Internal Status Controller

To design an adaptive analog front-end, the proposed $\Sigma\Delta$ ADC should provide various kinds of SNR values to reduce power dissipation according to the input amplitude. Various methods for achieving different SNRs have been proposed to date, including changing the clock frequency. By changing the clock frequency, the $\Sigma\Delta$ ADC achieves different SNR characteristics. However, adopting a high clock frequency complicates the design of an analog circuit, such as operational transconductance amplifier (OTA), since the unity-gain frequency of the OTA should be three or four times higher than the clock frequency of the $\Sigma\Delta$ ADC [32]. A high order $\Sigma\Delta$ modulator is an alternative approach to modify the SNR. In the case of more than a 3rd order $\Sigma\Delta$ modulator, however, the performance of the $\Sigma\Delta$ modulator seriously suffers from nonidealities due to the finite gain and bandwidth of the OTA and instabilities caused by saturation of the integrator [33].

One possible method to reduce power dissipation is to allow the internal attributes of the circuit to vary, depending on the task at hand [34]. If the structure of the system changes dynamically according to the control variables, which are generated externally or internally, the system will have programmable or adaptable characteristics. This is called the dynamic structure variation technique. A simple way to accomplish this technique is to apply multi-sub systems with different structure into the system. By selecting the appropriate structure among the multi-sub systems according to the control variables, the desired system can be implemented dynamically. However, designing these multi-sub systems occupies a great deal of chip-area and the chip cost is increased. Therefore, to adopt the adjustable system structure for low power dissipation without any area penalty, we introduce the adaptive-SNR technique to the $\Sigma\Delta$ modulator.

The proposed multi-order single-loop $\Sigma\Delta$ ADC topology is shown in Fig. 11.17. By adapting bypass signal path PA, proposed $\Sigma\Delta$ ADC offers a multi-order

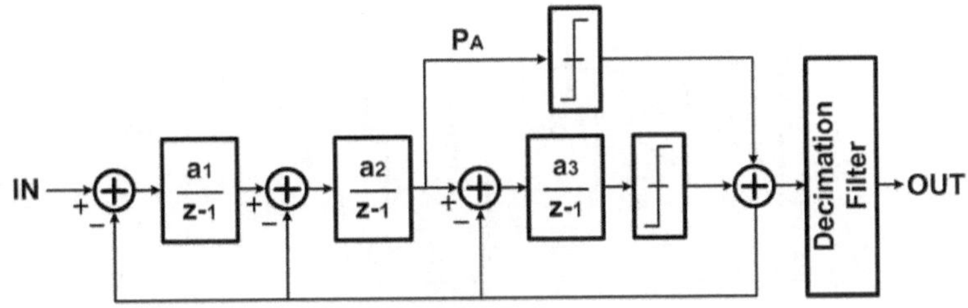

Fig. 11.17 Proposed multi–order single–loop $\Sigma\Delta$ ADC

topology. When the signal path PA is enabled, this ΣΔ ADC act as a 2nd order one by disabling 3rd orders integrator. Otherwise, this ΣΔ ADC operates as a 3rd order one by disabling the signal path P_A. The loop coefficients a_1, a_2, and a_3 are set to 0.3, 0.45, and 0.5, respectively.

Figure 11.18 shows High-level simulation results of the proposed ΣΔ ADC compared with those of the conventional ΣΔ ADC. Different kinds of SNR values can be generated according to the input amplitude.

The proposed ΣΔ ADC is described in Fig. 11.19. A switch SW_1 determines the order of the ΣΔ ADC between the 2nd and 3rd while SW_2 chooses its clock frequency. A combination of these switches allows the ΣΔ ADC to obtain four different kinds of SNRs. The control parameters of the ΣΔ ADC are externally stored and applied by the control register and the DSP for easy and convenient management. By selecting the proper parameters according to the input amplitude, the ΣΔ ADC obtains optimal SNR from in terms of power and performance.

Figure 11.20 shows the detailed architecture of the ΣΔ ADC. When the SW_1 is closed, it operates in 2nd order by bypassing the output from the 2nd integrator to the *OUTN* or *OUTP*. Because the resistance value of the conventional *N*-type switch varies according to the drain voltage, the 2nd integrator output degrades seriously

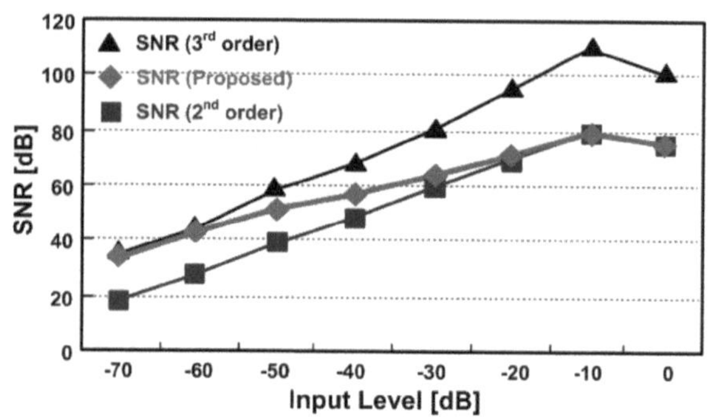

Fig. 11.18 Simulation results of the conventional vs. proposed ΣΔ ADC

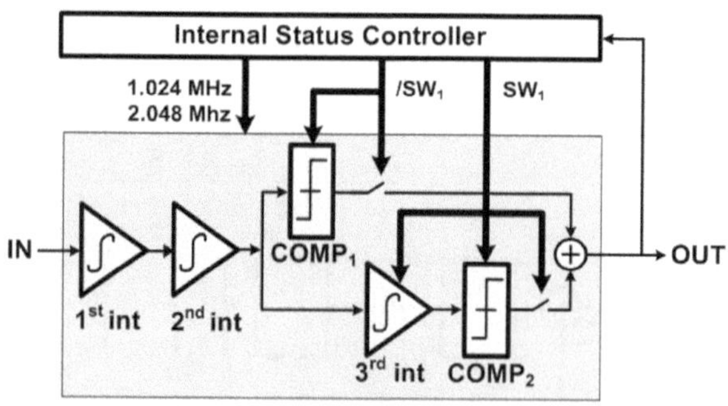

Fig. 11.19 Proposed ΣΔ ADC exploiting adaptive-SNR technique

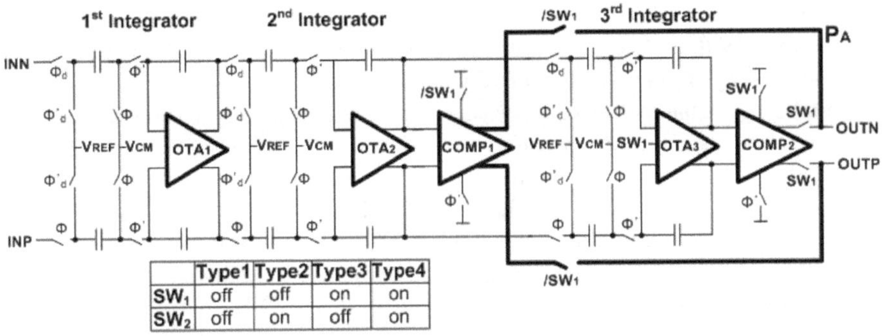

Fig. 11.20 Detailed $\Sigma\Delta$ modulator architecture

when it passes through SW_1. Hence, in order to prevent this distortion, a high performance switch is necessary so that its resistance value is constant under the entire range of drain voltage. However, the high performance switch is difficult to design and consumes additional power. Therefore, we adopts $COMP_1$, which converts the 2nd integrator output into a PWM signal and passes it through the $/SW_1$ without signal distortion. When the $COMP_1$ is activated, the 3rd integrator and $COMP_2$ are completely turned off so as to eliminate extra power consumption. On the other hand, if the SW_1 is opened, the 3rd integrator accepts the output of the 2nd integrator as an input and performs a 3rd order modulation. In this phase, $COMP_1$ is turned off to avoid extra power dissipation. By turning SW_2 on and off, the clock frequency is changed between 2.048 and 1.024 MHz, respectively. By changing SW_1 and SW_2 separately, four configurations of the $\Sigma\Delta$ ADC having have different kinds of SNR are obtained, as summarized in the table of the Fig. 11.21. With SW_1 open, type

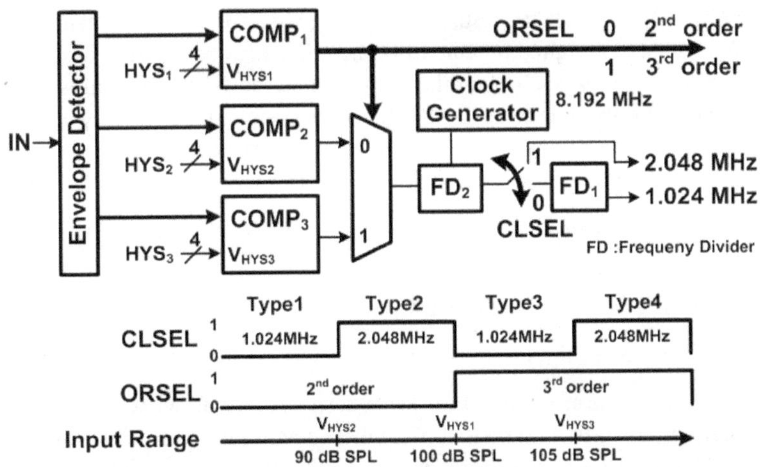

Fig. 11.21 Architecture and operation of the internal status controller

1 and type 2 2nd order $\Sigma\Delta$ modulators are achieved by turning SW_2 off and on, respectively. With SW_1 closed, type 3 and type 4 3rd order $\Sigma\Delta$ ADCs are realized by turning SW_2 off and on, respectively.

Another possible performance issue that is important to hearing aid users is the way audio characteristics can change abruptly in response to changes in the input level. To avoid the discontinuity problems that can occur when the order or clock frequency is changed, we designed the $\Sigma\Delta$ ADC so that it uses the gain control unit of the preamplifier. The fabricated gain control unit adopts the envelope of the output signal and modifies the preamplifier gain by controlling the V_{COM} value, thereby preserving the continuity of the SNR characteristics of the $\Sigma\Delta$ ADC.

Figure 11.21 shows the architecture and the operation of the fabricated internal status controller. Two control signals, the *ORSEL* signal and the *CLSEL* signal, are generated by the internal status controller, which senses the envelope of the $\Sigma\Delta$ ADC input signal to independently determine the order and clock frequency. The *ORSEL* signal selects between the second and third order of the $\Sigma\Delta$ ADC whereas the *CLSEL* signal chooses a reference clock frequency between 1.024 and 2.048 MHz. The combination of these two control signals produces four different states of the $\Sigma\Delta$ ADC for a fully internal adaptive-SNR analog front end.

The level of the output signal of the preamplifier is identified by the envelope detector and injected into the three comparators, each of which has a different hysteresis. The hysteresis value of $COMP_1$ is compared with the output signal of the envelope detector to generate the *ORSEL* signal. When the level of the input signal is less than 100 dB SPL, the *ORSEL* value changes to 0 to produce the second order $\Sigma\Delta$ ADC. On the other hand, when the level of the input signal is higher than 100 dB SPL, the *ORSEL* value changes to 1 to produce the third order $\Sigma\Delta$ ADC.

To change the clock frequency of the $\Sigma\Delta$ ADC, we adopted the *CLSEL* control signal. Furthermore, we determined the *CLSEL* value by comparing the hysteresis values of $COMP_1$ and $COMP_2$ with the output signal of the envelope detector. When the *ORSEL* signal is assigned a value of 0, the *CLSEL* value is either 0 or 1 depending on the value of V_{HYS2}. On the other hand, when the *ORSEL* signal is assigned a value 1, the *CLSEL* value is either 0 or 1 depending on the value of V_{HYS3}. The clock generator creates an 8.192 MHz internal system and, depending on the *CLSEL* signal, the internal system clock is divided into two reference clocks, with frequencies of 1.024 MHz and 2.048 MHz.

11.3.4.4 *Building* Block Circuits Design

Low Power Circuits for the Preamplifier

Figure 11.22 shows the structure of the MOS resistive circuit (MRC) of the proposed preamplifier and its basic two resistor model. The MRC is composed of four N-type transistors M_1. If all four transistors are operated in the triode region and the output voltage V_{o1} equals V_{o2}, the resistance from the MRC is as follows

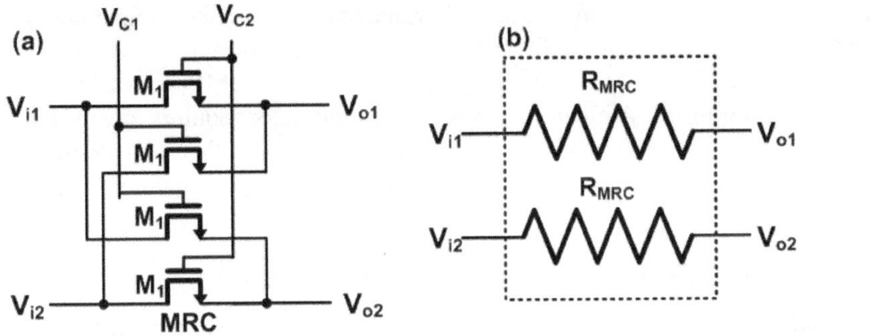

Fig. 11.22 (**a**) Structure of MRC (**b**) basic two resistor model

$$R_{\mathrm{MRC}} = \frac{V_{i1} - V_{i2}}{I_{i1} - I_{i2}} = \frac{L_1}{\mu_n C_{\mathrm{ox}} W_1 \left(V_{C1} - V_{C2}\right)} \tag{11.11}$$

where n is the carrier mobility, C_{OX} is the gate oxide capacitance per unit area, and W_I and L_I are the width and the length of the transistor M_1, respectively. From Eq.11.11, the resistance of the MRC can be changed by controlling the gate voltages V_{C1} and V_{C2} of transistor M_1.

In Fig. 11.23, a low power OTA for the proposed $\Sigma\Delta$ modulator is designed to have a compensated two-stage, which has an input stage with a cross-coupled active load and a class AB output. By using a PMOS input differential pair, the common mode level of the OTA is closed to ground level. This allows the use of small NMOS transistors for the switches of the $\Sigma\Delta$ modulator. Moreover, having the PMOS input differential pair minimizes the output noise due to the 1/f noise and optimizes the slew rate and unity-gain frequency [35].

The designed OTA shows 79 dB DC gain, 8.01 MHz unity gain bandwidth, and a 63 phase margin for a 3 pF load. For the proper operation of the $\Sigma\Delta$ modulator, the OTA dc gain is required to be of the order of 40–80 dB [36]. The two-stage OTA makes it possible to obtain sufficient dc gain. A comparator circuit is also designed

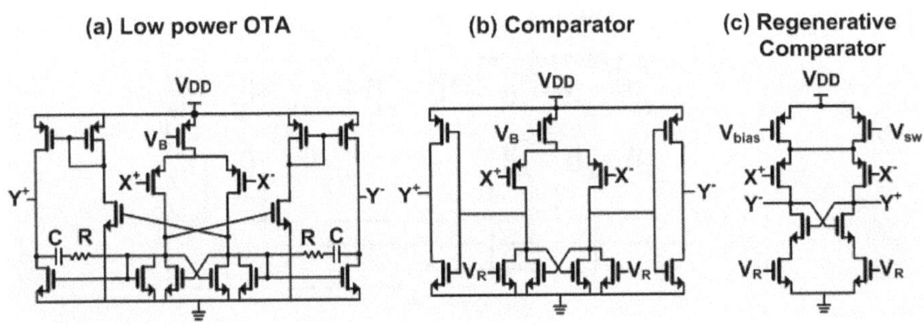

Fig. 11.23 (**a**) OTA (**b**) Comparator (**c**) Regenerative comparator

for a internal status controller. The LSB value of the fabricated comparator is 3.8 mV and the comparison time is 100 ns with a 1 μA bias current.

Figure 11.23 also shows the 1-bit quantizer, i.e., $COMP_1$ and $COMP_2$, with a clocked circuit [37]. A quantizer for the ΣΔ modulator requires low hysteresis and offset voltage. It means that good reset is more important than high resolution [38], and this regenerative comparator act as a good resetable comparator.

Hysteresis Comparator for the Internal Status Controller

A bandgap reference circuit is generally implemented to generate a bias current or voltage for an analog circuit, such as an amplifier or a comparator. Through a comparison with the input signal of the status controller, the reference voltage generated from the bandgap reference can be used to determine whether the *CLSEL* signal or the *ORSEL* signal is used for the internal status control signal. However, because the reference voltage can be changed in relation to the conditions of an individual user, the adoption of the bandgap reference circuit, which generates a constant reference voltage, limits the programmability of the proposed system. We therefore designed the hysteresis comparator for the purpose of producing a highly flexible hysteresis voltage for the internal status controller with low power consumption. Figure 11.24 presents the transistor level architecture of the designed hysteresis comparator. To control the reference voltage, we adopted 4 bit hysteresis control signals, namely $HYS_{1\sim3}$. To ensure the programmability of the *ORSEL* value, we enabled the HYS_1 signal to control the V_{HYS1} voltage with a dynamic range of 30 mV. In addition, the HYS_2 signal and the HYS_3 signal change the hysteresis values of the respective comparators, V_{HYS2} and V_{HYS3}, in order to change the mode of the ΣΔ ADC.

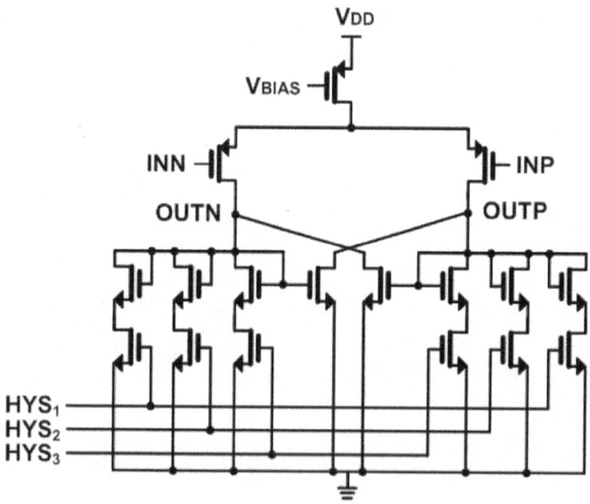

Fig. 11.24 Hysteresis comparator

Because our power supply voltage is very low, we need a suitable biasing method for the amplifier and the hysteresis comparator circuit so that we can enhance the performance and the power reduction of the analog front end. To design the amplifier and the hysteresis comparator without V_{th} stacking problems, we designed them for operating in the condition of $V_{\text{eff}} = V_{\text{GS}} - V_{\text{TH}} \approx 90$ mV. As a result, the proposed amplifier and the hysteresis comparator achieve low power consumption and low harmonic distortion with a moderate speed and area.

Clock Generator Design

In the design of our chip, we used an op-chip clock generator with three-stage delay elements. To achieve a wide tuning range, a moderate Q factor, and a low phase noise performance we fabricated a differential ring oscillator with an active load. Furthermore, to achieve a robust reference clock frequency without consideration of disturbances, we added an extra capacitor to the bias voltage node. In addition, we adopted an external control system so that the reference clock frequency could be adjusted to 8.912 MHz. For operation of the adaptive-SNR $\Sigma\Delta$ ADC, the generated reference clock is divided into 1.024 MHz and 2.048 MHz through a frequency divider in the internal status controller. The generated frequency is also used for the fabricated DSP up to kHz and for the heterogeneous $\Sigma\Delta$ DAC up to 2.048 MHz.

11.3.5 Low Power Digital Back-End

11.3.5.1 16 Channel IFIR DSP

Two of the major issues in digital hearing aid design are high flexibility and low power consumption. High flexibility accounts for a more accurate fitting that results in increased hearing clarity for the wearer. Lower power consumption prolongs battery lifetime. There have been active researches to achieve both high flexibility and low power consumption at once. One approach is to use a single configurable filter to divide only hi and lo bands [39, 40]. While this approach is simple and power efficient, it is only applicable to patients with specifically shaped audiograms and thus has limited flexibility. A more elaborate approach is to divide the frequency spectrum into many bands [41] [42]. Recent works have used up to 16 bands to provide better flexibility at the cost of higher power consumption. However, most of the previous works use general purpose digital signal processors (DSP) leading to high power consumption.

In this chapter, we propose an interpolated fixed impulse response (IFIR) filter-bank that uses one prototype filter and fifteen sub–filters to split the input signal into sixteen frequency components. In designing the filters, instead of using the general purpose DSP, hardwired digital FIR filters are used to minimize power consumption. Filter coefficients are approximated to reduce the complexity of multiplications. Many of the coefficients are set to 0.5, and can be reduced to simple shifting. For the small remaining number of multiplications, sub-expression elimination techniques are used to further reduce power consumption.

A characteristic of IFIR filters is its relatively small number of multiplications compared to its large number of delay elements. Thus less computation is required compared to a conventional filter of similar length. Also, this leaves an opportunity for further power reduction since consecutive delay elements not involved in multiplication can be implemented using a circular buffer, instead of shift registers.

System Overview

Figure 11.25 shows the architecture of a typical digital hearing aid system. The analog input signal from the transducer is converted by an analog to digital converter circuit. This signal is split into many components containing different parts of the frequency spectrum. The components are combined after being scaled using different gain values to achieve the desired frequency response. This signal is converted back to an analog signal using a digital to analog converter circuit. After being amplified the frequency compensated sound signal is output through a speaker.

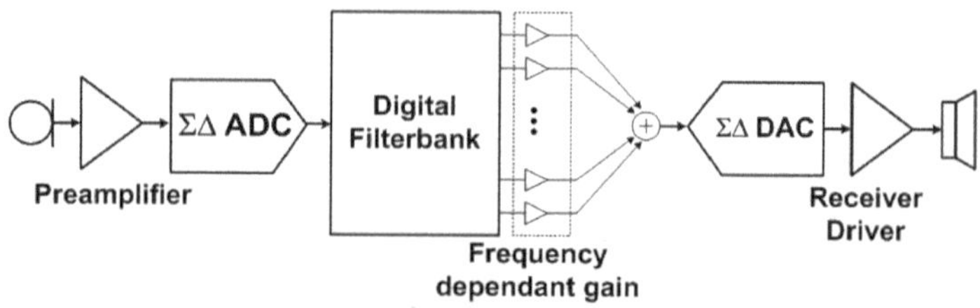

Fig. 11.25 Typical digital hearing aid

A big trend in hearing aids is towards miniaturization and this has led to the use of smaller sized batteries and thus stricter power consumption requirements. In the case of completely in the canal hearing aids the power consumption of the digital filterbank becomes more important since the power requirements of the speaker driver are reduced. To minimize the power consumption of the proposed digital filterbank, a completely hardwired parallel approach is chosen, with optimizations for reducing computational complexity where possible. Also finite impulse response filters are used for their phase-linearity and robustness using fixed length words.

Interpolated Finite Impulse Response Filter

The IFIR technique is used to derive a high order FIR filter from a model FIR filter by inserting zeros between each coefficient of the original filter. This has the effect of compressing, and then repeating the images of the original filter across the frequency spectrum as shown in Fig. 11.26. The number of repetitions can be controlled by the number of inserted zeros, L. The resulting filter has multiple pass–bands and stop-bands with narrower transitions than the original filter. While

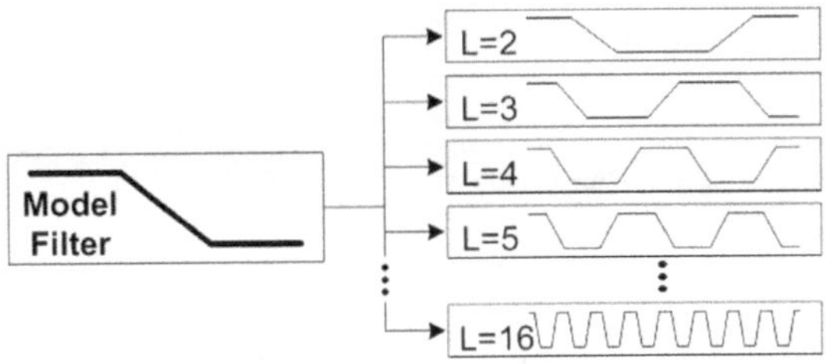

Fig. 11.26 Deriving interpolated finite impulse response filters

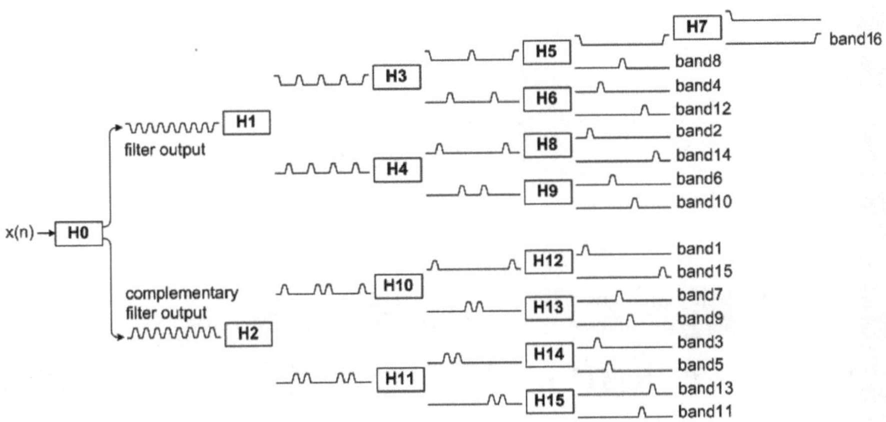

Fig. 11.27 Proposed filter bank structure

interpolation of a filter results in more delay elements, the number of multiplications remains the same. This is important since the number of multiplications determine the computational complexity.

The filterbank is composed of sixteen IFIR filters that each takes one input and put out two complementary output signals as shown in Fig. 11.27. The first filter in the tree, *H0*, splits the input signal into two prototype signals. Each pass-band in the prototype signals represents one of the desired frequency components. The job of the remaining filters is to separate each of the pass–bands into a separate component. Each filter uses a combination of one of the model filters and the interpolation factor L to separate the desired pass-bands. The filterbank is structured so that the most resource intensive operations are positioned near the top of the tree. This prevents costly filters from being duplicated in multiple nodes of the tree. Out of the seventeen resulting bands, band16 is used as an anti-aliasing filter.

Three model filters of different lengths are used to derive all of the FIR filters in the filterbank in Fig. 11.28. This approach has the advantage of requiring the design of just three sets of coefficients for the entire design. An optimization on one set of coefficients will affect all of the IFIR filters that share those coefficients. The model filter used for each filter is determined by the required sharpness of the transition. The model filters are defined by the Eqs. (11.12)–(11.14).

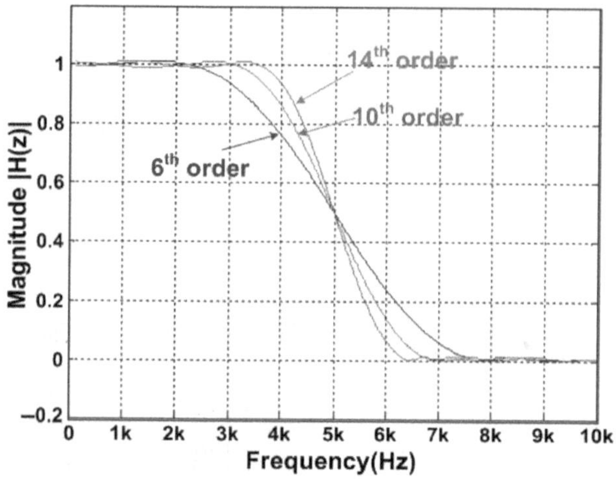

Fig. 11.28 Model half–band filters

- 6th order half-band low pass filter:

$$H(z) = h_0\left(1 + z^{-6}\right) + h_1\left(z^{-2} + z^{-4}\right) + 0.5z^{-3} \tag{11.12}$$

- 10th order half-band low pass filter:

$$H(z) = h_0\left(1 + z^{-10}\right) + h_1\left(z^{-2} + z^{-8}\right) + h_2\left(z^{-4} + z^{-6}\right) + 0.5z^{-5} \tag{11.13}$$

- 14th order half-band low pass filter:

$$H(z) = h_0\left(1+z^{-14}\right)+h_1\left(z^{-2}+z^{-12}\right)+h_2\left(z^{-4}+z^{-10}\right)+h_3\left(z^{-6}+z^{-8}\right)+0.5z^{-7} \tag{11.14}$$

The filter coefficients have a direct impact on the computational complexity of the filters. Special care was put into optimizing the coefficients to have the least amount of bits possible. Signed digit format was used to reduce the average number of bits in each coefficient. Then approximation was used to further reduce that number, while satisfying the specifications for the model filters. Optimization using only approximation resulted in coefficients with an average of 5 partial products. Optimization using both signed bit representation and approximation reduced that number to 3. The resulting coefficients are shown in Table 11.1.

Figure 11.29 shows the structure of an FIR filter that is derived from the 10th order model filter. The symmetry of the coefficients is exploited to reduce the number of multiplications in half. The complementary output is obtained with just one additional subtraction using the following equation.

Table 11.1 Optimized FIR filterbank coefficients

Filter	Coefficients	Optimized Coefficients (signed bit representation)
6th order	h0 = −0.049	h0 = −0.0488281 (0.0000100)
	h1 = 0.294	h1 = 0.294922 (0.01001100)
10th order	h0 = 0.0236	h0 = 0.0235596 (0.000010000001)
	h1 = −0.0758	h1 = 0.0761719 (0.0000001)
	h1 = 0.307	h2 = 0.306614 (0.01010001)
14th order	h0 = −0.0159	h0 = −0.0158691 (0.0000000000)
	h1 = 0.0372	h1 = 0.0371094 (0.00001010)
	h2 = −0.0882	h2= −0.0878906 (0.0001010)
	h3 = 0.312	h3 = 0.3125 (0.01010)

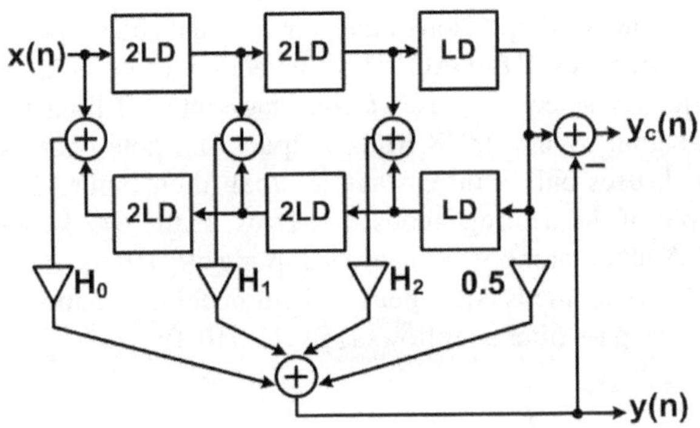

Fig. 11.29 IFIR filter derived from 10th order model filter

$$y_c(n) = x(n - (N - 1)/2) - y(n) \tag{11.15}$$

The delay blocks in the filter can get quite large with increasing L, and implementation using registers results in high power consumption. In order to alleviate this, a circular buffer is employed for delays greater than 7 cycles. In the circular buffer, the data remains stationary and only the pointer is incremented at each clock edge.

Mixer

The function of the mixer is to scale and combine the outputs from the filter bank to obtain the final full-band frequency gain compensation. The scaling is applied in powers of 2, allowing it to be implemented using a simple shift instead of multiplication.

11.3.5.2 Heterogeneous ΣΔ DAC

In a conventional ΣΔ DAC, the input signal and the sampling clock of the ΣΔ modulator should have the same frequency to guarantee accurate operation [43]. Because the ΣΔ modulator is operated with an oversampled clock frequency, the interpolation filter should upsample its input signal up to the sampling frequency of the ΣΔ modulator. Furthermore, because the oversampling ratio of the ΣΔ modulator is generally high, the interpolation filter dissipates a large amount of power to upsample the input signal frequency up to the sampling clock frequency. To reduce the power consumption of the interpolation filter and achieve propose proper operation, we implemented the proposed heterogeneous ΣΔ DAC.

Architecture of the Heterogeneous ΣΔ DAC

The heterogeneous ΣΔ DAC enables the adoption of different frequencies between the input signal (512 kHz) and the sampling clock (2.048 MHz) of the ΣΔ modulator. Figure 11.30 shows a comparison of the conventional and proposed interpolation filter of the heterogeneous ΣΔ DAC. The conventional interpolation filter, which is based on frequency selectivity, uses a five-stage internal filter to upsample the 512 kHz input signal to the 2.048 MHz output. The heterogeneous ΣΔ DAC, on the other hand, uses only a three-stage internal filter. Figure 11.31 also represents the principle of the heterogeneous ΣΔ DAC. If the 512 kHz input signal is injected into a ΣΔ modulator operating at a sampling clock frequency of 2.048 MHz (Fig. 11.31(a)), the outcome is equivalent to the transfer of the input signal through the fourth order low-pass filter as follows (Fig. 11.31(b)):

$$H(z) = 1 + z^{-1} + z^{-2} + z^{-3} \tag{11.15}$$

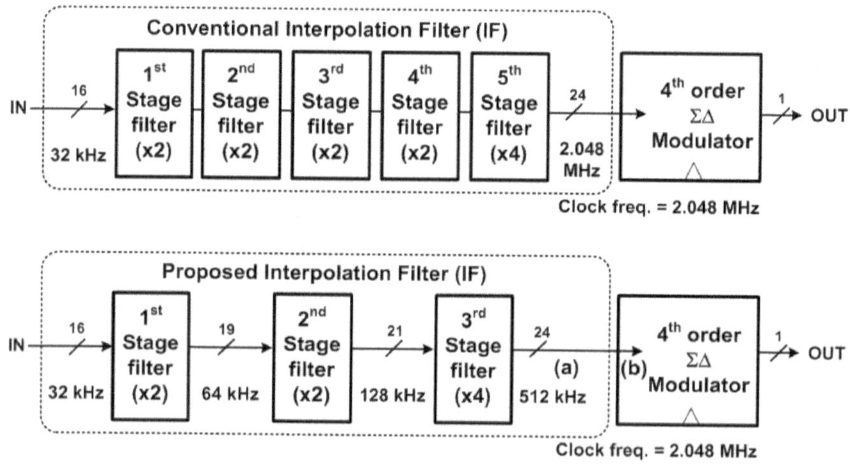

Fig. 11.30 Architecture of the heterogeneous ΣΔ DAC

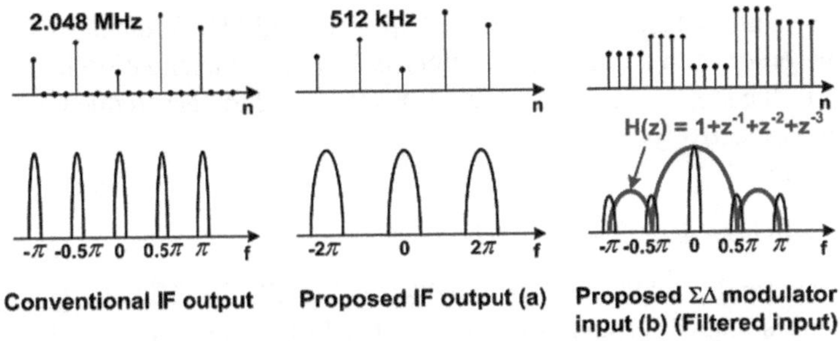

	Power (μW)	Area (μm^2)	Error
Conventional DAC	32.81	23653	1
Proportional DAC	22.23	16102	1.0016
Penalty	40.38 % ↓	40.51 % ↓	0.16 % ↑

Fig. 11.31 Principles of the heterogeneous ΣΔ DAC

Thus, the heterogeneous ΣΔ DAC is modeled as a conventional ΣΔ DAC that has the low–pass filtered output of an interpolation filter. The filtered input of the ΣΔ modulator suppresses the signal component outside of the pass band region. Because the ΣΔ DAC adopts a high oversampling ratio, the pass band region is wider than the bandwidth of the input audio signal. Most of the signal energy is therefore inside the fundamental pass band, and the signal is degraded by a tolerable level. The heterogeneous ΣΔ DAC reduces the power dissipation by 40.4% and the area by 40.5%, furthermore the reported error due to the filtered input is only 0.16% higher than that of the conventional ΣΔ DAC.

Interpolation Filter Design

The proposed heterogeneous ΣΔ DAC consists of a three-stage interpolation filter and a fourth-order ΣΔ modulator. After the 32 kHz DSP output signal is injected into the proposed ΣΔDAC, the interpolation filter increases the signal frequency to 512 kHz. To avoid the design difficulties created by a single-state interpolation filter and a high upsampling ratio, we divided the designed interpolation filter into three stages. The upsampling ratios of the interpolation filter are 2, 2, and 4, respectively. The first-stage interpolation filter upsamples the 32 kHz DSP output signal up to the 64 kHz of the first internal signal. The attenuation level from the pass band to the stop band of the twentieth-order first-stage interpolation filter is 50 dB. In the second-stage interpolation filter, the 64 kHz frequency of the first internal signal is upsampled up to 128 kHz of the second internal signal. The order of this second-stage interpolation filter is 12, and the attenuation level from the pass band to the stop band is 55 dB. We used a value of four for the upsampling ratio of the third-stage interpolation filter because the frequency selectively of this type of filter is lower than that of the previous two interpolation filters. Thus, the 128 kHz frequency of the second internal signal is upsampled up to the 512 kHz frequency of the output signal. The order of this interpolation filter is 16 and the attenuation level from the pass band to the stop band and near the 128 kHz and 256 kHz frequencies is 55 dB and 70 dB, respectively.

11.3.5.3 H-bridge as a Speaker Driver

The generated pulse width modulation (PWM) signal from the heterogeneous ΣΔ DAC output is loaded through a low pass filter. In this case, an H-bridge circuit is usually used to drive the receiver of the digital hearing aid. H-bridge alternates its output between 1 and 0 according to the PWM signal values and controls the current directions of the receiver.

11.3.6 Implementation and Measurement Results

Figure 11.32 shows a chip microphotograph of the fabricated hearing aid processor with the internal gain verification algorithm. The overall system power dissipation is 122 μW from a single 0.9 V supply and the active area occupies 5.4 mm^2 in a 0.18 μm standard CMOS technology. First of all, a high level simulation is accomplished to prove the proposed real time gain verification algorithm using closed loop feedback architecture. To corroborate a co-operation both the hearing aid and the internal gain verification algorithm gate level simulation and a transistor level simulation are also performed. At last, the hearing aid processor which includes a real time gain verification algorithm and the multi mode operation has been fabricated as a fully operational single hearing aid chip.

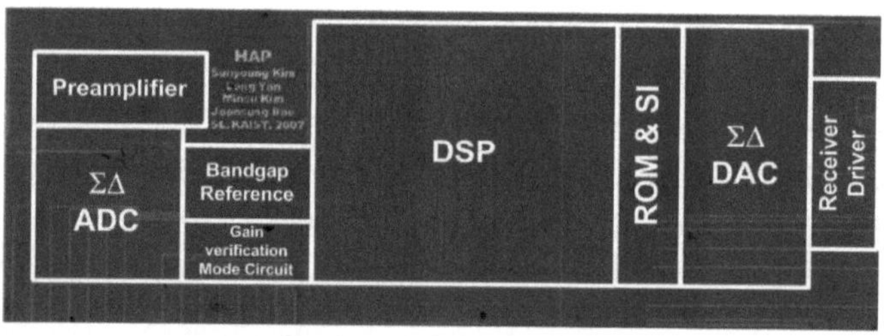

Fig. 11.32 Chip microphotograph of the digital hearing aid chip

In Fig. 11.33, variations of the measured threshold gains and the threshold knee points of the preamplifier are presented as a function of V_{VC} with V_{TH} as a parameter. The threshold knee point is determined by different values of V_{TH}. By reducing V_{TH}, this point is decreased simultaneously. Further, the threshold gain of the preamplifier is determined according to the value of V_{VC} with V_{TH}. By using these parameters, excessively high gain is prevented and the power dissipation is reduced.

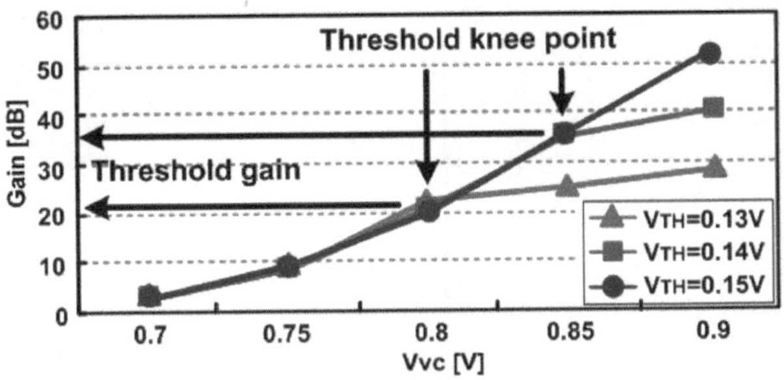

Fig. 11.33 Measured performance of the proposed preamplifier

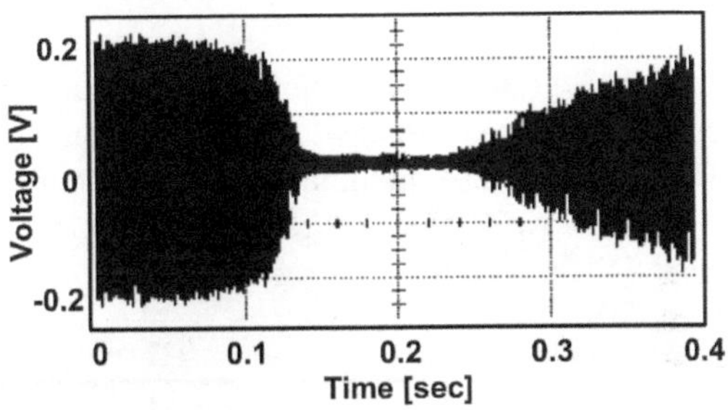

Fig. 11.34 Measured attack and release response

The measured attack and release response is shown in Fig. 11.34. The output signal is shown when the input voltage level suddenly drops by 25 dB. After a 0.1 sec delay, its output gain increases gradually according to the input level. The measured output of the preamplifier is shown in Fig. 11.35. The reported output dynamic range is from 0.4 to 0.82 V.

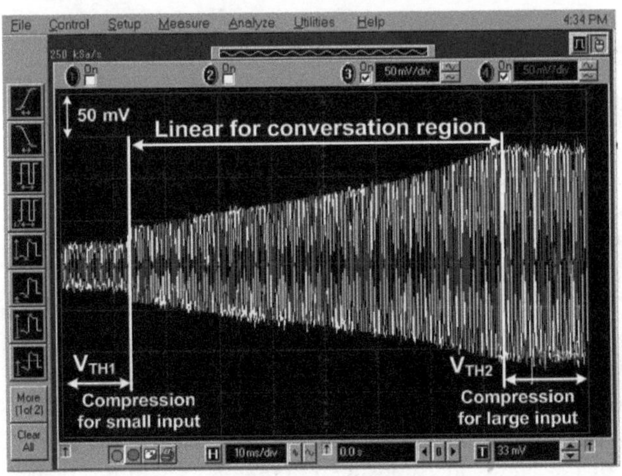

Fig. 11.35 Multi-threshold preamplifier output

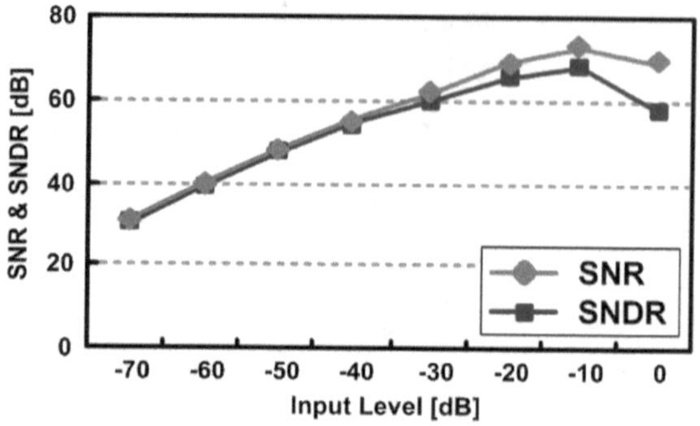

Fig. 11.36 Measured SNR/SNDR versus input amplitude

Figure 11.36 shows plots of the measured SNR and SNDR as a function of the input amplitude. The measured output voltage spectrum is presented in Fig. 11 37. The proposed ΣΔ modulator is a type 1 modulator with 2 kHz sinusoidal input signal and 1.024 MHz clock frequency. Under these this conditions, the measured peak SNR is 72 dB.

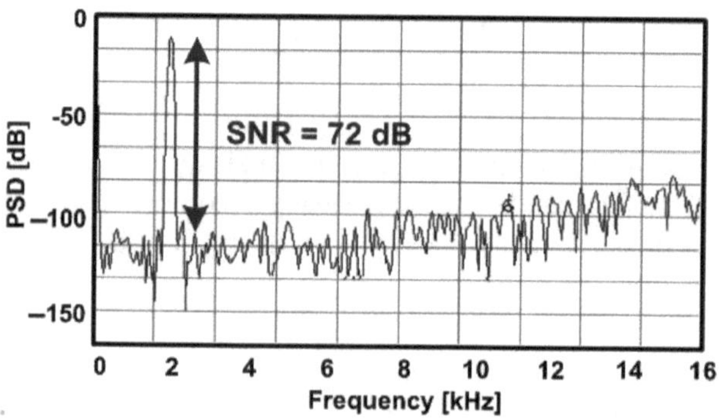

Fig. 11.37 Measured spectrum of the proposed ΣΔ modulator

Figure 11.38 shows the measurement results of the internal status controller. According to the input amplitude of the adaptive-SNR ΣΔ ADC, the *ORSEL* signal changes the order between two and three, while the *CLSEL* signal alternates the clock frequency between 1.024 and 2.048 MHz.

Figure 11.39 shows the output responses of the fabricated DSP according to each frequency range. In this case the DSP gain is selected as 1 to verify the frequency splitting performance of the filter back. The frequency compensated gain characteristics of the DSP are also measured and presented in Fig. 11.40. A typical ski slope loss audiogram is used to extract the compensated audiogram. To satisfy the requirements of the National Acoustics Laboratory for fitting of hearing aids the accuracy of the compensation curve should be achieved within 5 dB of the desired

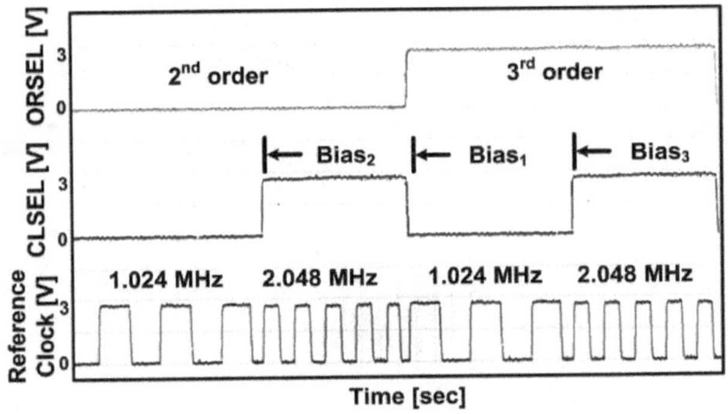

Fig. 11.38 Status controller output and reference frequency

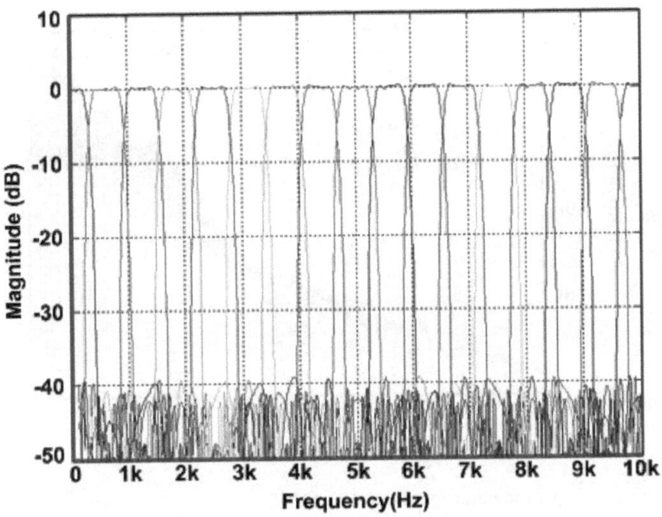

Fig. 11.39 IFIR filterbank output

curve across the frequency spectrum. Finally, the measured group delay of the filter bank is less than 4.5 ms which is small enough for use in hearing aids.

Figure 11.41 shows the FFT results of the heterogeneous $\Sigma\Delta$ DAC output when the frequency of the input audio signal is 1 kHz and the sampling frequency is 2.048 MHz. The peak SNR of the heterogeneous $\Sigma\Delta$ DAC is 86 dB.

Figure 11.42 shows a measurement setup of the proposed digital hearing aid chip with an internal gain verification algorithm. A plastic type ear canal is implemented to realize a human external ear canal which is 6 mm long, 9 mm wide and 3.5 cm length. To model the hearing aid in the external ear canal, two dummy patches are inserted into the plastic type ear canal. A space between two dummy patches forms

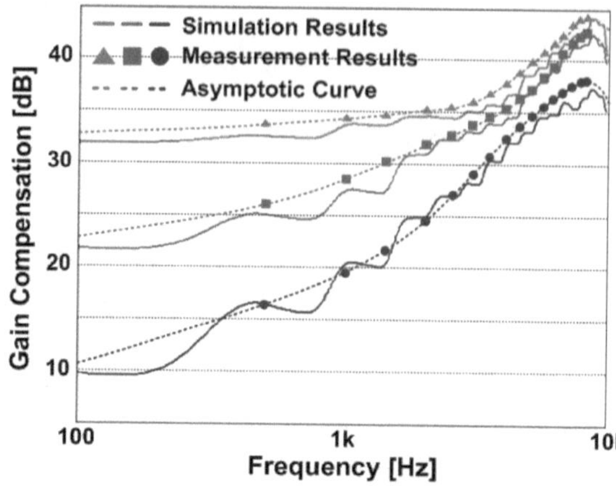

Fig. 11.40 Output responses of the DSP

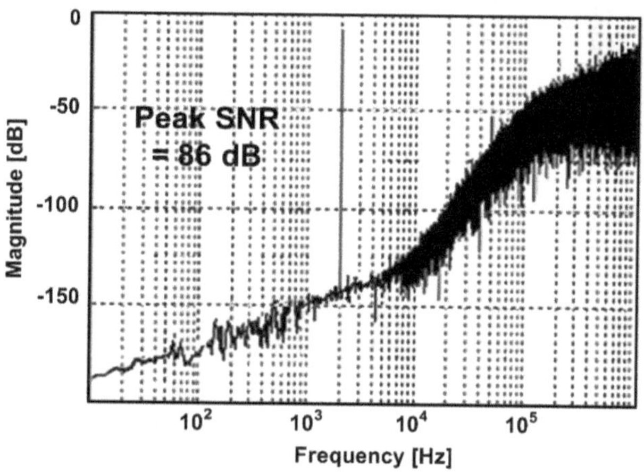

Fig. 11.41 FFT results of the heterogeneous ΣΔ DAC

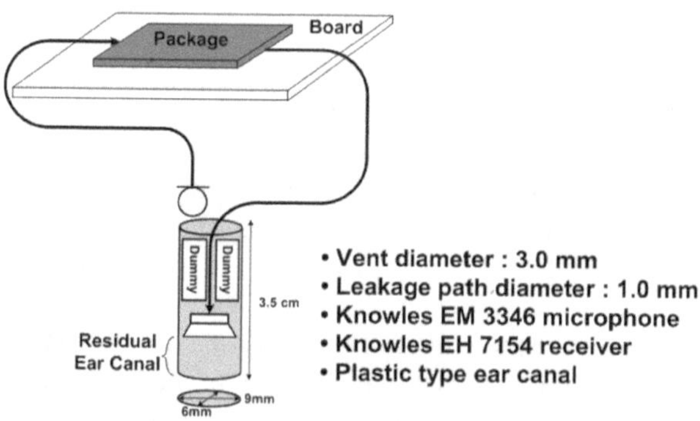

Fig. 11.42 Measurement setup

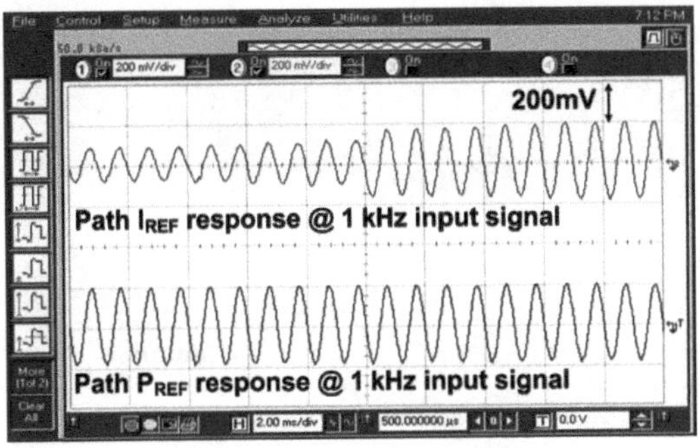

Fig. 11.43 Responses of the internal gain verification algorithm

a vent path which diameter is 3.0 mm and a space between a single dummy patch and an external ear canal wall forms a leakage path which diameter is 1.0 mm.

Figure 11.43 presents the measured responses of the internal gain verification algorithm. Because a signal S_{ID} from a ΣΔ DAC output has a smaller value compared with a signal S_{IM} from a feed backed signal via vent and leakage paths after the first iteration process, a DSP gain parameters are increased for augment a S_{ID} signal value in next iteration process. Therefore, a first DSP parameter value for a 1 kHz frequency range, P_{N_2}, is renewed continuously with repeated internal iteration processes. This process is ended when the two different signals S_{IM} and S_{ID} have same values. In this measurement, the input signal has 1 kHz frequency component. During the 0.5 second processing time for each frequency component, the acquired DSP gain parameters, P_{N_2F}, and the responses of the internal gain verifying operation due to the proposed algorithm are measured.

The measured characteristics of the mode conversion between the hearing aid mode and the smart earphone mode according to the external environment are shown in Fig. 11.44. After the control signal S_{MS} turns on, the output value of the hearing aid processor follows the speech audio signal instead of the external audio sound signal. Moreover the output value of the hearing aid processor becomes same level as the external audio sound when the external audio sound level is higher than the A_{REF} value.

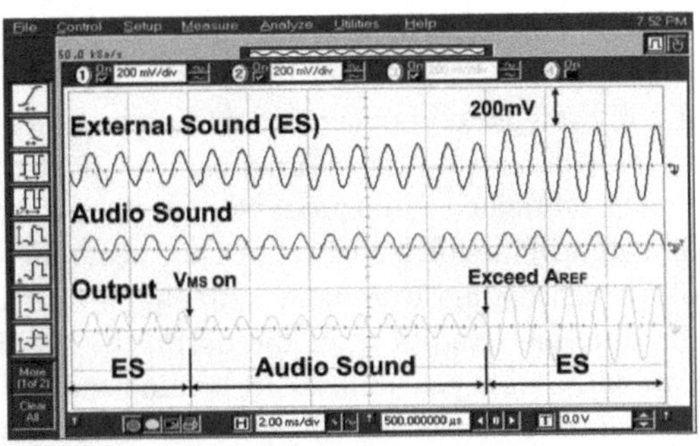

Fig. 11.44 Responses of the mode conversion

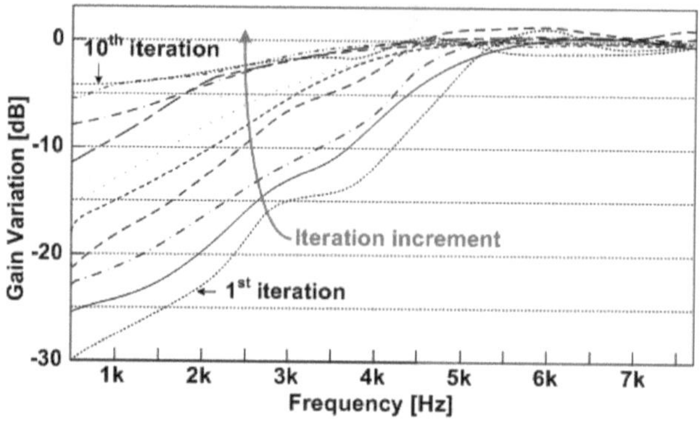

Fig. 11.45 Gain verification profile

Figure 11.45 shows a gain variation profile due to the DSP gain parameter sets of the hearing aid chip with the proposed gain verification algorithm. During the internal gain verification period, the test signal generator creates 16 pure sinusoidal test signals from 0.5 to 8 kHz frequency range in steps of 500 Hz. It reveals that the vent and the leakage path usually affect the low frequency gain of a hearing aid.

Table 11.2 summarizes the performance of the fabricated chip. The power consumption of the hearing aid processor chip is 122 μW at 0.9 V supply voltage. The peak signal-to-noise ratio (SNR) of the overall hearing aid system is 85 dB and the measured typical input referred noise is 3.9 μVrms. The number of classified DSP channel is 16 and each frequency channel has 0.5 kHz range. Therefore, the total 3 dB bandwidth of the hearing aid system is 8 kHz.

The performance comparison results are summarized in Table 11.3. In the reported results among the previous works connected to the hearing aid, the major design points are the specific design technique for enhancing the user conveniences and considering the human characteristics and the low power techniques. According to a reported result, the fabricated hearing aid processor consumes the lowest power with even includes the real-time internal gain verification algorithm.

11.3.7 Conclusions

A fully integrated human-like audio processor is proposed and implemented with physiological real time internal gain verification algorithm.

The fabricated audio processor has three major features: (a) Internal gain verification algorithm which enables fast and accurate on chip autonomous gain verification of the hearing aid to achieve suitable performance of the hearing aid according to the different conditions of the users. (b) Multi-mode operation which offers revolutionary new possibilities in hearing aid for everyone, not just people who are deaf or hard of hearing as a personal audio assistant device (c) Four different low

Table 11.2 Performance summary

Overall System Performance					
Supply voltage	0.9 V				
Peak SNR (Overall system)	85 dB				
Power dissipation	72 μW (Analog) / 50 μW (Digital)				
–3 dB bandwidth	8 kHz				
Human factor considered techniques	Internal gain verification algorithm				
Core area	$3.88 \times 1.44\ mm^2$				
CMOS process	0.18 μm CMOS technology				
Blocks Performance					
Multi-threshold preamplifier	Input referred noise	$3.9\ \mu W_{rms}$			
	Power consumption	31 μW			
	Area	$0.243\ mm^2$			
	Maximum gain	38 dB			
	Output dynamic range	0.4 – 0.82 V			
Adaptive-SNR ΣΔ ADC	Power consumption	26 – 38 μW			
	Type	1	2	3	4
	Peak SNR (dB)	75	85	77	89
Multi-channel IFIR DSP	Power consumption	25 μW			
	Number of channel	16			
	Area	$1.8\ mm^2$			
	Clock frequency	32 kHz			
Heterogeneous ΣΔ DAC	Input frequency	512 kHz			
	Clock frequency	2.048 kHz			
	Gate count	16 K			

power techniques which enables environment-aware power management with high flexibility.

The human factors considered design adopts an internal gain verification algorithm. The proposed gain verification algorithm uses a closed loop feedback

Table 11.3 Performance comparison

Parameter	[44]	[45]	[46]	This work
Supply voltage	1.3 V	1.1 V	1.0 V	0.9 V
Peak SNR	77 dBA	–	70 dB	85 dB
Power consumption	2 mW	297 μW	200 μW	122 μW
Type	Digital	Digital	Programming	Digital
# of DSP band	4		–	16
Design techniques based on human factors	No	No	No	Yes
CMOS technology	Low V_{TH} 0.8 μW	0.6 μW	1.2 μW	0.18 μW

mechanism to estimate an additional internal gain to a vent, a leakage path and a residual ear canal volume when a hearing aid is worn. By using the proposed algorithm a resonance gain error between the real ear canal and the test cavity can be minimized. In addition, additional gains due to the various internal signal paths also achieved with only internal iteration processes due to the closed loop feedback algorithm. To prove this algorithm with the co-operation of the hearing aid, a hearing aid chip implementation is also accomplished. By including various operation modes, a smart earphone operation and a location indication operation into the low power digital hearing aid, a fabricated hearing aid processor performs as a personal audio assistant device.

The fabricated audio processor also gives a multi-mode function; a hearing aid mode, a smart earphone mode and a location indication mode. The hearing aid mode offers a conventional hearing aid operation with accurate external sound amplification and transmission according to the user's demands. The smart earphone mode enables to receive an electrical sound signal from an external audio source such as MP3 player. Finally the location indication mode enables a discrimination of a sound source location which is essential to enhance an understanding of speech. The fabricated hearing aid processor introduces the possibilities of hearing aid as a personal audio assistant device to allow everyone to control and enhance the sounds around us.

Four different low power techniques include a multi-threshold preamplifier, an adaptive-SNR ΣΔ ADC, 16 channel IFIR DSP, and a heterogeneous ΣΔ DAC is proposed and implemented. To reduce the power dissipation of the human factored hearing aid processor, the multi-threshold preamplifier is designed. The dynamic range of the multi-threshold preamplifier exists from 0.4 to 0.82 V and dissipates only 31 μW from a single 0.9 V supply by using 2 MRC units with the gain amplifier. The newly designed internal status controller enables the adaptive-SNR analog front end to be operated in a fully internal mode. With the combination of two internal control signals, we can achieve four different types of the enhanced adaptive-SNR analog front end by independently controlling the order and the clock frequency. A low power 16 channel DSP with interpolated finite impulse response (IFIR) filterbank for use in digital hearing aids is developed. The filterbank integrates sixteen IFIR filters in a tree configuration to obtain a total of sixteen bands. The IFIR filters are derived from three model half-band pass filters with stop-band attenuation of 40 dB. Coefficient approximation and subsequent sub-expression elimination techniques are used to reduce the computational power requirement for each filter. Additionally, delay elements are implemented using circular buffers to minimize power dissipation from shifting of samples within the filters. Our implementation of the heterogeneous $\Sigma\Delta$ DAC reduces the power consumption with only a slight degradation in performance. It also reduces the power dissipation and the area by 40.4% and 40.5%, respectively, and the reported error is only 0.16% higher than a conventional $\Sigma\Delta$ DAC. The fabricated audio processor chip reports 122 μW power consumption at a 0.9 V supply voltage and 5.4 mm^{2}core area occupation at a 0.18 μm standard CMOS technology.

11.4 Cochlear Implant

11.4.1 Introduction of the Cochlear Implant

A cochlear implant (CI) is a surgically implanted electronic device that provides a sense of sound to a person who is profoundly deaf or severely hard of hearing. A cochlear implant is very different from a hearing aid. Hearing aids amplify sounds so they may be detected by damaged ears. Cochlear implants bypass damaged portions of the ear and directly stimulate the auditory nerve. That means, a cochlear implant receives sound from the outside environment, processes it, and sends small electric currents near the auditory nerve. These electric currents activate the nerve, which then sends a signal to the brain. The brain learns to recognize this signal and the person experiences this as "hearing". Signals generated by the implant are sent by way of the auditory nerve to the brain, which recognizes the signals as sound. Hearing through a cochlear implant is different from normal hearing and takes time to learn or relearn. However, it allows many people to recognize warning signals, understand other sounds in the environment, and enjoy a conversation in person or by telephone.

The research with this device began in the 1950s and the first commercial devices were approved by the FDA in the mid-1980. Several CI devices have been developed over the 20 years. According to the Food and Drug Administration (FDA), as of April 2009, approximately 188,000 people worldwide have received implants [47]. In the United States, roughly 41,500 adults and 25,500 children have received them [48].

At present, there are three major cochlear implant manufacturers including Advanced Bionics Corporation, USA (www.advancedbionics.com), Med-El Corporation, Austria (www.medel.com), and Cochlear Corporation, Australia (www.cochlear.com), with Cochlear being the dominating player controlling 70–80% of the cochlear implant market worldwide. Several startup companies are also developing advanced and low-cost multi-electrode cochlear implants, including Advanced Cochlear Systems (www.advcoch.com) in Seattle, Washington, Nurobiosys Corporation in Seoul, Korea, and Nurotron Biotechnology Inc. based in both Irvine, CA and Hangzhou, China (www.nurotron.com) [49] (Fig. 11.46).

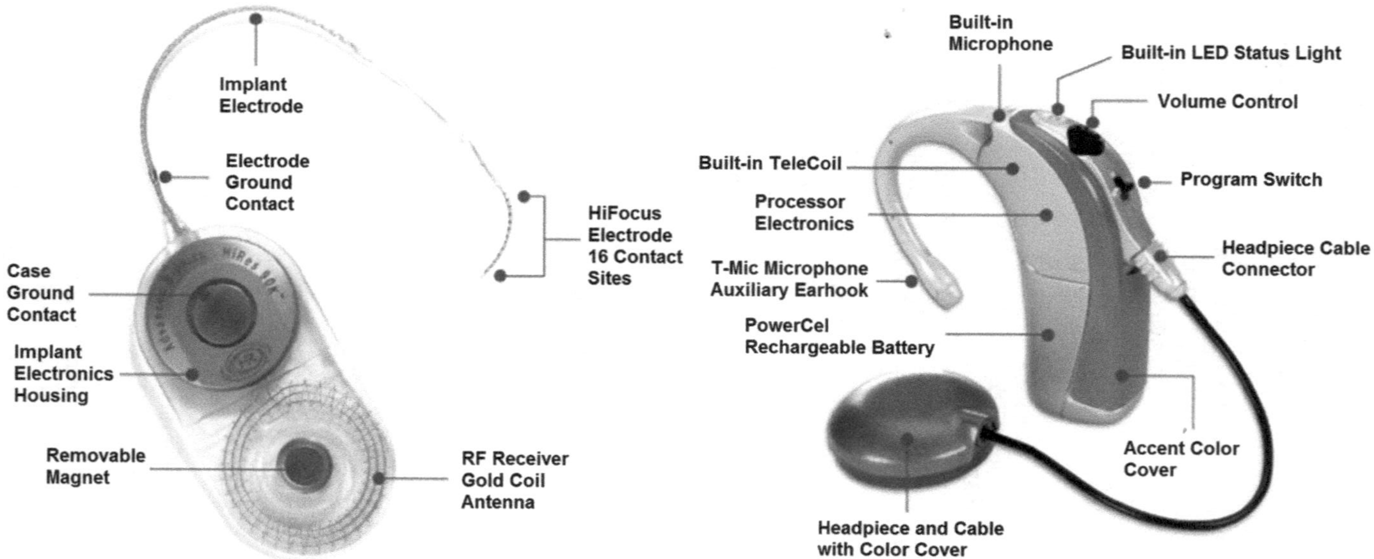

Fig. 11.46 Cochlear implant device (Source: Advanced Bionics Corporation)

11.4.2 Design of the Cochlear Implant

All the implant devices have the following features in common: a microphone that picks up the sound, a speech processor that converts the sound into electrical signals, a transmission system that transmits the electrical signals to the implanted electrodes with transcutaneous RF link, and an intra-cochlear electrode array (consisting of multiple electrodes) that is inserted into the cochlea by a surgeon in Fig. 11.47.

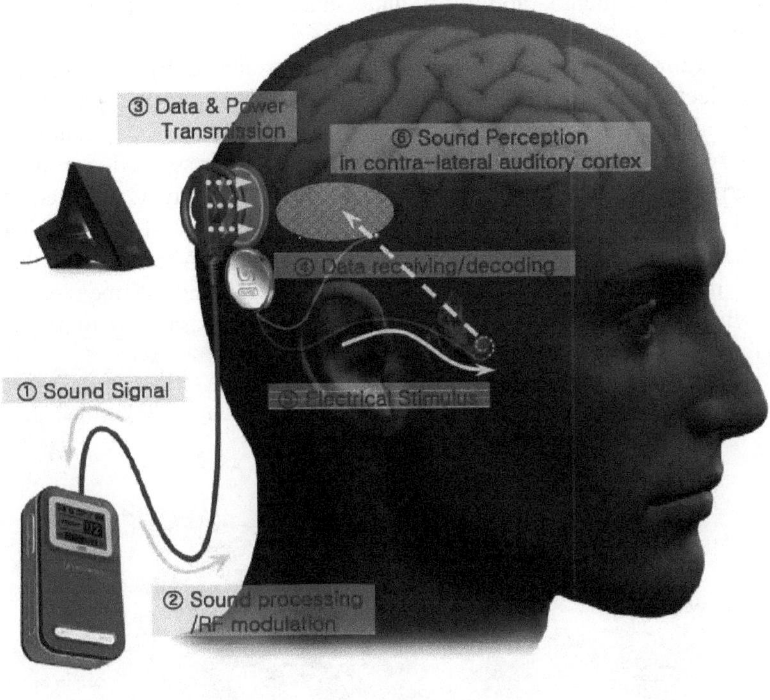

Fig. 11.47 Cochlear implant [50]

The CI system consists of an external speech preprocessor, an internal implantable unit, and an inductive telemetry link connecting the two. The external speech preprocessor consists of a microphone which picks up sound from the environment and a speech processor which selectively filters sound to prioritize audible speech and sends the electrical sound signals through a thin cable to the transmitter. The internal implantable unit which is consists of a receiver-stimulator IC secured in bone beneath the skin, which converts the signals into electric impulses and sends them through an internal cable to electrodes and an intracochlear array of electrodes which send the impulses to the nerves in the scala tympani and then directly to the brain through the auditory nerve system. The inductive telemetry link consists of circuits and coils for forward transmission of power and data, and backward transmission of indicators of IC and electrode function [51].

An electrode connected to the device is inserted into the inner ear. The electrode is simply a bundle of tiny wires that have open contacts spread out along the length of the cochlea. Thus, the electrical signals can be sent to different areas of the cochlea and represent different frequency sounds. State-of-the art cochlear implant devices now have up to 24 electrodes that stimulate the auditory nerve. These multi-channel implants have the advantage of stimulating many different nerve fibers individually, thereby transmitting more detailed information to the brain. The more information that reaches the brain, the greater the patient's ability to understand what is happening in his/her environment.

Direct stimulation of the auditory nerve is produced by currents delivered through electrodes placed in the scala tympani (ST), one of three fluid-filled chambers along the length of the cochlea. Figure 11.48 shows a anatomy of the implanted cochlea [52]. The three chambers (in the cross sections) and a partial insertion of an electrode array into the ST. The array is inserted through a drilled opening made by the surgeon in the bony shell of the cochlea overlying the ST and close to the base of the cochlea. Alternatively, the array may be inserted through the second flexible membrane of the cochlea, the round window membrane, which also is close to the basal end of the cochlea and ST.

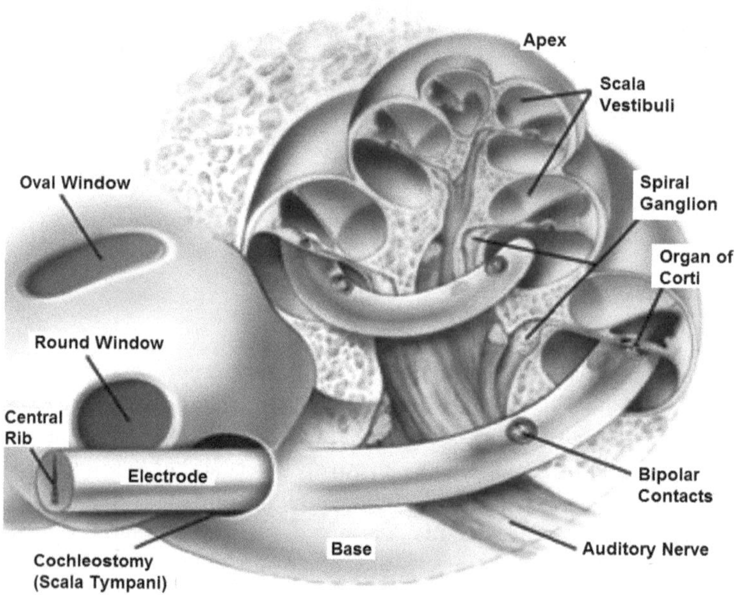

Fig. 11.48 Anatomy of the implanted cochlear [52]

11.4.3 Future of the Cochlear Implant

Cochlear implants are rarely used in ears that have a functional level of residual hearing. However, electric Acoustic Stimulation (EAS) devices, including the Hybrid "short-electrode" cochlear implant, have been developed that combine a cochlear implant with a sound amplifying hearing aid [53]. EAS devices have the potential to make cochlear implants suitable for many people with partial hearing loss. The sound amplifying component helps users to perceive lower frequency sounds through their residual natural hearing while the cochlear implant allows them to hear middle and higher frequency sounds. The combination enhances speech perception in noisy environments [54]. Further improvements for all patients might be produced by somehow increasing the number of effective channels supported by cochlear implants [55].

References

1. Cooper RA (2008) Quality-of-life technology. IEEE Eng Med Biol Mag 27(2)10–11, Apr 2008
2. Bitner M-Glindzicz (2002) Hereditary deafness and phenotyping in humans. Br Med Bull 63:73–94
3. Tekin M, Arnos KS, Pandya S (2001) Advance in hereditary deafness. Lancet 358:1082–1090
4. National Center for Health Statistics [Online]. Available: www.cdc.gov/nchs/
5. National Institute on Deafness and Other Communication Disorders (NIDCD), Statistics about hearing disorders, ear infections, and deafness [Online]. Available: http://www.nidcd.nih.gov/health/statistics/hearing.asp
6. Acoustical Performance Criteria, Design Requirements and Guidelines for Schools, ANSI Standard S12.60-2002.
7. Kochkin S (2001) MarkeTrak VI: The VA and direct mail sales spark growth in hearing aid market. Hearing Rev 8(12):16–24, 63–65
8. Seelman KD, Palmer CV, Ortmann A, Mormer E, Guthrie O, Milel J, Brabyn J (2008) Quality-of-life technology for vision and hearing loss. IEEE Eng Med Biol Mag 27(2):40–55, Apr 2008
9. The Royal National Institute for Deaf People (2006) Hearwear–the future of hearing. Exhibition at V&A victoria and albert museum, London, 26 Jul 2005–5 Mar 2006
10. Types of Hearing Aids [Online]. Available: http://www.hearingaidscentral.com
11. The Future of Hearing Aid Technology [Online]. Available: http://www.starkeypro.com
12. Looking at the History of Hearing Aids: the past, present, and future of hearing aid technology [Online]. Available: http://hearing-aids.americahears.com/aid-technology.html
13. Peluso V, Vancorenland V, Augusto Marques M, Steyaert MSJ, Sansen W (1998) A 900-mV Low-Power ΣΔ A/D Converter with 77-dB Dynamic Range. IEEE J. Solid-State Circuits 33(12):1887–1897, Dec 1998
14. Yao L, Michiel S,. Steyaert J, Sansen W (2004) A 1-V 140-μW 88-dB audio sigma-delta modulator in 90-nm CMOS. IEEE J. Solid-State Circuits 39(11):1809–1818, Nov 2004
15. Agnew J (1999) Digital signal processing in hearing aids. J Acoustical Society of Am 105(2):1210, Feb 1999
16. Kim S, Lee J-Y, Song S-J, Cho N, Yoo H-J (2006) An energy-efficient analog front-end circuit for a Sub-1-V Digital hearing aid chip. IEEE J. Solid-State Circuits 41(4):876–882, Apr 2006

17. Kim S, Cho N, Song S-J, Yoo H-J (2007) A 0.9 V 96 μW Fully operational digital hearing aid chip. IEEE J. Solid-State Circuits 42(11):2432–2440, Nov 2007
18. Stelmachowicz PG, Hoover B, Lewis DE, Brennan M (2002) Is functional gain really functional? Hearing J 55(11):38–42, Nov 2002
19. Kim S, Lee SJ, Cho N, Song S-J, Yoo H-J (2008) A Fully Integrated Digital Hearing Aid Chip With Human Factors Considerations. IEEE J. Solid-State Circuits 43(1):266–274, Jan 2008
20. Kim S, Lee SJ, Cho N, Song S-J, Yoo H-J (2007) A fully integrated digital hearing-aid chip with human-factors considerations. IEEE Int Solid State Circuits Conf Dig. Tech. Papers 154–155, Feb 2007
21. Stelmachowicz PG, Hoover B, Lewis DE, Brennan M (2002) Is functional gain really functional? Hearing J 55(11):38–42, Nov 2002
22. Dillon H (2001) Hearing aids, 1st edn, Boomerang Press
23. Kim S, Lee J-Y, Song S-J, Cho N, Yoo H-J (2006) An energy-efficient analog front-end circuit for a Sub-1-V digital hearing aid chip. IEEE J. Solid State Circuits 41(4):876–882, Apr 2006
24. Almadian M (2001) Transmission Line Matrix (TLM) modelling of medical ultrasound. Ph.D. dissertation, Jun 2001
25. Nakao K, Nishimura R, Suzuki Y (2006) Calculation of transfer function of acoustic feedback path for in-the-ear hearing aids with a correction for specific acoustic impedance of a tubule. Acoust Sci Tech 27(4):242–244
26. Stelmachowicz PG, Hoover B, Lewis DE, Brennan M (2002) Is functional gain really functional? Hearing J 55(11):38–42, Nov 2002
27. Thompson H, Thomas N (2005) A supersonic future, blueprint broadsides 2(2):1–20, Jun 2005
28. DUET^{TM} DIGITAL Advanced DSP System with FRONTWAVE, 2003 [Online]. Available: http://www.gennum.com.
29. Mangelsdorf CW (200) A variable gain cmos amplifier with exponential gain control. Dig. Symp. VLSI Circuits 146–149, Jun 2000
30. Hauptmann J, Dielacher F, Steiner R, Enz CC, Krummenacher F (1992) A low-noise amplifier with automatic gain control and anticlipping control in cmos technology. IEEE J. Solid-state Circuits 27(7):974–981, July 1992
31. Moulton D (1993) About the loudness of sounds and the risk of hearing damage, 1993 [Online]. Available: http://www.moultonlabs.com.
32. Gregorian R, Temes GC (1986) Analog MOS integrated circuits Wiley, Ch.5
33. Geerts Y, Steyaert M (2002) Design of multi-bit delta-sigma A/D converters. Kluwer academic publishers Ch.2
34. Tsividis Y, Krishnapura N, Palaskas Y, Toth L (2003) Internally varying analog circuits minimize power dissipation. IEEE Circuits Devices Mag 19(1):63–72, Jan 2003
35. David A, Johns, Martin K (1997) Analog integrated circuit design, Wiley 1997, Ch.10
36. Rabii S, Wooley BA (1997) A 1.8-V digital-audio sigma-delta modulator in 0.8-μm CMOS. IEEE J. Solid-state Circuits 32(6):783–796, June 1997
37. Kim S-E, Song S-J, Kim JK, Kim S, Lee J-Y, Yoo H-J (2004) A small ripple regulated charge pump with automatic pumping control schemes. Dig. IEEE Eur Solid State Circuits Conf 383–386, Sep 2004
38. Sauerbrey J, Tille T, Schmitt-Landsiedel D, Thewes R (2002) A 0.7-V MOSFET-Only switched-Opamp Σ-Δ modulator in standard digital CMOS Technology. IEEE J. Solid-state Circuits 37(12):1662–1669, Dec 2002
39. Kim S, Lee J-Y, Song S-J, Cho N, Yoo H-J (2006) An energy-efficient analog front-end circuit for a sub-1-V digital hearing aid chip. IEEE J Solid-State Circuits 41(4):876–882, Apr 2006
40. Yoo J, Kim S, Cho N, Song S-J, Yoo H-J (2006) A 10-μW digital signal processor with adaptive-SNR monitoring for a sub-1 V digital hearing aid. IEEE Int Symp Circuits Syst, May 2006
41. Brennan R, Schneider T (1998) A flexible filterbank structure for extensive signal manipulations in digital hearing aids. Dig. IEEE Symp Circuits Syst 6:569–572, May 1998

42. Lunner T, Hellgren J (1991) A digital filterbank hearing aid-design, implementation and evaluation. Dig. IEEE Int Conf Acoustics, Speech Signal Process 5:3661–3664, 1991
43. Norsworthy SR, Schreier R, Gabor C (1997) Temes, delta-sigma data converters—Theory, design, and simulation, IEEE Press Marketing
44. Neuteboom H, Kup BMJ, Hanssens M (1997) A DSP-Based Hearing Instrument IC. IEEE J. Solid-State Circuits 32(32):1790–1806, Nov 1997
45. Gata DG, Sjursen W, Hochschild JR, Fattaruso JW, Fang L, Iannelli GR, Jiang Z, Branch CM, Holmes JA, Skorcz ML, Petilli EM, Chen S, Wakeman G, Preves DA, Severin WA (202) A 1.1-V 270-μA Mixed-signal hearing aid chip. IEEE J. Solid-State Circuits 37(12):1670–1678, Dec 2002
46. Serra-Graells F, Gomez L, Huertas JL A true 1-V 300-μW CMOS-subthreshold log-domain hearing-aid-on-chip. IEEE J. Solid-State Circuits 39(8):1271–1281, Aug 2004
47. Peoria's first cochlear implant surgery has grandfather rediscovering life [Online] Available: http://www.pjstar.com/features/x876590686/Peoria-s-first-cochlear-implant-surgery-has-grandfather-rediscovering-life.
48. Eyeing smaller, faster, smarter ear implants [Online] Available: http://www.usatoday.com/news/health/2009-08-16-cochlear-implant_N.htm.
49. Zeng F-G, Rebscher S, Harrison W, Sun X, Feng H (2008) Cochlear implants: system design, integration, and evaluation. IEEE Rev Biomed Eng. 1:115–142
50. An SK (2009) Multi-channel Cochlear Implant. Bio-Medical IC Workshop
51. An SK, Park S-I, Jun SB, Lee CJ, Byun KM, Sung JH, Wilson BS,. Rebscher SJ, Oh SH, Kim SJ (2007) Design for a simplified cochlear implant system. IEEE Trans Biomed Eng 54(6):973–982, Jun 2007
52. Wilson BS,. Dorman MF (2008) Cochlear implants: current designs and future possibilities. J. Rehabil Res Dev 45(5):695–730, 2008
53. New hybrid hearing device combining advantages of hearing aids, implants [Online] Available: http://www.sciencedaily.com /releases/2008/04/080417100013.htm
54. Gantz BG, Turner CW (2004) Combining acoustic and electrical speech processing: Iowa/Nucleus hybrid implant. Acta Otolaryngol 124:344–347
55. Turner CW, Gantz BJ, Vidal C, et al. (2004) Speech recognition in noise for cochlear implant listeners: benefits of residual acoustic hearing. J. Acoust Soc Am 115(4):1729–1735, Apr 2004

Chapter 12
Cardiac Rhythm Management IC's

Erno Klaassen

12.1 Introduction

Cardiac rhythm management devices can be grouped into two broad categories: pacemakers and implantable cardioverter defibrillators (ICD's). These devices, over the last decades, have continued to grow in capability and complexity, and provide therapy for a wide range of cardiac rhythm disorders. The devices themselves are only one important part of the entire system, which includes device, leads, programmer, and the patient. This chapter will provide some background about the need for these devices, their function in the system, and details on their internal electrical design, focusing on the integrated circuits.

12.1.1 Anatomy of the Heart

The heart is composed of four chambers, as show in Fig.12.1. De-oxygenated blood from body tissues flows into the right atrium (RA), through the tricuspid valve, and is pumped from the heart into the lungs by the right ventricle (RV). Oxygen-rich blood from the lungs flows into the left atrium (LA), through the mitral valve, and is pumped out to body tissues by the left ventricle (LV). Electrical signals flowing on the surface of the heart synchronize the pumping of the four chambers. The electrical impulse for a normal heartbeat originates in the sinoatrial (SA) node. From the SA node, the electrical signal flows through both atria, and causes the cardiac muscles to contract and squeeze blood into the ventricles. The atria are electrically connected to the ventricles by the atrioventricular (AV) node, which conducts and slightly delays the electrical signal from the atria before passing it to the ventricles, where it conducts through a network known as left and right bundle branches. The delay through the AV node allows the atria to fully contract and the ventricles to fill

E. Klaassen (✉)
St. Jude Medical, Cardiac Rhythm Management Division, Sunnyvale, CA, USA
e-mail: eklaassen@sjm.com

H.-J. Yoo, C. van Hoof (eds.), *Bio-Medical CMOS ICs*,
Integrated Circuits and Systems, DOI 10.1007/978-1-4419-6597-4_12,

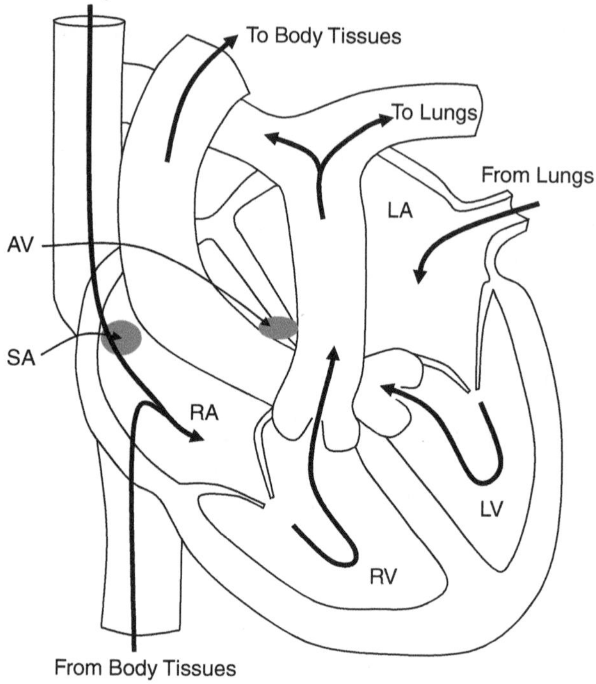

Fig. 12.1 Sketch of the four chambers of the heart, depicting direction of blood flow

before ventricular contraction starts. The ventricles then contract simultaneously, pumping blood into the lungs and body tissue.

An example of the well-known surface electrocardiogram (ECG) is shown in Fig. 12.2. The ECG is a recording of the electrical activity of the heart, and the various characteristic segments of the waveform are referred to by letters as P-Q-R-S-T. The P-wave represents depolarization (excitation) of the atria. The P-Q interval, typically about 200 ms, represents the conduction delay between the atria and ventricles through the AV node. The QRS originates from the ventricular depolarization, and

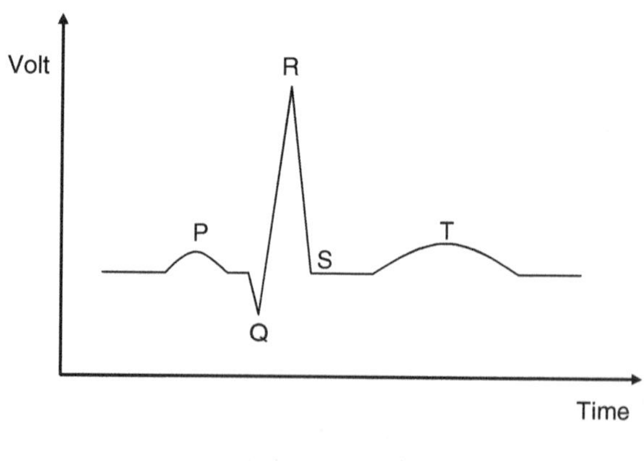

Fig. 12.2 Surface ECG showing the various components of the cardiac cycle

is followed by the T-wave which corresponds to ventricular repolarization, or return to the tissue resting state.

Arrhythmias of the heart include bradycardia, which manifests as an unusually slow heartbeat (<60 bpm), and tachycardia, which manifests as a regular but unusually high heartbeat (>100 bpm). Ventricular fibrillation is a dangerous condition that involves a rapid and irregular contraction of the ventricles, and prevents the ventricles from pumping blood efficiently. Ventricular fibrillation, if left unchecked, can lead to death within several minutes.

12.1.2 Pacemakers

The first implantable pacemakers were developed in the 1950s. The first implantation into a human was performed by Dr. Åke Senning in 1958 in Sweden, with a pacemaker built by Rune Elmqvist at a company called Elema-Schonander (which eventually became part of St. Jude Medical) [1]. Other early pioneers in this area included Chardack, Greatbatch and Bakken [2]. These early devices treated bradycardia by asynchronously stimulating the cardiac tissue of the ventricles at a fixed, predetermined rate. They required implantation through thoracotomy (a procedure where the patient's chest is surgically opened), and consisted of transistor oscillators built out of discrete bipolar transistors, resistors, and capacitors, powered by mercury-zinc or nickel-cadmium batteries, and sealed inside an epoxy resin. The need for thoracotomy was later eliminated through the use of catheter electrodes that could be passed into the right ventricle through a chest vein [3].

Later developments in pacemakers included the concept of demand pacing, introduced in the 1960s, where a pace pulse is only delivered when no intrinsic heart signal is sensed, dual chamber pacing, introduced in the 1970s, where sensing and pacing takes place through leads into both the right atrium and right ventricle, and rate-responsive pacing, where the pacing rate is adjusted based on physiological activity (often sensed through acceleration).

Pacemakers today have become sophisticated implantable medical systems that provide the doctor with a wide range of adjustable parameters and a wealth of diagnostic data, accessible through wireless telemetry.

12.1.3 Implantable Cardioverter Defibrillators

Another class of cardiac rhythm management devices is known as the implantable cardioverter defibrillator (ICD). ICD's typically include all the bradycardia capabilities of pacemakers, but also are able to detect ventricular tachycardia and provide high voltage cardioversion therapy. Because of the required extra circuitry, larger battery, and high voltage storage capacitors, ICD's are physically larger than pacemakers.

The ICD was conceived by Michel Mirowski in the 1960s, the first human implant took place in 1980, and was first approved by the FDA in 1985 for use in the United States [4]. Like with pacemakers, the first ICD's required a thoracotomy. The defibrillation leads, which have to deliver much greater energy than pacing leads, were epicardial patches, sutured to the outside of the heart. These early devices were implanted in the abdomen, due to their relatively large size. Transvenous defibrillation leads were developed later, and together with reductions in ICD size made possible pectoral implants and reduced the implant procedure from one requiring general anesthesia and thoracotomy to a much less invasive, often out-patient procedure.

12.2 Components of Pacemaker and ICD

12.2.1 Leads

The leads that connect the pacemaker or ICD to the heart are attached to a header on the device. In the past, most device manufacturers employed their own connector technology, which required the use of adapters when a lead from one manufacturer was used with the device from another manufacturer. These days, the connectors between the header and lead typically conform to a standard referred to as "IS-1" as determined by the International Standards Organization (ISO), which have a 3.2 mm diameter terminal pin [5]. The connector on the proximal end of the lead is inserted into the header and held in place by a set-screw, as shown in Fig. 12.3. Most modern leads are introduced transvenously into the heart, and employ either an active or passive fixation mechanism [6]. Myocardial leads, which attach to the outside of the heart, are still used, but typically only for young children whose veins are too narrow to accommodate a lead.

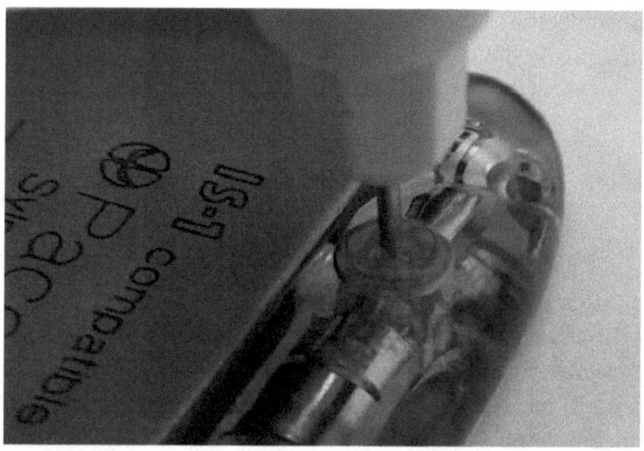

Fig. 12.3 Close-up of a pacemaker header showing the tool used to tighten the set-screw that holds the lead connector in place. Courtesy of St. Jude Medical

Leads come in a large variety of lengths and diameters, and can be divided into bipolar and unipolar varieties. Bipolar leads contain two conductors, and have two electrodes—typically a ring and a tip. Sensing and pacing through these leads occurs between these two electrodes. Unipolar leads have only one conductor and electrode, and sensing and pacing is done between this single electrode and another electrode in the system, usually the metal case of the device. Unipolar leads tend to be thinner, but they are more prone to inadvertent stimulation of the chest muscles, and are more susceptible to picking up myopotential interference from the electrical activity in surrounding muscle tissues. Leads are insulated in a layer of silicone or polyurethane.

Figure 12.4 shows a picture of a pacing lead with a passive fixation mechanism. The connector pin attaches to the device header. The distal end of this bipolar lead has ring and tip electrodes, as well as the fixation mechanism. The suture sleeve can be positioned at the appropriate location along the lead, and allows the physician to suture the lead in place.

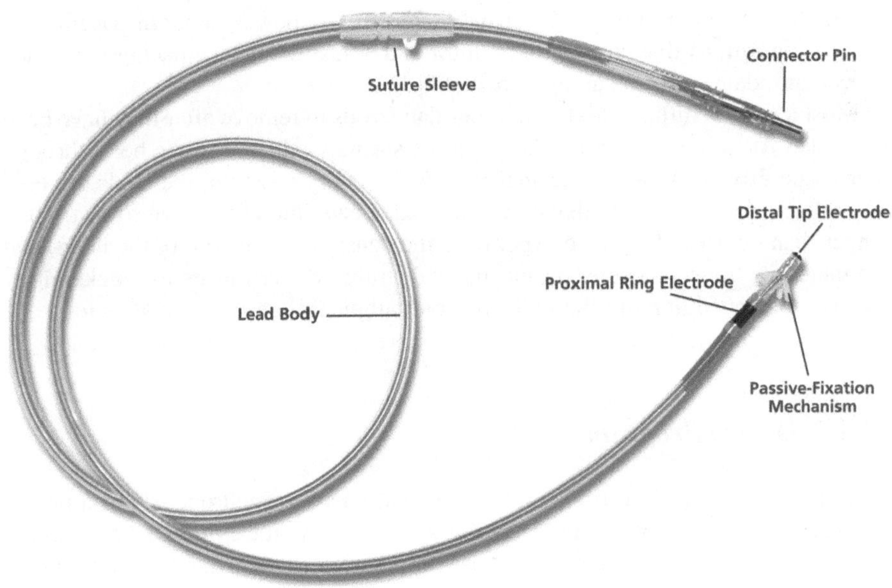

Fig. 12.4 Photo of a bipolar pacing lead. Courtesy of St. Jude Medical

Two important electrical properties of leads are the electrode resistance and polarization. Ideally the resistance of the electrode-tissue interface in a lead should be high, in order to minimize the amount of leakage current that flows during a pacing pulse of a given voltage amplitude. This resistance can be made large by reducing the electrode diameter. Polarization is a measure of a lead's susceptibility to surface charging due to positive or negative ions attracted to its surface. Such polarization inhibits the lead's ability to deliver pacing pulses. Electrodes with large surface area have lower polarization. In order to balance the seemingly contradictory requirements of small electrode diameter and large electrode surface area, modern lead electrodes have textured porous surfaces that increase total surface area without affecting electrode diameter.

At time of implant, a special device called a Pacing System Analyzer (PSA) is used to measure the sensing signal picked up by the lead (usually 6–10 mV for leads in the ventricles and >2 mV for leads in the atria [6]) and to determine the voltage threshold required to stimulate the heart. This threshold is also known as the "capture threshold". It is usually determined by using the PSA to deliver smaller and smaller amplitude pacing pulses until the heart is no longer stimulated. The capture threshold at implant is called the acute capture threshold. During the weeks following implant, the capture threshold typically increases at first, and then later diminishes, but usually not as low as the acute level. This is the acute-to-chronic capture threshold progression. Most modern leads have steroid-eluting tips that have been found to reduce the variation in capture threshold after implant. In order to provide a safety margin for proper stimulation of the heart, doctors often use a device setting of twice the measured acute capture threshold [6]. Since pacing at higher than required voltages is an inefficient use of power, modern pacemakers employ algorithms that constantly monitor and adjust the device pacing amplitude to accommodate shifts in capture threshold [7].

Most leads are difficult and sometimes dangerous to remove after they have been implanted for some time, since they become strongly adhered to the heart through scar tissue growth. When a pacemaker or ICD battery runs out, the leads are usually left in place when the device is replaced. Leads therefore have to last much longer than devices. They also experience the constant movement of the heart, and are therefore subject to constant mechanical strain. Modern leads use nickel alloy conductors wound in multifilar coils to avoid fatigue failures in the conductors.

12.2.2 Device Programmer

Early pacemakers had no programmability, and paced asynchronously at a predetermined rate. Later devices, in the sixties and early seventies, provided adjustment of the pacemaker function through the use of potentiometers that could be accessed transcutaneously with needles, or by cutting a resistor housed in the end of a pacemaker tail [8]. The first digitally programmable pacemaker, built by Cordis, allowed the physician to transmit settings to the device by means of a series of magnetic pulses [9].

For modern devices, at time of implant, and during subsequent patient checkups, communication with the device is achieved with a purpose-built programmer, as shown in Fig. 12.5. The programmer can establish communication with the implanted device through an inductive or RF telemetry link. More details on these telemetry protocols will be provided later in this chapter. Through the programmer's interface, the physician is able to enable or disable features in the device, program a large variety of device settings, run diagnostic tests, update device firmware, or download patient diagnostic data such as ECG recordings. It also allows live

Fig. 12.5 Photo of a device programmer. Courtesy of St. Jude Medical

monitoring of the ECG waveforms and other device measurements. Programmers and the communication protocols used in the telemetry with the implanted devices are proprietary to each device manufacturer; no standards exist yet that allow the programmer from one device company to program devices from another company, or that allow for the concept of a "universal programmer".

12.2.3 Device Subsystems

The implantable device can be divided broadly into a number of subsystems. The volume of an ICD device is very roughly divided in four equal parts as 25% battery, 25% electronics, 25% header, and 25% high voltage capacitors. In a pacemaker, there are no high voltage capacitors, and the battery normally occupies a proportionally higher fraction of the total device volume.

Each of these subsystems will now be discussed separately.

12.2.4 Case, Feedthrough and Header

Modern pacemakers and ICD's are encased in a hermetically sealed titanium shell. Titanium offers an extremely high strength-to-weight ratio and is biocompatible. The case is typically sealed in a laser-welding step. The metallic case forms one of the electrodes in the pacing/sensing system. For unipolar pacing, it is typically used as anode (positive electrode). In ICD's, it can be used as one of the shocking electrodes.

The electrical connections to the outside of the can are made through a hermetic feedthrough. The connections from the feedthrough are located inside a molded silicone header that accommodates the lead connections.

The outside of the case is laser-marked with the manufacturer name, serial number, and with a special code called the North American Society of Pacing and Electrophysiology / British Pacing and Electrophysiology Group (NASPE/BPEG) code [10]. The code refers to the function and capabilities of the implanted device, and is shown in Table 12.1.

Table 12.1 NASPE/BPEG device code

Position	I	II	III	IV	V
Category	Chamber(s) Paced	Chamber(s) Sensed	Resp. to Sensing	Rate Modulation	Multisite Pacing
	O = None	O = None	O = None	O = None	O = None
	A = Atrium	A = Atrium	T = Triggered	R = Rate Modulation	A= Atrium
	V = Ventricle	V = Ventricle	I=Inhibited		V = Ventricle
	D = Dual (A/V)	D = Dual (A/V)	D = Dual (T/I)		D = Dual (A/V)

The first position refers to the chambers the device can pace (atrium, ventricle, or both). The second position refers to the chambers the device can sense (atrium, ventricle, or both). The third position is the device operation for providing pacing pulses. "Inhibited" means the device will inhibit a pacing pulse if a sensed event is detected. "Triggered" means the device will provide a pacing pulse upon a sensed event. The fourth position refers to whether the device is capable of rate modulation. Rate modulation adjusts the pacing rate in response to some measurement of patient activity. This can be done in a manner of different ways, which includes accelerometers, heart impedance measurements, or temperature [6]. The fifth and last position refers to whether the device has multi-site pacing or sensing capability, either in the atrium, ventricle, or both. Modern devices can be programmed in a variety of ways, but the device code will always refer to the maximum capability of the device.

Figure 12.6 shows a photograph of an ICD device that shows some of the common markings on the outside of the can, and the polyurethane header with its set-screws. The lead connectors are electrically connected to the circuitry inside the case through conductors that run through the header, through a hermetic feedthrough, into the device.

12.2.5 Battery

Batteries in the first implantable pacemakers used nickel-cadmium or mercury-zinc chemistries, and some were inductively rechargeable. The cells in these batteries

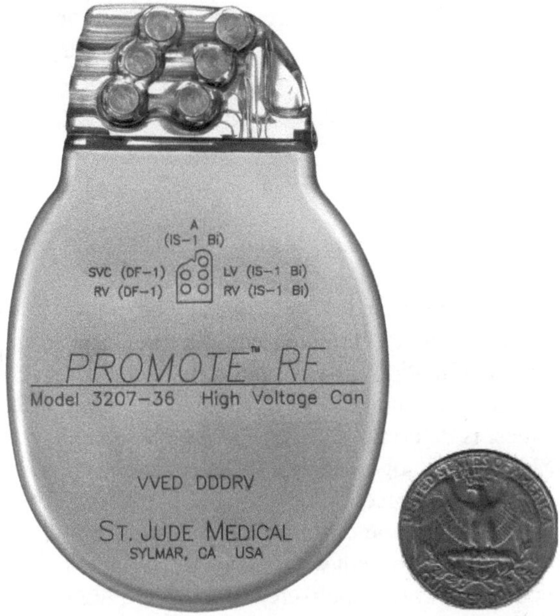

Fig. 12.6 Photo of an implantable cardioverter defibrillator. Courtesy of St. Jude Medical. U.S. quarter showing the approximate relative size of the device

suffered from high internal leakage that limited the longevity of early pacemakers. In the search for better device longevity, nuclear-powered pacemakers were developed and implanted. These devices were powered by radioisotopes, and generated electrical energy through thermoelectric or beta-voltaic effects [11]. While such devices provided excellent longevity, the various challenges of using radioactive material in an implantable medical device meant that these devices never gained widespread use.

In the 1970s, the lithium-based power cell, first developed by Wilson Greatbatch [12], became the most common type of battery for implantable pacemakers. Varieties of lithium-based chemistries remain the dominant type of batteries in use today. Most pacemakers use lithium iodine cells, which have an extremely low self-discharge rate, and a stable voltage throughout most of the cell's useful life, with a predictable drop in voltage towards the end of life [7]. Pacemaker batteries have a relatively high equivalent series resistance, on the order of tens of kilo-ohms at the end of service (EOS). They are therefore not capable of sourcing high currents, even if those currents flow for a short period of time. This poses an additional challenge on the circuit design in the pacemaker, especially for higher-power diagnostics, sensor, or telemetry functions. While the average current for such functions can be made low through duty-cycling, the limited current available from a pacemaker battery also constrains the maximum peak current drawn from the battery while the function is active. For ICD's, which draw high currents from the battery during the charging of the high-voltage capacitors, other lithium-based battery chemistries are used, such as Lithium/Silver Vanadium Oxide (SVO) [13]. Current draw during the charging of the high voltage capacitors in an ICD reaches peaks of several Amperes for many seconds, and these types of batteries are able to sustain such currents, but still have sufficiently low internal discharge rates to provide excellent device longevity.

12.2.6 ICD Capacitors

The electrical energy used for defibrillation therapy in an ICD is temporarily stored on high voltage capacitors inside the device. When ventricular fibrillation is detected, these capacitors are charged to a target voltage using a high voltage boost converter. Once charging completes, the capacitors are switched across the desired lead combination, and the defibrillation energy is imparted on the heart. Usually, at time of implant, the defibrillation threshold (DFT) for the patient, measured in Joules, is determined for by purposely inducing ventricular fibrillation, and delivering a series of shocks at different energy levels until a safe level is determined that ensures successful therapy. ICD's have features that allow induction of ventricular fibrillation during implant to accommodate this DFT testing. Most devices have several different ways in which such induction can be achieved. Common ways include DC current delivered to the heart, high frequency pacing, or specially timed high voltage pacing pulses that coincide with the T-wave. The defibrillation energy level programmed in the device is increased by a safety margin to account for variation in DFT over time. Defibrillation waveforms are usually delivered as biphasic waveforms, as shown in Fig. 12.7. Biphasic waveforms have been found to allow for lower DFT's [14]. Modern ICD's can deliver energies on the order of 40 J. In order to accommodate such high energy levels, the internal capacitors are charged to voltages close to 1000 V.

The high voltage capacitors are charged only when the device detects ventricular fibrillation, and are thus left discharged when not in use. In order to maintain fast charge times, the capacitors are periodically charged to maximum voltage, in a process known as capacitor reformation, or capacitor maintenance. The ICD handles this automatically, and the interval with which this is done is usually around

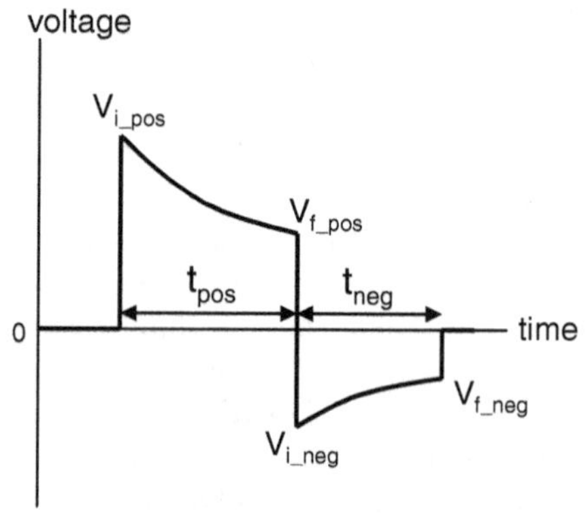

Fig. 12.7 Biphasic defibrillation waveform

3 months. As the capacitor is charged up, local leakage sites in the capacitor dielectric material conduct current which leads to a chemical reaction that re-forms the dielectric material at the leakage site, and thus eliminates much of the leakage. This ensures that, when the capacitor is charged during an actual detected episode of ventricular fibrillation, the leakage does not adversely impact the charging time. At the end of a capacitor maintenance cycle, the charge on the capacitors is bled off internal to the device.

12.3 Electronics

The electronics within the pacemaker or ICD are built on one or more hybrid boards, that contain a mix of integrated circuits, discrete semiconductors, resistors, capacitors, and sensors. The dynamic range of signals within the electronic circuitry spans that of sensing signals with a resolution of μV to high-voltage therapy of close to 1 kV. While there is a desire for the highest level of integration possible, the need for high voltage tolerance (both from an internally generated defibrillation pulse as well as from external defibrillation), large capacitance values (on the order of tens of μF), as well as the integration of various types of sensors, makes integration of all the functionality of a pacemaker, and certainly that of an ICD, not yet feasible.

12.3.1 Basic Pacemaker Functions

Even the most basic pacemakers these days tend to be complex embedded systems, with many programmable features, diagnostics, and modes of therapy. At their most basic, the two externally visible functions of pacemakers are the capability of sensing cardiac electrical signals through the leads, and the capability of delivering, when necessary, pacing pulses back through those same leads.

A block diagram showing some of the most common basic pacemaker functions is shown in Fig. 12.8.

Pacemaker batteries have capacities on the order of 900 mA-hrs. This translates into approximately 100 μA-years. So if a pacemaker draws, on average, 10 μA of battery current, the device longevity would be approximately 10 years. Actual device longevity is strongly dependent on device settings, what features are activated, electrode resistance, degree of telemetry use, etc. This simplified example highlights that, for a device with a longevity of about 10 years, a sustained additional current draw of 1 μA in the electronics would reduce longevity by a very significant one year, and 100 nA reduces longevity by over one month. The management of battery current is therefore of utmost priority in the design of pacemaker electronics, and at the IC block design level, every nano-ampere tends to deserve scrutiny, especially in blocks that cannot benefit from a low power-on duty cycle.

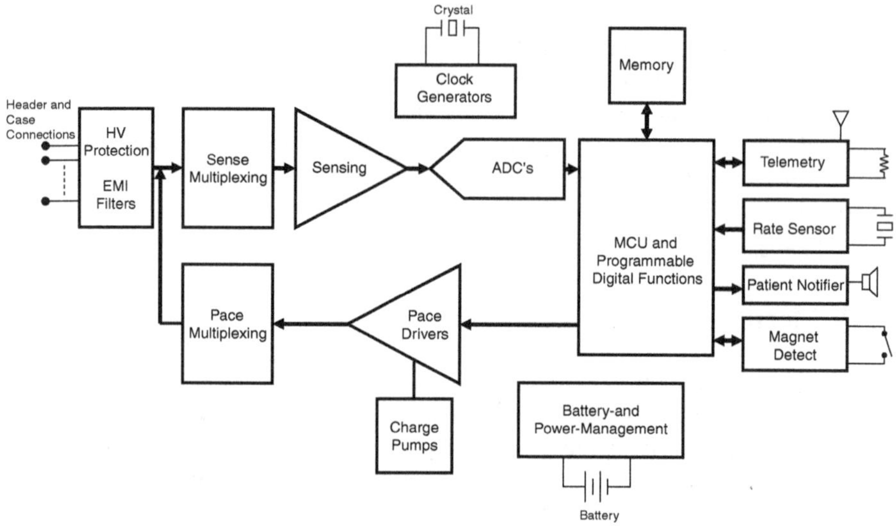

Fig. 12.8 Simplified block diagram of the electronics in a basic pacemaker system

12.3.2 Sensing Circuits

Today's pacemakers are usually dual-chamber devices that can pace and sense in both the atrium and the ventricle, and commonly have provisions for three bipolar leads, one in the atrium, and one in each of the two ventricles. The case of the pacemaker is also used as an electrode (in unipolar configurations). In order to accommodate programmable sensing vectors, the inputs from the electrodes are usually multiplexed into a set of sensing channels, where the cardiac signal can be differentially sensed.

The frequency spectrum of interest for cardiac signals is between tens of Hertz and several hundred Hertz. Myopotential signals associated with muscle movement, which represent interference for cardiac sensing, have a bandwidth from about 100 Hz to several kHz. Sensing channels for pacemakers are therefore typically implemented as band-pass channels, with one or more low-frequency poles at tens of Hertz, and one or more high-frequency poles at several hundred Hertz. This filtering can be done in the analog domain, before the ADC, or implemented with digital filters after the ADC. Digital filtering can be more versatile, allowing greater degrees of flexibility in pole locations, filter profiles, etc., but the digitization of a wider bandwidth analog signal requires a faster, and thus more power hungry, ADC.

The sense channels are typically electrically disconnected from the lead during a pacing pulse, to prevent saturation of the sense amplifiers by the pacing waveform, which is much larger in amplitude than the sensing signals. Since the sensing inputs connect electrically to the device header, and through there to the leads inside the heart, special care must be taken to protect these inputs from electrical overstresses that can couple into the device through the leads. Such overstresses can come from external defibrillation, such as from an automated external defibrillator (AED) or from one applied by paramedics or doctors. Most devices also employ an EMI filter, ideally located as close as possible to the device header, to low-pass filter the signals coming into the device.

Sensing signals contain features that can be as small as several tenths of mV for smaller components like the atrial P-wave to tens of mV in amplitude for the largest features in a QRS complex. Depending on the full-scale-range available, this implies a required sensing chain gain, before ADC conversion, of around 26–40 dB. Since sense amplifiers need to remain powered continually, throughout the entire cardiac cycle, the power dissipation in this function is extremely critical. Sense amplifiers with supply currents of several 100 nA have been reported [15–17]. For these types of amplifiers and filters, switched capacitor topologies are commonly used, often running at much lower rates than found in other, higher-power and wider bandwidth applications. Switched capacitor circuits run at low frequencies can pose design challenges when the leakage through the switches is sufficiently high to affect circuit operation [18].

12.3.3 ADC

Following amplification and filtering, the sensing signals are digitized. Even until quite recently, some pacemakers did not actually digitize the sensing signals, but used analog-based threshold-detection techniques to sense intrinsic cardiac events. Obviously the flexibility of such sensing techniques is quite limited, and almost all modern pacemakers digitize the sensed signals, and use digital means to detect atrial or ventricular activity. The resolution required of such ADC's is usually no less than 8 bits [18], and the conversion rate is determined by the Nyquist frequency to be at least twice as high as the bandwidth of the incoming analog signal. If the low-pass filtering of 100–200 Hz is accomplished in the analog domain, before the analog-to-digital converter (ADC), that sets a conversion rate at 200–400 samples per second. Since, like the rest of the sensing channel, the ADC's typically operate during the entire cardiac cycle, extreme care is taken to ensure they are optimally designed. Some examples of different approaches to such designs can be found in [18–20].

The power requirements for the ADC's are extremely low, and represent a very extreme corner of data converter design. Fortunately, the conversion rates required are also relatively low compared to many other applications. In fact, the conversion rates cited above are off the charts, on the low end, on some recent ADC survey papers [21–23]. ADC's have been published with energy/conversion numbers as low as 31 pJ per conversion for an 8-bit converter [23]. An ADC specifically designed for digitization of cardiac signals was reported at 300 pJ per conversion for an 8-bit converter [18]. In a noise-limited converter, the extra power required per bit would increase 4× per bit, while trends in data converter publications reflect an increase of closer to 2× per bit for moderate to low resolution converters [22].

In sensing and processing cardiac waveforms, it can be useful to employ a data converter with adjustable resolution, to trade either conversion rate or power with resolution. Successive approximation register (SAR) ADC's lend themselves to a tradeoff of resolution and conversion rate, as they can terminate the conversion before the least significant bits are calculated. This tradeoff only saves net power if the standby current is small. Sigma-delta converters can also provide such a tradeoff [24, 25], but these types of converters can pose additional design challenges to allow the implementation of signal multiplexing [26] if the sensing architecture requires that.

12.3.4 Pace Driver and Mux

The output path of the pacemaker system shown in Fig. 12.8 consists of the pace drivers and multiplexers. Pacing pulse parameters are usually programmable, allowing the clinician to set the pacing amplitude and pace pulse width. Pacing pulse amplitudes required depend on the capture threshold, as described earlier, and are usually programmable in increments of around 100 mV, up to maximum voltages of 7–10 V. The amount of energy delivered to the patient by the pace pulse sets a theoretical lower limit on device current draw. For example, if a device is pacing with an amplitude of 2.5 V and a pulse width of 0.5 ms, at a rate of 70 pulses per minute (ppm) with 100% pacing into one chamber (no pulses are inhibited by the detection of intrinsic cardiac activity), into a 500 Ω lead resistance, the average current draw is given by:

$$I_{\mathrm{AVE}} = \frac{V_{\mathrm{PA}}}{R_L} \cdot \frac{T_{\mathrm{PW}}}{T_{\mathrm{PP}}} = \frac{2.5\mathrm{V}}{500\Omega} \cdot \frac{0.5\mathrm{ms}}{0.857\mathrm{s}} = 2.92\mu\mathrm{A} \qquad [12.1]$$

This represents roughly 30% of the battery capacity of the example 900 mA-hrs device described earlier. If lead impedance is lower, capture threshold is higher (necessitating a higher pace pulse amplitude), or if the pulse width needs to increase, the current number above will grow accordingly.

The interaction between pace pulse amplitude and duration is reflected in the so-called strength-duration curve [6] as depicted in Fig. 12.9. Two reference points on the strength duration curve are rheobase and chronaxie. Rheobase is defined as the minimum voltage required to capture the heart, regardless of how long a pulse width is used. Chronaxie is the pulse width required to capture the cardiac tissue at a pulse amplitude of twice the rheobase. The energy per pulse grows as the curve approaches the asymptote of rheobase, due to rapid increase in the required pulse width. As the pulse width is reduced below the chronaxie point, the required energy also increases, due to the rapid increase in required pulse amplitude. An optimum energy level is found between these two extremes, near the chronaxie point.

Battery voltages are typically in the range of 2–4 V, depending on the battery type and depletion level of the battery. For pacing voltages above the battery voltage, a charge pump, or some other type of boost converter, is required. An example

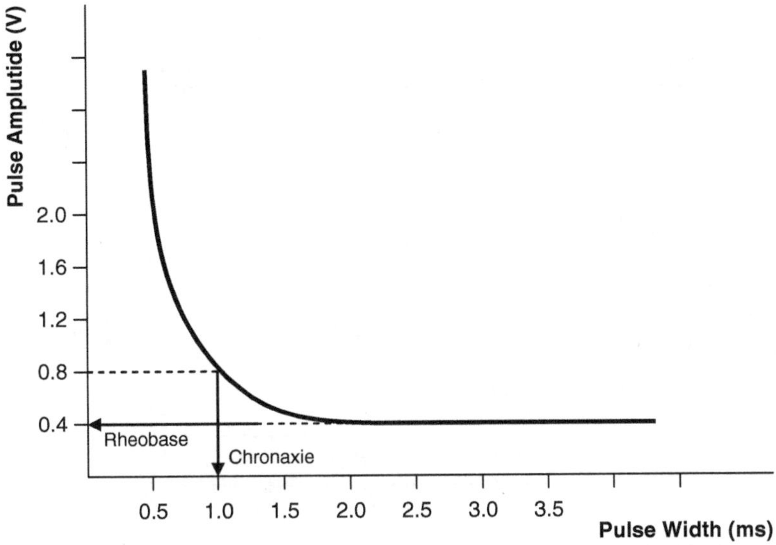

Fig. 12.9 Strength-duration curve for stimulation of the heart, showing the pulse amplitude required as a function of pulse width

of such a charge pump, providing an integer multiple of the battery voltage, is shown in [18]. At its most basic, such charge pumps circuits charge one or more capacitors in parallel, and then place them in series across the load. Ripple from the charge pump is filtered out by a suitably large reservoir capacitor. The current drawn from such charge pumps is limited by the allowed voltage droop, the switching rate, and values of the intermediate capacitors. The inefficiency of such charge pumps is limited mainly by resistive losses in the switches and the dynamic current used to charge the parasitic capacitances (mainly gate capacitance in the switches). Due to size limitations of on-chip capacitors, some, or even all of the capacitors in a charge pump may need to be implemented with discrete off-chip components.

The pace output drivers deliver the pacing pulse either directly to the pace outputs, or can be multiplexed through a programmable switching array. Examples of output driver designs can be found in [18] and [27]. A schematic of the type of amplifier described in [18] is shown in Fig. 12.10. Core-voltage transistors are represented in the schematic with a thin line for the gate, while high-voltage transistors are represented with a thick line. A core-voltage differential pair (M1/M2) serves as the input to the amplifier. This pair serves as a transconductor, amplifying the differential input into a differential current. The differential current is mirrored into a two separate current-mirror loads, one made of core-voltage transistors, the other of made up of high voltage transistors.

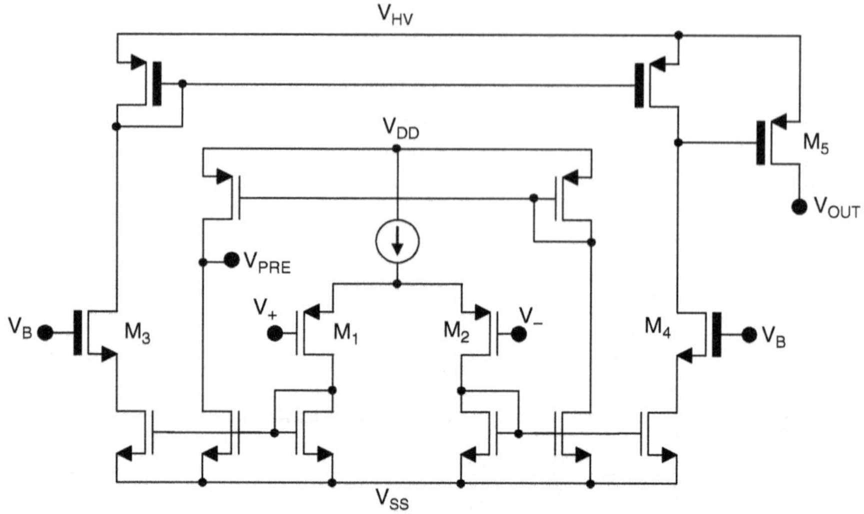

Fig. 12.10 Simplified schematic of a high voltage driver that can be used to generate pacing waveforms [18]

Cascode transistors M3/M4 protect the low-voltage transistors from exposure to the large voltage swings on the high voltage stage. The high-voltage stage is followed by an output transistor M5, which is simply a common-source PMOS stage. The frequency compensation of the amplifier is not shown for simplicity.

The amplifier is used in [18] in a switched-capacitor DAC as shown in Fig. 12.11. The output of the low voltage stage (V_{PRE}) is used to turn that stage into a unity gain buffer during the first phase (Φ_1). During that phase, a reference voltage (VREF) and the input offset of the low voltage amplifier are stored on C_1. The value of C_1 is adjustable, and the trim for that capacitor is the digital input of the DAC. During

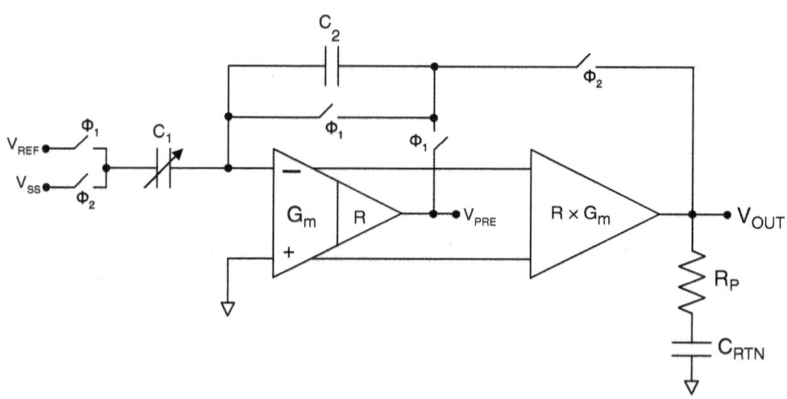

Fig. 12.11 Switched capacitor DAC that can be used to generate pacing output waveforms [18]

the second phase (Φ_2), a capacitive feedback path around the high voltage output is closed, and the output voltage is driven to a value given by

$$V_{OUT} = \frac{C_1}{C_2} \cdot V_{REF} \qquad [12.2]$$

The resistive load R_P in Fig. 12.11 is the patient load, which is approximated here as an ohmic resistance. C_{RTN} is a charge-balance capacitor, the purpose of which will be described below.

Since the load can be as low as several hundred Ohms, and can vary significantly among different patients, the output impedance of the driver needs to be in the range of tens of Ohms, or lower. Pacing outputs are always relative to a pace return, which can be chosen by the clinician. In bipolar leads, the return is usually chosen as the other electrode in the lead (tip or ring). One important requirement for pacing stimuli is that of charge neutrality. If a unipolar pacing pulse with the parameters shown in Eq. 12.2 is applied to the heart, 2.92 μA of net DC current flows out of the positive lead electrode, through the cardiac tissue, and into the return electrode. At sufficiently high levels of DC current, in the hundred micro-ampere range, such current can cause arrhythmias such as ventricular fibrillation. Even at lower levels, on the order of hundreds of nano-amperes, DC current can corrode the lead electrode material in some leads, or cause polarization of the lead-tissue interface that interferes with pacing and sensing.

One way to achieve pacing charge neutrality is to effectively AC couple the pacing circuitry with a capacitor sufficiently large to not filter the pacing pulse in a clinically significant way. A very simplified form of such a circuit is shown in Fig. 12.12. The associated pacing waveform applied across the tissue is shown in Fig. 12.13. During the pacing pulse itself, switch S1 is closed, and S2 is open. The pacing voltage, developed on the pacemaker IC as the difference between the Pace_out and Pace_rtn nodes, drops across the tissue load R_p. The pacing current that flows, I_{pace}, charges up the pace return capacitor C_{rtn}. C_{rtn} is chosen large

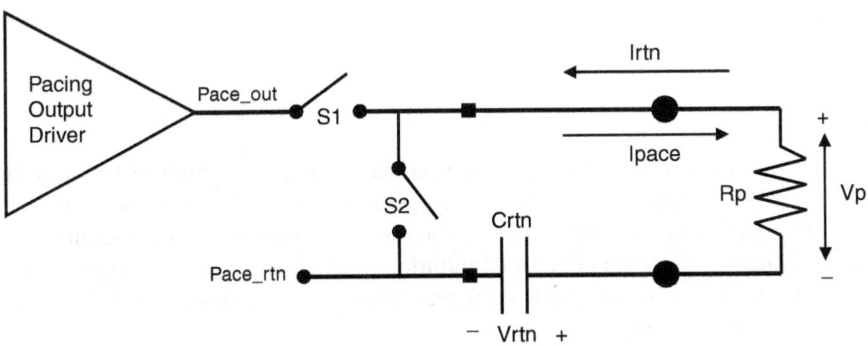

Fig. 12.12 Schematic representation of passive charge balancing

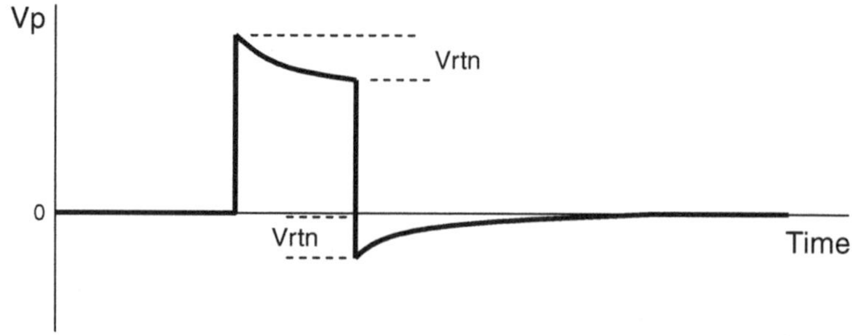

Fig. 12.13 Pacing voltage Vp across pacing load, showing pacing pulse and charge balancing

enough to not cause an excessively large voltage to be developed across it, since that voltage V_{rtn}, starts dropping the value of V_p during the pace pulse.

After the end of the pace pulse, switch S1 is opened, and S2 is closed, which discharges the pace return capacitor C_{rtn}. This part of the cycle is responsible for the negative portion of the pacing waveform shown in Fig. 12.13. Once the pace return capacitor is fully discharged, the charge that has flowed through the lead electrodes (the integrated current I_{pace} minus the integrated current Irtn) will be the ideal net-zero.

In actual implementations, this situation is made more complicated by the need for different pace outputs and returns, which are each connected to each other through various lead electrodes and tissue resistances, but the fundamental concept remains similar. This type of charge balancing approach can be considered as a passive charge balance. The benefit is that the return current I_{rtn} comes fully from the pace return capacitor, and therefore does not require battery current. Drawbacks include the need for a large value capacitor, and the time spent during the passive discharge (several time-constants of $R_p \times C_{rtn}$). Alternative ways of achieving charge neutrality involve active charge balancing, where an active circuit is used to drive a return current that balances the integrated pacing current. Such techniques can be faster, may avoid the need for a capacitor, but typically come with the penalty of additional current drawn from the battery.

12.3.5 MCU

Another important function in modern pacemakers is the function that makes the determination of when (or even if) a pacing pulse is needed. In the block diagram of Fig. 12.8 this function is represented by the block labeled MCU and Programmable Digital Functions. Not only does the inhibition of an unnecessary pacing pulse save power in the device, it has also been found that intrinsic cardiac activity is usually preferable to pacing [6].

Detection of sensed events (intrinsic cardiac activity) is, in the simplest implementation, based on a threshold detection of the digitized sensing signals from the

ADC. Signal edges above the programmable detection threshold are detected, those below it are ignored. If the threshold is set too low, this can lead to "oversensing", and the device will inappropriately inhibit pacing. If the threshold is set too high, this can result in "undersensing", and the device will inappropriately pace. Such detection is accomplished in the digital domain for a pacemaker like the one shown in the block diagram in Fig. 12.8, but it could also be accomplished without digitization using analog comparators [16].

The interval between a sensed event or pacing output pulse and the subsequent sensed event or pacing output pulse is known as the escape interval. The first part of the escape interval, just following the sensed or paced event, is known as the pacemaker refractory period. During the refractory period, the device is not trying to detect intrinsic activity, since there are usually too many signals that could lead to oversensing. These signals include remaining energy from the pacing pulse, evoked response from the cardiac tissue in response to the pacing pulse, or the T-wave that results from repolarization of the ventricles. Following the refractory period is the alert period, where the pacemaker will try to detect intrinsic cardiac activity. If a given amount of time expires from the beginning of the escape interval (last paced or sensed event), the pacemaker will deliver a pacing pulse. The detection of a sensed event or delivery of a pacing pulse will reset the timing cycle. In a single chamber pacemaker, for example a device running in VVI mode (per the mode definitions in Table 12.1), it will sense in the ventricle and, if not inhibited by the detection of a sensed event during the alert period, will pace in the ventricle. Dual chamber pacemakers that sense and pace in both the atria and ventricles have a larger numbers of states in the basic pacing cycle. In such a pacing cycle, there are four main states of pacing, which are determined by the permutations of sensed vs. paced events in the atrium and the ventricle [6]. These states are designated by letter combinations where "A" indicates an atrial paced event, "P" indicates an atrial sensed event, "V" indicates a ventricular paced event, and "R" indicates a ventricular sensed event. "AR", for example, designates an atrial paced event followed by a ventricular sensed event. One of the main benefits to dual chamber pacing is to maintain synchrony between the atria and the ventricles, which means that these chambers of the heart are pumping blood in a coordinated fashion. A detailed discussion of the timing cycles for dual chamber pacing is beyond the scope of this chapter, but can be found elsewhere [6, 28]. One of the challenges is to ensure that the ventricles are not paced too fast in response to atrial fibrillation. This is accomplished in modern pacemakers by mode switching, where the device switches to another mode of pacing when high atrial activity rates are detected.

12.3.6 Sensor I/O

Another important function found in most modern pacemakers is the rate detector. Some patients have a condition known as chronotropic incompetence, which means that their intrinsic heart rate does not increase enough with the physiological demand for more oxygen. Different types of sensors have been employed to determine the desired rate of pacing. Some of these involved pH measurements of the blood, or blood temperature. Those types of measurements provide good indicators of physiologic demand, but they were found to be too slow to respond. They also require an electrical connection to the outside of the device. Most of today's pacemakers use activity sensors that indirectly measure physiologic demand by measuring patient motion with an accelerometer. Most of these types of devices measure acceleration in the axis normal to the chest of the patient (so normal to the front face of the device). The processed acceleration signal is used to adjust the pacing rate to adjust for changes in activity level.

The accelerometers used in this application can be MEMS-based devices made by the micromechanical processing of silicon or devices fabricated with conventional techniques. The typical sensing mechanism is based on detecting the motion of a proof mass suspended by a spring, either through piezoelectric, piezoresistive, or capacitive means.

Another type of sensor used in most pacemakers is a magnetic sensor. These are used to detect the presence of a magnet held near the device, external to the patient's body. Detection of such a magnet causes different system behavior, depending on device manufacturer. Some common magnet-mode behavior of pacemakers include the device switching to asynchronous (DOO or VOO) pacing without rate response, changing to a default base pacing rate, and/or switching to a default pace pulse amplitude. Magnetic sensors that can be used for this application include mechanical reed switches, hall sensors, anisotropic magneto-resistive (AMR) and giant magneto-resistive (GMR) sensors. Reed sensors provide a digital output signal, depending on the state of the relay, while the other types of sensors provide an analog output signal. The biasing to and output from these sensors are connected to circuitry that provide a magnet detect signal to the digital logic. In the case of a reed sensor, this includes de-bouncing or other filtering circuitry, and in the case of the sensors with an analog output will usually include a threshold detection circuit.

12.3.7 Telemetry

While the magnet detection system provides a simple yet limited way for the device to interact with the environment outside the patient's body, the telemetry subsystem forms the more versatile interface that allows physicians to interrogate and program the implanted device, using a programmer like the one shown in Fig. 12.5. Most modern pacemakers use a form of inductive telemetry, but RF telemetry is starting to become available.

There are no industry standards for these inductive telemetry links – all device manufacturers tend to use their own proprietary protocol. Since the telemetry is the only "window" into the device once the can is sealed, it is not only used after implant, but also during the final stages of device manufacturing, and prior to implant as the device is configured outside the patient's body.

Inductive telemetry systems work at relatively short distances. The clinician places an inductive wand over the device, on the patient's chest. The wand contains a large coil that transmits to, and receives signals from, the device. A second much smaller coil inside the device receives the signals from the wand, and transmits signals back to it. Since telemetry is used for only a relatively small fraction of the lifetime of a device, most of the circuits used in this block can remain powered down. A telemetry link is initiated only from the programmer side, so the device only needs to periodically "sniff" for a wakeup signal from the programmer. Since that circuitry needs to run continuously, the power requirements for the wakeup sniffing are stringent. The rest of the telemetry circuitry can dissipate, for the short time that it is on, whatever power is necessary to reliably transmit and receive the data. Since the coupling in the inductive link is not very strong, the transmit circuitry in the device must be capable of driving relatively large currents into its coil. The receive circuitry also needs to detect the relatively small induced voltages on the coil from the external wand. There are several challenges in the design of such circuitry. One important one is the fact that the amplitude of the receive signal on the device telemetry coil can vary tremendously depending on the positioning of the external wand. This is strongly affected by how much body tissue is present between the implant, and the relative angles of the device and the wand. The circuitry needs to tolerate such variation without impacting the telemetry link. Another challenge is the need to maintain a reliable telemetry link while other electrically noisy functions are active inside the device. This is a bigger challenge for ICD's, where capacitor charging can happen, but is still a design challenge inside pacemakers as well.

RF telemetry is starting to become available in pacemakers (and has been available in ICD's for several years). Two common frequencies for RF communication are in the Medical Implant Communications Service (MICS) band, or in unlicensed parts of the RF spectrum. MICS is a standard defined in 1999 specifically for implantable medical devices [29]. It operates in a frequency range from 402 to 405 MHz. The only other types of devices licensed by the FCC in this band are weather balloons [30]. The standard specifies a number of different channels within this band, each with a bandwidth of 300 kHz. The maximum radiated power allowed is 25 μW, and a typical intended range of a communications link is a few meters. The MICS standard specifies aspects of the communication protocol that include a clear-channel assessment (CCA), which involves a listen-before-transmit. CCA requires that a device about to transmit first determines received signal strength on a given channel, which needs to fall below a certain threshold before it is allowed to transmit. The standard allows for emergency device transmission without a CCA.

There are numerous benefits to RF telemetry over inductive communication. Some of these benefits revolve around the longer range of communication enabled by RF. During implant, for example, the device programmer can be located further away from the patient, and therefore outside of the sterile operating field. RF telemetry also allows for home monitoring, by means of devices such as bed-side monitors. These systems allow for remote follow-ups, so that the device can be interrogated remotely, saving the patient a trip to a clinic. The bed-side monitor connects through a phone line to a secure central database that provides access for the physician. RF telemetry systems also support higher data rates than inductive ones, which speeds up interrogation, or the download of new settings or firmware into the implanted device. Some of the technical challenges with implementing MICS telemetry systems involve the variability in antenna matching and losses through the patient's tissue. Losses can be as high as 40–45 dB [31]. Peak currents also have to be limited to a level the high internal resistance battery can accommodate; usually less than several mA.

12.3.8 Clock Generator and Power Management

A clock generation subsystem provides the clocks used throughout the system. At least one of these clocks needs to allow for accurate real-time-clock and pacing frequency generation, and is typically based off a 32.768 kHz (2^{16} Hz) crystal. Such crystals can provide frequency references with tens of part-per-million accuracy. The crystal is driven at its resonance frequency with a sustaining circuit. This relatively slow frequency reference is constantly running, and can provide timing for other duty-cycled functions throughout the system. Faster clocks, for digital circuitry for example, can be generated as a derived clock using a phase-locked-loop (PLL) or by other means. Care must be taken in limiting how widely even the slower time-base clock is routed around the system. The dynamic current drawn by this clock driving a capacitive load is given by

$$I_{\mathrm{DYN}} = C_L \cdot V_{\mathrm{DD}} \cdot f \qquad [12.3]$$

where C_{L} is the load capacitance, V_{DD} is the power supply voltage, and f is the clock frequency. If the supply voltage is 1.8 V, and the load capacitance is 10 pF, the dynamic current is 589 nA, which equates to almost 6 months of device longevity for the example battery capacity shown above.

An important housekeeping function in a pacemaker system is the battery and power management subsystem. This function includes regulators that take the battery voltage and regulate it down (or potentially up) to a stable voltage that suits the needs of the different subsystems. It also contains means to measure the battery voltage to indicate when the battery has reached the elective replacement interval (ERI) or end of service (EOS) points. Some battery chemistries have a gradual and relatively predictable decline in battery voltage as the battery capacity is depleted, but others can have more sudden drops that only occur near the EOS. For the latter case, devices can use a battery current monitor that integrates the battery current over the life of the device, so that ERI and EOS can be predicted more accurately [18]. Multiple supplies can be used inside the pacemaker system, since different functions can typically have competing supply voltage requirements. Some analog functions, such as the pace drivers, benefit from the highest supply possible, to provide a larger signal swing, and to provide high gate drive to analog switches. Digital functions, on the other hand, benefit from a lower supply voltage in order to reduce standby leakage and dynamic power dissipation. The regulation of supplies may also be handled dynamically, to enable low power standby modes that allow a function to retain its state with a lower voltage during a standby period, while providing sufficient supply to allow for dynamic operation when the function is active.

The patient notifier function shown in Fig.12.8 is included in some devices, and provides a way for the device to warn the patient about a device condition that requires a followup, such as the battery voltage reaching pre-set ERI or EOS limits. Patient notifiers typically use either vibratory or auditory means of enunciating the device condition.

12.4 Basic ICD Functions

Early ICD's only provided defibrillation detection and therapy functions, but most ICD's these days also offer pacemaker functionality. The simplified block diagram of basic electrical function in an ICD are shown in Fig. 12.14.

As the block diagram shows, it is effectively a superset of the pacemaker blocks. The blocks that are shared between ICD's and pacemakers may have different specification in the two different devices, depending on differences in battery voltages, required regulated supply levels, filtering coefficients, ADC resolution, specific clocks required, amount of memory required, MCU speed, etc. Those details will not be discussed here, but the functions specific to only ICD's will be described.

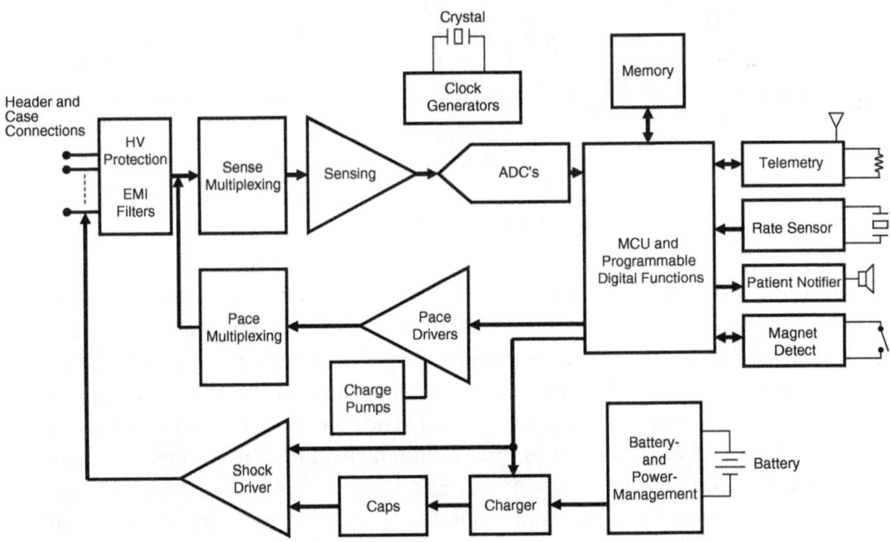

Fig. 12.14 Simplified block diagram of basic electrical functions in an ICD

A distinction can be made between sensing and arrhythmia detection. Sensing refers to determination of a threshold crossing, while arrhythmia detection is the determination of a ventricular tachycardia or ventricular fibrillation episode. Normal cardiac waveforms are relatively large and stable, but the signature of a ventricular tachyarrhythmia event can have very irregular, low amplitude signals. In order for these to be sensed, an ICD needs to have a sufficiently low sensing threshold. One of the challenges is ensuring that such low thresholds do not result in oversensing, which could lead to false detection and inappropriate therapy. Most sensing algorithms are based on adjusting the sensitivity settings based on the most recent sensed amplitudes. A particular sensing algorithm used in St. Jude Medical devices uses a dynamically adjusted sensing threshold [4]. In this algorithm, a rectified version of the sensed signal is used. The sensing of a ventricular signal begins a refractory period, during which the device measures the amplitude of the largest peak. That peak amplitude is used to set a starting threshold level for sensing after the refractory period ends. This starting threshold is usually set at a programmable level around 50% of the peak amplitude. The threshold is then linearly reduced down to a maximum sensitivity limit. This is graphically shown in Fig. 12.15. The linear reduction in threshold can optionally be delayed to temporarily keep a lower sensitivity. This is a very simplified example of one particular implementation of a sensing algorithm.

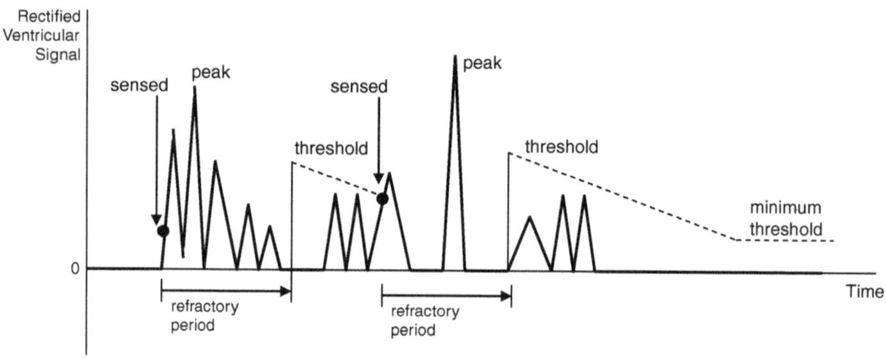

Fig. 12.15 Simplified example of an ICD sensing algorithm

Detection of arrhythmias (again in the example of St. Jude Medical devices) is based on binning of the timing of the sensed events. This binning is based on a comparison of an instantaneous interval period to a moving average. ICD's can be programmed to detect fibrillation only, or determine different tachycardia zones. These tachycardia zones can be programmed. The detection of a tachycardia is made based on the number of counts in the various tachycardia zone bins. The number of counts required can be programmed, and need to fall within a certain number of consecutive events [4]. An additional challenge to proper detection is that the device should not deliver therapy in cases of super-ventricular tachycardia (SVT), which is any tachycardia originating above the ventricles (such as atrial fibrillation). A variety of algorithms are used to discriminate between such conditions and actual ventricular arrhythmias [4].

Many modern ICD's offer tiered therapy. The levels of therapy start with anti-tachycardia pacing (ATP), step up to cardioversion, or low energy shocks, and top out with defibrillation, or high energy shocks. ATP is only effective in certain types of ventricular tachycardias, but it has the benefit of being relatively painless. It involves the delivery of several pacing pulses at a rate faster than the tachycardia to break the electrical cycle causing the arrhythmia. For cardioversion or defibrillation therapy, the high voltage capacitors in the device begin charging when the fibrillation is detected. Some devices, known as "committed" ICD's, will follow through with therapy once the charging process has started. Others, known as "uncommitted" ICD's, have the ability to terminate therapy by aborting the shock if normal sinus rhythm is restored before therapy is delivered. The charge on the capacitors is then bled off internal to the device, typically using the same mechanism used for capacitor maintenance.

Unless an IC process with sufficiently high voltage transistors (as high as 1000 V) is used, charger circuitry usually employs a hybrid of integrated and discrete circuitry. Charging needs to happen as quickly as possible, and depending on target voltage usually takes no more than about 10 seconds, depending on the state of the battery. This process can draw as much as several amperes from the battery, and since the type of charger used is typically a high voltage boost converter, charging creates a lot of electrical noise that can pose challenges for other functions in the system. Once the capacitors are charged, and unless the device aborts therapy, the high voltage energy is delivered through the selected leads. As shown in Fig. 12.7, devices typically deliver a biphasic waveform, since these require lower voltages than monophasic waveforms. Devices can usually be programmed to deliver a fixed shock time, or a fixed "tilt". Tilt is defined as the percentage drop from leading edge to trailing edge. The tilt of the first phase of the waveform in Fig. 12.7 is therefore given by:

$$\text{Tilt} = \frac{V_{i\_pos} - V_{f\_pos}}{V_{i\_pos}} \cdot 100\% \qquad [12.4]$$

where V_{i_pos} is the leading edge amplitude and V_{f_pos} is the trailing edge amplitude. The benefit of fixed tilt instead of fixed pulse duration is that, since the shock energy is delivered from an initial voltage on a capacitor (known stored energy), the tilt will determine what fraction of that energy is dissipated into the heart (delivered energy). Since the resistance, and therefore the time constant of the discharge, is variable, specifying a fixed duration will result in a delivered energy that will vary. Shock parameters, at the clinician's level, can be programmed in Volts or in Joules. Most devices also allow the programming of various shocking vectors, which include the RV coil, a lead in the superior vena cava (SVC), and the ICD case.

Figure 12.16 shows a die photograph of a mixed-signal IC used in an ICD.

Fig. 12.16 Die photograph of a mixed-signal IC used in an ICD. Courtesy of St. Jude Medical

12.5 IC Process Technology

12.5.1 Process Technology

Since the early days of pacemakers that were built using discrete transistors, incorporation of IC's in these devices have made possible a size reduction and longevity improvement, while vastly increasing the number of clinical features available. Process technology needs for pacemakers and ICD's are driven by some of the main functions inside the device.

Pacing requires transistors with voltage ratings on the order of 10 V. Charging and shocking functions, if they were to be integrated into the IC's, would require transistors with 1000 V ratings (or more). Even if such functions are implemented using external high voltage transistors, the voltages required to drive such devices are in the 10's of Volts. The on-resistance of transistors in pacing and charging circuitry should also be relatively low, in order to limit series resistance of analog switches, and provide sufficient output drive strength.

Digital functions require small standard cell sizes, and thus drive a need for small line-width transistors. With standard CMOS scaling, supply voltages and dynamic power dissipation are decreased. The various leakage currents in the transistors quickly increases to levels that are unacceptable for designs where nano-amperes matter. This tradeoff happens for virtually every type of integrated system, but for pacemakers and ICD's the CMOS node where this becomes an issue is much larger than for higher power and higher speed systems. Depending on the digital complexity, and the amount of current budgeted for digital leakage, standard processes below 0.25 μm tend to have too high an off-state leakage (I_{off}) to allow for traditional design techniques. The predominant mechanism leading to high I_{off} is sub-threshold leakage from drain to source. In any standard CMOS logic gate, when the gate's output is at either the supply or ground, there should be, to first order, no static current path between supply and ground for the gate. This is accomplished by all the current paths having at least one transistor with a gate-source voltage (V_{gs}) below the threshold voltage (V_t) of the transistor (ideally $V_{gs} = 0$). Even with $|V_{gs}| < |V_t|$, however, some finite amount of current can flow through the channel with an applied drain-source voltage (V_{ds}). As CMOS geometries shrink, threshold voltages are being intentionally reduced in order to improve speed and dynamic power dissipation. Secondary effects like drain-induced barrier lowering (DIBL) reduce the threshold voltage even further [32]. Some design techniques that can be used to reduce the effect of high transistor I_{off} currents will be presented later. Some IC processes have process options where all or some of the core transistors can receive a threshold adjust implant that raises the threshold voltage. Such implants, usually referred to a high-Vt (HVT), low-leakage (LL), or ultra-low-leakage (ULL) process options, can help reduce I_{off} in the transistors, and consequently reduce the leakage current of a digital block. But even with such threshold adjust implants, which usually still target more mainstream (and higher speed) designs, I_{off} is usually still an issue for large digital functions in a pacemaker or ICD for small geometry processes. In that case, other than process customization, only design-based approaches can help solve the issue. Standard processes below 0.18 μm, in addition to I_{off}, have gate leakage currents that start to become significant at these power levels.

Even the entire worldwide market for pacemakers and ICD's, at this point, is not sufficiently large to drive the enormous financial investment required for the development of fully customized IC processes. But despite the rather unique requirements for these processes, there is some synergy with requirements for other, higher volume markets like power management and disk drive IC's.

12.5.2 Low Power Design Techniques

Most of the types of design techniques well-known for low power design can be applied to the design of pacemaker or ICD electronics. The only distinction might be that, because of the extreme drive for low power, the cost/benefit looks sufficiently different to make a technique that has an unacceptable tradeoff for other applications (e.g. increased area, design time, etc.) look more attractive.

Low power design starts with a proper architecture-level design. At this level, things like power supply voltages, I/O architecture, functional partitioning, and IC floor-planning must carefully take into account power tradeoffs. It is also vital not to over-specify performance targets for various parts of the system, since that will often result in unnecessarily high power dissipation.

A well-known analog design technique for low power is CMOS sub-threshold design. First used in watch applications [33], sub-threshold CMOS design involves biasing circuitry in such a way that the transistors are in weak inversion. In this regime, the small-signal drain current of the transistor becomes exponentially dependent on the small-signal gate-source voltage, much like a bipolar transistor. A full description of sub-threshold design techniques is beyond the scope of this chapter, and much more detail can be found elsewhere [33, 34]. Some of the challenges of sub-threshold design are that current mismatch (such as in current mirrors) can be extremely large, and that CMOS process models do not always accurately fit silicon operation in this region of operation.

Many design techniques for low-power digital design can also be found in literature [35, 36]. Some of these techniques include clock gating for reduction of dynamic current, power gating for dynamic and static (standby) current reduction, dynamic supply biasing, and body-biasing. Clock gating involves careful design of clock trees and their distribution, and gating clocks to sections of the design that are not being used. This eliminates unnecessary toggling of gates, and thus saves dynamic power. In power gating, different power domain islands are created inside the design, and only those sections that are needed are powered up. Care must be taken to also gate off any signals coming from a powered-down section, to avoid floating logic nodes from causing shoot-through current in gates they connect to in adjacent blocks that remain powered. Dynamic supply biasing is a technique where the power supply for a digital block is raised or lowered, depending on what speed the block needs to run at. If temporarily, high speed is required, the supply is raised. When the block goes into a standby (either low-speed operation or a static condition where only the internal states need to be retained) the supply is dropped to a lower value. This is an intermediate point to full power-gating, and can allow the preservation of minimal-speed operation and/or static logic states inside the block. Both power gating and dynamic supply biasing can reduce leakage, but have the extra penalty of an activation charge required to bring the supply back up to a normal level. Body-biasing is a way to introduce a non-zero source-bulk voltage in the PMOS and/or NMOS transistors to effectively increase the threshold voltage of the transistors through the body effect. This threshold voltage increase reduces the sub-threshold leakage, and hence the static current of the logic blocks. Body-biasing can be accomplished by biasing the wells of the transistors (only available for NMOS transistors in dual-well processes) above the supply (for PMOS) or below ground (for NMOS), as shown in Fig. 12.17. The well supply and ground are VDD_W and VSS_W, and the source supply and ground are VDD_S and VSS_S. Digital logic swings are between VDD_S and VSS_S. A similar effect, sometimes referred to as source biasing, can be accomplished by generating an additional supply just above ground (for the NMOS) and just below Vdd (for the PMOS). In the case of source biasing, no special NMOS well is required. Source biasing also reduces the supply voltage across the logic gate, which further helps decrease leakage, and also decreases dynamic power due to the reduced logic swing. Drawbacks of this technique are an increased standard cell size, the need for a customized standard-cell library, and the need to somehow generate these additional

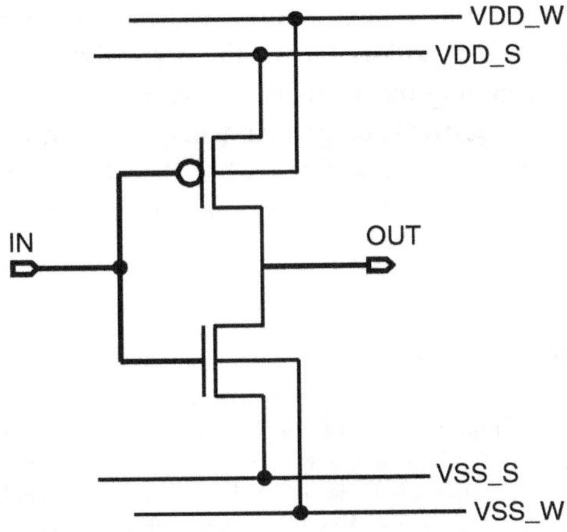

Fig. 12.17 Schematic of a body-biasing scheme for reducing leakage current

supplies. It is not necessary to apply body or source biasing to both the NMOS and PMOS side.

If the transistors leak equally, and the prevalence of logic "1" and "0" states is equal, applying the technique to only one side reduces static leakage by 50%.

12.6 Future Trends

It is inevitable that the development trends from the last several decades continue into the future. This means that devices will become increasingly smaller, contain more clinical features, and become more complex. A big challenge currently is to, despite such increased complexity, make the devices easier to use. Trends in this direction have been to include more automaticity, and this is likely to continue.

Another possible future development is that devices maintain feedback to hemodynamic efficiency, through the use of different types of sensors, and the detection, and ultimately treatment of other related (comorbid) medical conditions.

Much work is currently ongoing to strive for body-area networks, some of which can accommodate a variety of implanted and externally worn medical devices. It is possible that implanted cardiac rhythm management devices act as a node in such networks.

RF telemetry, as it has increased the data rates and physical distance of communications with the implanted device, could start enabling the remote processing of algorithms. Once device data is uploaded into an external device, or even into a secure medical network, the limited computational power inside an implanted device that might preclude certain algorithms, is no longer a limitation. Any such algorithms that do not require real-time feedback to patient therapy can be processed remotely. This requires that the processing power and benefits of the algorithms outweigh the power required for the device telemetry.

Another potential future advance lies in the area of energy scavenging. This area has been explored since the early sixties [8], but could hold future promise. Such scavenging mechanisms could convert bio-energy (mechanical, chemical, or other) into electrical energy that helps power the device. The challenge is to develop a scavenging technique that, volumetrically, is more efficient than an equivalent amount of additional battery capacity, over the intended life of the device.

References

1. Elmqvist R, Senning A (1960) An implantable pacemaker for the heart. In: Smythe CN (ed) Medical Electronics, Proceedings of the Second International Conference on Medical Electronics, Paris, 1959, Tliffe and Sons, London
2. Geddes LA (1990) Historical highlights in cardiac pacing. IEEE Eng Med Biol 9:12–18
3. Parsonnet V, Zucker IR, Asa MM (1962) Preliminary investigation of the development of a permanent implantable pacemaker using an intracardiac dipolar electrode. Clin Res, 10:391
4. Kenny T (2006) The nuts and bolts of ICD therapy. Blackwell, Malden MA
5. International Standards Organization (2000) Implants for surgery—cardiac pacemakers—Part 3: Low-profile connectors (IS-1) for implantable pacemakers ISO 5841-3:2000
6. Kenny T (2005) The nuts and bolts of cardiac pacing. Blackwell, Malden MA
7. Mallela VS, Ilankumaran V, Rao NS (2004) Trends in cardiac pacemaker batteries. Indian Pacing Electrophysiol J 4(4):201–212
8. Jeffrey K, Parsonnet V (1998) Cardiac pacing, 1960–1985: A quarter century of medical and industrial innovation. Circulation 97:1978–1991
9. Parsonnet V, et al. (1973) A permanent pacemaker capable of external non-invasive programming. Trans Am Soc Artif Intern Organs 19:224–228
10. Bernstein AD, Daubert JC, Fletcher RD, et al. (2000) The revised NASPE/BPEG generic code for antibradycardia, adaptive-rate and multisite pacing. Pacing Clinic Electrophysiol 25:260–264
11. Huffman FN, Migliore JJ, Robinson, WJ, Norman, JC (1974) Radioisotope powered cardiac pacemakers. IEEE Trans Nucl Sci NS21(1):707–713
12. Greatbatch W, et al. (1971) The solid-state lithium battery: A new improved chemical power source for implantable cardiac pacemakers. IEEE Trans. Biomed Eng. 18:317–324
13. Takeuchi ES, Quattrini PJ, Greatbatch W (1988) Lithium/silver vanadium oxide batteries for implantable defibrillators. Pacing Clinic Electrophysiol 11(11):2035–2039
14. Bunch TJ, Hayes DL, Friendman PA, (2008) Clinically relevant basics of pacing and defibrillation. In: Hayes DL, Friedman PA (eds) Cardiac pacing, defibrillation and resynchronization, 2nd edn. Wiley-Blackwell, West Sussex U.K
15. Daliri M, Maymandi-Nejad M (2008) A 0.8 V 420 nW CMOS switched-opamp switched-capacitor pacemaker front-end with a new continuous-time CMFB. Int Conf Electron Circuits Syst, ICECS 2008, Aug 31–Sept 3, pp. 758–761
16. Silveira F, Flandre D (2004) Low power analog CMOS for cardiac pacemakers. Kluwer, Dordrecht, The Netherlands
17. Gerosa A, Maniero A, Neviani A (2004) A fully integrated dual-channel log-domain programmable preamplifier and filter for an implantable cardiac pacemaker. IEEE Trans Circuits Syst 15(10):1916–1925
18. Wong LSY et al. (2004) A very low-power CMOS mixed-signal IC for implantable pacemaker applications. IEEE J Solid-State Circuits 39(12):2446–2456
19. Gerosa A, Neviani A (2005) A 1.8 μW sigma-delta modulator for 8-bit digitization of cardiac signals in implantable pacemakers operating down to 1.8 V. IEEE Trans Circuits Syst II:Express Briefs 52(2):71–76

20. Ryan JG, Carroll KJ, Pless BD (1989) A four chip implantable defibrillator/pacemaker chipset. IEEE 1989 Custom Integrated Circuits Conference
21. Walden RH (1999) Analog-to-digital converter survey and analysis. IEEE J Selected Areas Commun 17(4):539–550
22. Murmann B (2008) A/D converter trends: power dissipation, scaling and digitally assisted architectures. IEEE 2008 Custom Integrated Circuits Conf 105–112
23. Scott MD, Boser BE, Pister KSJ (2003) An ultralow-energy ADC for smart dust. IEEE J Solid-State Circuits 38(7):1123–1129
24. Nys OJAP, Dijkstra E (1993) On configurable oversampled A/D converters. IEEE J Solid-State Circuits 28(7):736–742
25. Gulati K, Lee HS (2001) A low-power reconfigurable analog-to-digital converter. IEEE J Solid State Circuits 36(12): 1900–1911
26. Markus J, Silva J, Temes GC (2004) Theory and applications of incremental delta sigma converters. IEEE Trans Circuits Syst 51(4):678–690
27. Swaroop P, Vasani AJ, Ghovanloo M (2006) A high-voltage output driver for implantable biomedical stimulators and I/O applications. MWSCAS '06, 49th IEEE Int Midwest Symp Circuits Systems (1):566–569
28. Hayes DL, Wang PJ, Asirvatham SJ, Friedman PA, (2008) Pacemaker and cardiac resynchronization timing cycles and electrocardiography. In: Hayes DL, Friedman PA (eds) Cardiac pacing, defibrillation and resynchronization, 2nd edn. Wiley-Blackwell, West Sussex U.K
29. FCC rules and regulations (1999) MICS Band plan. Part 95, 47 CFR 95.601-95.673 Subpart E
30. Savci HS, Sula A, Wang Z, Dogan NS, Arvas E (2005) MICS transceivers: regulatory standards and applications. IEEE Southeast Con 179–182
31. Bradley PD (2006) An ultra low power, high performance medical implant communication system (MICS) transceiver for implantable devices. IEEE Biomed Circuits Syst Conf 158–161
32. Roy K, et al. (2003) Leakage current mechanisms and leakage reduction techniques in deep-submicrometer CMOS circuits. Proc IEEE, 91(2):305–327
33. Vittoz EA (2006) Origins of weak inversion (or sub-threshold) circuit design. In: Wang A, et al. (eds) Sub-threshold design for ultra low-power systems. Springer, New York
34. Vittoz EA, Fellrath J (1977) CMOS analog integrated circuits based on weak inversion operation. IEEE J Solid-State Circuits 12(3):224–231
35. Wang A, et al. (eds) (2006) Sub-threshold design for ultra low-power systems. Springer, New York
36. Soudris D, Piguet C, Goutis C (eds) (2002) Designing CMOS circuits for low power. Kluwer, Boston MA

Chapter 13
Neurostimulation Design from an Energy and Information Transfer Perspective

David A. Dinsmoor, Robert W. Hocken, Jr., Wesley A. Santa, Jalpa S. Shah, Larry Tyler, and Timothy J. Denison

13.1 Introduction

Neurostimulation—defined as electrical charge delivery for the purpose of affecting the behavior of nervous tissue—is one of the fastest growing applications in biomedical engineering. In the United States alone, neurostimulation products represented a $628 million market in 2006 with an expected annual growth rate of 20% [1]. Example applications include neurostimulation for pain control, incontinence, hearing loss, epilepsy and essential tremor. Even more exciting for engineers, researchers and venture capitalists are the nascent and under-developed applications of neurostimulation—particularly neurostimulation to restore function lost to neurological diseases or injury. At the heart of any such system is a circuit which drives neural tissue with electricity. An example radiograph showing the key elements of a neuromodulation system is shown in Fig. 13.1 these elements include the energy source, neurostimulation circuitry, mechanical packaging and stimulating electrodes. All neuromodulation systems have these general elements.

This chapter provides an overview of the circuit systems required for a state-of-the-art central or peripheral neurostimulator—very specialized applications such as cochlear, retinal implants or other sensory prostheses are treated elsewhere [18, 17]. Section two outlines a complete neurostimulation system including all aspects of energy flow, from the battery to the tissue. Sections three and four discuss the details of the tissue-electrode interface and highlight key safety considerations. The fifth section presents two emerging areas which may be considered for future neurostimulator design: closed-loop, adaptive stimulation architectures and optogenetic stimulation. The intent of this chapter is to provide a comprehensive overview, through design tutorials and examples, of how a practical system is architected and implemented. It is important to note that any implanted medical device requires extensive, formal verification and validation. This chapter alone should not be construed as sufficient for designing implant grade medical electronics.

D.A. Dinsmoor (✉)
Medtronic Neuromodulation Integrated Circuit Technology, Minneapolis, MN 55432, USA
e-mail: david.a.dinsmoor@medtronic.com

H.-J. Yoo, C. van Hoof (eds.), *Bio-Medical CMOS ICs*,
Integrated Circuits and Systems, DOI 10.1007/978-1-4419-6597-4_13,

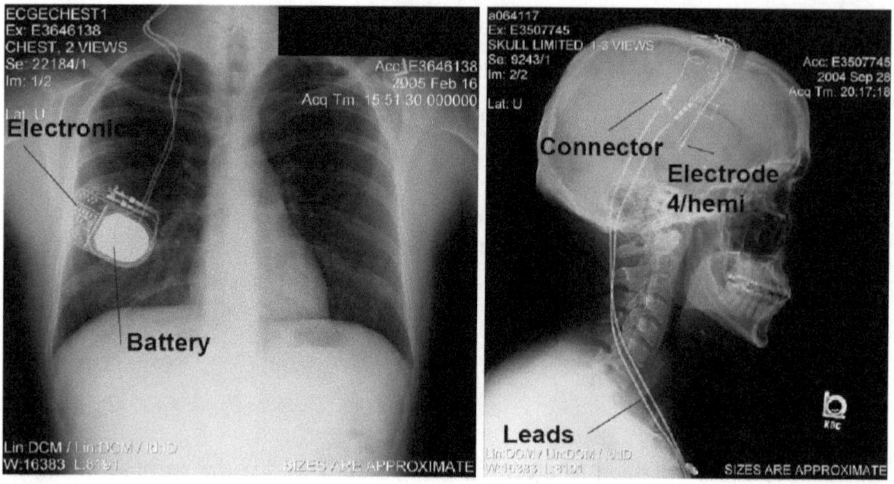

Fig. 13.1 Anterior / posterior (*left*) and lateral radiographs showing implantable neurostimulator and stimulating electrodes targeting the anterior nucleus of the thalamus

13.2 Overview of Challenges and System Requirements

At the core of any neurostimulation system are the circuits which transfer energy from the charge storage element to the tissue. The role of the implantable neurostimulator (INS) circuit designer is to ensure that the charge delivery circuit (hereinafter the "stimulation engine") is capable of interacting with all elements of the stimulation system—the target tissue, lead, other elements in the INS and external programmer—so as to meet the goals of safety, efficiency and efficacy.

The interface between the stimulating electrode(s) for an INS and the target tissue forms a fairly complex electrochemical interface [2] which remains the subject of active research. Paramount to the design of any chronically implanted INS is safe delivery of charge across this interface to the target tissue while still generating the desired physiologic effect. Over the years, researchers have elucidated various electrode designs and stimulation patterns that avoid or limit unintended tissue damage and electrode corrosion during charge delivery [3]. Balanced, biphasic charge delivery which does not exceed application dependent stimulation frequency, net DC transfer, charge density and charge-per-phase levels generally results in neurostimulation with limited tissue and electrode damage. These levels for many neurostimulation applications (such as deep brain stimulation (DBS) and spinal cord stimulation) have been presented in other references and are not discussed further here [4]. To this end, most commercially available INS's are generally capable of delivering charge in a multi-programmable manner similar to that shown in Fig. 13.2.

The key elements of the pulse train shown in Fig. 13.2 are anodic and cathodic charge delivery half-phases, for which the total charge delivered during the anodic phase reverses most—if not all—of the charge delivered in the cathodic phase. The

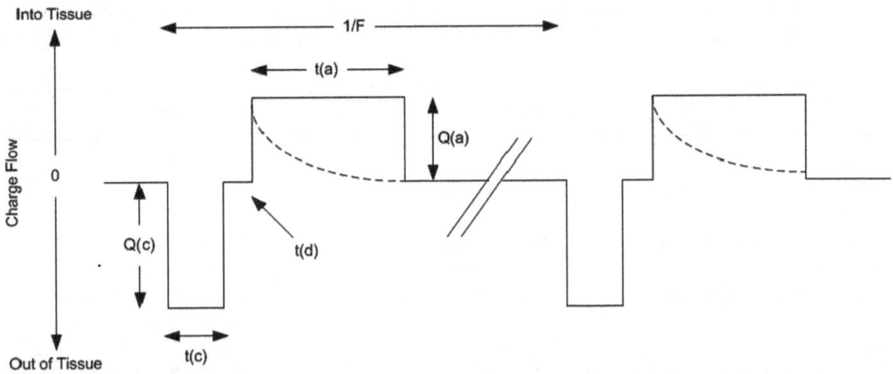

Fig. 13.2 Generalized stimulation pulse train. *Q*(*c*) is cathodic charge flow out of the tissue; *t*(*c*) is the duration of the cathodic charge flow; *t*(*d*) is the interval between cathodic and anodic phases; *Q*(*a*) is the anodic charge flow into the tissue; *t*(*a*) is the duration of the anodic charge flow; *F* is the frequency of charge delivery. The solid line for *Q*(*a*) indicates "active" anodic charge recovery versus the "passive" charge recovery illustrated with the dotted line. Adapted from [5]

anodic phase may consist of passive or active recharge phases, or a combination thereof. Passive recharge is enabled by the presence of a coupling capacitor which is often placed in series between the INS output and the electrode. One function of this capacitor is to avoid a potential path for DC current flow into the tissue. During passive recharge, charge is allowed to flow into the tissue in a decaying exponential fashion set by the product of the electrode impedance and the coupling capacitor. Passive recharge is the most efficient way to generate the anodic half-phase, but is only possible if coupling capacitors are used. During active recharge, charge is actively injected across the electrode-tissue interface for the purpose of rapidly reversing DC potentials on the electrode so as to enhance electrical sensing, limit tissue damage, or allow for balanced-charge high rate stimulation.

The aforementioned INS design considerations are not new to engineers and scientists. The origins of clinically successful implantable pulse generators (IPG's) for neurostimulation applications lay in cardiac electrical stimulators which have been in use clinically since the 1950s. In the late 1960s, Medtronic, Inc. began leveraging their cardiac electrical stimulation technology to develop a spinal cord stimulator. Although the fundamental concept of charge delivery to a particular target tissue is similar across cardiac and neurostimulation applications, certain key distinctions do exist for neurostimulators; namely, the need for broader stimulation parameters, more electrodes and capability for rechargeable power sources. These added constraints drive engineers to not only adapt pre-existing technologies, but develop new circuits and design practices as well. Of interest is a comparison of example INS design constraints versus those of cardiac IPG as presented in Table 13.1. Other general requirements of the system are embodied in this table as well. Very specialized neurostimulation applications such as subretinal stimulators have their own unique set of design requirements [6].

Table 13.1 Example stimulation constraints for an INS versus a cardiac IPG (Medtronic EnPulse DR). For the cardiac IPG, currents are assumed for a 500 Ω load 200 μs after the leading edge of the pace

	INS	Cardiac IPG
Frequency	1–250+ Hz	0.5–3.5 Hz
Cathodic / anodic current (peak)	250 μA–15 mA	1–15 mA
Duration of cathodic current	30 μs–2 ms	120 μs–1.5 ms
Number of electrodes	16 or more	2 (atrial / ventricular)
Sensed signal amplitude	>1 μV, 12–40 Hz (beta band)	>150 μV, 13–30 Hz (atrial depolarization)
Battery type	Secondary or Primary	Primary
Nominal stimulation current	63 μA (3 mA, 210 μs, 100 Hz)	5.6 μA (7 mA, 400 μs, 1 Hz), 100% A+V pacing
Example regulatory standard differences	ISO 14708-3 (INS), EN 60118-13 (Cochlear Implant)	EN 45502-2-1, ISO 14708-2

13.3 Completing the Energy Transfer Circuit: From Battery to Body

In many ways, the expanded design envelope and the higher current consumption of the INS over the cardiac IPG cement the stimulation engine as the true heart of the INS—a role typically reserved for the microprocessor in most embedded systems [7]. This "stimulation engine-centric" paradigm is presented graphically in Fig. 13.3.

The essence of the INS is energy translation as is evidenced in Fig. 13.3. When a battery is used to power the INS electronics, chemical energy is stored in the battery. Alternatives to battery power include direct inductive coupling, a super capacitor, biogalvanic cells or the like. The redox reactions between the battery electrolyte and the anode / cathode pair result in the battery terminal voltage available to power the rest of the system. In the case of a secondary battery, the electrochemical potential of the battery must be periodically replenished. This is typically accomplished via inductive charging by an external recharger. The stimulation engine in turn utilizes the potential of the battery to drive across the electrochemical phase boundary of the electrode / tissue interface and subsequently modulate the neural activity encoded in the activation patterns of the target nervous tissue. The energy delivered to the tissue may be titrated to meet the unique needs of the particular neurostimulation application, much like certain pharmaceuticals are titrated in a dose-dependent fashion to elicit a specific response. Technological advancements are gradually steering energy titration from physician or patient-mediated "open-loop" control to "closed-loop" control, automatically integrating and feeding back any number of sensed parameters to the stimulation engine for optimal response.

With Fig. 13.3 as a reference, we will explore a prototype INS where energy flow is traversed starting from the chemical energy stored in a secondary, inductively

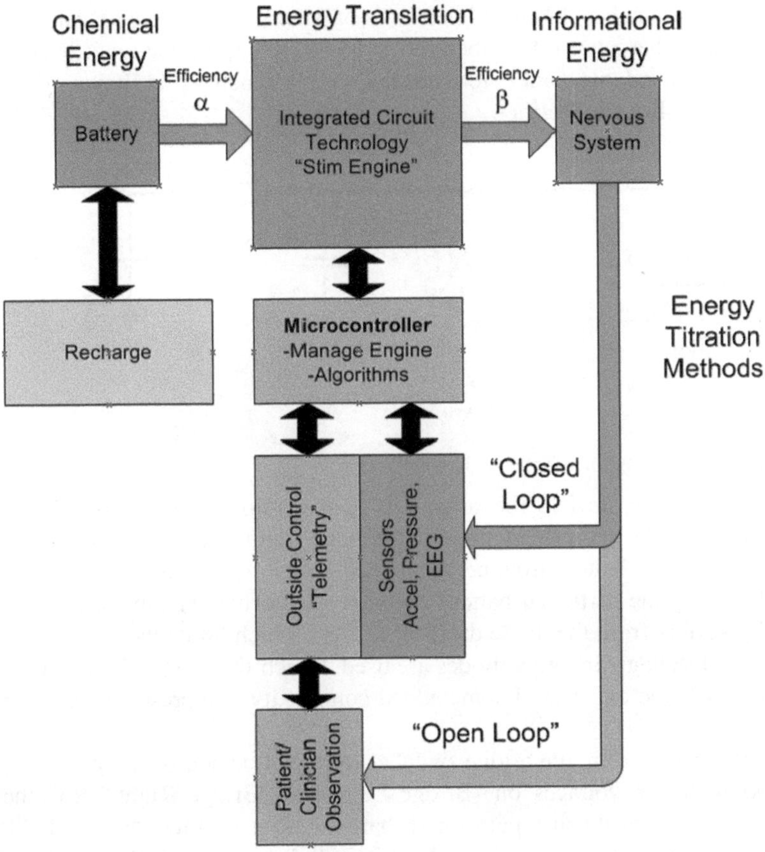

Fig. 13.3 Signal flow model for neurostimulators illustrating the modes of energy translation from the battery to the tissue [8]

recharged battery and culminating in charge delivery to the target tissue at the tip of the stimulating electrode. Specific areas of discussion below include recharging and interfacing with the battery, boosting the battery voltage to potentials suitable for therapeutic use, generating an appropriate stimulation signal, and delivering the stimulation signal to the tissue in a safe and effective manner. We conclude the chapter with a discussion on future directions for neurostimulation circuitry with alternative methods of energy transfer.

13.3.1 Secondary Cell Recharge

The first INS circuit design challenge is managing recharge; that is, energy flow into the battery when rechargeable (secondary) cells are used as the power source for the INS. Secondary cells are typically recharged via an inductively coupled link at low-RF frequencies (10's of kHz to 10's of MHz). Other methods such as ultrasonic recharge have been described in the literature [9] but have not been used in commercially available devices to-date. The simplest method, as shown in Fig. 13.4, is a standard full-wave rectifier.

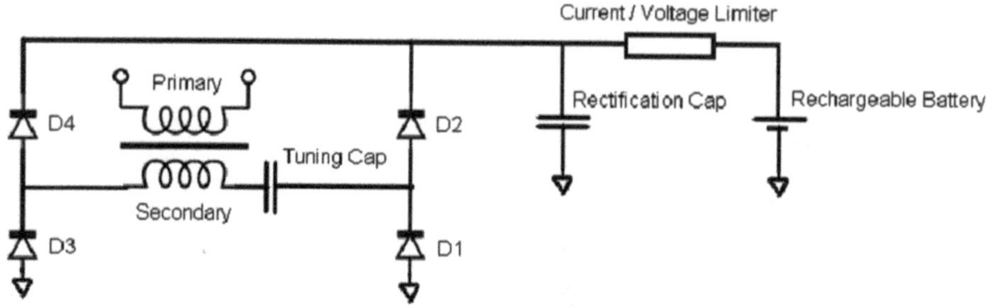

Fig. 13.4 Full wave rectifier for secondary cell recharge

In this example, an INS recharger acts as the primary with respect to the secondary coil inside the INS. A current / voltage limiter acts as a variable resistor which prevents the battery from being charged at a higher voltage or charge rate than that allowed by the particular battery chemistry. Unfortunately, the design shown in Fig. 13.4 suffers from the diode drops of D1–D4 which limits its usefulness unless low forward-voltage Schottky diodes are used. Synchronous rectifiers eliminate the diode drops at the expense of some added complexity; a representative example is shown in Fig. 13.5.

As seen in Fig. 13.5, the bridge switches (M1–M4) synchronously open or close in response to the voltages on "Bridge Left" and "Bridge Right." Switches M1 and M4 close to establish a path for recharge current to flow when node "Bridge Right" flies high. Similarly, when "Bridge Right" is low and "Bridge Left" flies high, switches M2 and M3 are on and conducting. Regardless of the method of rectification used, care must be taken in the following areas:

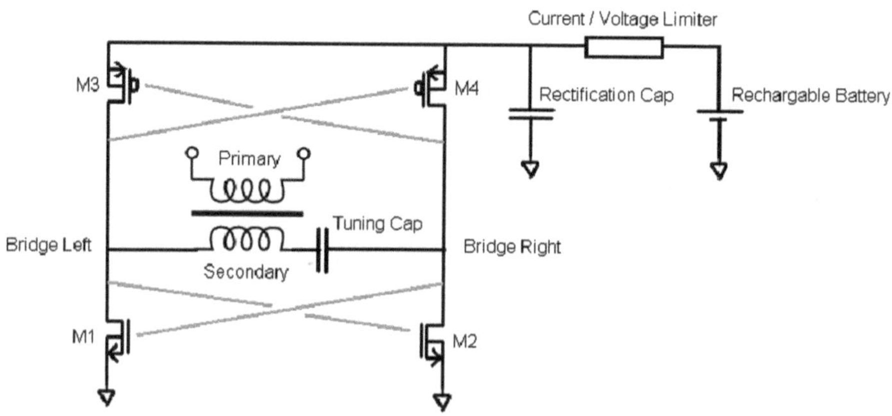

Fig 13.5 Synchronous rectifier for secondary cell recharge

1. Optimization of inductive link efficiency,
2. Current and voltage management to avoid damage to the INS or interference with other INS elements, and;
3. Thermal management to avoid excessive heating during inductive coupling, either intended or not.

The final point above is of special concern because it involves potential risk to the patient should the temperature of the INS rise sufficiently high to cause tissue damage during recharge. Standards such as EN 45502-1: 1997 §17.1 provide guidance on allowable thermal excursions for an INS. The literature is replete with other examples of methods and techniques scientists, engineers and researchers have used to optimally recharge or power their INS's and other medical devices [10, 11], As such, recharge is not discussed further here.

13.3.2 Energy Source Characteristics

The characteristics of the INS energy source are critical to the energy flow problem in the INS and provide key constraints on circuit design. As seen in Table 13.1, secondary cells are frequently used in INS's versus cardiac IPG's given the higher overall current draw in the former versus the later. Simply put, primary cells are precluded in high-rate stimulation applications as the higher current draws would necessitate too frequent device change-outs. The discharge curve for an example 3 mA-h Li (Ni, Co, Al)O_2 (LNCAO) secondary cell (the Quallion QL0003I) is shown in Fig. 13.6. The term "depth-of-discharge" refers to the fraction of the total electrical energy stored in a battery recoverable by discharging at a certain point of time. The "C" annotation on each curve refers to the discharge rate as a function of the nominal battery capacity; for instance, the 2 C curve shows the battery discharge curve with a 6 mA load whereas 0.2 C indicates a 0.6 mA load. These cells feature a low output impedance (<10 Ω) to support high-rate charge delivery. This may be contrasted with the much higher impedance (<10,000 Ω) of the low-rate Li/I_2-polyvinylpyridine primary cells used in cardiac IPG's [12]. Other chemistries such as $Li_4Ti_5O_{12}/LiCoO_2$ have an even flatter discharge curve with a nominal operating voltage of 2.5 V.

The efficiency of the chemical to electrical current conversion (shown as "Efficiency α" in Fig. 13.3) conferred by the battery may be viewed two different ways; the first being the charge or coulombic efficiency (mA-h) and the other being the power efficiency (W). In a secondary battery, coulombic charge efficiency may be defined as the following ratio when an equal number of coulombs are used to recharge the battery and then pulled out as a load:

$$\Delta \text{ Depth of Discharge (Recharge)} / \Delta \text{ Depth of Discharge (Load)}$$

Coulombic charge efficiency is excellent in Li-Ion chemistries and is typically in excess of 99% [14]. The power efficiency is use-dependent and is a function of the

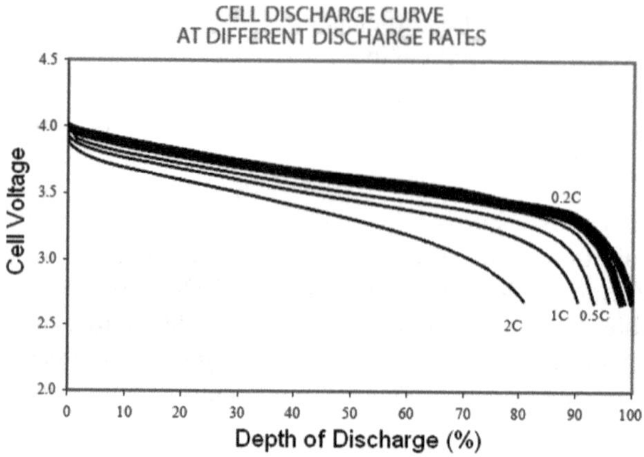

Fig. 13.6 Discharge curve for Quallion QL0003I [13]

ohmic losses—I^2R—from the series resistances in the recharge and load paths of the system. Good circuit design practices which minimize these resistances generally result in negligible power losses (<1%).

13.3.3 Boosting the Voltage—Providing Overhead for the Stimulation Engine

For a constant current stimulation engine, the secondary batteries described above provide special challenge with respect to maintaining enough headroom to source the specified stimulation current across the expected range of tissue impedances. One method is capacitor stacking. Capacitor stacking involves a two-phase process where a number of capacitors are pre-charged in the first phase to either a fraction of, or a full battery voltage. In the second phase, these capacitors are connected in series to provide the headroom for the stimulation engine. Although the circuit implementation is simple, this design practice has limitations as it necessitates large, very low impedance switches to series-connect the capacitors. The pre-charge phase also leads to a large load current as the capacitors are charged, which may lead to supply droop in systems without very large battery decoupling capacitors and/or low impedance battery chemistries. A final disadvantage is that the stacked capacitors need to be very large to support the droop or ripple requirements of the boosted supply. By means of example, consider a neurostimulation application where a 4x battery supply is needed to provide sufficient headroom. If a 10 μF boosted supply bypass capacitor is required, for instance, the stacked capacitors need to be at least 40 μF each.

Another method for achieving a higher voltage is a switch-mode power supply such as a boost converter. These methods are typically very efficient, especially when a ferrite core inductor is used. However, inductors generally take up a significant volume in the space-constrained environment of an INS. Even more problematic are the ferrites, which can lead to unintended movement, thermal injury from RF induction heating or image artifacts during magnetic resonance (MR) scanning. Industry trends are pushing implantable medical devices in the direction of "MR-Safe" or "MR-Conditional" where the use of ferrites is generally avoided. Nevertheless, some INS's do use ferrites [15].

In many cases, a charge pump may be used to generate sufficiently high supply with efficiencies approximating those of an equivalent boost converter. A circuit often used that is capable of generating a 1X–5X multiple of the battery supply is shown in Fig. 13.7. This circuit consists of four pump capacitors (C1–C4), two hold capacitors (C_{HOLD1}, C_{HOLD2}) and related switches. The two separate hold capacitors are used to maintain separate supplies for up to two different stimulation programs, noting that one of the programs may be used for the active recharge pulse. Separate supplies help maximize efficiency as the lowest headroom supply may be selected for each program. The capacitors may also be optionally shorted together if desired.

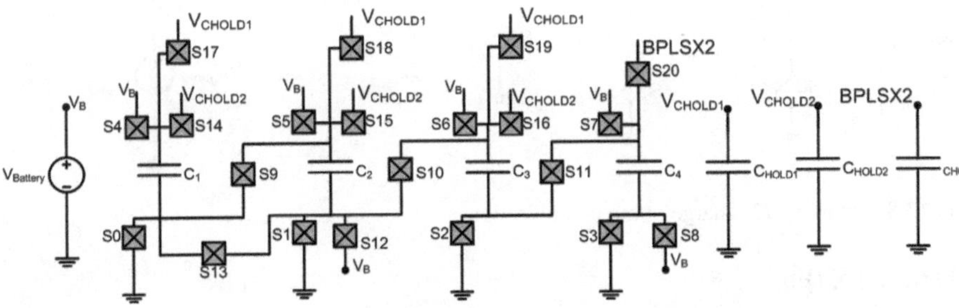

Fig. 13.7 Charge pump capable of generating (Vbattery 1X–Vbattery 5X)

A 2X battery supply (hereinafter the "supply pump") is also maintained on a separate hold capacitor and is available for use in other parts of the system. The switches for generating this 2X battery supply is multiplexed with the other charge pump switches (S3, S7, S8 and S20).

Maximum efficiency of the charge pump can be achieved by generating the minimum voltage necessary—plus some margin—to source the specified current for a given tissue impedance. A feedback loop can be used to indicate whether the voltage at the hold capacitors is too high or too low. To reduce or to increase the voltage at the hold capacitors according to the feedback, certain configurations are set for different multiples of the battery voltage—this feedback also includes adjusting the timing of a variable non-overlapping clock generator which is used to clock the charge pump switches. It is important to note that the tissue impedance changes in only a quasi-DC fashion in most common neurostimulation applications [16]. As such, adjustments of the charge pump battery multiple or operating frequency are generally infrequent or unneeded unless stimulation settings are intentionally changed. Changes may also be needed as the battery depletes between recharge intervals.

So as to ensure the lowest on-resistance for all switches, the gates of the switches should be driven at the highest voltage available. The three possible voltage sources include the voltages on the BPLSX2 hold capacitor, C_{HOLD1} and C_{HOLD2}. A high-voltage comparator may be used to continuously monitor these three voltages and select the highest voltage as the output (V_{MAX}). V_{MAX} may be used as the high-side supply for the level shifters and buffers driving the charge pump switches. The high-voltage comparator should utilize both DC and AC hysteresis to avoid chattering and inappropriate mode switching.

Described below are the configurations for the 1X and 1.5X multiples of the battery voltage with respect to the charge and pump phase. The other pump combinations are not explicitly described but are straightforward extensions of the architecture shown in Fig. 13.7.

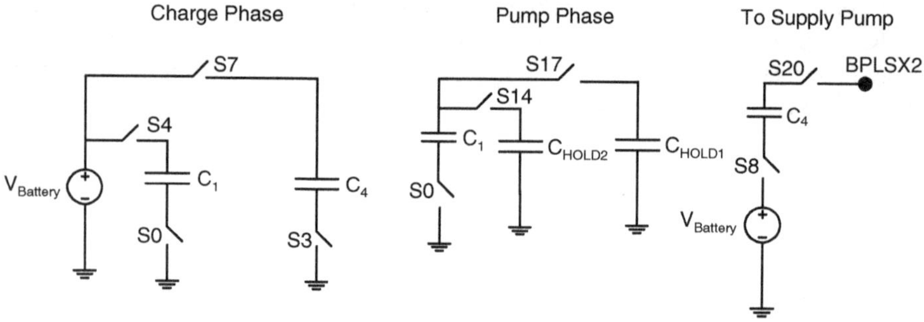

Fig. 13.8 Vbattery 1X charge pump configuration

Vbattery 1X (Fig. 13.8)

- During the charge phase, pump capacitors C1 and C4 are connected in parallel to Vbattery through switches S4/S0 and S7/S3, respectively. The pump capacitors are subsequently charged to Vbattery 1X.
- During the pump phase, C1 is connected in parallel to one or both of hold capacitors C_{HOLD1} and C_{HOLD2} via S0, S14 and S17. Simultaneously, capacitor C4 (for the supply pump) is connected on the low side to the battery via S8 and on the high side to the BPLSX2 via S20.

Vbattery 1.5X (Fig. 13.9)

- During the charge phase, pump capacitors C1 and C2 are connected in series via switches S1, S4, S9. The battery voltage splits across C1 and C2 so both capacitors are charged to Vbattery 0.5 X. C4 is connected to the battery via S3 and S7 as part of the supply pump.
- During the pump phase, C1 and C2 are connected in parallel with each other (via S13) and in series with the battery (via S12). The parallel combination of C1 and C2 are connected in parallel—via S14, S15, S17 and S18—to one or both of the hold capacitors so as to pump to 1.5X Vbattery. Simultaneously, capacitor C4 (for the supply pump) is connected on the low side to the battery via S8 and on the high side to the BPLSX2 via S20.

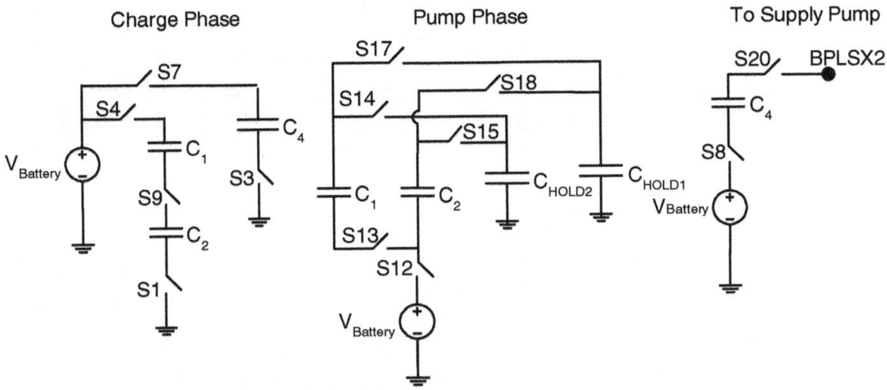

Fig. 13.9 Vbattery 1.5X charge pump configuration

13.3.4 Generating the Stimulation Signal

Once a suitably boosted battery voltage is made available for use by the stimulation engine, the next step is to transform the energy into the stimulation signal (as illustrated in Fig. 13.2) used to modulate the target nervous tissue. The two key attributes of the stimulation signal are that it is of an appropriate pattern and is delivered to the appropriate nervous structure. In commercially released INS's to-date, the stimulation pattern is usually at a fixed frequency (i.e., closed loop or adaptive stimulation techniques are not yet utilized). Retinal [17] and cochlear [18] implants are an exception, however, which is consistent with their functional role as a sensory prostheses; these devices vary their stimulation signal in response to sensed visual or auditory stimuli, respectively.

Shown below in Fig. 13.10 is a subset of a scaleable, constant current stimulation engine with active electrodes capable of fractionally regulating currents in bipolar or unipolar mode. In unipolar mode, the case of the INS usually acts as the current sink or source. The stimulation current is shown passing through discrete AC coupling capacitors before flowing into the target tissue (drawn as a 1 kΩ resistor). While coupling capacitors are used in the example below, they are not always needed or feasible to include—some ultra-space constrained sensory prostheses have insufficient space to include any coupling capacitors. Special care must be taken with respect to failure-mode analysis and relevant patient safety standards should the stimulating electrodes directly couple to the tissue. The current trend—where discrete AC blocking capacitors are not used—is to include monolithic coupling capacitors plus other active circuitry to prevent charge accumulation and destructive faradic current across the electrode / tissue interface [19].

Each of the active electrodes are configured as a sink, source, or disabled for the stimulus pulse. If all enabled electrodes are configured as either a sink or source, unipolar mode is presumed and the CASE switch is used as the source/sink path for the stimulus pulse.

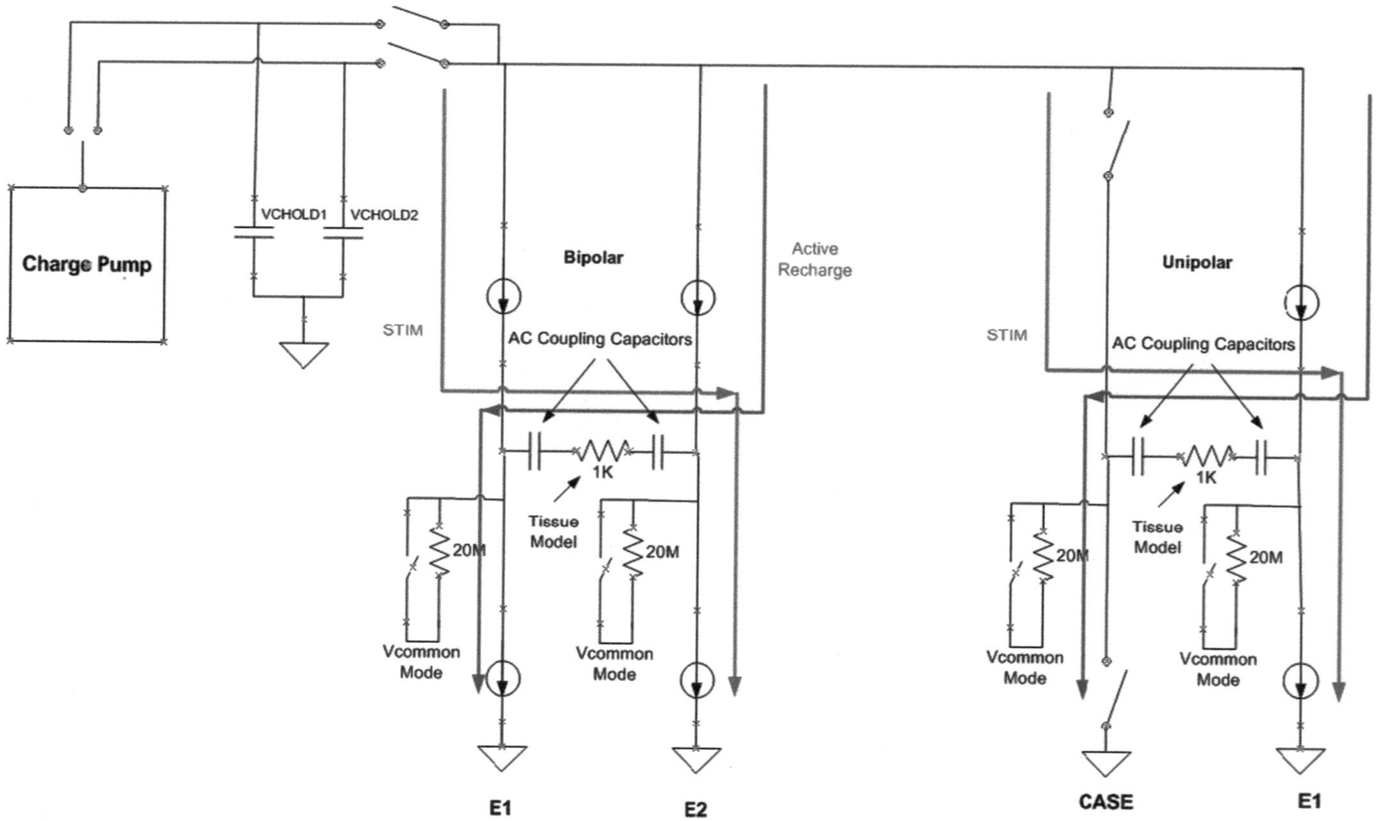

Fig. 13.10 Constant current output stage. Both unipolar (where the INS case acts as the current source / sink) and bipolar stimulation configurations are shown. The target tissue (shown as a 1 kΩ load) is AC coupled to the stimulation engine. A common mode voltage—an arbitrary voltage greater than or equal to ground—is provided when passive recharge is used for the anodal pulse

For both bipolar and unipolar modes, if multiple electrodes are configured as source and/or sink, the field can be regulated via the amount of current provided by each electrode in fractionalized increments. For example, if two electrodes are configured as sources (E0, E1) with a third electrode as a sink (E2), the field can be regulated by programming one of the two source electrodes (E0) with 1/4th current and the other source (E1) as 3/4th current. This results in 25% of the programmed current going through E0 and 75% of the programmed current in E1. All of the current would then flow into programmed sink electrode E2.

Two modes of recharge are often provided for flexible use when tissue coupling capacitors are used. Passive recharge, when enabled, discharges residual electrode voltage to the common mode voltage—in this example, an arbitrary voltage greater than or equal to ground. Active recharge provides a stimulus pulse in the opposite polarity of the programmed cathodic pulse.

A reference current must be created as the basis for the stimulation pulse. In some designs, a master reference current is either sourced or sunk into a reference resistor. The voltage generated across the reference resistor is then tapped to each electrode sink/source output stage. Each electrode may use its own tap to facilitate trimming which further improves matching between electrode sink/source combinations. Taps may also have a coarse selection used to select a fraction of the amplitude for active recharge.

Attention to stimulation engine supply and headroom is crucial for designs where maximum efficiency is desirable. The high voltage rail of the electrode circuitry shown in Fig. 13.10 is selectable, and can be driven by either the voltage reserved on C_{HOLD1} or C_{HOLD2} (as described previously). Each electrode contains two (source/sink) comparators. These comparators are used to determine when the electrode is out of compliance and/or runs out of headroom. So as to maximize efficiency, the headroom is kept as low as possible while maintaining compliance. In addition, the current sinks and sources need to remain accurate while requiring minimal headroom.

It is important to note that there is an upper bound on the amount of energy that any INS stimulation engine can deliver to the tissue. We define E_{PULSE}, or $Q(c)t(c)$ from Fig. 13.2, as the amount of cathodic charge delivered per pulse. E_{PULSE} may alternately be written as $C \times dV_{STIM} \times t(c)$ where C refers to the effective series capacitance in the delivery pathway and dV_{STIM} is the stimulation voltage generated across the capacitance during the stimulation pulse. The maximum total cathodic energy deliverable into the tissue is subsequently $C \times (dV_{STIM})^2 \times t(c) \times F_{STIM}$. dV_{STIM} can approach the maximum voltage that may be generated by the charge pump or stack, but is limited by voltage losses in the delivery pathway due IR drop across switches and/or current source headroom. The maximum frequency is limited by the time required between stimulation pulses to replenish the hold capacitor or the capacitor stack. Thus, the stimulation engine does act to gate maximal charge delivery into the tissue; the stimulation engine must be designed to account for the requirements of the particular neurostimulation application.

The following sections provide relevant details of these circuits at the transistor level. These include the reference current generator, the active source and sink designs, sharing the reference resistor across all electrode sources and sinks, output regulation with a reference resistor, distribution of current through the electrodes and a final section elaborating how these pieces are tied together to form a complete stimulation engine.

13.3.4.1 Reference Current Generator

The reference current generator supplies the reference current for the electrode block, which in turn is used to set the stimulation current at the output nodes. The reference current is split into two equivalent source and sink reference currents; these currents are gained up to a full scale current for the downstream electrode output interface.

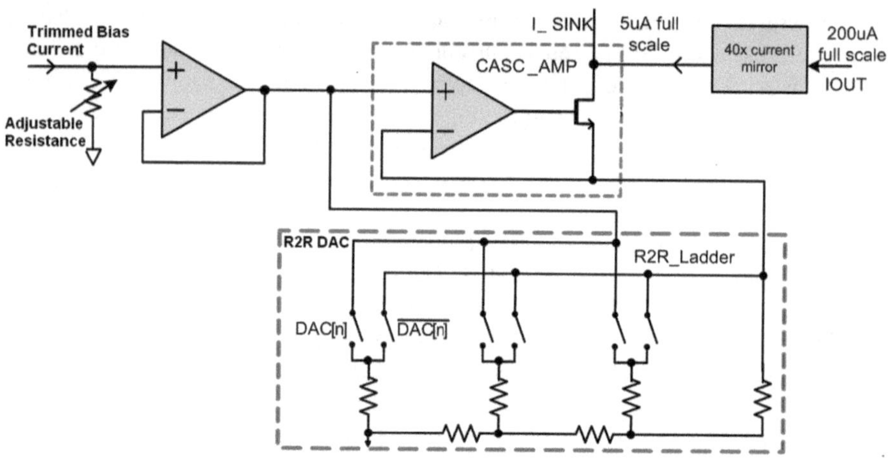

Fig. 13.11 Reference current generator (with R2R DAC and current multiplication)

The elements of a good reference current generator are accuracy and flexibility to cover the wide design space required by the INS. An example reference current generator is shown in Fig. 13.11. This current generator consists of a trimmed bias current and adjustable resistance to set a reference voltage, a low-offset buffer which follows the aforementioned reference voltage, an R2R based digital-to-analog converter (DAC), an amplifier which drives the current source cascode (CASC_AMP) and a 40 × current mirror. Based on the stimulation current requirements, the reference generator can be set to provide anywhere from 20 nA to 5 μA of current. The current (I_SINK) is further multiplied by a factor of 40 via the 40 × current mirror to provide 200 μA of full-scale output current to the stimulation circuits.

13.3.4.2 Active Sources and Sinks

The current sinks and sources must accurately drive the reference current into the target tissue. In this design example, the reference current described above is sampled, then amplified and passed to the tissue using an active mirror system. Since the

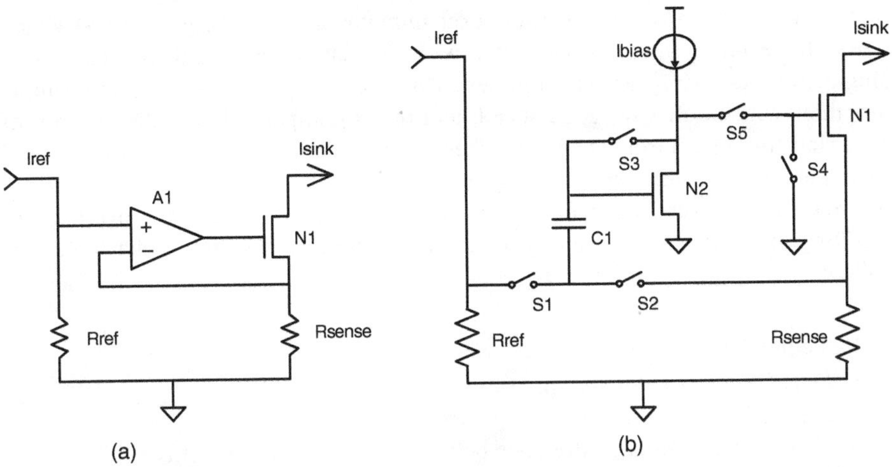

Fig. 13.12 Feedback sense-resistor based constant current output stage (sink). **(a)** is a generic sense resistor based current mirror, whereas **(b)** shows a two-phase design with a single transistor as an amplifier

current source and sink circuits are complementary, only the description for the sink is described below. In many designs, a feedback sense resistor based architecture (as seen in Fig. 13.12) is utilized for two primary reasons:

(1) The architecture allows the current matching to be driven by resistors rather than transistors. This results in better matching in most integrated circuit processes; and,
(2) The architecture eliminates the need to keep any output transistors in saturation, reducing voltage headroom requirements to improve efficiency under many conditions of neurostimulation.

The generic sense-resistor based architecture shown in Fig. 13.12(a) has the following design characteristics and principle of operation.

- Iref (from the reference current generator) generates a voltage on Rref.
- Amplifier A1 is driven by the voltage on Rref with feedback from Rsense. A1 drives the gate of N1 until the voltage on Rsense matches the voltage on Rref, creating an active current mirror.
- The Isink/Iref ratio can be controlled by the Rref/Rsense ratio.
- N1 can be driven into triode and still maintain accuracy, so the saturation voltage of the overall current sink can be as low as Isink × (Rsense + the "on" resistance of N1). Said another way, the circuit benefits from increasing N1's output resistance by the loop gain.
- The current matching of this circuit depends on the resistor matching as well as amplifier offset.

Fig. 13.12(b) shows a transistor level implementation of Fig. 13.12(a) which works by using a sampling capacitor with precharge and output phases, plus a single transistor (N2) as an amplifier. This combination creates a very simple amplifier with small size, good speed, and offset compensation at the expense of requiring timing control for the switches. In essence, correlated sampling is used to implement the stimulation pulse. Since the stimulation pulses are inherently discrete events—versus other types of continuous-time signals—the overhead of clocking the switches is generally acceptable. Details of the circuit phasing are as follows:

- In the precharge phase:
  - Only S1, S3, and S4 are on.
  - N2 is diode-connected, carrying Ibias.
  - C1 stores the voltage difference between Rref and the gate of N2.
  - N1 is off.
- In the output phase:
  - Only S2 and S5 are on.
  - The capacitor C1 retains the voltage acquired in the precharge phase.
  - When the voltage on Rsense is exactly equal to the earlier voltage on Rref, the stored voltage on C2 biases the gate of N2 properly so it balances Ibias.
  - If, for example, the voltage on Rsense is lower than the original Rref voltage, then the gate of N2 is pulled lower, allowing Ibias to pull up on the gate on N1 so it will pass more current to Rsense. This is the feedback mechanism to mirror the voltage on Rref to Rsense in the output phase.
- The matching of this current mirror depends primarily on resistor matching, but the circuit designer should note that charge injection, leakage, and gain can cause systematic offsets as with any switched capacitor circuit

This basic architecture is advantageous for multiple reasons, as it may be expanded upon and enhanced by utilizing the following techniques—all of which are discussed in detail below.

1. A shared reference resistor for optimal matching and expansion,
2. Output regulation with a reference resistor, and;
3. Output regulation with sensing resistor

13.3.4.3 Scaling Considerations for Electrode Sinks and Sources

Expandability is an important part of any architecture, but is especially critical in the design of an INS with 16 channels or more. By sharing the reference resistor and only replicating the output stage (as shown in Fig. 13.13), the architecture shown in Fig. 13.12 may be easily expanded as electrode counts increase. Additionally, the matching is improved between sink outputs and less power is consumed because

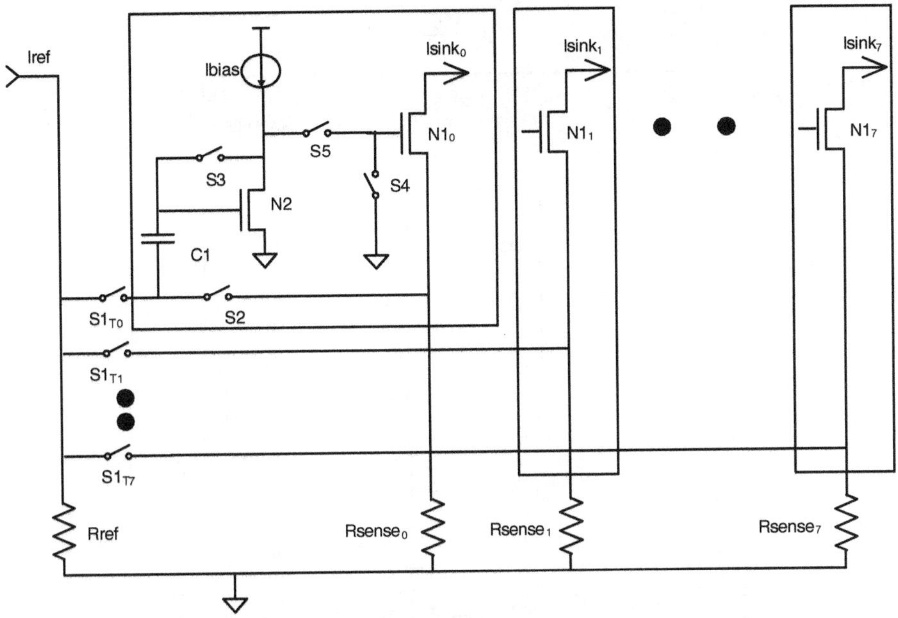

Fig. 13.13 Eight electrode constant current sink with shared reference resistor

only one reference resistor is pre-charged. In the example below, an eight electrode system is assumed.

The reference resistor expands to provide multiple taps, one for each of the eight electrodes. This is illustrated in Fig. 13.13 by breaking switch S1 into multiple taps ($S1_{T0}$–$S1_{T7}$). Each tap is then used to pre-charge the capacitor, C1, within each sink output. With this architecture, the Iref current is enabled and establishes a voltage on the reference resistor. Then, for each sink output stage that is enabled, the capacitor is pre-charged. If the sink output for a given electrode is not enabled, power may be saved by not pre-charging the capacitor for that electrode.

Since all sink outputs use the same reference resistor, the design's matching is mainly dependent on the resistive matching between sense resistors (R_{sense0}–R_{sense7}). Gain variation results if the reference resistance is not well matched to the R_{sense} resistor. This variation, along with sink to source matching, may be controlled by discrete trimming as discussed in the next section.

13.3.4.4 Output Regulation with a Reference Resistor

There are three ways to regulate the output current, I_{sink}: (1) I_{ref} current; (2) Reference Resistor, R_{ref}; (3) Sensing Resistor, R_{sense}. Using the reference current, I_{ref}, is the most straightforward control of the output current. The second method is to have programmable taps off of the reference resistor (as seen in Fig. 13.14).

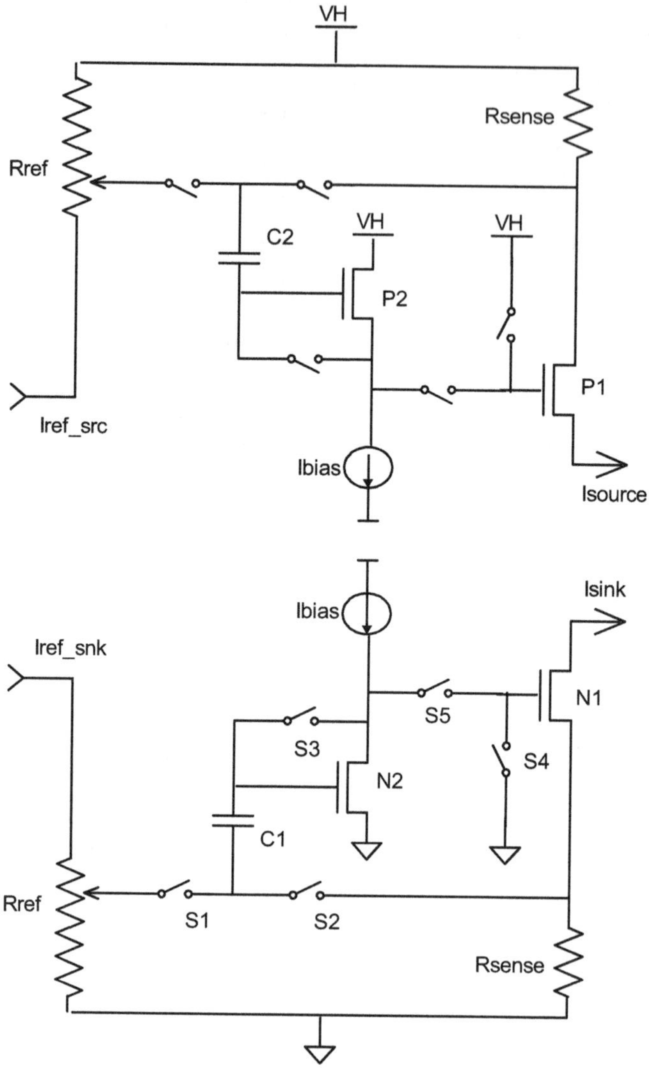

Fig. 13.14 Reference resistor voltage taps

Taps on the reference resistor provide several key features:

- Fine trim to improve matching across all 8 sink output stages (similarly across all eight source stages)
- Fine trim to improve matching between source and sink output stages
- Coarse trim to vary the resulting output current (gain).

An advantage of using one reference resistor with programmable taps is that it requires one pre-charge operation. The resulting output current is then regulated by the tap selected to perform the pre-charge. By selecting a reference resistor at some multiple of Rsense, a lower gate drive on N2/P2 during the stimulation phase results in a higher gate drive on N1/P1, and a subsequently higher output current. Manipulating the Rref to Rsense ratio therefore adjusts the gain and resulting output current.

13.3.4.5 Fractional Current Regulation Through Electrodes

As stated above, the gain through the output stage can be controlled by making the sense resistor (R_{sense}) some divisor or multiple of the reference resistor (R_{ref}). Like R_{ref}, R_{sense} can be scaled down to affect gain using taps. Similarly, placing multiple instances of the output transistor (N1) and sense resistor (R_{sense}) would provide a gain equal to the number in parallel. In this tutorial, R_{ref} is four times R_{sense}; with 16 N1/R*sense* combinations in parallel, the resulting full scale output current for the design in Fig. 13.15 is:

$$I_{sink} = 4 \times 16 \times I_{ref}$$

This assumes perfect matching between the reference and sense resistors.

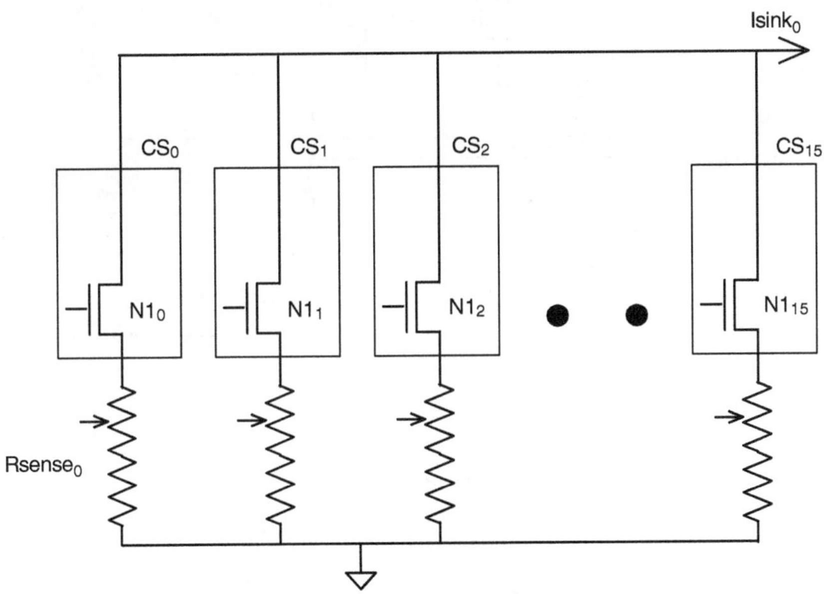

Fig. 13.15 Sensing resistor and current steering

This complete output stage architecture allows for multiple methods for regulating the output current. Placing N1/Rsense combinations in parallel allow for an easy way to regulate the current between electrodes. This action also provides the mechanism for accurate distribution of current among the electrodes.

13.3.4.6 Tying It All Together: A Complete Stimulation Engine

In Fig. 13.16, the two-phase feedback sense-resistor output sink is tied together with the single shared reference resistor, programmable reference resistor taps and sense resistor taps to implement an eight channel constant-current sink. A similar current source would also connect in parallel but is not shown for clarity.

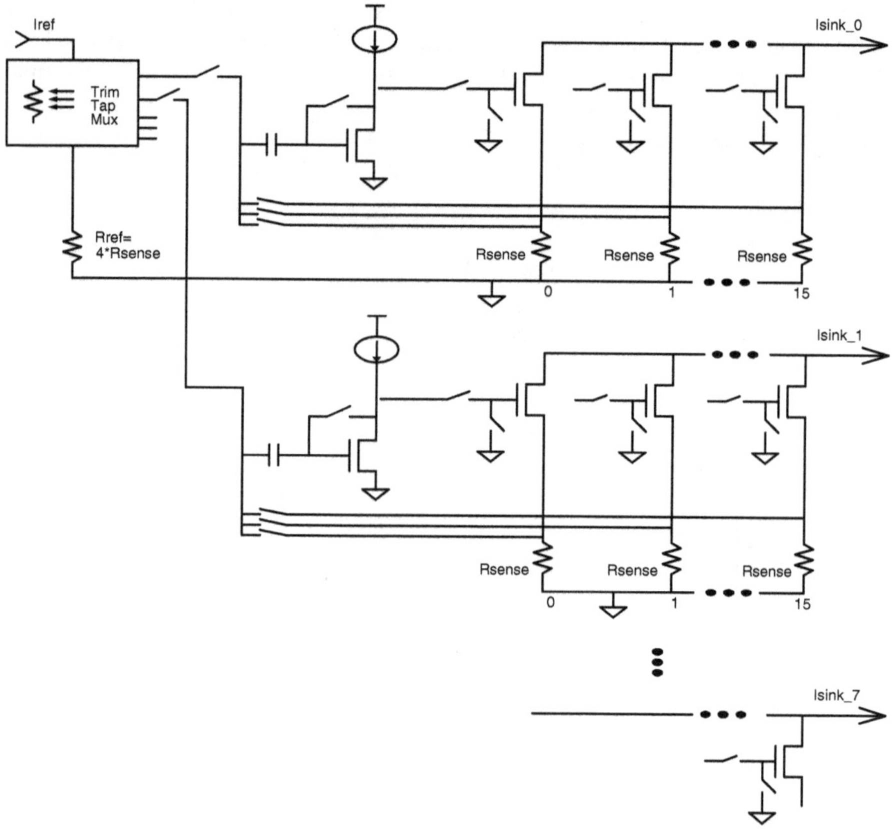

Fig. 13.16 Complete current sink implementation

In summary, the amplitude for this tutorial circuit is regulated by controlling Iref from the reference current generator. The reference resistor is used to trim for matching and regulating output for active recharge. The N1/Rsense parallel combinations are used for current steering. The stimulation pulse from a circuit built on a 0.8 μm BiCMOS process using a design incorporating elements of the aforementioned architecture is shown in Fig. 13.17. A performance summary is included in Table 13.2.

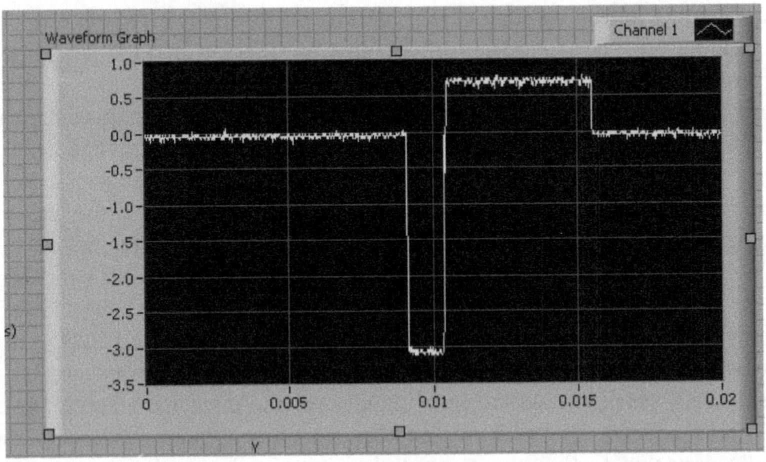

Fig. 13.17 Representative biphasic stimulation pulse oscilloscope capture. In this example, the stimulation engine is programmed to a deliver a 1 ms, 3 mA cathodic pulse followed by a 5 ms, 600 μA anodic pulse. The interphase delay is set to 80 μs. A 1 kΩ load is placed between the two stimulating electrodes

Table 13.2 Summary performance specifications for aforementioned stimulation engine

Die size	<1 cm^2
Quiescent current consumption (includes system microprocessor, supplies, etc...)	<10 μA
Number of electrodes	Scaleable
Total pulse amplitude	0.250 mA to 15 mA, 0.05 μA step size
Supply voltage (battery)	2.3 V to 3.2 V
Electrode-to-electrode matching	<0.5%
Source / sink current partitioning increments	4-bit fractionalized
Typical load stimulation current consumption (3 mA cathodic pulse, 210 μs pulse width, 2.5 V battery, $f = 100$ Hz, 1 kΩ load, passive recharge); includes background current	105 μA
Net efficiency for stimulation example above (corresponds to "Efficiency β" in Fig. 13.3).	72% (Battery to target tissue)

13.4 The Tissue Interface and General Safety Considerations

Once the stimulation engine has transformed the battery energy into the stimulation signal, the final step is delivering the energy to the target tissue via the stimulating electrode. As mentioned previously, the current sinks and sources are generally AC coupled to the tissue. The actual choice of capacitance value depends on a number of factors including capacitor leakage, standoff voltage and physical size. For some stimulation engine designs, the degree of allowable stimulation pulse "tilt" may also be a driving factor. Tilt is defined as the percentage drop in the stimulation current from the leading to trailing edge of the cathodic pulse, and is inversely proportional to the product of the capacitance of the coupling capacitor and the DC impedance of the target tissue.

A critical design element of a practical system includes components necessary for patient and INS safety. To this end, present on each electrode are filter capacitors and voltage / current clamping structures; these elements are intended to handle defibrillation or electrocautery transients, high frequency EMI, ESD and the like without damaging the INS. Electrocautery is commonly used as part of the device implantation procedure. The filter capacitors are often integrated as part of the hermetic electrode feedthrough. Filtering requirements for INS's have not been standardized to the level and maturity currently defined for cardiac IPG's and ICD's in documents such as EN 45502-2-1. The reasons for this discrepancy are attributable to the historic lack of sensing, standard functionality set and inter-manufacturer compatibility in INS's. Generally speaking, feedthrough capacitors are made large enough to handle expected EMI without interfering with lead impedance measurements or other circuits that sense signals from the lead. A representative example of the aforementioned electrode clamping structures is shown in Fig. 13.18.

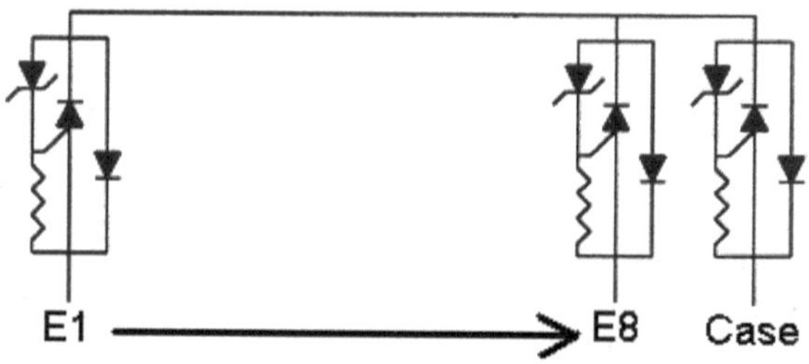

Fig. 13.18 Electrode clamping structures

For Fig. 13.18, the forward conduction path (shown in Fig. 13.19) is straightforward via the forward biased diode incorporated within the structure.

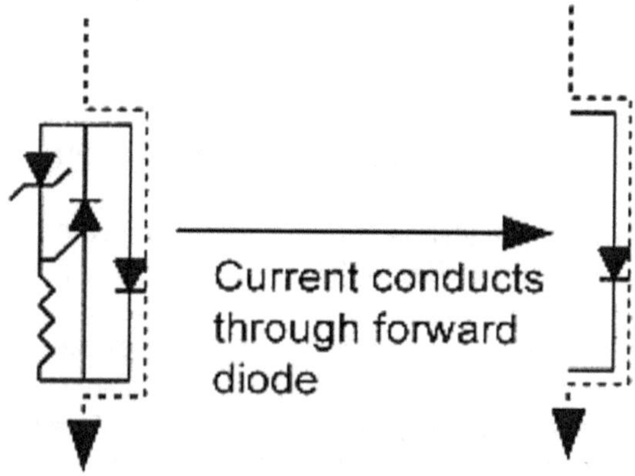

Fig. 13.19 Forward conduction path

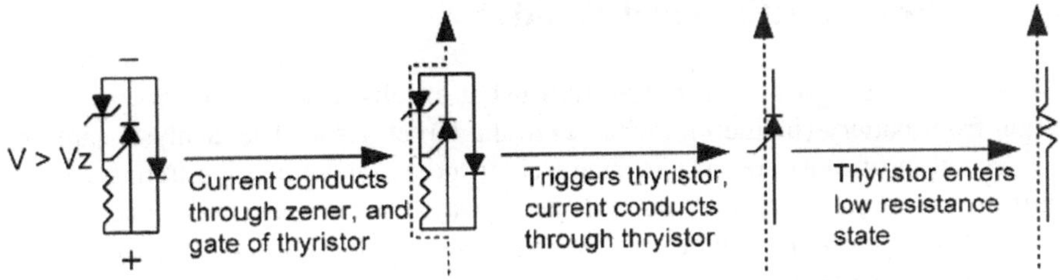

Fig. 13.20 Reverse conduction path

When current flow is in the reverse direction (as seen in Fig. 13.20), the current flows through a zener-triggered thyristor structure. If a voltage greater than the zener voltage is applied across the terminals, the thyristor is triggered. This causes the device to enter a low resistance state for current flow in that direction. Compared to just a zener diode, this low impedance state reduces the power needed to be dissipated. By means of example, 1 A through a 10 V zener diode results in ten times as much power dissipation as 1 A through a 1 ohm resistor.

In Summary, a representative charge interface path between the tissue and stimulation engine is shown in Fig. 13.21. Charge from the stimulation engine flows through the 4.7 μF AC coupling capacitors prior to driving across the lead interface and into the tissue (drawn as a resistor). Also shown are 1.5 nF filter capacitors and protection structures. Overall, the tissue interface is similar to that used in cardiac IPG's. Key differences do exist, however, in the trip points of the thyristor due to higher potential lead voltages during neurostimulation versus those of the cardiac IPG.

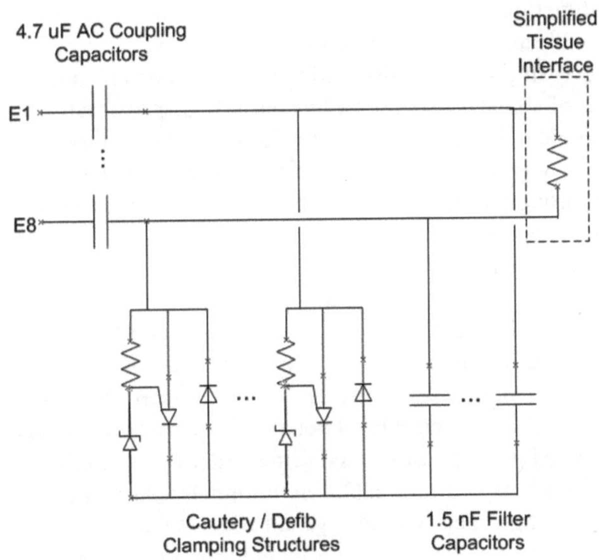

Fig. 13.21 Complete tissue interface path

13.5 Future Directions and Trends

The last three sections analyzed an INS using a holistic schema of energy translation from battery (including recharge) to the target tissue. This analysis captures the operation of most approved systems as of today. Two emerging trends offer to shift the paradigm of stimulation in different ways, and the circuit designer should be mindful of these developments. The first is the use of sensors and control algorithms to achieve closed loop, adaptive stimulation of neural networks. This involves an evolution of devices to act more as a supplemental neural circuit than a simple actuator, much as cardiac IPG's evolved in the late 1960s to include sensing for intrinsic cardiac events. The second trend is a new transduction mechanism for activating neuronal circuits through the use of optogenetic stimulation. Each of these trends builds on the circuit concepts presented so far, and will be discussed briefly here.

13.5.1 Closed-Loop, Adaptive Stimulation

Although we have considered transistor-level circuit design of a neurostimulator, a circuit designer up until this point might still rightfully ask: how does stimulation interface functionally within the context of the neural circuit, and how do we make that functional interface better? These concepts are well established in sensory prosthestics but are relatively new in neuromodulation for DBS, spinal cord stimulation and the like.

The first consideration is the nature of disease and its relation to circuit design. Pathologic dysfunction of the nervous system may take a number of forms, and in accordance with an information processing framework, may be viewed as an information processing failure. Information might be corrupted due to noise or the intermittent loss of signal, or it may be lost entirely due to a transmission failure or lesion of central elements as occurs with infarction due to stroke. The information transfer functions may be corrupted due to many factors including the loss of individual neurons throughout the brain or the failure of various biochemical reactions affecting cellular processes.

A particular form of information processing failure is increasingly being investigated as a causal agent in numerous brain pathologies such as epilepsy, Parkinson's disease, bipolar disorders and obsessive compulsive disorders. This failure occurs when the largely uncorrelated firing of individual neurons throughout a region of brain tissue devolve into a coherently organized synchronous oscillation. In this state, the normal, transiently correlated behavior of individual elements throughout the network is forced into a phase-locked firing pattern that significantly reduces the mutual information between afferent/efferent signals and completely disrupts the information processing capacity of the system as a whole.

The precise mechanism(s) of action of stimulation in deep brain tissue and other neural systems remains controversial. However, there are several ways in which the mutual information between input/output of firing neurons might be affected by applying high frequency electrical pulses to disrupted systems. In one case, it is postulated that DBS actually induces an "information lesion" [20] in which portions of dysfunctional brain structures are effectively disabled, allowing other, redundant, systems to communicate without error. From an information theory point of view, the rapid stimulation rate reduces the entropy of the neural circuit by greatly restricting its available states and thereby its ability to carry information. From a therapeutic standpoint, the benefit of suppressing the pathological information content within the circuit is posited to provide similar benefits as a lesion [20]. Another hypothesis [21] is that high frequency DBS is acting to desynchronize overly coherent oscillation, which, in turn, increases the information capacity of the dysfunctional region.

A major difficulty in deciphering neural dynamics is the barrier to extracting information from the brain circuit. Scientific tools that monitor neural dynamics are needed to uncover the basic principles of function, the therapeutic effects of stimulation, and to provide the observability needed for adaptive neuromodulation. Systems for accomplishing these tasks are becoming practical, as we learn enough about brain coding to architect devices for practical sensing and stimulation.

Adding sensing technology to a neurostimulator—along the lines of cardiac IPG with rate- and motion-responsive pacing—could provide several benefits with respect to the effect of neurostimulation on modulating the information content of the neural circuit. The scientific benefit is driven by the need for better understanding basic network dynamics, information flow, and mechanisms of action for neurological therapies. From a clinical standpoint, there is interest in using sensing of neurological activity to help provide closed-loop therapy based on therapeutically relevant biomarkers. The goals of closed-loop therapy—also known as adaptive modulation—are to improve therapeutic outcomes and potentially increase device longevity by entering low-energy states when stimulation is not required. The addition of sensing can also provide quantitative diagnostics to aid in therapy titration in "open" and closed loop modes as seen in Fig. 13.3. The addition of sensing, however, requires thoughtful engineering to be practical, effective, and safe. Example systems and methods for closed-loop DBS applications leveraging local neural field potential sensing have been described in the literature [22], [23]. As our understanding of biomarkers and algorithms improves, the functional interface of the neurostimulator will evolve from simple pulsatile excitation to modulated information flow to and from the nervous system.

13.5.2 Optogenetic Neuromodulation

While serving as the core technology of many neurological therapies, electrical stimulation still suffers from several drawbacks. Constraints on electrode geometry and placement can result in an inability to modulate specific neural populations, and stimulation of non-target networks may cause undesirable side-effects. Conducting electrodes in tissue may also restrict the level of tolerated electromagnetic exposure from modalities like MRI and electrosurgery, and large stimulation currents may undermine the ability to simultaneously sense underlying neural activity when implementing a closed-loop therapy system [22]. These drawbacks motivate the need for alternative techniques to therapeutically modulate neural network activity.

"Optogenetic" neuromodulation is a rapidly evolving technology achieved by activating light-sensitive channel proteins (opsins) embedded in neuronal populations. Opsins are expressed on neuronal membrane by lentiviral-based delivery of their genes, allowing for direct cellular targeting through genetic mechanisms as opposed to reliance on electrode positioning. Optical stimulation can also eliminate the galvanic path that can limit EMI/MRI and sensing performance. Until recently, optogenetic implantable devices were of modest interest because the inefficiency of opsin stimulation required roughly two-orders of magnitude higher power than electrode-based technology. Recent advances in molecular engineering of Channelrhodopsin-2 have potentially reduced these requirements significantly, making this technology attractive for translational research for therapy devices [24, 25]. This fact allows for the development of a scaleable implantable optogenetic stimulation system capable of chronically delivering required light intensities for modulation of modified opsin channels. The block diagram for such a system is shown in Fig. 13.22.

This approach to neuromodulation is a fundamental paradigm shift from electrical techniques. As a disruptive technology, it is difficult to predict the impact of the

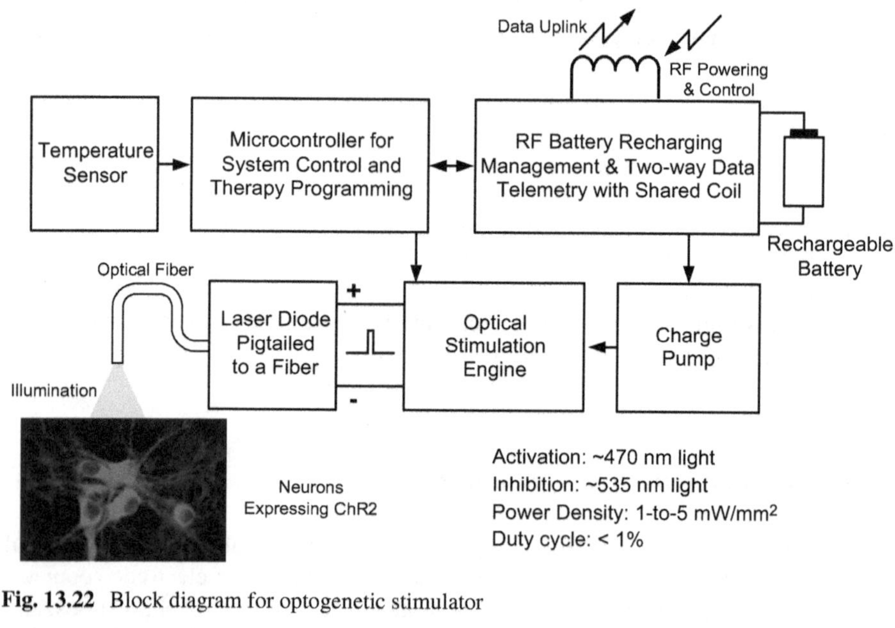

Fig. 13.22 Block diagram for optogenetic stimulator

technique for therapy applications, but the use of this stimulation is already making in-roads for scientific exploration and the circuit designer should be aware of this technology as the optogenetic technology continues to evolve.

13.6 Conclusion

This chapter presented the design of a neurostimulator from the viewpoint of the energy translation from battery (including recharge) to tissue, which we believe is the most appropriate method to analyze the circuit design. The first three sections covered battery considerations, power-translation circuits, stimulator output stages, and safety considerations relevant to a typical design representative of most approved systems as of today. In addition, two emerging trends were discussed that are threatening to disrupt the established techniques of neurostimulation and modulation. The first is the use of sensors and control algorithms to achieve closed loop, adaptive stimulation of neural networks. This involves an evolution of devices to act more as a supplemental neural circuit than a simple actuator. The second trend is a fundamentally new transduction mechanism for activating neuronal circuits through the use of optogenetic stimulation. Each of these trends builds on the neural modulation concepts established today, but with the potential for technical and therapeutic improvements if the clinical data matches the optimism of leading researchers.

At a minimum, our hope is that this overview will provide a circuit designer with the tools needed to critically analyze existing neurostimulator architectures, as well as inspire future designs with improved performance for the benefit of patients suffering chronic disease.

References

1. Windhover Information Inc. Neurostimulation market expanding. Web. Jan 2007 http://sis.windhover.com/buy/abstract.php?id=2007400004
2. Roblee, LS, Rose TL (1990) Electrochemical guidelines for selection of protocols and electrode materials for neural stimulation. neural prostheses fundamental studies. In: Agnew WF, McCreery DB (eds) Englewood Cliffs: Prentice Hall, pp 26–66
3. Durand DB (2006) Electrical stimulation of excitable systems. biomedical engineering fundamentals. In: Bronzino JD, Raton B (eds). CRC Press, pp 28–1 to 28–21
4. Merrill DR, et al (2005) Electrical stimulation of excitable tissue: design of efficacious and safe protocols. J Neurosci Methods 141:171–198
5. Troyk PR, Cogan SF (2005) Sensory neural prostheses. Neural engineering. He B (ed), Kluwer, New York p 7
6. Sivaprakasam M, et al. (2005) Architecture tradeoffs in high-density microstimulators for retinal prosthesis. IEEE Trans Circuits Sys–I: Regular Papers 52(12), Dec. 2005
7. Catsoulis J (2005) Designing embedded hardware, 2nd edn. Sebastopol, O'Reilly
8. Jensen S, et al (2008) Information, Energy, and entropy: design principles for adaptive, therapeutic modulation of neural circuits. In: 2008 European Solid-State Circuits Conference (ESSCIRC 2008), Edinburgh, UK. 32–39
9. Chavan A, et al (2005) Implantable and rechargeable neural stimulator. U.S. Patent 7,616,990, filed October 24, 2005, and issued November 10, 2009

10. Harrison R (2007) Designing efficient inductive power links for implantable devices. In: Proc. 2007 IEEE Intl. Symposium on Circuits and Systems (ISCAS 2007), New Orleans, LA, pp 2080–2083
11. Gord JC, et al. (2008) Method and apparatus for efficient power/data transmission. U.S. Patent 7,379,774, filed October 14, 2004, and issued May 27, 2008
12. Webster JG (1995) Battery. design of cardiac pacemakers. Webster JG (ed) New York: IEEE Press, 161–170
13. http://www.quallion.com/images-pdf/BION.pdf
14. Smith AJ, et al. Precision measurements of the coulombic efficiency of lithium-ion batteries and of electrode materials for lithium-ion batteries. submitted to J Electrochem Soc
15. Schulman J, et al. (2006) An implantable bionic network of injectable neural prosthetic devices: the future platform for functional electrical stimulation and sensing to restore movement and sensation. Biomedical engineering fundamentals. Bronzino JD, Raton B (eds) CRC Press 34-1 to 34-17
16. Abejon D, Feler C (2007) Is Impedance a parameter to be taken into account in spinal cord stimulation? Pain Physic 10:533–540
17. Rothermel A, et al. (2009) A CMOS chip with active pixel array and specific test feature for subretinal implantation. IEEE J Solid-State Circuits 44(1), Jan 2009
18. McDermott HJ (2006) Cochlear implants. Wiley Encyclopedia of Biomedical Engineering. Hoboken, Wiley
19. Sooksood K, et al. (2009) Recent advances in charge balancing for functional electrical stimulation. IEEE EMBC 2009, Minneapolis, MN, Sep 2009
20. Grill WM, et al. (2004) Deep brain stimulation creates an informational lesion of the stimulated nucleus. Neuroreport. 2004 May 19;15(7):1137–1140
21. Wingeier B, et al. (2005) Intra-operative STN DBS attenuates the prominent beta rhythm in the STN in Parkinson's disease. Experimental Neurology
22. Avestruz A, et al. (2008) 5 μW/Channel spectral analysis IC for chronic bidirectional brain-machine interfaces. IEEE J Solid-State Circuits 43(12), Dec 2008
23. Shoeb A, et al. (2009) A micropower support vector machine based seizure detection architecture for embedded medical devices. IEEE EMBC 2009, Minneapolis, MN, Sep 2009
24. Berndt A, et al. (2008) Bi-stable neural state switches. Nature Neurosci 12:229–234
25. Gradinaru V, et al. (2009) Optical deconstruction of parkinsonian neural circuitry. Science 324:354–359

Chapter 14
Artificial Retina IC

Jun Ohta

14.1 Introduction

Artificial retina or, in general, artificial vision, is a prosthesis device to regain vision for the blind. The similar sensory prosthesis device is an artificial cochlea, which has been successfully developed and widely used in many deaf patients in the world-wide to regain sound. Now in the world, a number of research and development on artificial reina [17] are progressing and commercial products will be produced commercially in the near future.

This chapter is organized as follows; First, fundamentals of artificial retina are presented. The structure of human eye and blindness are described in the section. Artificial retina devices are classified in terms of the implantation place of the device. There are three types of artificial retina devices; ep-retina implantation, sub-retina implantation and supra-choroidal implantation. They are explained with some description of comparison among them. In the next section, basic circuits for artificial retina devices are introduced. First, retinal stimulator circuits are mentioned. Requirements for safety operation in human body are addressed such as AC powering and charge balancing. Then, photosensors which are important circuits for sub-retinal implantation are described in details. In the last section, the case study of an artificial retina device is introduced.

Before starting the first section, the terminology of retinal prosthesis is briefly addressed. There are several terms for retinal prostheis; artificial vision, artificial sight, visual prosthesis, artificial retina and so on. As this chapter focuses on the stimulation of retina by CMOS technologies, an artificial retina will be used in this chapter.

J. Ohta (✉)
Graduate School of Materials Science, Nara Institute of Science and Technology, Nara, Japan
e-mail: ohta@ms.naist.jp

H.-J. Yoo, C. van Hoof (eds.), *Bio-Medical CMOS ICs*,
Integrated Circuits and Systems, DOI 10.1007/978-1-4419-6597-4_14,

14.2 Fundamentals for Artificial Retina

14.2.1 Retina and Blindness

We gain a large portion of information in external world through vision. The front-end of visual information is retina. The human retina is a thin, layered tissue with a thickness of 0.1–0.4 mm attached to the inner surface of the eyeball [55] as shown in Fig. 14.1. The retina has a layered structure with photoreceptor cells for light detection in the bottom layer and ganglion cells for output in the top layer. The retina plays an important role in visual information collection and processing, and so its dysfunction can result in blindness. Figure 14.2 shows the prevalence of blindness in the world [56], USA [12], and Japan [26]. Among these diseases, retinitis pigmentosa (RP) and age-related macular degeneration (AMD) have no effective remedies at present. In both cases, the photoreceptors gradually become dysfunctional, so that the patient eventually becomes blind.

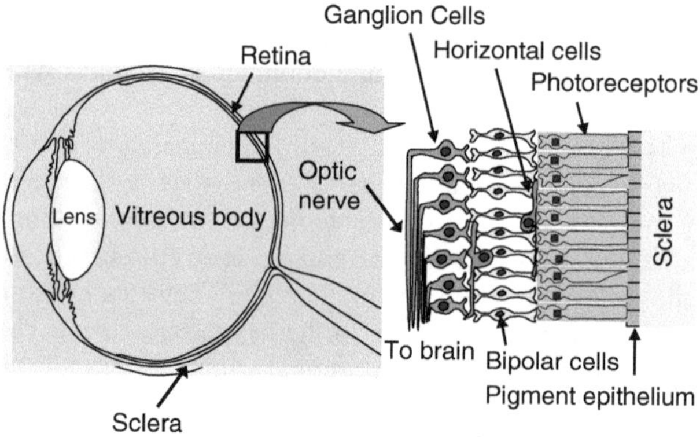

Fig. 14.1 The schematic structure of humane eye and retina

14.2.2 Principle of Artificial Retina

In RP and AMD, photoreceptor cells are dysfunctional, but most of the other retinal cells, such as ganglion cells, are still alive, unless the disease is in the terminal stage [43, 15]. Consequently, by stimulating the remaining retinal cells, visual sensation or phosphene can be evoked. This is the principle of the retinal prosthesis or artificial retina. Based on this principle, an artificial retina device stimulates retinal cells with a patterned electrical signal so that a blind patient may sense a patterned phosphene, or something like an image.

According to the site at which the retinal stimulator is placed, the artificial retina device is classified into three categories: epi-retinal stimulation [16, 23, 18, 49, 39, 44, 14, 36, 45, 46], sub-retinal stimulation [5, 58, 40, 18, 49], and suprachoroidal

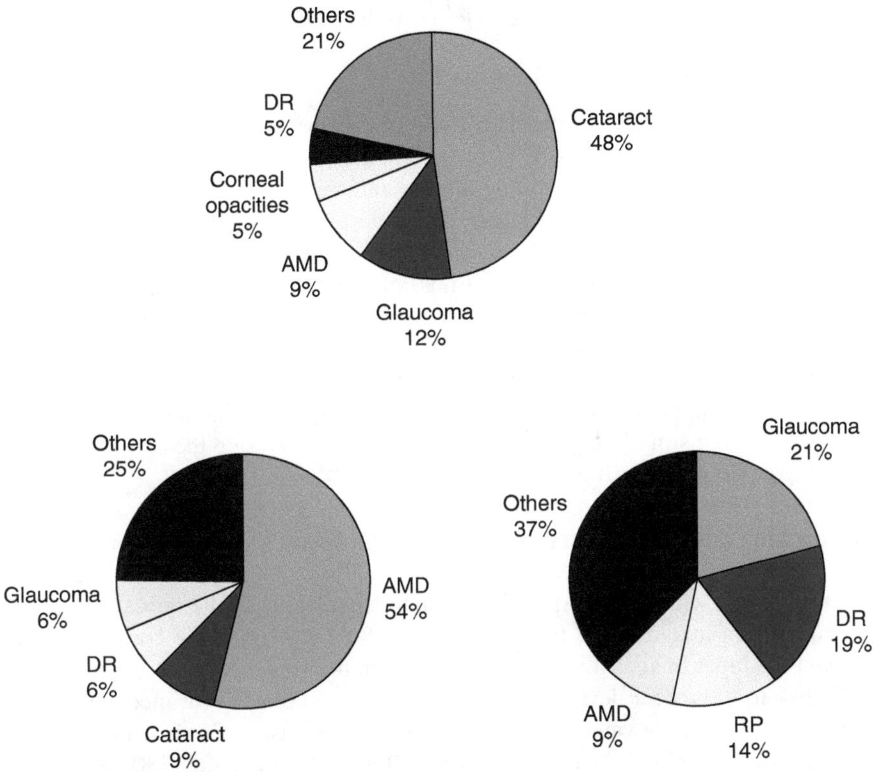

Fig. 14.2 The causes of the blindness in (**a**) worldwide [56], (**b**) USA [12], and (**c**) Japan [26]

transretinal stimulation (STS) [21, 20, 10, 33, 34, 47, 48], which has recently been developed. They are described in Sect. 14.2.3.2.

The stimulation site may be located not only in retinal cells, but also in the pathways to the brain, such as the optic nerves [54, 42, 4], which are the transmission lines of visual information, and, of course, in the visual cortex [2, 7, 30], which is in charge of the terminal of the visual information processing. Figure 14.3 shows the arrangement of these elements. They are classified to extraocular retinal outside of eyeball prostheses, while the former ones to intraocular retinal prostheses.

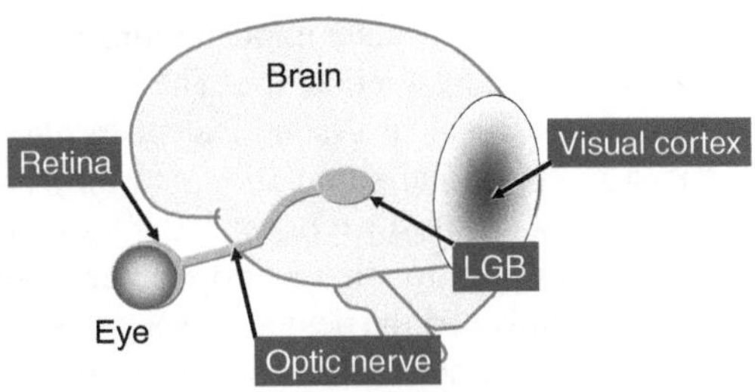

Fig. 14.3 Stimulation places in a visual pathway. LGB: Lateral geniculate body

14.2.3 Classification of Artificial Retina

14.2.3.1 Extraocular Artificial Retina

The extraocular retinal prosthesis, which stimulates the visual cortex or optic nerve electrically, can be applied to patients with no retinal cells. This means that these methods can be applied to any disease related with blindness, including RP and AMD. In the case in which the visual cortex is stimulated, the stimulator is implanted in the surface of the visual cortex by opening a scull. This method has been applied to the first human trials among retinal prostheses by Brindley and Lewin [2] and has been successfully applied in some patients by the Dobelle Institute over a long period of time [7]. Although coritial implant can widely be applied to any blind diseases, it has disadvantages such as difficult surgical operation, infection due to the electrical wire through the skull, and retinotopy.

The method of stimulating the optic nerve involves covering the optic nerve with a cuff-type electrode to stimulate the nerve [54].

Both of these methods require difficult surgical operations because the surgical sites are related to the nervous system or the brain. These methods have only been performed in limited human trials. In addition, these methods must deal with retinotopy, which is the spatial correspondence between the retinal image and the recognition image in the brain. It is difficult to determine the correspondence between the input image and the electrode site on the visual cortex.

Thanks to the recent development of wireless technologies for medical implant devices [57], the disadvantage of the infection is released. Cortical implantation with wireless power and data transmission system will be emerged [30]. Regarding the optic nerve stimulation, another method with easier surgical operation has been proposed [42].

14.2.3.2 Intraocular Artificial Retina

The research of retinal prosthesis at present is mainnly focused on intraocular artificial retina. The reason is stimulation of retinal cells involves an easier surgical procedure and is possibly less affected by retinotopy because the stimulation points are located near the retina. As mentioned previously, this method is classified into three types according to the stimulator implantation site: epi-retinal stimulation [16, 23, 18, 49, 39, 44, 14, 36, 45, 46], sub-retinal stimulation [5, 58, 40, 18, 49], and STS [21, 20, 10, 33, 34, 47, 48]. Figure 14.4 describes these three types of retinal implantation. In epi-retinal implantation, a retinal stimulator is attached to the retina surface and fixed by a retinal tag. The system of epi-retinal implantaion can be used as an artificial cochlear system except for the stimulator. There are several groups engaged in epi-retinal stimulation and they are mostly advanced in clinical trials. The disadvantage of this type is the fixation of the stimulator. In addition, the electrodes may happen to stimulate optic nerve instead of ganglion cells. In this case, the patient implanted the device may sense streak-like phosphen. The fixation of the stimulator sometimes causes this type of phosphen.

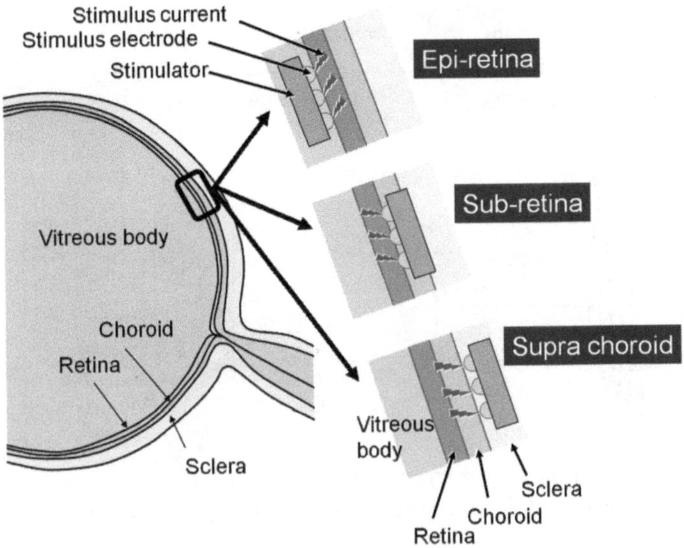

Fig. 14.4 Three types of retinal prosthesis

STS has the same advantage of epi-retinal implantation. In addition, the surgical operation is the easiest among retinal prostheses. The eye is not needed to be open, because a stimulator is inserted into a pocket produced in a sclera. The implantation in a sclera pocket means the electrode has a distance from a retina, so that the threshold current to elicit the phosphen may be higher than the other methods. Acute clinical trials hav shown that the threshold is 200–1000 μA, which is a little higher of than that of in the other methods and the patients suffered from RP sensed phophens when stimulated [20].

In sub-retinal implantation, a retinal stimulator is inserted underneath the retina as shown in Fig. 14.4, so that the device is fixed naturally. The electrodes are attached to the retina, especially photoreceptors and thus the retinotpy is probably maintained. Some sub-retinal devices are integrated with stimulus electrodes and photo-sensors in the same plane, so that they act like photoreceptors. Thus such a sub-retinal device is ideal for artificial reitna. However, it is difficult to develop such an integrated device for the sub-retinal implantation. The details of such an integrated device are described in the next section.

14.2.4 *Artificial Retina System*

In epi-retial and STS systems, the power supply and stimulus pattern data generated from input image data are transmitted wirelessly by electromagnetic coupling of the primary coil, which is placed outside the body, and the secondary coil, which is placed inside the eye, as shown in Fig. 14.5 [28].

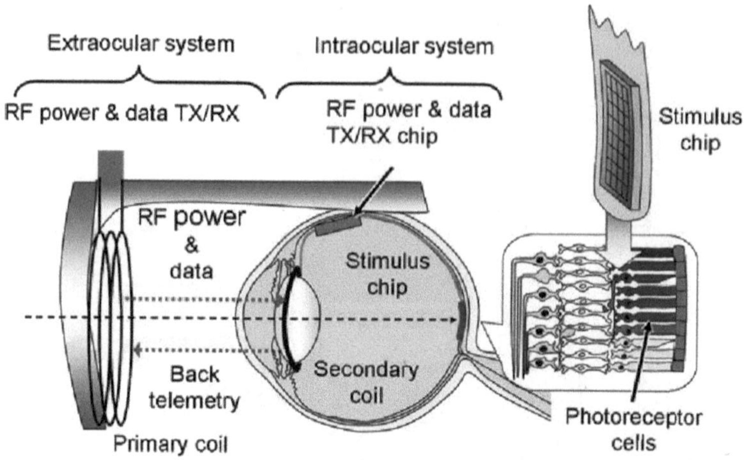

Fig. 14.5 Typical configuration of STS-based retinal prosthesis system. The pirmary coil is installed in an eye glass and the secondary coil is implanted in a crystalline lens. Source: [28] with permission, © IEEE

The total functional block diagram of a typical artificial retina system is shown in Fig. 14.6. Outside a body, the system consists of a camera system and a wireless transmission system with a battery. The image data taken by a camera is processed to the data suitable for retinal stimulation, such as binning and thresholoding. A wireless receiver system including a coil, a stimulus current generator and a multiplxser (MUX) are implanted inside a body. One of electrodes is selected by the MUX. The stimulus current flows from one electrode to the return electrode. The details will be exlained in the next Sect. 14.3.

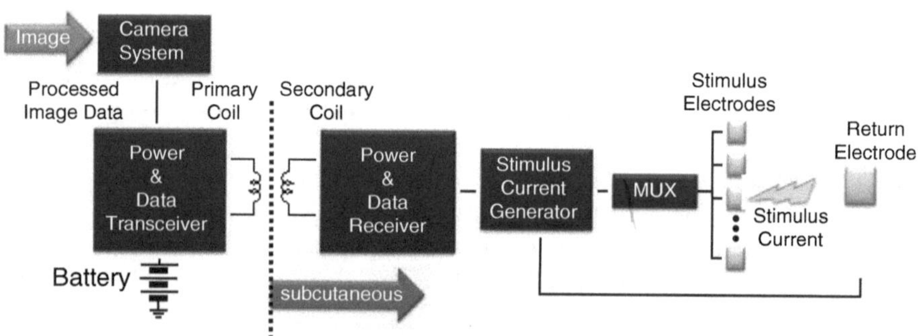

Fig. 14.6 The total functional block diagram of a typical artificial retina system

Figures 14.7 and 14.8 show an illustration of STS system and its mock-up developed by the consortium of Osaka University, Nara Institute of Science and Technology, and Nidek, Co. Ltd. In these cases as shown in Figs. 14.5–14.8, the coil system is located behind the ear like an artificial cochlear system. The place

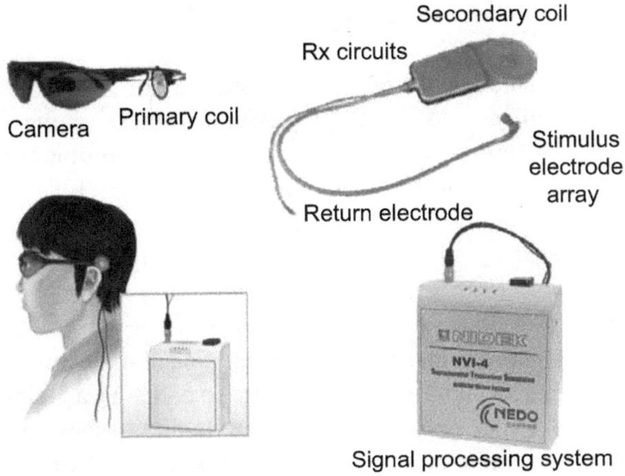

Fig. 14.7 Illustraion of STS-based retinal prosthesis system. Courtesy of Nidek Co. Inc

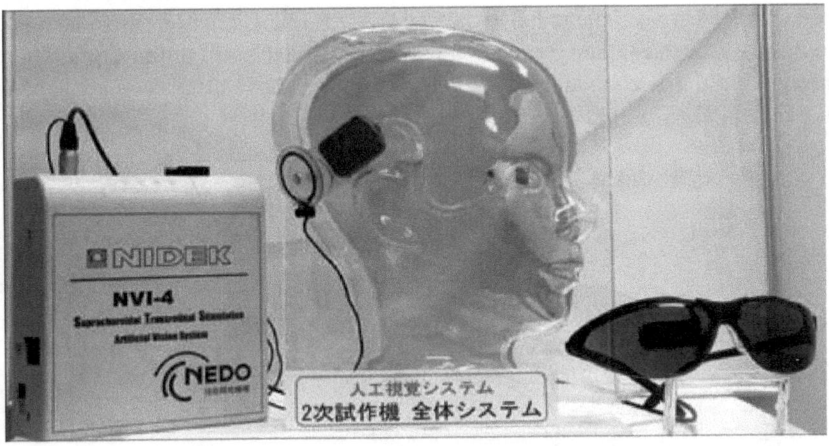

Fig. 14.8 A mock-up system of STS-based retinal prosthesis. Courtesy of Nidek Co. Inc

where the secondary coil is installed is an important issue and will be discussed in the next section.

14.3 Basic Circuits for Artificial Retina

In this section, some basic circuits for artificial retina are described. Before circuits description, the basic concept of stimulation of retinal cells is mentioned. Then stimulator circuits and photosensors are explained with some examples reported.

14.3.1 Stimulation of Retinal Cells

First of all, let us consider the stimulation of neural cells in, for example, artificial cochlear and retinal prostheses. In these applications, stimulation is achieved by extra cellular stimulation, in which the stimulus current affects neural cell's state flowing through the body fluid or electrolyte [27]. Thus we need to evaluate the system composed of metal electrode in an electrolyte solution which has positive and negative mobile ions as shown in Fig. 14.9. When a voltage is applied to a stimulus electrode, mobile ions are gathered to the electrode, and eventually an electric double layer is produced near the electrode. The thickness of this layer is very thin so that the associated capacitance value C_{ed} is large; typically 10–20 μF/cm^2 [38]. This capacitance is also called as a Helmholtz capacitance. Resistance R_S also exists between the electrolyte and tissues, and is called as spreading resistance. Its value is typically about 10 kΩ. Finally, if oxidization–reduction reactions in the interface between the electrolyte and the metal occur when voltage is applied, then electrons are transferred, producing "Faradaic current". The Faradaic current is divided into two categories; reversible and irreversible. This Faradaic current raises impedance Z_F between the metal and electrolyte. If redox reactions are not so large, then this resistance can be ignored and this is the usual case in retinal stimulation in a safety region. The equivalent circuits based on the above discussion are shown in Fig. 14.9(b) [9].

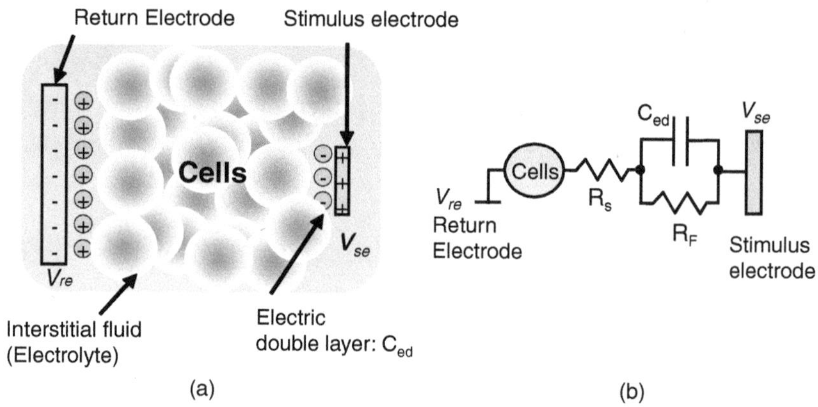

Fig. 14.9 Stimulation of cells (**a**) and its equivalent circuits (**b**). R_S is a series resistance of electrolyte, Z_F is an impedance associated with Faradaic current, and C_{ed} is a capacitance associated with an electric double layer

14.3.2 Stimulator

A stimulator in an artificial retina IC is composed of an array of metal electrodes and a substrate. The impedance between the electrode and the cells may change if the distance between the electrode and the cells changes. Thus, constant current

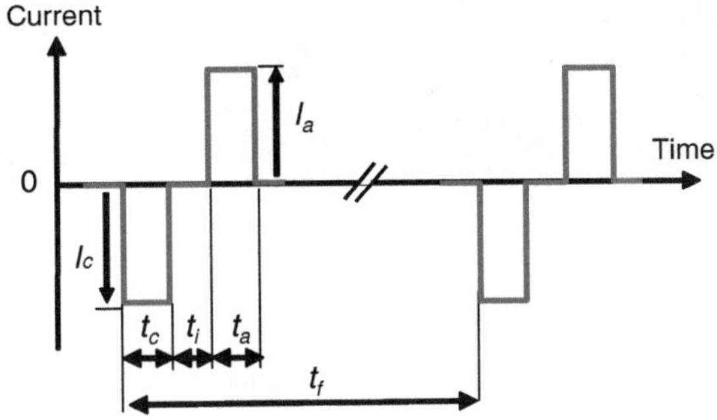

Fig. 14.10 Pulse parameters of stimulus biphasic current. The anodic current I_a and the cathodic current I_c are generated in current sources and injected into the electrolyte as a biphasic pulse

stimulation is generally used. In addition, for the purpose of achieving charge balance, a biphasic pulse is usually used to ensure long-term safety. Figure 14.10 shows the biphasic current waveforms and the parameters of the stimulus current pulse.

Generally cathodic first (CF) current pulse is used for evoking neural cells, while anodic current pulse is used for charge balancing [27]. The interval time between cathodic and anodic pulses t_i is waiting time for evoking cells. It is noted that anodic first (AF) current pulse can also evoke neural cells [27]; actually which current pulse pattern is more effective depending on the implantation place, stimulation current level, and so on. Figure 14.11 shows two types of biphasic stimulus current sources [6]. In Fig. 14.11(a), current sink and source are used, while in Fig. 14.11(b), biphasic current is produced by switching the direction of current flow from the current generator.

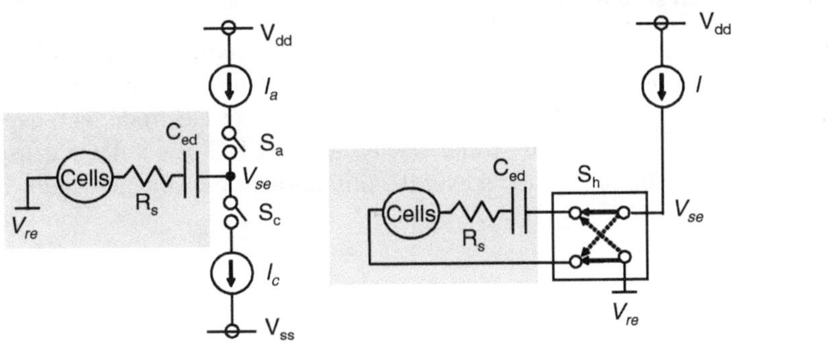

Fig. 14.11 The biphasic current circuits for stimulating neural cells. (**a**) two current generators for cathodic and anodic currents, and (**b**) one current generator with switching circuits [6]

When stimulus current is injected into neural cells, the voltage swing must be keep within a "voltage safety window" of the electrode materials used [38] as well as a power supply voltage. The voltage window is the region in which only reversible reactions occur. If the voltage at the electrode interface exceeds this value, irreversible chemical reactions occur following bubbling, pH change, producing

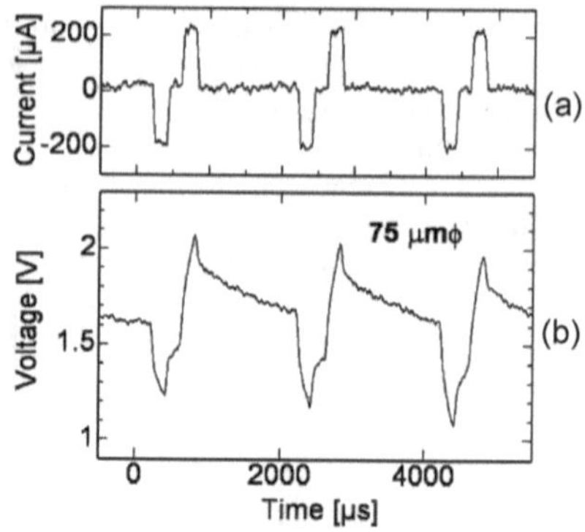

Fig. 14.12 Injection current pulse wave into a saline solution from a Pt electrode with the size of 75 μ (**a**) and its associated voltage waveform (**b**). Source: [53] with permission

materials and so on. Such situation is sometimes harmful for biological tissues. Thus the voltage swing must be inside this window. For example, the voltage window of Pt electrode is 0.9 and/or −0.6 V *vs. SCE* [38]. Figure 14.12 shows an experimental results of current injection pulse waveform into a saline solution and its associated voltage waveform [34]. The electrode is made of Pt with a diameter of 75 μ. It is noted that the first voltage decrease is "IR drop" associated with the resistance of electrolyte solution. By estimating the voltage window, you need to subtract "IR drop". By accounting this "IR drop" value, the voltage siwng is inside the voltage window as well as the power supply voltage of 5 V. If the voltage exceeds the safety window, then electrolysis may occur as shown in Fig. 14.13 [34].

14.3.2.1 Charge Balance

The charge balance is of a great important issue in retinal prosthesis in long term operation in a viewpoint of safety. If the charge balance in a biphasic current is not maintained, some amount of charges are stored in an electrode, and eventually a DC bias voltage can appear on the electrode. Such DC bais voltage causes a high voltage during stimulation to be over the limitation of the voltage window.

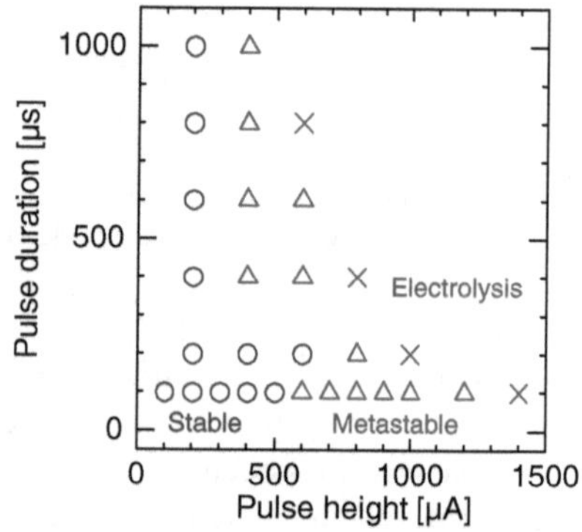

Fig. 14.13 The region of reversible and irreversible condition when Pt electrode with a diameter of about 75 μ is used to inject biphasic current pulse into a saline solution. Source: [34] with permission

The circuits in Fig. 14.11(a) cannot keep charge balancing if current mismatch appears in the two current generators, while the circuits in Fig. 14.11(b) can maintain the charge balancing because both cathodic and anodic currents are produced by the same current generator. Figure 14.14 shows passive and active circuit methods to keep charge balancing in retinal stimulation.

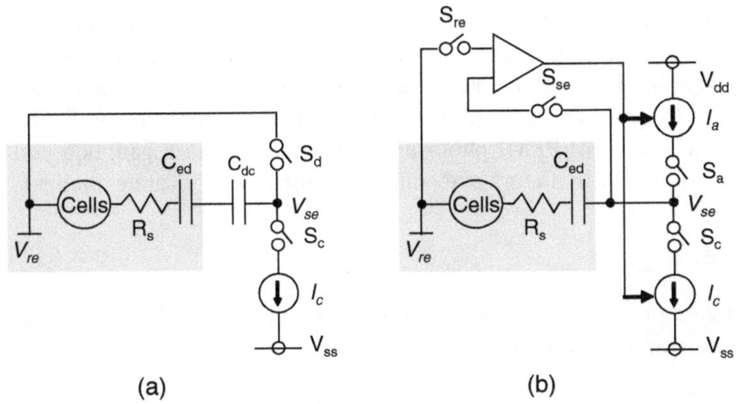

Fig. 14.14 Charge balance circuits for stimulator. Passive type (**a**) [6] and active type (**b**) [36]

Figure 14.14(a) is a passive charge balance circuit where a capacitance C_{dc} is inserted in front of an stimulus electrode to block DC current [6]. In the first cathodic pulse phase, the current generator sinks current I_c from neural cells. After some interval time t_i, the blocking capacitor C_{dc} including neural cells is electrically shorted by switching S_d on. This shorting process discharges the stored charge in the neural cells and thus results in charge balancing in this stimulator. In the passive method, in the anodic phase instead of a rectangular pulse shape, a decay curve appears. It is noted that the blocking capacitor generally requires a large value of about 0.25 μF when $I_c = 500\,\mu A$ and $t_c = 500\,\mu s$, and the voltage drop across C_{dc} is 1 V. This means that anodic phase time requires about 2.5 msec when $R_s \sim 1\,k\Omega$.

To avoid this, an active charge balancing is proposed and shown in Fig. 14.14(b) [36]. In the circuits, the voltage at the electrode V_{se} is monitored and if $|V_{se}|$ exceeds some value V_{th}, then a small amount of anodic or cathodic current pulse is added repeatedly according to the polality at V_{se} until $|V_{se}| < V_{th}$.

14.3.3 Photosensor

Human retina has photoreceptors to convert light intensity to electrical signals. For artificial retina, a photosensor acts like a photoreceptor and is mainly a pn junction diode or a photodiode (PD) with some circuits. Artificial retina devices using PD like as a photoreceptor are mentioned. In the following sections, a basic operation of a PD is described. Next, four types of photosensors are introduced for artificial retina device. First type is a stimulator using micro-photodiode array (MPDA). In this type, a photocurrent is directly used to stimulate retinal cells. Next, active pixel sensor (APS) is introduced. APS is using a PD with some amplification blocks which are widely used in a pixel circuit for a CMOS image sensor. When an image sensor device is implanted, low voltage operation and low power consumption are required. CMOS image sensors match these requirements. In addition, one can integrate some signal processing circuits on a CMOS image sensor chip. This integration capability is a great advantage because the total system can be reduced in its physical size, which is suitable for implantable devices. As third type, a log-sensor whose photorespose is logarithmic like a photoreceptor cell is introduced, and finally a pulse frequency modulation (PFM) photosensor which acts like a ganglion cell will be briefly mentioned. Stimulators embedded with these sensors are explained.

14.3.3.1 Photodiode

The operation principle of the pn-junction PD is quite simple. In a pn-junction diode, the forward current I_F is expressed as

$$I_F = I_{\mathrm{diff}}\left[\exp\left(\frac{eV}{nk_BT}\right) - 1\right], \tag{14.1}$$

where n is an ideal factor and I_{diff} is the saturation current or diffusion current. k_B and T are the Boltzman constant and absolute temperature, respectively.

Light, incident on the PD, produces a photocurrent I_{ph}. The photocurrent I_{ph} is expressed as

$$I_{\mathrm{ph}} = R_{\mathrm{ph}}P, \tag{14.2}$$

where R_{ph} is a photo-sensitivity and P is an incident light power. The maximum photo-sensitivity $R_{\mathrm{ph,max}}$, where the quantum efficiency is 100%, is determined as,

$$R_{ph,max} = \frac{e\lambda}{hc}, \tag{14.3}$$

where e, *lambda*, h and c are the electron charge, the wavelength of the incident light, Plack constant and light speed in vaccuum, respectively.

The total current from the pn-junction photodiode, I_L, under bright condition is expressed as follows:

$$\begin{aligned} I_L &= I_{\mathrm{ph}} - I_F \\ &= I_{\mathrm{ph}} - I_{\mathrm{diff}}\left[\exp\left(\frac{eV}{nk_BT}\right) - 1\right]. \end{aligned} \tag{14.4}$$

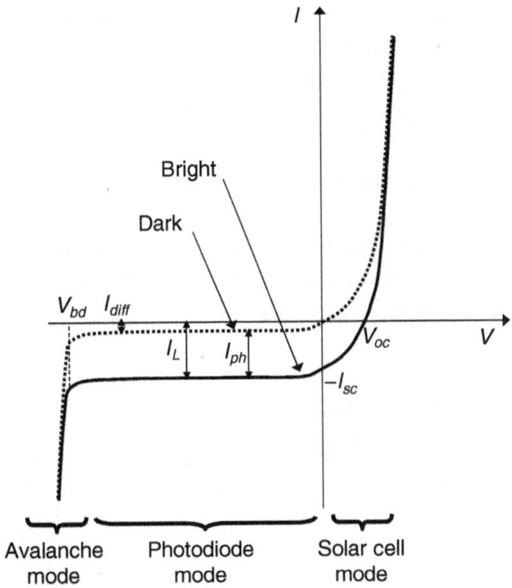

Fig. 14.15 PD I–V curves under dark and bright conditions. There are three operation modes; solar cell mode, photodiode mode, and avalanche mode

Figure 14.15 illustrates the I–V curves of a pn-PD under dark and illuminated conditions. There are three modes for bias conditions: solar cell mode, PD mode, and avalanche mode, as shown in Fig. 14.15.

Solar Cell Mode

In the solar cell mode, no bias voltage is applied to the PD. Under light illumination, the PD acts as a battery, that is it produces a voltage across the pn-junction. In Fig. 14.15, the open circuit voltage V_{oc} is shown. In the open circuit condition, the voltage V_{oc} can be obtained from $I_L = 0$ in Eq. 14.4, and thus

$$V_{oc} = \frac{k_B T}{e} \ln \left(\frac{I_{\mathrm{ph}}}{I_{\mathrm{diff}}} + 1 \right). \quad (14.5)$$

This shows that the open circuit voltage does not linearly increase in proportion to the input light intensity.

PD Mode

The second mode is the PD mode. When a PD is reverse biased, that is $V < 0$, the exponential term in Eq. 14.4 can be neglected, and thus I_L becomes

$$I_L \approx I_{\mathrm{ph}} + I_{\mathrm{diff}}. \quad (14.6)$$

This shows that the output current of the PD is equal to the sum of the photocurrent and diffusion current. Thus, the photocurrent lineally increases according to the input light intensity.

14.3.4 Photosensor Array in Artificial Retina IC

This section describes photosensing circuits to convert input light pattern into electrical signals in artificial retina IC. Light-sensing is incorporated in artificial retina device as shown in Fig. 14.16. Image is taken by the array of photosensors, or an imager.

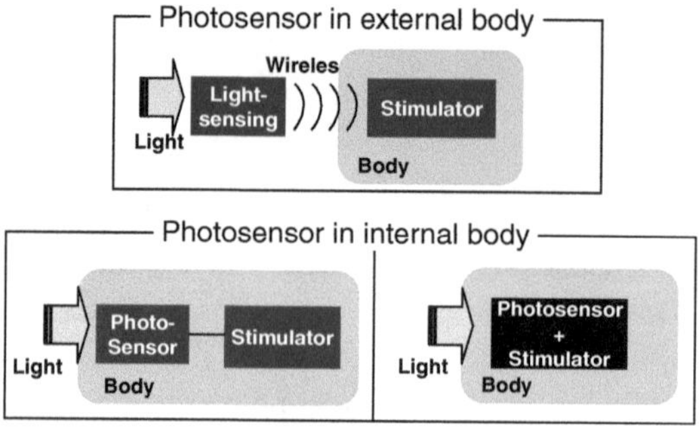

Fig. 14.16 Pixel circuit of a log CMOS image sensor

Usually in eppretinal implantation and STS, an imager is placed outiside of the body as shown in Fig. 14.16. A conventional camera system can be used in this case. The image data is transmitted into the stimulator implanted in a eye. Wireless transmission is widely used. The transmission data rate is high when the pixel count is large.

It is reported that a very tiny camera or a microcamera is implanted into a rabbit eye and thus a whole system can be implanted into a body [4], also shown in bottom of Fig. 14.16. A CMOS image sensor is suitable in this case, because it features low power consumption. In this case, a power is transmitted wirelessly.

In subretinal implantation, a pixel integrated with a photosensor and a stimulus electrode is used. The most simple case is to use a photodiode (PD) in a solar cell mode [5]. The stimulus current is a photocurrent itself. In a solar cell mode, current flows without an external bias voltage as mentioned in Sect. 14.3.3.1. Thus no external power supply is needed. The disadvantage of this device is its small amount of stimulus current.

The other cases are powered by external power supply. An APS (active pixel sensor), which uses the pixel of a CMOS image sensor, can be utilized for stimulating retinal cell effectively. Log-sensor and pulse modulation frequency (PFM)-based photosensor are also used in subretinal implantation.

The types of photo-sensors and their relation with implantation are summarized in Table 14.1.

Table 14.1 Types of retinal prosthesis using photosensor

Type	Method	Implantation	Project	Ref.
Implanted micro camera		Optic nerve	C-Sight	[4]
MPDA	Solar cell mode	Sub.	Optobionics	[5]
	IR conversion	Sub.	Stanford U.	[1, 37]
CMOS pixel with electrodes	Log sensor with adaptation function	Sub	Retina Implant	[8, 40]
	3-d integration chip	Epi	Tohoku U.	[46]
	PFM- APS-based photosensor	Sub, STS	NAIST/Osaka U./Nidek	[33, 34, 19, 11, 53]

14.3.4.1 Micro PD Array

A MPD (micro photodiode array) is an array of PDs and is used on a retinal stimulator; the topmost electrode of a photodiode is utilized as a stimulus electrode as shown in Fig. 14.17. A MPDA is implanted into a subretinal space for clinical trials and it evokes some phosphene. It is noted that a MPDA in a solar cell mode cannot deliver enough stimulus current to evoke retinal cells under normal light condition. When the area of PD is 20 × 20 μm, and the sensitivity of PD is 0.5 A/W, then the light intensity to evoke retinal cells is estimated to be 5 W/cm^2, which is too large for the natural scene. Also a MPD cannot maintain charge balancing, because it only produces one direction polarity of DC current. It is said that such a weak but constant stimulus current sometimes activates the weaken cells due to neuroprotective effect, and thus some subjects with MPDA implanted can sense phosphen. At present, it is generally recognized that an external power is needed to evoke retinal cells efficiently.

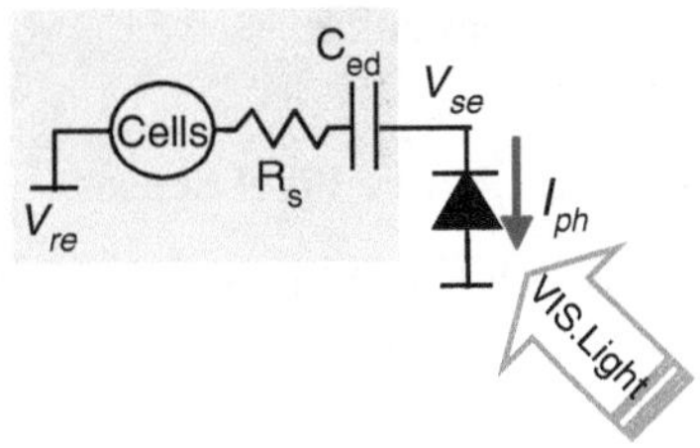

Fig. 14.17 A photodiode in a solar cell mode as a reitnal simulator

To overcome the problem, the strong intensity of IR light pattern is illuminated to the MPDA. In this case, a novel structure of electrode is developed so as to contact the retinal cell more tightly. In such structure, the stimulus current drastically decreased to about 1 μA [37].

In addition, the PD is operated with synchronizing the IR light source, so that it can achieve charge balancing as shown in Fig. 14.18 [37]. First the MPD is forward-biased at V_d which is very close to the on-voltage like the second graph of Fig. 14.18 (b). Thus the current does not flow to charge V_{se} into negative voltage. When the

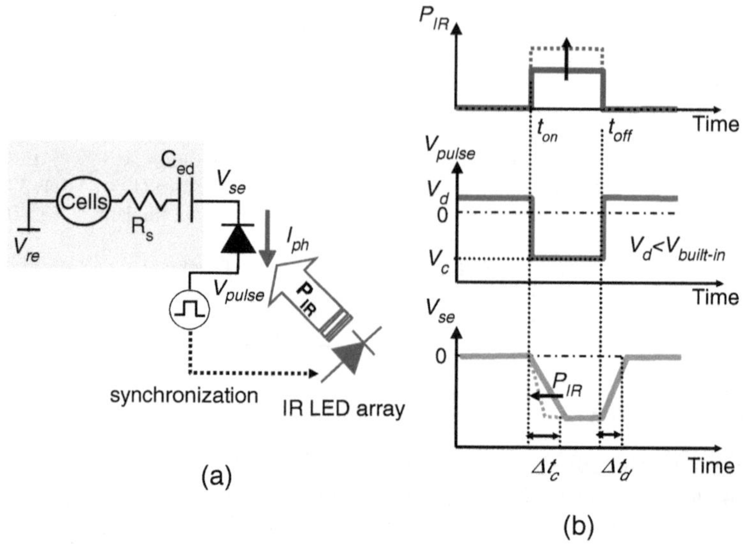

Fig. 14.18 Active type of a photodiode as a retinal stimulator. Strong IR light produces photocurrent

IR light is on, then the voltage decreases down to V_c. The MPD is reversely biased and the corresponding photocurrent flows. When the IR light turns off, then the bias voltage to MPD increases to V_d. In this case, because the C_{ed} is charged and the V_{se} is negative, and thus the diode is forward-biased to flow the current. This current flows until the C_{ed} is completely discharged for Δt to make V_{se}. equal to 0. Of course this is a very ideal case, and it may be difficult to realize such operation in large number of electrodes. Also in this scheme, synchronization data and power supply are required.

14.3.4.2 Active Pixel Sensor

A PD in a CMOS image sensor is usually operated in accumulation mode [31]. In this mode, the PD is electrically floated and when light illuminates the PD, photocarriers are generated and swept to the surface due to the potential well in the depletion region. The potential voltage decreases when electrons accumulate. By measuring the voltage drop, the total intensity of light power can be determined. It should be noted that the accumulation of electrons is interpreted as the process of discharge from the charged capacitor by generated photocurrent as shown in Fig. 14.19. In this figure, a photodiode is expressed as an equivalent circuits which consist of a light-controlled current source and a capacitor C_{PD}. The capacitor originates from the depletion width of the pn junction in the photodiode. First, the photodiode is reset to V_{dd}, and then floated. The capacitor C_{PD} is charged by this reset process. When light with the power of P is impinged on the photodiode, photocurrent I_{ph} is produced. This photocurrent I_{ph} discharges the capacitor C_{PD}. Thus the node

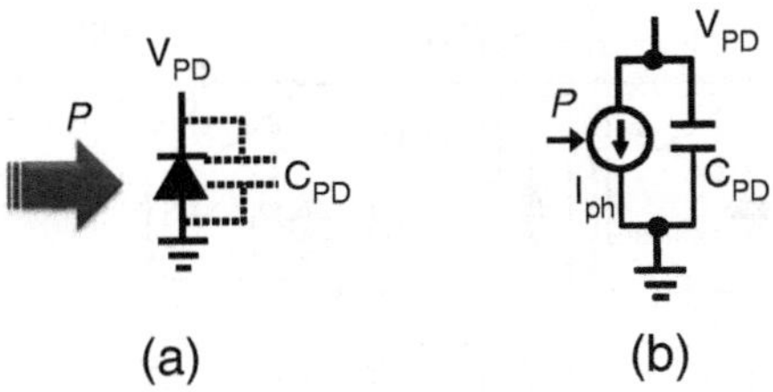

Fig. 14.19 Floated photodiode circuits (**a**) and equivalent circuits (**b**). A photodiode is expressed as a parallel connection of a light-controlled current source and a capacitor C_{PD} in a depletion region of a photodiode. P is a light intensity, I_{ph} is a photocurrent

voltage of the anode in the photodiode, V_{PD} decreases in proportional to the input light intensity, P. The detailed explanation is appeared in [31].

Let us consider, using a simple but typical case, why the accumulation mode is required in a CMOS image sensor. We assume the following parameters: the sensitivity of the PD $R_{ph} = 0.3$ A/W, the area size of the PD $A = 1000$ lux, and the illumination at the PD surface $L_o = 100\ \mu\text{m}^2$. Assuming that 1 lux roughly corresponds to 1.6×10^{-7} W/cm^{-2} [13], the photocurrent I_{ph} is evaluated as

$$\begin{aligned} I_{ph} &= R_{ph} \times L_o \times A \\ &= 0.3\ \text{A/W} \times 100 \times 1.6 \times 10^{-7}\ \text{W/cm}^{-2} \times 100\ \mu\text{m}^2 \\ &\approx 10\ \text{pA}. \end{aligned}$$

While it is possible to measure such a low photocurrent, it is difficult to precisely measure photocurrents of the same order at a video rate from a two-dimensional array composed of a large number of points.

Figure 14.20 shows the structure of a CMOS image sensor. The vertical scanner accesses the pixels in the same row, and the outputs from the pixels of the same row are simultaneously accessed to be fed into the vertical column lines and stored in the capacitor C_{SH} located in the columns. Finally, the stored charges in C_{SH} are addressed by a horizontal scanner and read out to the outside chip.

In this architecture, each pixel has a buffer amplifier to enable the PD to operate in an accumulation mode. Actually this buffer amplifier is a source follower; a current load is employed in a column line. This type of a pixel is called as an active pixel sensor (APS) [31]. Figure 14.21 shows the circuits of APS. In the figure, the gray box is equivalent to a pixel, which has one photodiode and three transistors (3T). Thus this type of APS is called as 3T-APS. M_{RST} is a reset transistor which charges the pn junction capacitor C_{PD}. When C_{PD} is charged to about V_{dd}, M_{RST} turns off and the node V_{PD} is floated. Incident light produces photocurrent, which discharges C_{PD}, and thus V_{PD} gradually decreases. V_{PD} appears in the vertical output line by the source follower, M_{SF} with the current load through the select transistor, M_{SEL}.

The advantage of a CMOS image sensor is its low voltage operation and low power consumption compared with the CCD (charge coupled device) image sensor. This advantage is suitable when it is implanted into the eye. In [4], a micro CMOS camera is implanted in the crystalline lens as shown in Fig. 14.22. The output of

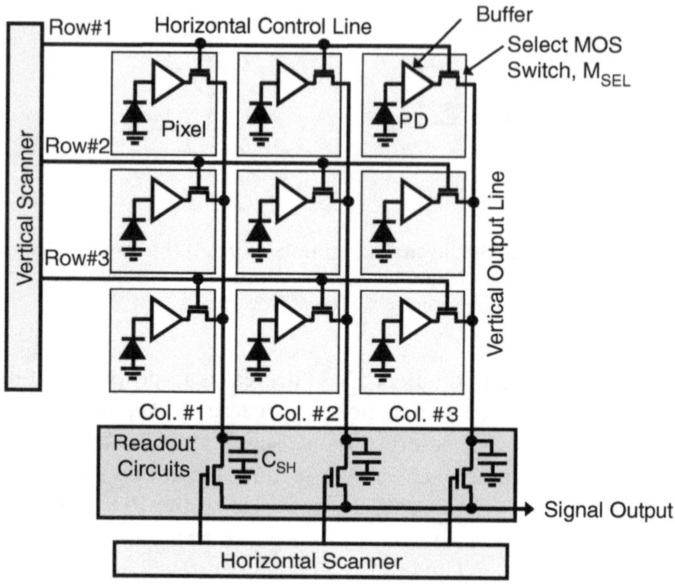

Fig. 14.20 The architecture of a CMOS image sensor. Each pixel has a PD and a buffer amplifier. The vertical scanner accesses the pixels in a row line, and the horizontal scanner accesses the sample and hold capacitor C_{SH}, which stores the output signal from each pixel

Fig. 14.21 The schematic of an active pixel sensor (APS)

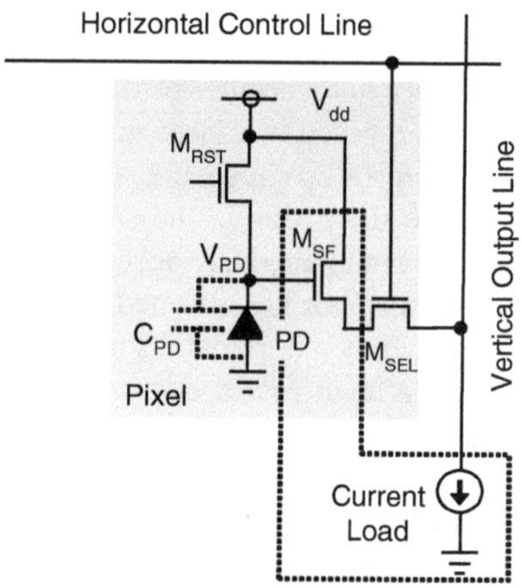

the CMOS image sensor is connected to the stimulator which stimulates the optic nerve.

14.3.4.3 Log Sensor

A conventional image sensor responds linearly to the input light intensity. A log sensor is based on the subthreshold operation mode of MOSFET [31]. A log sensor pixel uses the direct current mode, because the current mirror configuration is a

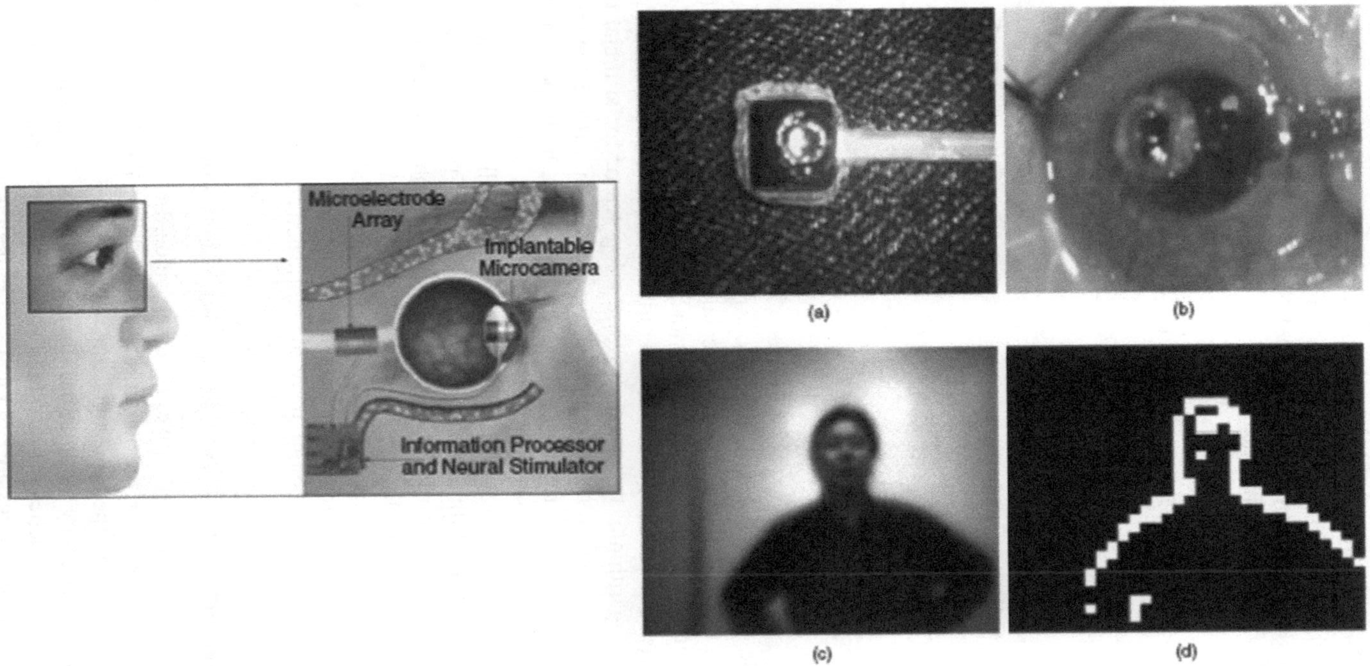

Fig. 14.22 A micro CMOS image sensor implanted in the crystalline lens. Capture image data is processed in edge data, which is used to stimulate optic nerve. Source: [4] with permission, © IEEE

log sensor structure when the photocurrent is so small that the transistor enters the subthreshold region. Another application of log sensors is for wide dynamic range image sensors. Figure 14.23 shows the basic pixel circuit of a logarithmic CMOS image sensor. In the subthreshold region, the MOSFET drain current I_d is very small and exponentially increases with gate voltage V_g:

$$I_d = I_o \exp\left(\frac{e}{mk_BT}\left(V_g - V_{\mathrm{th}}\right)\right), \tag{14.7}$$

where I_o and m are constant values.

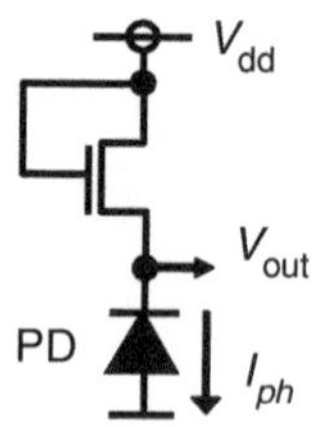

Fig. 14.23 Pixel Circuit of a log CMOS Image Sensor

Although a log sensor has a wide dynamic range over 100 dB, it has some disadvantages, such as low photosensitivity especially in the low illumination region, slow response due to subthreshold operation, and a relatively large variation of the device characteristics due to subthreshold operation. As the response of the human photoreceptor is logarithmic, it is natural to use a log-sensor for artificial retina. The disadvantages mentioned above are not so harmful when it is used in artificial retina, because at present the resolution is not so high and a high speed response is not required. Figure 14.24 shows pixel circuits employed with a log-sensor. In this architecture, one log-sensor is used to obtain ambient illumination level, and the other measures the local illumination level. The output of pixel is the difference between the local illumination level and the ambient or global illumination level [8, 40]. This process mimics the response of human retina where the response is adaptive to the ambient illumination level. As such the artificial retina achieves ultra wide dynamic range covering from light from a star at night to midsummer daylight. Figure 14.25 shows the microphotograph of the fabricated chip.

14.3.4.4 Photosensor Based on Pulse Frequency Modulation

The output in a conventional APS pixel is analog value. In PFM (Pulse frequency modulation) based photosensor, the output is a stream of digital pulses and the pulse frequency is propotional to the input light intensity. Figure 14.26 shows basic circuits of PFM (pulse frequency modulation). In a photosensor based on PFM, when the accumulation signal reaches the threshold value, the output pulse signal is produced, the accumulated charges are reset and accumulation starts again. Repeating this process, the output pulse signals continue to be produced. The frequency of the

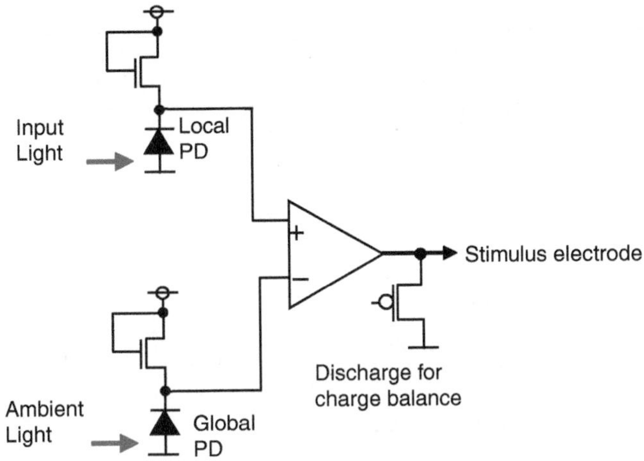

Fig. 14.24 Pixel circuit of a log CMOS image sensor [8]

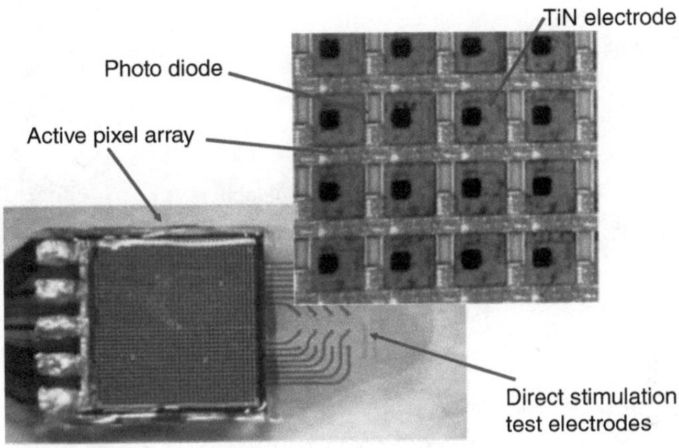

Fig. 14.25 Pixel circuit of a log CMOS image sensor. Courtesy of Retina Implant AG. Source: [41] with permission © IEEE

output signal production is proportional to the input light intensity. PFM-like coding systems are found in biological systems [29], which have inspired the pulsed signal processing [25, 22]. Beside the similarity with biological systems, PFM may be effective to adjusting the intensity of evoking neural cells. When pulse amplitude increases, the region of evoking retinal cell may increase; this possibly results in the larger size of phosphen and not in the stronger intensity of phosphen. On the contrary, when the pulse frequency is higher, then the intensity of phosphen may be stronger. Figure 14.27 is a block diagram of a retinal stimulator embedded in a PFM-based photosensor. The input light intensity is covnerted into the pulse frequency by

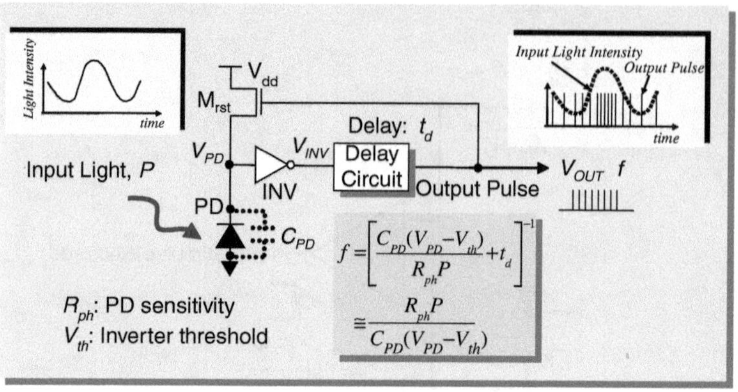

Fig. 14.26 Pixel circuit of a PFM-based photosensor [35]

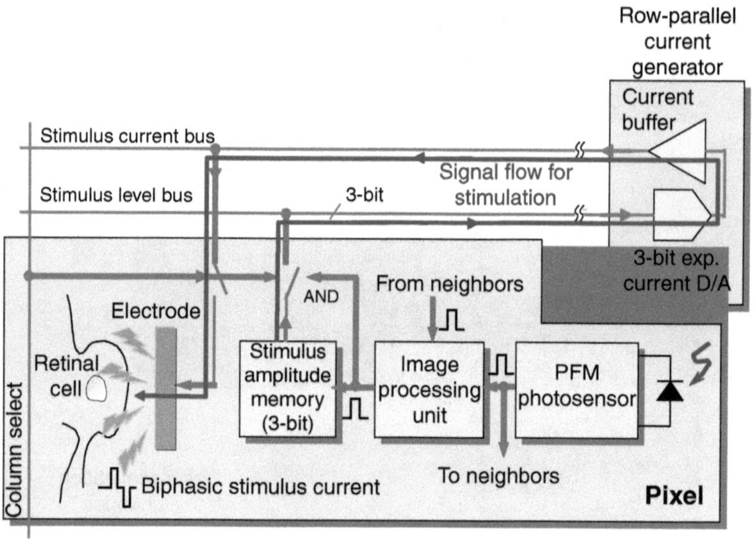

Fig. 14.27 Block diagram of a pixel with a PFM-based photosensor and signal processing [35]

a PFM-based photosensor, and the votage pulse train is processed in a biphasic current pulse train for stimulating retinal cells effectively. Figure 14.28 shows the chip photograps of the artificial reina IC.

This artificial retina IC is used to evoke retina detached from a frog. The Pt stimulus electrode is formed on top of the pad region made of Al. The detached retina is placed on the chip and IR light is incident on the chip through the retina. It is noted that the chip responds to the IR light but the retina does not respond to the IR light. The retinal cells are monitored by inserting a recording wire electrode. The whole system is immersed in a Ringer solution. The chip is protected with an expoxy resin against the water. Figures 14.29(a) and (b) shows the experimental

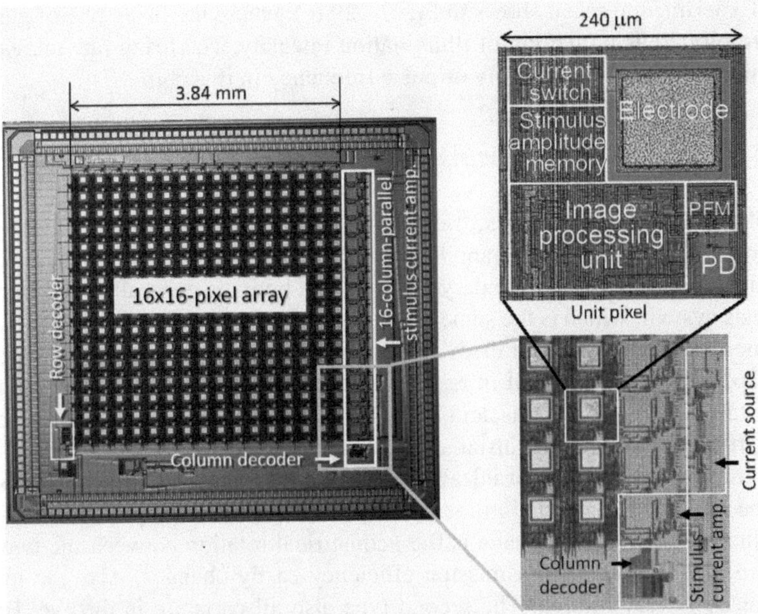

Fig. 14.28 A chip photo of a PFM-based artificial retina IC [35]

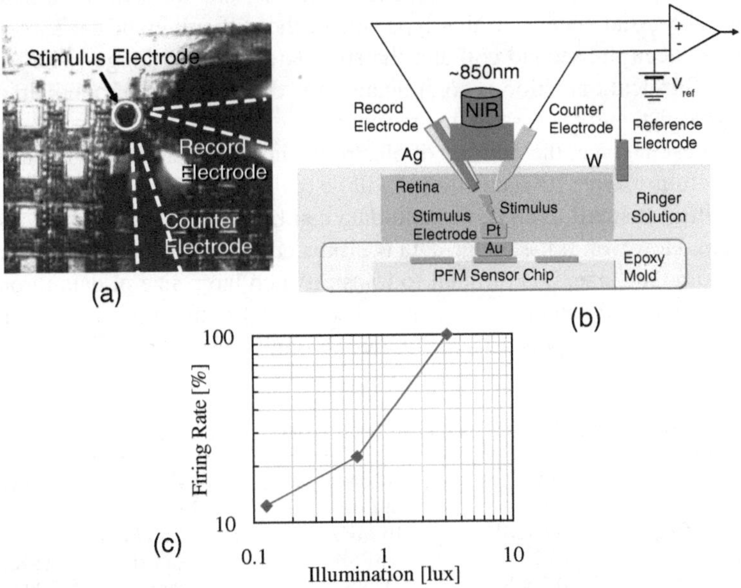

Fig. 14.29 Setup and results of *in vitro* experiments using detached frog's retina on a stimulator embedded with a PFM-photosensor. (**a**) shows the photograph of the top view in electrophysiological setup, (**b**) illustrates the experimental setup, and (**c**) experimental results. Source: [11] with permission

setup. Experimental result shows in Fig. 14.29 (c), where the firing rate of ganglion cells are plotted as a function of illumination intensity. The firing rate increases in proportional to the light intensity or pulse frequency in this chip.

14.3.5 Power and Data Transmission

In implantable medical devices, transmitting power and data from external body inside of eye is of great important. The inductive coupling of two coils, a primary coil outside a body and a secondary coil inside a body, is generally used in retinal prosthesis system, which is the same as in artificial cochlear system. However, there are some differences between them. First, there are three candidates in the installation place of a secondary coil in retinal prosthesis; the three candidates are inside the crystalline lens, on an episclera space, and in a subcateneous space behind the ear. In artificial cochlear, the installation place is the last one; a subcateneous space behind the ear. These types of installation have advantages and disadvantages. The first type realizes a complete intraocular implantation; all of the parts are implanted inside the eye. The disadvantage is the geometrical relation between the two coils varies so that the power transmission efficiency easily changes. Also the implantation place is very limited. The second type also all parts are in the eye. But the space for implantation is relatively alleviated compared the first type. This type has the same disadvantage like the first type, varying power transmission efficiency. In the third type, the place of the secondary coil is fixed so that the efficiency is constant. The disadvantage of this type lies in the difficulties of the surgery. The distance between the second coil and the stimulator is much longer than the other two types. Surgical operation is complicated; the collaboration of ophthalmologist, otolaryngologist, and neurosurgeon are required.

In the present stage, the number of electrodes in retinal prosthesis are quite few, but in near future, over 1000 electrodes will be realized. If the amount of the data to be transmitted is small, then power and data can be transmitted by using the same coil system. However, if the image data is produced by the image taken by a camera with medium data size, it is difficult to transmit such large size of data through the same coil system. Another solution is to use optical communication for the data transmission, while the power is transmitted through coils.

Table 14.2 Power/data transmission system in retinal prosthesis

Institute/Project	Boston	SecondSight	IMI	EPIRET
Type	Sub	Epi	Epi	Epi
Power carrier freq.	13.56 MHz	16 MHz	13.56 MHz	13.56 MHz
Modulation	ASK	DBPSK	Optical	ASK
Data rate	50-700bps	2Mbps	1Mbps	200 kHz
Supplied power	±2.5 V	±1.7 V	±10 V	~ 100 μA
Secondary coil	Episclera	SC space behind ear	Episclera	Lens
Back telemetry	NA	Yes	Yes	NA
Reference	[49]	[24]	[14]	[39]

Table 14.2 shows specifications of reported transmission transmission systems for retinal prostheses.

14.4 Case studies: Artificial retina Device for over 1000 Electrodes

14.4.1 Multiple Microchip Architecture

In order to realize better vision through a retinal prosthesis, over 1,000 electrodes would be preferable [3]. When increasing the number of electrodes, we are faced with problems associated with interconnection between electrodes and external lead wires with good mechanical flexibility. Specifically, the stimulator must be bent to match the curvature of the eyeball.

Figure 14.30 shows the methods used to realize a stimulus electrode array [34]. A direct connecting method, which is commonly used in retinal prosthesis devices, is shown in Fig. 14.30(a), where each electrode is directly connected by a lead wire. It is difficult to increase the number of electrodes in this method.

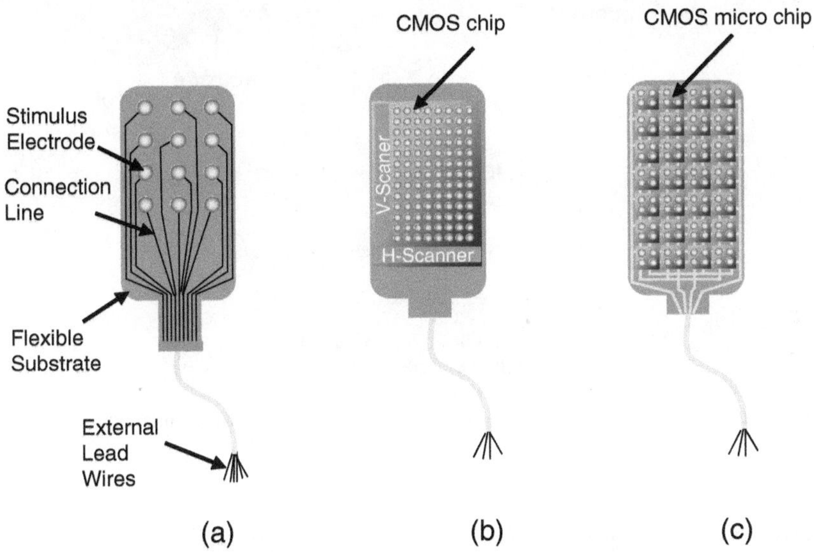

Fig. 14.30 Methods for increasing stimulus electrodes in retinal prosthesis device. (**a**) direct connection, (**b**) introducing scanners on a CMOS chip, (**c**) multiple microchip architecture. Source: [34], wth permission

It is a good idea to introduce a CMOS-based chip in the stimulator because scanning circuits (scanner) can be integrated in order to reduce the amount of wiring, as shown in Fig. 14.30(b). Random access can be implemented using decoder circuits instead of scanners. For implantation, the thin and flexible CMOS-based stimulator is preferable to be thin and flexible in order to fit the eye and to avoid damaging tissue. However, silicon is rigid, and thinning of the CMOS chip increases the risk of breakage.

To solve this problem, a smart stimulator that consists of a number of CMOS-based microchips distributed on a flexible substrate, as shown in Fig. 14.30(c) is proposed and demonstrated [52], [33, 53].

Figure 14.31 shows the concept of multiple microchip architecture and its cross-section. Each microchip incorporates several stimulus electrodes, which can be externally controlled to turn on and off through an external control circuit. In addition to solving the interconnection issue, CMOS-based stimulators offer several advantages, such as signal processing. To allow flexibility, we place several microchips on a flexible substrate in a distributed manner. This stimulator has been developed primarily for STS, and also can be applied to other methods, such as sub-retinal stimulation, which is mentioned in Sect. 14.4.2.

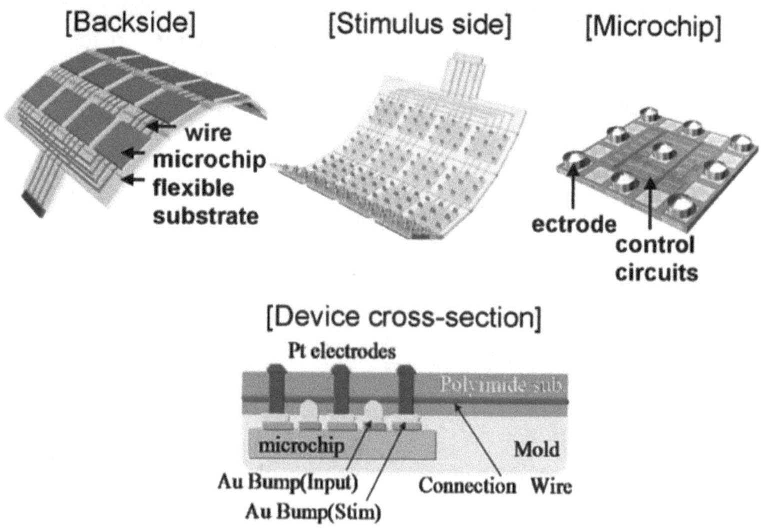

Fig. 14.31 Conceptual schematics of multiple microchip based retinal stimulator

14.4.1.1 Microchip Specification

The microchip architecture has nine stimulation pads and four input lines, including the power supply lines. A block diagram of the chip is shown in Fig. 14.32(a) [34]. Each stimulation pad is assigned a unique four-bit address that can selectively activate one of the nine electrodes on the microchip. The four input lines are VDD, GND, CTLR, and STIM. The VDD and GND lines are used for the power supply (VDD = 5 V), and control and stimulation can be achieved with only two lines, namely, CTLR and STIM. Each of the stimulation electrodes can be selected with the number of the pulses applied on the CTLR line. This is achieved by the

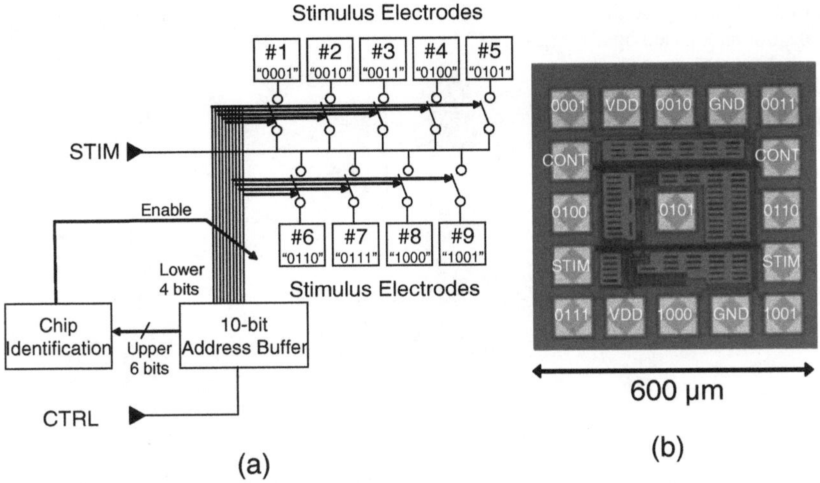

Fig. 14.32 Block diagram of a microchip. Source: [34], with permission

microchip counting the pulses applied on the CTLR line using a 10-bit address buffer. As shown in Fig. 14.32(a), the lower four bits of the address buffer are used for electrode selection, and the upper six bits are used for chip identification. The stimulation current is provided from outside of the chip and is fed into the STIM terminal.

One of the stimulation electrodes is selected depending on the value in the lower four bits of the address buffer. The six-bit address space for microchips facilitates the control of an arbitrary number of microchips (up to 64) using only one set of input lines. Consequently, the multi-chip stimulation device platform can configure a 64-chip device with 576 stimulation electrodes. In order to ensure flexibility, the microchip array is assembled at a pitch of 1,000 to 1,200 μm. The microchips are diced from a mother chip, which is fabricated using 0.35 μm standard CMOS technology. The mother chip contains 16 microchips. Figure 14.32(b) shows microphotographs of a mother chip and a microchip measuring 600 × 600 μm.

14.4.1.2 Stimulator Specificaton

Figure 14.33 shows the fabricated retinal stimulator based on the multiple microchip architecture [53]. The four microchips are placed on a flexible polyimide substrate using flip-chip bonding technology. The thickness of the chip is approximately 50 μm. The total thickness of the stimulator is approximately 200 μm. On the front side, nine Pt bulk stimulus electrodes are formed on one microchip, so that, in this case, 36 stimulus electrodes are used in the stimulator. The electrode is formed on an Al pad of the microchip using stud bump technology. With the exception of the stimulus electrodes, the surface of the stimulator is covered with epoxy resin or

Fig. 14.33 The fabricated retinal stimulator based on multiple microchip architecture. Source: [50], with permission

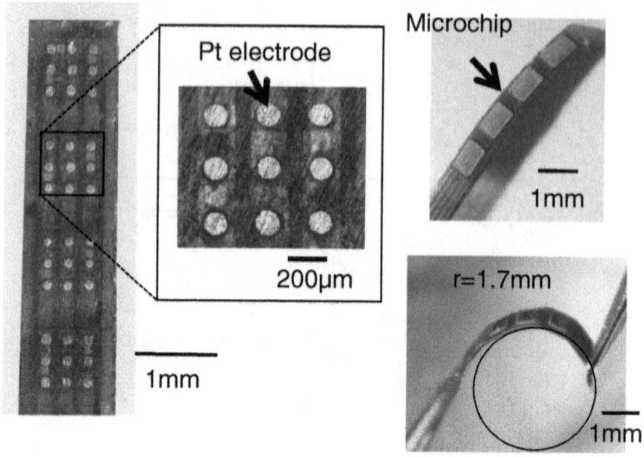

palylene. As shown in Fig. 14.33, the fabricated stimulator can be bent easily, and the bendig radius in this case is approximately the same as that of the rabbit eyeball, i.e., 1.7 mm.

14.4.1.3 In vivo experiment

In vivo experiment, we implanted the fabricated stimulator into the scleral pocket of a rabbit eye using an operation procedure described in [34]. The rabbit is anesthetized. Figure 14.34 shows the experimental setup.

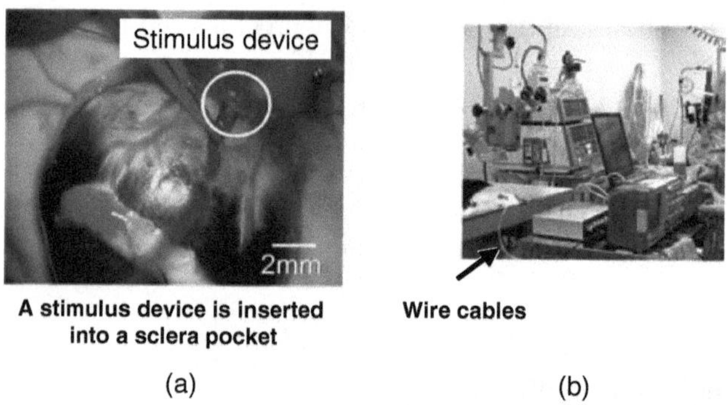

Fig. 14.34 *In vivo* experiment using the fabricated retinal stimulator. (**a**) the stimulator inserted into the sclearal pocket of a rabbit eye. (**b**) The experimental setup. Courtesy of Osaka University

The recording electrode used to measure the electrical evoked potential (EEP) is a stainless-steel screw. The electrode is screwed into the skull at the area of the visual cortex so that the tip touches the dura mater. The reference electrode is screwed into the bregma. The stimulator is inserted into a pocket formed in sclera. A return

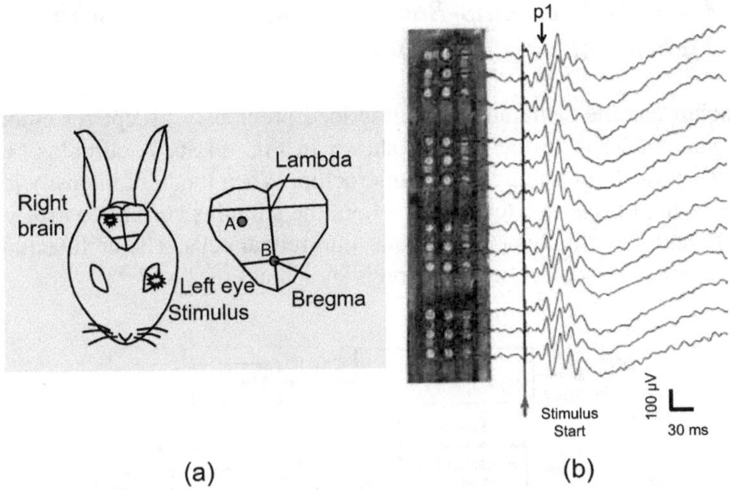

Fig. 14.35 EEP signals obtained by the retinal stimulator based on a multiple microchip architecture. Source: [50], with permission

electrode was inserted into the vitreous cavity. Monophasic 0.5-ms-duration pulses with anodic polarity are used to elicit the EEPs.

Figure 14.35 shows the experimental results for the EEPs, where "p1" is concluded to be the EEP signal from the ganglion cells because of the latency of the signal. Based on p1, the threshold is approximately 100 μA ($< 1\text{mC}/cm^2$). For each stimulus electrode, a clear EEP signal is obtained. These experimental results clearly demonstrate that any one electrode among the 36 electrodes can be assigned to be the stimulus electrode, which can be used to stimulate retinal cells.

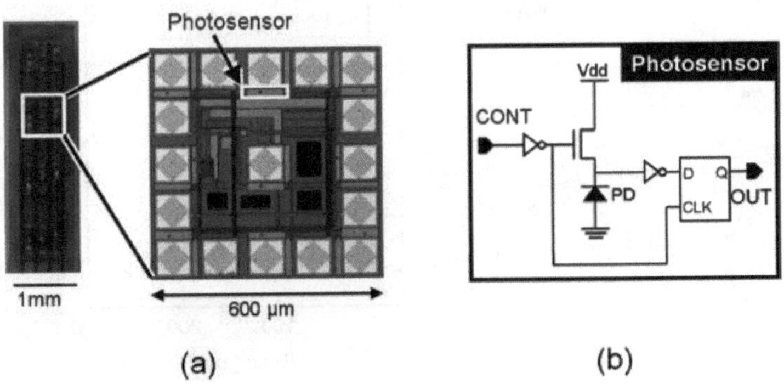

Fig. 14.36 A photograph of a microchip with light controlled function (**a**) and the circuits of the photosensor (**b**). Source: [51], with permission © IEEE

14.4.2 *Multiple Microchip-Based Retinal Stimulator with Light-Controlled Function*

A microchip has the same structure described previously except for embedded a light-sensing function in the chip as shown in Fig. 14.36. A stimulus current is controlled by the light impinged on a microchip. When the light intensity reaches a threshold value of the inverter in Fig. 14.36, the stimulus current control switch is turned on, and the stimulus current flows into retinal cells. The architecture of the microchip with light-controlled function is shown in Fig. 14.37.

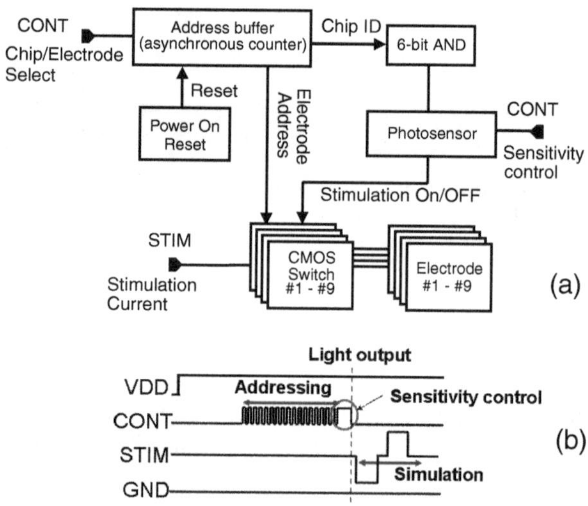

Fig. 14.37 A block diagram of a microchip with light-controlled function (**a**) and timing diagram of the microchip (**b**) [32]

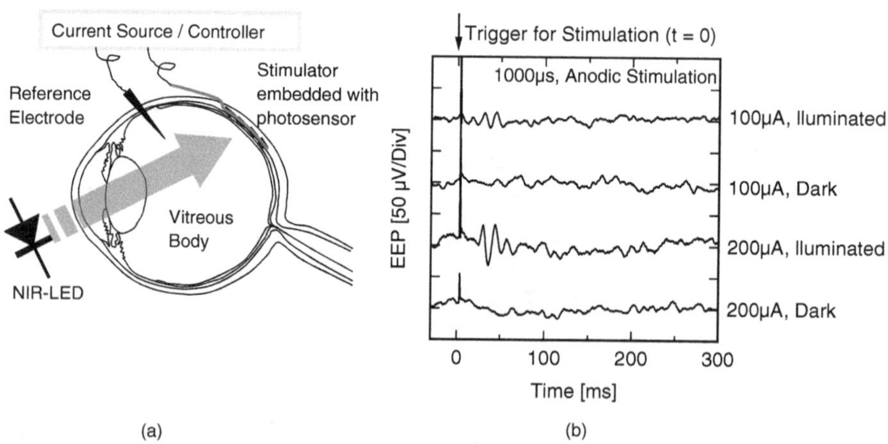

Fig. 14.38 *In vivo* experiment for EEP. (**a**) Experimental setup. (**b**) EEP signals obtained in the experiment. Source: [51], with permission © IEEE

The fabricated stimulator is implanted into a pocket formed in the sclera of a rabbit eye as shown in Fig. 14.38. Near infrared (NIR) LED array is illuminated on the eye where the stimulator is implanted. It is noted that NIR light cannot evoke photoreceptors and can penetrate the epithelium and some thickness of the sclera of a rabbit. EEP signal is measured through the screw electrodes set in a visual cortex. After implanting the stimulator, we confirm that VEP (visual evoked potential) signal is not measured by NIR light used in this experiment before the measurement of EEP signal. When NIR light incidents on the eye, a clear EEP signal can be obtained as shown in Fig. 14.38, and thus the stimulation of retinal cells is successfully demonstrated. The threshold of the stimulus current is about 100 μA, which is the same as in the previous experiment.

Acknowledgments The part of this research was supported by the Strategic Research Program for Brain Sciences, MEXT, Japan, by the Asahi Glass Foundation, and by a Health and Labour Sciences Research Grant, Japan. I would like to thank late Prof. Yasuo Tano and Porf. Takashi Fujikado of Osaka University, the member of Vision Institute of Nidek Co. Ltd., and Prof. Takashi Tokuda of Nara Institute of Scinece and Technology.

References

1. Asher A, Segal AW, Baccus AS, Yaroslavsky PL, Palanker D (2007) Image processign for a high-resolution optoelectronic retinal prosthesis. IEEE Trans Biomed Eng. 54(6):993–1004, July 2007
2. Brindley GS, WS Lewin (1968) The visual sensations produced by electrical stimulation of the medical occiital cortex. J Physiol 194(2):54–59, Feb 1968
3. Cha K, Horch WK, Normann AR (1988) Mobility performance with a pixelized vision system. Vision Res 32(7):1367–1372
4. Chaia X, Li U, Wu K, Zhou C, Caho P, Ren Q (2008) C-sight visual prostheses for the blind—optic nerve stimulation with penetrating electrode array. IEEE Eng Med Bio Mag 27(5):20–28, Sep–Nov 2008
5. Chow YA, Chow YV, Packo K, Pollack J, Peyman G, Schuchard R (2004) The artificial silicon retina microchip for the treatment of vision loss from retinitis pigmentosa. Arch. Ophthalmol 122(4):460–469
6. Demosthenous A, Triantis FI, Liu X (2008) Circuits for implantable neural recording and stimulation, chapter 11. Artech House, Inc., Norwood, MA
7. Dobelle HW (2000) Artificial vision for the blind by connecting a television camera to the visual cortex. ASAIO J.(Am Soc Artif Inter Organs J.), 46:3–9
8. Dollberg A, Graf GH, H¨offlinger B, Nisch W, DJS Spuentrup, Schumacher K (2003) Zrenner E. A Fully Testable Retinal Implant. In: Proceedings of International Conference on. Biomedical Engineering pp 255–260, Salzburg, June 2003
9. Franks W, Schenker I, Schmutz P, Hierlemann A (2005) Impedance characterization and modeling of electrodes for biomedical applications. IEEE Trans Biomed Eng 52(7):1295–1302, July 2005
10. Fujikado T, Morimoto T, Kanda H, Kusaka S, Nakauchi K, Ozawa M, Matsushita K, Sakaguchi H, Ikuno Y, Kamei M, Tano Y (2007) Evaluation of phosphenes elicited by extraocular stimulation in normals and by suprachoroidal transretinal stimulation in patients with retinitis pigmentosa. Graefes Arch Clin Exp Ophthalmol 245:1411–1419
11. Furumiya T, Ng DC, Yasuoka K, Kagawa K, Tokuda T, Nunoshita M, Ohta J (2006) Functional verification of pulse frequency modulation-based image sensor for retinal prosthesis by in vitro electrophysiological experiments using frog retina. Biosensors Bioelectron, 21(7):1059–1068, Jan 2006

12. The Eye Diseases Prevalence Research Group. Causes and prevalence of visual impairment. among adults in the United States. Arc Ophthalmol 122(4):477–485, April 2004
13. Hopkinson RG, Goodman MT, Prince RS (2004) A guide to the use and calibration of detector arrya equipment. SPIE Press, Bellingham, Washington
14. Hornig R, Laube T, Walter P, Velikay-Parel M, Bornfeld N, Feucht M, Akguel H, R¨ossler G, Alteheld N, Notarp DL, Wyatt J, Richard G (2005) A method and technical equipment for an acute human trial to evaluate retinal implant technology. J Neural Eng 2(1): S129–S134
15. Humayun SM, Prince M, de Juan E, Barron Y, Moskowitz M, Klock BI, Milam HA (1999). Morphometric analysis of the extramacular retina from postmorten eyes with retinitis pigmentosa. Invest Ophthalmol Visual Sci 40:143–148
16. Humayun SM, Weiland DJ, Fujii YG, Greenberg R, Williamson R, Little J, Cimmarusti V, Boeme VG, Dagnelie G,de Juan Jr. E (2003) Visual perception in a blind subject with a chronic microelectronic retinal prosthesis. Vision Res 43:2573–2581
17. Rizzo FJ III, Snebold L, Kenny M (2007). Development of a visual prosthesis: a review of the field. Human Press Inc., Totowa, NJ
18. Rizzo FJ III, Wyatt J, Loewenstein J, Kelly S, Shire D (2003) Methods and perceptual Thresholds for short-term electrical stimulation of human retina with microelectrode arrays. Invest Ophthalmol Visual Sci 44(12):5355–5361, Dec 2003
19. Kagawa K, Yasuoka K, Ng DC, Furumiya T, Tokuda T, Ohta J, Nunoshita M (2004). Pulsedomain digital image processing for vision chips employing low-voltage operation in deepsubmicron technologies. IEEE Selct Topic Quantum Electron 10(4):816–828, July 2004
20. Kamei M, Fujikado T, Kanda H, Morimoto T, Nakauchi K, Sakaguchi H, Ikuno Y, Ozawa M, Kusaka S, Tano Y (2006) Suprachoroidal-transretinal stimulation (sts) artificial vision system for patients with retinitis pigmentosa. Invest. Ophthalmol Visual Sci 47:E-Abstract 1537
21. Kanda H, Morimoto T, Fujikado T, Tano Y, Fukuda Y, Sawai H (2004). Electrophysiological studies of the feasibility of suprachoroidal-transretinal stimulation for artificial vision in normal and RCS rats. Invest Ophthalmol Visual Sci 45(2):560–566
22. Lehmann T, Woodburn R (1999). Biologically-inspired learning in pulsed neural networks. In: Cauwenberghs G, Bayoumi AM (eds) Learning on silicon: adaptive VLSI neural Systems. Kluwer Academic Pub, Norwell, MA, pp 105–130
23. Liu W, Humayun SM (2004). Retinal Prosthesis. In: Dig. Tech. Papers Int'l Solid-State Circuits Conf. (ISSCC), pp 218–219, San Francisco, CA, Feb 2004
24. Liu W, Vichienchom K, Clements M, DeMarco CS, Hughes C, McGucken E, Humayun SM, de Juan E, Weiland DJ, Greenberg R (2000). A neuro-stimulus chip with telemetry unit for retinal prosthetic device. IEEE J. Solid-State Circuits 35:1487–1497, Oct 2000
25. Maass W (1999). In: Bishop MC (ed). Pulsed neural networks. The MIT Press, Cambridge, MA
26. Research Report by Ministry of Health, Labour and Welfare of Japan, 2005
27. Mortimer TJ (1999) Electrical excitation of nerve, chapter 3. Prentice-Hall, Inc., Englewood Cliffs,NJ
28. Ng DC, Furumiya T, Yasuoka K, Uehara A, Kagawa K, Tokuda T, Nunoshita M, Ohta J (2006) Pulse frequency modulation-based cmos image sensor for subretinal stimulation. IEEE Trans Circuits Sys II, 53(6):487–491, June 2006
29. Nicholls GJ, Martin RA, Wallace GB, Fuchs AP (2001). From neuro to brain, 4th edn. Sinauer Associates, Inc., Sunderland, MA
30. Normann RA, Greger BA, Paul House, Romero SF, Fernandez FPE (2009) Toward the development of a cortically based visual neuroprosthesis. J Neural Eng 6(2):1–8
31. Ohta J (2007) Smart CMOS image sensors and applications. CRC Press, Boca Raton, FL
32. Ohta J, Tokuda T, Hiyama K, Sawamura S, Sasagawa K, Nishida K, Kitaguchi Y, Kamei M, Fujikado T, Tano Y (2009). Retinal stimulator embedded with light-sensing function in distributed microchip architecture for subretinal implantation. In: International Image Sensor Workshop, Bergen, Norway, June 2009

33. Ohta J, Tokuda T, Kagawa K, Furumiya T, Uehara A, Terasawa Y, Ozawa M, Fujikado T, Tano Y (2006). Silicon LSI-based smart stimulators for retinal prosthesis. IEEE Eng Med Biol Mag 25(5):47–59, Oct 2006
34. Ohta J, Tokuda T, Kagawa K, Sugitani S, Taniyama M, Uehara A, Terasawa Y, Nakauchi K, Fujikado T, Tano Y (2007). Laboratory investigation of microelectronics-based stimulators for large-scale suprachroidal transretinal stimulation (STS). Neural J Eng.,4(1): S85–S91
35. Ohta J, Tokuda T, Kagawa K, Terasawa Y, Ozawa M, Fujikado T, Tano Y (2007) Large-scale integration-based stimulus electrodes for retinal prosthesis. Springer, New York, NY
36. Ortmanns M, Rocke A, Gehrke M, Tiedtke H-J (2007) A 232-channel epiretinal stimulator ASIC. IEEE J Solid-State Circuits 42(12):2946–2959, Dec 2007
37. Palanker D, Huie P, Vankov A, Asher A, Baccus S (2005). Towards high-resolution optoelectronic retinal prosthesis. BIOS, 5688A
38. Robblee SL Rose LT (1990) Electrochemical guidelines for selection of protocols and electrode materials for neural stimulation, chapter 2. Prentice-Hall, Inc., Englewood Cliffs, NJ
39. Roessler G, Laube T, Brockmann C, Kirschkamp T, Mazinani B, Goertz M, Koch C, Krisch I, Sellhaus B, Trieu KH, Weis J, Bornfeld N, Ro"othgen H, Messner A (2009) In: Mokwa W, Walter P (eds). Implantation and explantation of a wireless epiretinal retina implant device: observations during the EPIRET3 Prospective Clinical Trial Invest Ophthalmol Visual Sci 50(6):3003–3008, June 2009
40. Rothermell A, Liu L, Aryan PN, Fischer M, Wuenschmann J, Kibbel S, Harscher A (2009). A CMOS chipwith active pixel arrya and spceific test features for subretinal implantation. IEEE J. Solid-State Circuits, 44(1):290–300, Jan 2009
41. Rothermell A, Wieczorek V, Liu L, Stett A, Gerhardt M, Harscher A, Kibbel S (2008) A 1600-pixel subretinal chip with dc-free terminals and ±2v supply optimized for long lifetime and high stimulation efficiency. In Digest Technical Papers Internationall Solid-State Circuits Conference (ISSCC), San Francisco, CA, Feb 2008
42. Sakaguchi H, Kamei M, Fujikado T, Yonezawa E, Ozawa M, Cecilia-Gonzalez C, Ustariz-Gonzalez O, Quiroz-Mercado H, Tano Y (2008) Artificial vision by direct optic nerve electrode (AV-DONE) for a blind patient with retinitis pigmentosa. Invest Ophthalmol Visual Sci 49:S4044
43. Santos A, Humayun SM, de Juan, Greenburg JR, Marsh JM, Klock BI, Milam AH (1996) Preservation of the inner retina in retinitis pigmentosa; a morphometric analysis. Arc. Ophthalmol 114:40–46, 1996
44. Schwarz M, Hauschild R, Hosticka JB, Huppertz J, Kneip T, Kolnsberg S, Ewe L, Trieu KH (1999) Single-Chip CMOS Image sensors for a retina implant system. IEEE Trans.Circuits Sys II, 46(7):870–877, July 1999
45. Suaning JG, Lovell HN (2001) CMOS neurostimulation ASIC with 100 channels, scaleable output, and bidirectional radio-frequency telemetry. IEEE Trans Biomed Eng 48(2):248–260, Feb 2001
46. Tanaka T, Sato K, Komiya K, Kobayashi T, Watanabe T, Fukushima T, Tomita H, Kurino H, Tamai M, Koyanagi M (2007) Fully implantable retinal prosthesis chip with photodetector and stimulus current generator. In: Technology in Digest International Electron Devices Meeting (IEDM), pp 1015–1018, Washington, DC, December 2007
47. Terasawa Y, Tashiro H, Uehara A, Saito T, Ozawa M, Tokuda T, Ohta J (2006) The development of a multichnanel electrode array for retinal prosthesis. J Artif Organs 9(4):263–266, 2006
48. Terasawa Y, Uehara A, Yonezawa E, Saitoh T, Shodo K, Ozawa M, Tano Y, Ohta J (2008). A visual prosthesis with 100 electrodes featuring wireless signals and wireless power transmission. IEICE Electronics Express, 5(15):574–580, 2008
49. Theogarajan SL (2008). A low-power fully implantable 15-channel retinal stimulator chip. IEEE J. Solid-State Circuits, 43(10):2322–2377, Oct 2008

50. Tokuda T, Asano R, Sugitani S, Taniyama M, Terasawa Y, Nunoshita M, Nakauchi K, Fujikado T, Tano Y, Ohta J (2008) Retinal stimulation on rabbit using CMOS-based multi-chip flexible stimulator toward retinal prosthesis. Jpn J Appl Phys 47(4B):3220–3225, April 2008
51. Tokuda T, Hiyama K, Sawamura S, Sasagawa K, Terasawa Y, Nishida K, Kitaguchi Y, Fujikado T, Tano Y, Ohta J (2009) CMOS-based multichip networked flexible retinal stimulator designed for image-based retinal prosthesis. IEEE Trans Electron Devices 56(11):2577–2585
52. Tokuda T, Pan YL, Uehara A, Kagawa K, Nunoshita M, Ohta J (2009) Flexible and extendible neural interface device based on cooperative multi-chip CMOS LSI architecture. Sensors & Actuators A, 122(1):88–98, July 2005
53. Tokuda T, Sugitani S, Taniyama M, Uehara A, Terasawa Y, Kagawa K, Nunoshita M, Tano Y, Ohta J (2007) Fabrication and validation of a multi-chip neural stimulator for in vivo experiments toward retinal prosthesis. Jpn J Appl Phys 46(4B):2792–2798, April 2007
54. Veraart C, Wanet-Defalque MC, Gerard B, Vanlierde A, Delbeke J (2003) Pattern recognition with the optic nerve visual prosthesis. Artif Organs, 11:996–1004, 2003
55. Wandell AB (1995) Foundations of vision. Sinauer Associates, Inc., Sunderland, MA
56. http://www.who.int/mediacentre/factsheets/fs282/en/index.html
57. Wise DK, Anderson JD, Hetke FJ, Kipke RD, Najafi K (2004) Wireless implantable microsystems:high-density electronic interfaces to the nervous system. IEEE Proc 92(1): 76–97, Jan 2004
58. Zrenner E (2002) Will retinal implants restore vision? Science 295:1022–1025, Feb 2002

Index

A

AC coupling, 138, 143, 319, 463, 475
Acknowledgment packet (ACK), 345
Acoustic telemetry, 225
Acoustic wave, 220, 224–225
Action potential measurement, 127
Active pixel sensor (APS), 492, 494, 496–498
Acute capture threshold, 426
Adaptive dynamic range optimization (ADRO), 374
Adaptive modulation, 477
Adaptive noise reduction, 387
Adaptive-SNR technique, 388, 391–392
Adaptive stimulation, 453, 463, 476, 479
Adaptive threshold rectifier (ATR), 285, 354–355, 358–359, 362–363, 368
ADC, *see* Analog-to-digital converter (ADC)
ADSL, 246
Advanced Bionics Corp, 101, 413
Advanced Cochlear Systems, 413
AES-128 algorithm, 208
AES enable signal (AESEn), 206
AFM lithography, 102
Age-related macular degeneration (AMD), 482
Alternate-mark-inversion (AMI), 241
ALU processing, 205
Ambulatory monitoring, 59, 68, 76, 117, 191, 294, 346
American Heart Association, 69
American National Standards Institute, 372
Americans with Disabilities Act, 371
Amplitude shift keying (ASK), 227, 231, 241–246, 259, 263, 266, 268, 272, 344–345, 355–356, 362
AM *vs*. FM, 235
Analog front-end (AFE), 304
Analog modulation, 229, 235
Analog multiple channel modulation, 238
Analog-to-digital converter (ADC), 3, 5, 8, 153, 157–189, 239, 251, 284, 333, 342, 355–356, 361–362, 379, 382, 385, 388, 391–394, 396–397, 406, 412, 432–433, 439, 443
  comparator-based asynchronous binary search, 172–182
    A/D-converter timing, 176
    calibration, 181–182
    clock generation, 176
    embedded threshold, 176–178
    encoding, 176–178
    feedback D/A converter, 180–181
    operating principle, 173–175
    passive track-and-hold, 178–180
    power breakdown, 182
  specifications, 158–161
    ideal, 158–159
    implementation, 161
    practical, 159–160
Analog *vs*. digital modulation, 247
Anisotropic magneto-resistive (AMR), 440
Anodic first (AF) current, 489
Antenna efficiency, 254–255
Anti-aliasing filter, 399
Antitachycardia pacing (ATP), 445
Arrhythmia, 24, 191–192, 195, 197, 202–203, 212, 282, 284, 339, 444–445
Arrhythmia detection, 192, 202, 444
Artifact problems, 32, 72–73, 77, 109
Artificial retina, 481–511
  basic circuits, 487–505
    active pixel sensor, 496–498
    log sensor, 498–500
    micro PD array, 495–496
    photosensor array, 494–504
    photosensor, 491–494
    power and data transmission, 504–505
    pulse frequency modulation, 500–504

H.-J. Yoo, C. van Hoof (eds.), *Bio-Medical CMOS ICs*,
Integrated Circuits and Systems, DOI 10.1007/978-1-4419-6597-4,

Artificial retina (*cont.*)
stimulation of retinal cells, 488
stimulator, 488–491
case studies, 505–511
multiple microchip architecture, 505–510
multiple microchip-based retinal stimulator, 510–511
fundamentals for, 482–487
classification of, 484–485
extraocular artificial retina, 484
intraocular artificial retina, 484–485
principle of, 482–483
retina and blindness, 482
intraocular, 484
ASIC, 151–153, 286, 291, 342
ASK, *see* Amplitude shift keying (ASK)
Association for the Advancement of Medical Instrumentation (AAMI), 107–108
Asystole, 203
Atrial fibrillation, 439, 445
Attenuation, 32
Augmented electrode, 25
Automated external defibrillator (AED), 433
Automatic gain control (AGC), 388–390
Automatic repeat request (ARQ), 252
Autonomic nervous system, 293–294, 298
Axbo® system, 291
AZTEC algorithm, 198

B

Backscattering, 356
Bandpass filter, 141
Band-pass operational spectrum, 333
BAN, *see* Body area network (BAN)
Barkhausen criterion, 233
Base line drift, 32
Baseline wonder, 32
Beat detection algorithm, 287
BER, *see* Bit error rate (BER)
Beta-voltaic effects, 429
Beta wave, 26
Bias current tolerance, 111
Bigeminy, 203
Binary search algorithm, 167, 181
Bio-ceramic, 102
Biocompatibility, 32–33, 43, 48, 51–53, 253
Bio-compatible package, 300
Biocompatible polymer, 102
Bioelectricity, 13–29
biomedical device, 24–28
electrocardiography, 24–25
electroencephalography, 25–27
electromyography, 27–28
electrical properties, 14–23
cell membrane, 14–15
equivalent circuit model, 17
membrane potential, 15–19
synaptic transmission, 21–23
equivalent circuit model, 23–24
research trends, 28–29
Biomarkers, 477
Bio-medical CMOS IC, 3–6
architecture of, 3–6
categories of, 3
Biomedical electrodes
electrode design, 66–71
electrode-skin interface, 34–66
electrode-electrolyte interface, 34–54
skin, 54–66
electrode standards, 107–117
biosignal monitoring electrodes, 107–112
stimulation electrodes, 113–117
implant electrodes, 91–107
electrode designs, 96–101
microelectrodes, 101–107
modern disposable electrodes, 71–74
solid conductive adhesive electrodes, 74–76
wearble electrode, 76–91
external electrostimulation electrodes, 78–91
Biomedical telemetry, 220–228, 259–272
commercial biomedical transmitter, 270–272
Nordic NR24L01+, 272
Zarlink ZL70101, 270–272
Zarlink ZL70250, 272
conduction, 225
monitoring and identification, 269–270
portable heart rate monitoring, 269
RFID, 269
near-field coupling, 225–227
near-field *vs.* far-field, 227–228
nerve implant monitoring, 268–269
cochlear implants, 268
retinal prosthesis, 269
orthopedic implant monitoring, 263–268
distraction nail driver, 263–264
hip prosthesis fixation, 264–266
telemetry IC design, 266–268
physiological monitoring, 259–263
bladder pressure, 259
pacemaker monitoring, 261–262
textile integrated breathing, 261

wireless capsule endoscopy, 263
wireless ECG, 259–261
wireless endoscopy, 263
wave propagation, 220–225
acoustic wave, 224–225
EM wave, 221–224
Biomedical transmitter, commercial, 270
Bio-medicine, 2
Biopotential acquisition, 125–132, 137–138, 151, 153
biopotential electrodes, 127–129
biopotential signals, 126–127
examples of ICS, 151–153
interference the, 129–131
noise considerations, 131–132
Biopotential electrode, 127, 127–129, 131, 138, 142–144, 150, 350
Biopotential fiber sensor (BFS), 350
Biopotential monitoring, 127, 132, 135, 143, 150, 209
Biopotential signal, 125–126
Biosignal distortion, 32–33
Biosignal monitoring, 32, 52, 62, 64–65, 75, 78, 80, 88, 92, 108, 117
Biotelemetry, 219–220, 224, 226–227, 231, 233, 235, 242–243, 246–247, 249–250, 253–254, 257–258
Biphasic waveform, 48, 430
Bipolar disorder, 476
Bipolar return-to-zero (BRZ), 241
Bisping, 96–97
Bisping transvenous screw, 96–97
Bit error rate (BER), 240, 246, 251, 253, 317–318, 329, 346
Bladder pressure monitoring, 259
Block code error correction, 253
Blocking electrode, 43
Blood glucose tester, 29
Bluetooth, 3, 220, 290, 303–304, 332, 334, 340–341, 357
Bode plot, 40–41, 44
Body area network (BAN), 3, 5–6, 8, 225, 280, 282, 291, 295, 297, 303–304, 341, 344
Body Channel Communication (BCC), 3, 5, 303–335
channel characteristics, 305–308
human body, 304–305
system operation demonstration, 331–335
design architecture, 332–333
realization, 333–335
WBS receiver AFE, 326–327
WBS transceiver, 327–331
wideband signaling communication, 308–314
wideband signaling transceiver, 314–325
all-digital quadratic CDR circuit, 322–324
direct digital transmitter, 324–325
WBS receiver AFE, 319–322
Body sensor network (BSN), 1, 284, 339–343, 347–349, 354, 363–366, 368
Bodywire channel, 304, 307, 317, 330, 335
Boltzman constant, 16, 492
Bootstrapping, 179
BPSK, 243–244, 246, 259, 261–262, 269
Bradcardia, 203
Brain-computer interfaces (BCI), 8
Brain-machine interface, 118
BSN, *see* Body sensor network (BSN)
Butler-Volmer equation, 39, 50–51

C
CABS, *see* Comparator-based asynchronous binary search (CABS)
Calibration algorithm, 181, 184
Capacitive feedback amplifier, 142
Capacitively coupled dispersive electrode, 117
Capacitor reformation, 430
Capacitor stacking, 460
Capillary electrometer, 66–67
Capsule endoscopy, 4, 248, 263
Capture threshold, 101, 426, 434
Carbon-filled silicone rubber electrode, 89
Cardiac electrode, 92
Cardiac mapping electrode, 77
Cardiac pacing, 31, 48, 51, 63, 81, 83–85, 91, 93–94, 96–97, 117–118
Cardiac Society of Great Britain, 69
Cardinal Health, 290
Cardionet, 284
Cardio-stimulating drugs, 93
Carrier frequency selection, 253–258
antenna size *vs*. bandwidth, 256–257
regulations *vs*. bandwidth, 258
tissue absorption *vs*. antenna size, 253–255
Carson's rule, 234
Cathodic first (CF) current, 489
Chardack's differential current density, 106
Chardack's electrode, 95
Charge balanced waveform, 48
Charge coupled device (CCD), 497
Charge-sharing successive approximation, 157, 161
Charge transfer mechanism, 32
Charge transfer resistance, 50

Chemical injury, 47
Chip micrograph, 183, 326, 348, 362, 404
Chip package, 300
Chopper modulating amplifier, 140, 142
Chopper modulation, 139, 141
Chopping frequency, 141, 361
Chopping ripple, 142
Chronotropic incompetence, 439
Circuit phasing, 468
Classification processing, 202–206
  ECG classification algorithm, 202–203
  RISC, micro architecture, 203–206
Clear-channel assessment (CCA), 441
clkSAR, 176
Clock gating, 213, 448
Clock generator design, 397
Clock recovery, 242, 268, 316–317, 334
Closed loop feedback algorithm, 412
Closed loop healthcare, 5
Closed loop therapy, 477–478
CMOS IC technology, 3
CMOS image sensor, 492, 494, 496–498, 500–501
  architecture, 498
CMOS processing, 102
CMOS technology, 6, 127, 161, 173, 304, 326–327, 347, 362, 388, 404, 413, 507
CMRR instrumentation amplifiers, 131
Cochlear Corporation, 413
Cochlear implant (CI), 5, 8, 91, 101, 226, 248, 252, 268, 413–417
  anatomy of, 416
  design of, 415–416
  electric acoustic stimulation (EAS), 417
  future of, 417
  hybrid, 417
Code center, 159
Coding performance, 201
Colpitts oscillator, 233, 266
Common-mode gain, 136
Comparator-based asynchronous binary search (CABS), 172–187, 189
Comparator noise, 170
Complex impedance, 39, 41, 44, 58, 224
Compression enable signal (COMPEn), 206
Compression ratio (CR), 197–199, 201, 207, 212, 249–251, 346, 389
Conductive electrode, 83, 90, 116–117
Constant phase angle impedance, 41
Continua Healthcare Consortium, 2
Conventional rectifier, 357
Corkscrew electrode, 95
Corneocytes, 55
Correction algorithm, 170–171
CORTES algorithm, 198
Corventis, 282, 284
Coulombic adsorption, 37
Coulombic efficiency, 459
Coupling capacitor, 130, 143, 455, 463, 474
Coupling gel, 225
Coupling scheme, 138, 304
Crossbow, 281
Cross-coupled CMOS switches, 141
Crosstalk, 239
Cuff electrode, 105–106
Current balancing instrumentation amplifier, 144, 148–151
Current density consideration, 82
Current dispersing, 85, 90
Current mode amplifiers, 147–151
  chopper modulated current balancing, 150–151
  closed-loop current mode, 148–149
  open-loop current mode, 147–148
Current response, 46
Currents of linearity, 49
Current steering, 471–472
Curvus, 284
Custom-built electrode tester, 112
Cyberonics, 106
Cytoplasm, 14, 17, 21, 23

D

Data compression, 248–251
  loss-less compression algorithm, 249–250
  lossy compression algorithm, 250–251
Decoupling capacitance, 183
Deep brain stimulation (DBS), 8, 91, 151, 454, 476–477
Defibrillation, 31, 33, 54, 82–84, 86–87, 91, 108, 110, 113–114, 117, 424, 430–431, 433, 443, 445, 474
  electrode, 80, 82
  recovery, 110, 114
  threshold (DFT), 430
  waveform, 430
Delta coding, 192, 199, 206
Delta waves, 27
Demodulation, 227
Depolarization, 18, 20–21, 42, 422
D-flipflops resampling, 188
Diabetic care, 7
Diamond like carbon (DLC), 102, 106
Differential current density (DCD), 99–100, 107

Differential encoding, 242
Differential non-linearity (DNL), 159, 168, 171–172, 184–185
Differential PSK (DPSK), 244
Different pulse modulation encoding, 236
Digital audio streaming, touch-based, 332
Digital band-pass filters, 152
Digital hearing aid, 371–404
  acoustic responses, 379
  architecture of, 376
  autonomous internal gain verification, 380
  chip with on chip, 374–386
    internal gain verification algorithm, 376–377
    low power analog front-end, 387–397
    low power digital back-end, 397–404
    multi mode audio processor, 384–386
  design issues, 374
  ear simulator, 375
  future of, 372–373
  population trends, 371–372
  types of, 373–374
    behind the ear type (BTE), 373
    completely-in-canal (CIC), 374
    in the canal type (ITC), 373
    in the ear type (ITE), 373
Digitally-controlled oscillator (DCO), 318
Digital PM phase shift keying (PSK), 242
Digital pulse modulation encoding, 239
Digital signal processing (DSP), 246, 397
Digital-to-analog converter (DAC), 245, 333, 466
Direct-coupled interface (DCI), 304
Direct inductive coupling, 456
Direct signal compression, 198
  See also Lossy algorithm
Discrete cosine transform (DCT), 250–251
Discrete Fourier transform (DFT), 246
Distortion, 32, 54, 108, 140–142, 159–160, 171, 179, 236, 239–240, 340, 389, 393, 397
Distortion free-monitoring, 54
DNL, *see* Differential non-linearity (DNL)
Double balanced modulator, 243
Double sampling technique, 137
Double-sideband suppressed-carrier (DSB-SC), 229–230, 235, 239
Drain-induced barrier lowering (DIBL), 447
Drain-to-source current, 132
Driven-right-leg (DRL), 131
Dry electrode, 60, 299–300, 343, 350, 354, 363
DSSS technique, 335
Dynamic structure variation technique, 391

E
ECG analysis algorithms, 287
ECG classification algorithm, 202–203
ECG, *see* Electrocardiogram (ECG)
ECG signal compression algorithm, 198
ECG signal processor, 192–194
  algorithm, 192–193
  architecture, 193
  hardware implementation, 193–194
ECoG, *see* Electrocorticogram (ECoG)
Edge effect, 85
EEGASIC, 152
EEG, *see* Electroencephalography (EEG)
Effective number of bits (ENOB), 160
Effective resolution bandwidth (ERBW), 160–161, 171, 179, 185–187
E-healthcare, 1–2
E-hospital, 1
Elective replacement interval (ERI), 442–443
Electrical coupling, 5
Electrical evoked potential (EEP), 508
Electrical impedance tomography (EIT), 61
Electrocardiogram (ECG), 24, 31, 127, 259, 341, 350, 354, 422
Electrocardiography recording, Leads in, 25
Electrocautery transients, 474
Electrocorticogram (ECoG), 26, 126, 128, 132, 134, 138–139, 141, 143, 146, 151
Electrode contact impedance, 116
Electrode gels, 60
Electrode-gel-skin system, 56
Electrode impedance, 351
Electrode metal, 51
Electrode sinks, scaling considerations, 468
Electrode-skin impedance, 33, 59, 62
Electroencephalography (EEG), 4, 13, 25–27, 29, 70–71, 96, 117, 126–128, 132–133, 138–139, 141, 145–147, 151–153, 225, 239, 247–248, 290–291, 350
Electrolysis of water, 44
Electromagnetic interference (EMI), 129, 224
Electromotive force (EMF), 129
Electromyogram (EMG), 127, 350
Electromyography, 13, 27–28
Electrooculogram (EOG), 350
Electropuncture, 79, 92–93
Electrostatic coupling, 304, 307
Electrostatic generator, 78–79
Electrostatic interference, 129
Electrostatic transducers, 357
Electrostimulation electrode, 54, 80, 82, 88
Electrosurgery, 78, 83–84, 108, 117, 478

Elema-Schonander, 423
Elimination technique, 398, 413
Emotion monitoring, 293
EM waves, 8, 220–222, 224–225, 255, 259
Encryption key register (EKIR), 205
End of service (EOS), 429, 442–443
End of the block (EOB), 207
Energy scavenging, 450
Epi-retinal stimulation, 481–482, 484
ERBW, *see* Effective resolution bandwidth (ERBW)
Error correction, 240, 251–253, 272
 block codes, 252–253
 convolutional codes, 253
Escape interval, 439
Exponential gain control (EGC), 388
External gain fitting verification, 381
Extracellular fluid (ECF), 23
Extracellular interstitial, 15
Extracellular matrix, 14
Extraocular retinal prosthesis, 484

F
Fabric area network (FAN), 342
False negative (FN), 198
False positive (FP), 198
FAN algorithm, 198
Faradaic current, 79, 488
Farado-puncture, 93
Feature register (FTR), 205
Feedback D/A converter, 173–175, 180–183, 187
Feedback loop, 131, 138, 148–149, 151, 232, 360, 461
Feedback oscillation, 379, 382
Feedback reduction algorithm, 382
Feedback transconductance, 149
Ferroelectric capacitor-based rectifier, 358
Fick's law, 15
FIFO, 207
Figure of merit (FoM), 161, 171
Filterbank, 397–399, 401
 coefficients, 401
 structure, 399
Filter coefficients, 400
Filtering, 32, 138, 143, 150–151, 192–194, 198, 230, 432–433, 440, 443
Finite impulse response (FIR), 195
Fisher mapping, 295, 297
Flicker noise, 127, 132–134, 137–139, 141–143, 147, 150–151, 153, 323
Flip-chip bonding technology, 507
Focused ion beam patterning, 102
Food and Drug Administration (FDA), 413, 424
Forward error correction (FEC), 252, 270
Forward reaction, 48
Fourier transform, 199
Fraunhofer, 227, 290
Frequency division multiplexing (FDM), 238, 246
Frequency-scanning power meter, 234
Frequency shift keying (FSK), 241–246, 257, 262–264, 269, 272, 304
Functional electrical stimulation (FES), 19, 226
Functional gain verification, 375, 377

G
Gain amplifier, 135
Gain-bandwidth, 147–148, 312
Gain variation, 469
Gain verification algorithm, 376–377, 382–383, 404, 410–411
Galvanism, 78
Galvano-puncture, 93
Gap junctions, 22
Garment electrode, 91
Gastro-intestinal (GI) tract, 7, 219, 263
Gastro-intestinal investigation, 248
Gates toggling, 448
Gaussian distribution, 206
General purpose register file (GPR), 205
Giant magneto-resistive (GMR), 440
Gilbert cell, 243
Goldmann equation, 16, 20
Graphite-loaded polyester, 54

H
Hand-held device, 29
Hand-held electrodes, 79, 81
Hartley oscillator, 233
Hartley-Shannon law, 315
H-bridge circuit, 404
Headroom, 143–145, 147, 375, 460–461, 465, 467
Hearing aid, *see* Digital hearing aid
Heart anatomy, 421–423
Heart rate monitoring, portable, 269
Helmholtz capacitance, 488
Heterogeneous $\Sigma\Delta$ DAC, 402–403
 architecture of, 402
 interpolation filter design, 403
 principle of, 403
High-pass filter, 138, 150–151, 309
Hip prosthesis fixation, 264
Holter system, 199

Howling, 379, 391
See also Feedback oscillation
Huffman coding, 192–193, 199, 206–207, 249–250
Human body communication (HBC), 304, 332
Hunter-Roth intramyocardial electrode, 94
Hydrating, 64, 75, 89
Hydration, 38, 55–56, 60–61
Hydrogel, 60, 63–64, 74–75, 90
Hyperpolarization, 20–21
Hypnogram, 291–293
Hysteresis comparator, 396–397

I
ICD, *see* Implantable cardioverter defibrillator (ICD)
Impedance mismatch, 33
Impedance tomography, 33
Implantable cardioverter defibrillator (ICD), 5, 8, 262, 272, 421, 423, 424, 426–430, 441, 443–447, 474
Implantable neurostimulator, 454
Implantable pulse generator (IPG), 455
Induction telemetry, 226
Industrial, Scientific and Medical (ISM), 258
Infinite impulse response (IIR), 195
Information lesion, 477
INL, *see* Integral non-linearity (INL)
Insertion gain verification, 375, 377
Instrumentation amplifier (IA), 5, 125–127, 129–138, 140, 142–153, 286, 342, 355, 359, 362, 368
Instrumentation amplifier topology, 132–135
Integral non-linearity (INL), 159, 168–169, 171–172, 184–185
Inter-cellular communication, 23
Interfacial electrochemistry, 53
Interference theory, 129
Internal feedback verification, 380–381
Internal gain verification algorithm, 377, 381, 404, 407, 409–411
International Commission on Non-Ionizing Radiation Protection (ICNIRP), 219, 308
International Standards Organization (ISO), 424
Interpolated finite impulse response (IFIR), 397–398, 412
Inter-symbol interference (ISI), 317
Inverse discrete Fourier transform (IDFT), 246
In vivo experiment, 508, 510
IQ modulation, 245–246
IR drop, 465, 490

J
Jesuit missionaries, 92

K
Karaya gum, 63, 74
Karhunen-Loeve transform, 199

L
Large signal impedance, 113
Leadlock, 73
Least significant bit LSB, 158
Lempel Ziv algorithm, 250
Lie detector, 58
Life Shirt, 77
LifeWatch, 284
Limit voltages, 49, 51
Linear discriminant classification, 295
Linearity
limit current of, 49
limit voltage of, 51
Line encoding, 240–241, 268
Line-of-sight (LoS), 341
Link design parameters, 310
Link parameters, design specification of, 316
Local field potential (LFP), 126
Localized electrization, 79–80
Log sensor, 498, 500
Look-up table (LUT), 252
Loss-less compression, 248–250
Lossy algorithm, 198, 249–250
Low energy technique, 209–214
clock gating, 212
heterogeneous processor integration, 209
low supply voltage operation, 209–210
on-chip memory reduction, 212–214
segmentation-based pipelined, 211–212
Lower-power BAN radio, 5
Low-pass filter (199), 139–140, 143, 195, 323–324, 402–403, 433
LZW algorithm, 198

M
Magnetic resonance (MR) scanning, 461
Malleable electrode, 81
Manchester encoding, 242
Matlab, 256
Mean deviation (MD), 196
Med-El Corporation, 413
Media access controller (MAC), 270
Medical Device Radiocommunications Services, 258

Medical Implant Communication Services (MICS), 224, 248–249, 258, 270, 272, 344, 354–355, 359, 362–363, 441–442
Medtronic EDGE^{TM}, 87
MEMS processing, 102
Metal-electrolyte interface, 42, 106
Metal plate electrode, 68, 80–81
Michigan Probe, 102–103
Microelectrode, 101
Micro-fabrication technique, 102, 106
Micro PD array, 495
Micro photodiode array (MPDA), 492, 495
MICS, *see* Medical Implant Communication Services (MICS)
Miniaturization, 7, 219, 253, 282, 299, 398
Minimum-shift key (MSK), 263
Miro wireless endoscopic capsule, 225
MIT Media Lab, 294
Mobile Cardiac Outpatient Telemetry^{TM} (MCOT^{TM}), 284
Modulation methods, 228–248
  analog modulation, 229–235
    AM, 229–231
    discussion on, 235
    FM and PM, 231–235
  analog pulse modulation, 235–239
    analog multiple channel, 238–239
    PAM, 236
    PFM, 237–238
    PPM, 237
    PWM or PDM, 237
  data rates, 247–248
  digital modulation, 242–246
    ASK, 243
    digital multiple channel, 244–246
    FSK, 243
    PSK, 243–244
  digital pulse modulation, 239–242
    line, 240–242
    pulse code, 239–240
Monitor biosignal, 33, 52
Monophasic anodic pulse, 48
Moores Law, 3
MOSFET, 498, 500
MOS resistive circuit (MRC), 394, 390
Motion artefact, 59–60, 65
Motion-responsive pacing, 477
Motor points, 79, 83
Multi-disciplinary approaches, 29
Multielectrode silicon probe, 103
Multi mode audio processor, 384
Multiple bit modulation, 244
Multiplxser (MUX), 317, 486
Multi-point electrode, 69
Multi-site pacing, 428
Myopotential signal, 432

N

National Acoustics Laboratory, 406
NEF, *see* Noise efficiency factor (NEF)
Nemesysco, 294
Nernst equation, 16, 36, 53, 59
Nerve implant monitoring, 268
Nested chopper, 141, 354, 356, 360, 363
Neural probes, 126–129, 138, 249
Neuronal circuit, 476, 479
Neurostimulation, 453–479
  challenges and system requirement, 454–456
  energy transfer circuit, 456–463
    secondary cell recharge, 457–459
    energy source characteristics, 459–460
    voltage boosting, 460–463
    stimulation signal, 463–473
  tissue interface, 473–475
    closed-loop, adaptive stimulation, 476–477
    optogenetic neuromodulation, 477–479
Neurostimulator, 453, 476–477, 479
Neurotransmitter release, 22–23
Nike+, 2
Nintendo Wii Fit, 2
Noble metals, 43, 48, 51–52
Noise efficiency factor (NEF), 132, 136–137, 144–145, 147, 149, 152
Noise oscillator, 257
Nonrapid-eye-movement (NREM), 27
Non-return to zero (NRZ), 240–241, 316–317
Nyquist converter, 160
Nyquist frequency, 137, 160–161, 185, 433

O

Obsessive compulsive disorder, 476
Offset instability, 109, 113
Offset voltage, 33, 36, 109–110, 114, 141, 152, 396
Ohmic resistance, 254–255, 257, 437
Ohm's law, 16, 42
On-off keying (OOK), 224, 243
Operational transconductance amplifier (OTA), 138–139, 143, 391, 395
Opsins, 478
Optical wave propagation, 224
Optogenetic neuromodulation, 477–478
Optogenetic stimulation, 453, 476, 478–479

Orthogonal frequency multiplexing (OFDM), 245–246
Orthopedic monitoring, 266
Oscilloscope screen, 32
OTA, *see* Operational transconductance amplifier (OTA)
Outer Helmholtz plane (OHP), 37

P

Pacemaker, 8, 92, 94, 101, 251, 261–262, 272, 421–450
  artificial, 93
  basic ICD functions, 443–446
  communication principle, 262
  components, 424–431
    battery, 428–430
    case, feed through and header, 427–428
    device programmer, 426–427
    device subsystems, 427
    ICD capacitor, 430–431
    leads, 424–426
  electronics, 431–443
    ADC, 433–434
    basic functions, 431–432
    clock generator and power management, 442–443
    MCU, 438–439
    pace driver and Mux, 434–438
    sensing circuits, 432–433
    sensor I/O, 439–440
    telemetry, 440–442
  IC process, 446–449
    low power design, 447–449
    process technology, 446–447
  implantable cardioverter defibrillators, 423–424
  modern, 95, 426, 433, 438–440
  monitoring, 261
  refractory period, 439
Pacing equipment, 81
Pacing impulse, 93
Pacing pulse, 425–426, 428, 430–432, 435, 437–439, 445
Pacing system analyzer (PSA), 426
Paradoxical sleep, 27
Parkinson's disease, 91, 101, 476
Passive metals, 51
Passive recharge, 455, 465
Peak detector (PD), 385
Peak-to-peak differential, 188
Percentage root mean square difference (PRD), 201
Performance-degrading effect, 169
Perimeter effect, 85
  See also Edge effect
Personal area network (PAN), 3
Personalized health care, 76
P-FCD, *see* Planar-fashionable circuit board (P-FCB)
PFM, *see* Pulse frequency modulation (PFM)
Phase-locked firing, 476
Phase-locked-loop (PLL), 230, 232, 243, 246, 263, 323, 442
Photodiode (PD), 491, 494
Photo-lithography, 102
Photoreceptor, 482, 485, 491–492, 500, 511
Photosensor, 491–492, 494–495, 500, 503, 509
Physiological parameter, 7, 78, 219, 239
Piezoelectric harvester, 357
Planar-fashionable circuit board (P-FCB), 284–285, 349–355, 359, 361, 363–364, 368
Planck constant, 492
Plasma enhanced chemical vapour deposition (PECVD), 106
PLL, *see* Phase-locked-loop (PLL)
PMOS devices, 179
Point-to-multipoint communication, 335
Polarization, 42–43, 47, 51–52, 99, 110, 127, 138, 142–145, 147, 150–151, 153, 425, 437
Polarization voltage, 127, 138, 142–145, 147, 150–151, 153
Polysomnography test, 290
Post-processor, 206–208
  AES-128, 208
  Huffman coding, 206–207
Potential measurement, 35, 110, 134
Power consuming-decoding algorithm, 253
Power efficient instrumentation, 135–147
  chopper modulating amplifiers, 142–144
  chopper modulation, 139–142
  noise-efficiency factor, 144
  off-the-shelf amplifier, 135–137
  pseudo resistors, 137–139
Power spectral density (PSD), 140, 159–160, 363–364, 368
PPM, *see* Pulse position modulation (PPM)
PQRST wave, 195
PRBS, *see* Pseudo random binary sequence (PRBS)
Preamplifier gain, 361, 378, 394
Pregelled electrode, 82, 112
Prepping, 77

Pre processing, 194–202
ECG skeleton, 198–201
feature extraction, 195–198
filtering, 195
segmentation memory, 202
Pressure sensitive adhesive, 73
Programmable digital functions, 438
Programmable gain amplifier (PGA), 359
Programmable taps, 469–470
Propagation constant, 379
Proximity effect, 255
Pseudo capacitance, 41, 53, 56
Pseudo random binary sequence (PRBS), 317–318, 325–327, 329–330
Pseudo resistors, 137, 140, 145
Psychophysiological stimuli, 58
Pulse amplitude modulation (PAM), 236, 238
Pulse code modulation (PCM), 236, 239–240
Pulse frequency modulation (PFM), 236–238, 492, 494, 500–503
Pulse modulation, 228, 236, 239, 242, 251, 494
Pulse position modulation (PPM), 236–238, 262, 335
Pulse width modulation (PWM), 236–239, 262, 266, 268, 393, 404
P-wave, 195, 422, 433
PWM, *see* Pulse width modulation (PWM)

Q
Q factor, 256, 397
Quad level vector (QLV), 192, 194, 196–197, 199–200, 202, 207
Quadratic sampling, 323, 335
Quadratic sampling algorithm, 324–325
Quadratic sampling phase detector (QSPD), 323–325
Quantization, 249–250
Quantizer threshold, 158
Quantizing, 239, 251
Quasar, 281
Quinton Quick-Prep®, 65

R
Radiation resistance, 224, 254–255
Radio frequency identification (RFID), 226, 269
Radiofrequency (RF) telemetry, 219
Rapid-eye-movement (REM), 27, 292
Reactive ion etching (RIE), 102
Real-ear aided gain, 380
Real-ear occluded gain, 380
Real-time gain verification algorithm, 382–383
Real-time multimedia, 316, 333
Redo current response, 46
Redo voltage response, 45
Redox®, 65
Redundancy, 169, 248–249, 251–252
Reference current generator, 465–467, 472
Reference resistor, 465–466, 468–472
Refractory period, 20, 439, 444
Resistive electrode, 88
Retinal implants, 252, 453
Retinal prosthesis, 102, 238, 269, 481–482, 484–487, 490, 495, 504–505
Retinitis pigmentosa (RP), 482
Reverse reaction, 48
RF communication, 224, 441
RF telemetry, 449
RF wave propagation, 221
Ringer solution, 502
RISC, 193, 202–206, 212
Romeros algorithm, 287
R-on-T, 203
R peak detection, 197–198
R point, 195
R2R, 466
RR interval, 195, 197
Run-length-encoding (RLE), 249

S
Saltatory conduction, 21–22
Sampling capacitor, 134, 137, 166, 168–169, 180, 468
Sandwich electrode, 90
SAR algorithm, 162, 165–166, 169–170
SAR-CC quantization, 180
Scala tympani (ST), 415–416
Schmitt trigger, 317, 319–320, 326, 329
Schottky diode, 357, 458
Sense resistor, 467, 469, 471–472
Sensing algorithm, 444
Sensing capability, 428
Sensing resistor, 471
Sensium™, 281, 284, 342
Serial parallel interface (SPI), 194
SFDR, *see* Spurious-free dynamic range (SFDR)
Shannon theorem, 236, 247
Shimmer research, 281
Short pulse duration, 47–48
Sigma-delta converter, 434
Signal attenuation, 32, 108
Signal to noise and distortion ratio (SNDR), 159–160, 171, 185–187, 406
Signal to noise ratio (SNR), 131, 289, 316, 372, 410

Silicon microtechnology, 102
Single side band (SSB) modulation, 230
Sinks, 465–466, 473, 491
Sino-atrial node, 24
Sinusoidal wave, 306
Skeleton algorithm, 199, 202
Skeleton encoding, 194
Skeleton equation, 196
Skin abrasion, 63, 66, 111
Skin irritation, 59, 63, 66, 73–74, 77–78, 88–90, 117, 350–351, 354, 363
Skin potential, 59
Skin preparation technique, 65
Skin Rasp®, 65
Skin, schematic representation, 56
Skin stripping, 59, 65–66
Skipped beat, 203
Sleep disorder, 289–290
Sleep spindles, 27
SleepTracker®, 291
Slender wire electrode, 93
Small signal impedance, 113
SNDR, *see* Signal to noise and distortion ratio (SNDR)
S/N ratio, 232, 235, 237, 239–240, 242–244, 247, 251, 268
Social computing, 8
Spatial summation, 18–19
Specific adsorption, 38
Split-phase encoding, 242
Spreading resistance, 488
Spurious-free dynamic range (SFDR), 159–160, 185–186
Square wave, 19, 139, 153, 305
Standard deviation (STD), 196
Standard hydrogen electrode (SHE), 35, 37
Standing wave ratio (SWR), 222, 256–257
Status register (STR), 205
Steroid-eluting electrode, 100
Stimulation threshold, 52, 94, 96, 98, 100
St. Jude Medical devices, 444
Streaming potential, 70
Strength-duration curve, 434
Stress testing, 59–60, 62, 65–66, 69, 73–74, 75, 112
String galvanometer, 66, 68
Strongarm, 176–177
Stroop test, 295–296
Stud bump technology, 507
Subcarrier modulation (SCM), 238–239
Sub-retina implantation, 481
Sub-retinal stimulation, 482, 484, 506
Successive approximation register (SAR), 434
Successive approximation (SAR), 161–166
  basic operation principle, 163–166
    first block diagram, 166
    input sampling, 163–164
    MSB, 164–165
    MSB-1, 165
    remaining bits, 165
Suction electrode, 69
Sudden infant death syndrome (SIDS), 261
Superior vena cava (SVC), 445
Super-ventricular tachycardia (SVT), 444
Supply pump, 461–462
Supra-choroidal implantation, 481
Supra-choroidal transretinal stimulation (STS), 482–483
swaroTEST device, 116
Switched capacitor (SC) architecture, 136
Synaptic transmission, 21–23

T

Tab electrode, 75–76
Tachycardia, 203, 423, 444–445
3T-APS, 497
Target tip electrode, 100
Telemedicine, 1
Telemetry IC Design, 266
Telemonitoring implant, 226
Temperature measurement method, 115
Temporary memory (TM), 205
TENS, *see* Transcutaneous electrical nerve stimulation (TENS)
Test chip, 326, 328
Textile integrated breathing, 261
Therapeutic effect, 31, 47, 78, 83, 86, 477
Therapeutic waveform, 31
Thermal noise, 132, 141, 171, 173, 177
Thermoelectric effect, 429
Thermometer decoder, 324
Threshold triggering, 20
Tilt, 445, 473
Time division multiplexing (TDM), 238, 244
Tissue coupling capacitor, 465
Tissue damage, 44–45, 47, 454–455, 459
Toumaz, 281–282, 284, 341–342
Track-and-hold simulation, 186
Transconductance stage, 144, 147, 150
Transcutaneous electrical nerve stimulation (TENS), 31, 33, 54, 83–85, 88–89, 91
Transcutaneous energy transfer (TET), 226
Transduction mechanism, 476, 479
Transformation compression, 198
  See also Lossy algorithm

Transient response, 44
Transmission matrix, 379
Trier social stress test (TSST), 296
Trigeminy, 203
Trigger, 20, 174, 295, 309, 313, 315
Triggering, 237, 296, 319–320, 326
Trigger zone, 20
True positive (TP), 198
Tuning capacitance, 179
Turning point algorithm, 198
T-wave, 24, 33, 195, 423, 430, 439

U

UBTL test, 108
U-Healthcare, 1
Ultra-low-leakage (ULL), 447
Ultrasonic recharge, 458
Uncommitted ICDs, 445
Unipolar return-to-zero (RZ), 241
Universal-function electrode, 114

V

Vacuum tube radiofrequency generator, 94
Vacuum-tubes, 68
Van der Waals, 38
Variable gain amplifier (VGA), 342
Variable length coding (VLC), 249
VCO, *see* Voltage controlled oscillator (VCO)
Velcro®, 65
Ventricular assisting device (VAD), 251
Ventricular fibrillation, 423
Verification algorithm, 375–377, 404
Verification gain, 379, 381
Visual evoked potential (VEP) signal, 511
Viterbi algorithm, 253
Vocal patterns, 293
Voltage controlled oscillator (VCO), 232, 235, 237–239, 243, 263, 347
Voltage divider effect, 32
Voltage gain, 142, 314–315
Voltage response, 45
Voltage swing, 147, 436, 489–490
Voltage window, 489–490

W

Warburg, 41
Wavelet transform, 198
WBS communication link, 308–309, 335
Wearable Healthcare System, (WEALTHY), 1, 9, 77, 339–365
  example of, 343
  fabric circuit board, 349–354
    dry electrodes, 350–352
    inductors, 352–353
  health monitoring chest band, 364–365
  issues on using BSNs, 340–341
  previous works snapshots, 341–343
  reliability and low cost, 343–349
    adaptive power transmission, 346–347
    network controller soc, 347–348
    self-configured wearable BSN, 343–346
  wirelessly powered adhesive bandage sensor, 364
  wirelessly powered sensor, 354–364
    design, 355–356
    form factor, 354–355
    readout front-end, 359–362
    wireless power transmission, 356–359
Wearable monitoring, 77
Welch cup electrode, 69
Wheeler bandwidth, 257
Wideband signaling, 304, 332
Wireless body area network (WBAN), 280–300
  ambulatory cardiac monitoring, 282–289
    IMEC low-power ambulatory ECG, 286–289
    state-of-the-art, 284–285
    trends, 282–283
  challenges, 298–300
    dry electrode, 299–300
    multi-parameter sensor, 299
    ultra-low-power technology, 298
  emotion monitoring, 293–298
    IMEC wireless ANS monitoring, 294–298
    state-of-the-art, 293–294
    trends, 293
  wireless sleep monitoring, 289–293
    IMEC wireless sleep, 291–293
    state-of-the-art, 290–291
    trends, 289–290
Wireless capsule endoscopy, 263
Wireless ECG Monitoring, 259–261
Wireless endoscopic capsules, 226, 263
Wireless endoscopy, inductive power and data transmission, 263
Wireless LAN (WLAN), 220, 246, 281, 340–341, 357
Wireless personal area network (WPAN), 280
Wireless sensor network (WSN), 1
World Health Organization, 280, 371
Wristband sensor, 7

X

Xtratek electrode tester, 112

Z

Zener-triggered thyristor, 475
ZigBee, 3, 5, 220, 340–341, 357
Zip compression algorithm, 250

Kolibri 360 GmbH
Ferdinand-Jühlke-Straße 7
99095 Erfurt, Deutschland
produktsicherheit@kolibri360.de